A User's Guide to Measure Theoretic Probability

This book grew from a need to teach a rigorous probability course to a mixed audience – statisticians, mathematically inclined biostatisticians, mathematicians, economists, and students of finance – at the advanced undergraduate/introductory graduate level, without the luxury of a course in measure theory as a prerequisite. The core of the book covers the basic topics of independence, conditioning, martingales, convergence in distribution, and Fourier transforms. In addition, there are numerous sections treating topics traditionally thought of as more advanced, such as coupling and the KMT strong approximation, option pricing via the equivalent martingale measure, and Fernique's inequality for Gaussian processes.

In a further break with tradition, the necessary measure theory is developed via the identification of integrals with linear functionals on spaces of measurable functions, allowing quicker access to the full power of the measure theoretic methods.

The book is not just a presentation of mathematically rigorous theory; it is also a discussion of why some of that theory takes its current form and how anyone could have thought of those clever ideas in the first place. It is intended as a secure starting point for anyone who needs to invoke rigorous probabilistic arguments and to understand what they mean.

David Pollard is Professor of Statistics and Mathematics at Yale University in New Haven, Connecticut. His interests center on probability, measure theory, theoretical and applied statistics, and econometrics. He believes strongly that research and teaching (at all levels) should be intertwined. His book, *Convergence of Stochastic Processes* (Springer-Verlag, 1984), successfully introduced many researchers and graduate students to empirical process theory.

CAMBRIDGE SERIES IN STATISTICAL AND PROBABILISTIC MATHEMATICS

This series of high quality upper-division textbooks and expository monographs covers all aspects of stochastic applicable mathematics. The topics range from pure and applied statistics to probability theory, operations research, optimization and mathematical programming. The books contain clear presentations of new developments in the field and also of the state of the art in classical methods. While emphasizing rigorous treatment of theoretical methods, the books also contain applications and discussions of new techniques made possible by advances in computational practice.

Already Published

1. *Bootstrap Methods and Their Application,* by A. C. Davison and D. V. Hinkley
2. *Markov Chains,* by J. Norris
3. *Asymptotic Statistics,* by A. W. van der Vaart
4. *Wavelet Methods for Time Series Analysis,* by Donald B. Percival and Andrew T. Walden
5. *Bayesian Methods: An Analysis for Statisticians and Interdisciplinary Researchers,* by Thomas Leonard and John S. J. Hsu
6. *Empirical Processes in M-Estimation,* by Sara van de Geer
7. *Numerical Methods of Statistics,* by John F. Monahan

A User's Guide to Measure Theoretic Probability

DAVID POLLARD
Yale University

CAMBRIDGE UNIVERSITY PRESS
Cambridge, New York, Melbourne, Madrid, Cape Town, Singapore,
São Paulo, Delhi, Dubai, Tokyo, Mexico City

Cambridge University Press
32 Avenue of the Americas, New York, NY 10013-2473, USA

www.cambridge.org
Information on this title: www.cambridge.org/9780521002899

First published 2002
7th printing 2010

A catalog record for this publication is available from the British Library.

Library of Congress Cataloging in Publication Data

Pollard, David, 1950–
A user's guide to measure theoretic probability / David Pollard.
p. cm. – (Cambridge series in statistical and probabilistic mathematics)
Includes bibliographical references and index.
ISBN 0-521-80242-3 – ISBN 0-521-00289-3 (pbk.)
1. Probabilities. 2. Measure theory. I. Title.
II. Cambridge series in statistical and probabilistic mathematics.
QA273 .P7735 2001
519.2 – dc21 2001035270

ISBN 978-0-521-80242-0 Hardback
ISBN 978-0-521-00289-9 Paperback

Contents

Preface

This book began life as a set of handwritten notes, distributed to students in my one-semester graduate course on probability theory, a course that had humble aims: to help the students understand results such as the strong law of large numbers, the central limit theorem, conditioning, and some martingale theory. Along the way they could expect to learn a little measure theory and maybe even a smattering of functional analysis, but not as much as they would learn from a course on Measure Theory or Functional Analysis.

In recent years the audience has consisted mainly of graduate students in statistics and economics, most of whom have not studied measure theory. Most of them have no intention of studying measure theory systematically, or of becoming professional probabilists, but they do want to learn some rigorous probability theory—in one semester.

Faced with the reality of an audience that might have neither the time nor the inclination to devote itself completely to my favorite subject, I sought to compress the essentials into a course as self-contained as I could make it. I tried to pack into the first few weeks of the semester a crash course in measure theory, with supplementary exercises and a whirlwind exposition (Appendix A) for the enthusiasts. I tried to eliminate duplication of mathematical effort if it served no useful role. After many years of chopping and compressing, the material that I most wanted to cover all fit into a one-semester course, divided into 25 lectures, each lasting from 60 to 75 minutes. My handwritten notes filled fewer than a hundred pages.

I had every intention of making my little stack of notes into a little book. But I couldn't resist expanding a bit here and a bit there, adding useful reference material, spelling out ideas that I had struggled with on first acquaintance, slipping in extra topics that my students have seemed to need when writing dissertations, and pulling in material from other courses I have taught and neat tricks I have learned from my friends. And soon it wasn't so little any more.

Many of the additions ended up in starred Sections, which contain harder material or topics that can be skipped over without loss of continuity.

My treatment includes a few eccentricities that might upset some of my professional colleagues. My most obvious departure from tradition is in the use of linear functional notation for expectations, an approach I first encountered in books by de Finetti. I attempt to explain the virtues of this notation in the first two Chapters. Another slight novelty—at least for anyone already exposed to the Kolmogorov interpretation of conditional expectations—appears in my treatment of conditioning, in Chapter 5. For many years I have worried about the wide gap between the free-wheeling conditioning calculations of an elementary probability course and the formal manipulations demanded by rigor. I claim that a treatment starting from the idea of conditional distributions offers one way of bridging the gap, at least for many of the statistical applications of conditioning that have troubled me the most.

The twelve Chapters and six Appendixes contain general explanations, remarks, opinions, and blocks of more formal material. Theorems and Lemmas contain the most important mathematical details. Examples contain gentler, or less formal, explanations and illustrations. Supporting theoretical material is presented either in the form of Exercises, with terse solutions, or as Problems (at the ends of the Chapters) that work step-by-step through material that missed the cutoff as Exercises, Lemmas, or Theorems. Some Problems are routine, to give students an opportunity to digest the ideas in the text without great mental effort; some Problems are hard.

A possible one-semester course

Here is a list of the material that I usually try to cover in the one-semester graduate course.

Chapter 1: Spend one lecture on why measure theory is worth the effort, using a few of the Examples as illustrations. Introduce de Finetti notation, identifying sets with their indicator functions and writing $\mathbb{P}$ for both probabilities of sets and expectations of random variables. Mention, very briefly, the fair price Section as an alternative to the frequency interpretation.

Chapter 2: Cover the unstarred Sections carefully, but omitting many details from the Examples. Postpone Section 7 until Chapter 3. Postpone Section 8 until Chapter 6. Describe briefly the generating class theorem for functions, from Section 11, without proofs.

Chapter 3: Cover Section 1, explaining the connection with the elementary notion of a density. Take a short excursion into Hilbert space (explaining the projection theorem as an extension of the result for Euclidean spaces) before presenting the simple version of Radon-Nikodym. Mention briefly the classical concept of absolute continuity, but give no details. Maybe say something about total variation.

Chapter 4: Cover Sections 1 and 2, leaving details of some arguments to the students. Give a reminder about generating classes of functions. Describe the construction of $\mu \otimes \Lambda$, only for a finite kernel Λ, via the iterated integral. Cover product measures, using some of the Examples from Section 4. Explain the need for the blocking idea from Section 6, using the Maximal Inequality to preview the idea of a stopping time. Mention the truncation idea behind the version of the SLLN for independent, identically distributed random variables with finite first moments, but skip most of the proof.

Chapter 5: Discuss Section 1 carefully. Cover the high points of Sections 2 through 4. (They could be skipped without too much loss of continuity, but I prefer not to move straight into Kolmogorov conditioning.) Cover Section 6.

Chapter 6: Cover Sections 1 through 4, but skipping over some Examples. Characterize uniformly integrable martingales, using Section 6 and some of the material postponed from Section 8 of Chapter 2, unless short of time.

Chapter 7: Cover the first four Sections, skipping some of the examples of central limit theorems near the end of Section 2. Downplay multivariate results.

Chapter 8: Cover Sections 1, 2, 4, and 6.

If time is left over, cover a topic from the remaining Chapters.

Acknowledgments

I am particularly grateful to Richard Gill, who is a model of the constructive critic. His comments repeatedly exposed weaknesses and errors in the manuscript. My colleagues Joe Chang and Marten Wegkamp asked helpful questions while using earlier drafts to teach graduate probability courses. Andries Lenstra provided some important historical references.

Many cohorts of students worked through the notes, revealing points of obscurity and confusion. In particular, Jeankyung Kim, Gheorghe Doros, Daniela Cojocaru, and Peter Radchenko read carefully through several chapters and worked through numerous Problems. Their comments led to a lot of rewriting.

Finally, I thank Lauren Cowles for years of good advice, and for her inexhaustible patience with an author who could never stop tinkering.

David Pollard
New Haven

February 2001

Chapter 1

Motivation

SECTION 1 offers some reasons for why anyone who uses probability should know about the measure theoretic approach.

SECTION 2 describes some of the added complications, and some of the compensating benefits that come with the rigorous treatment of probabilities as measures.

SECTION 3 argues that there are advantages in approaching the study of probability theory via expectations, interpreted as linear functionals, as the basic concept.

SECTION 4 describes the de Finetti convention of identifying a set with its indicator function, and of using the same symbol for a probability measure and its corresponding expectation.

*SECTION *5 presents a fair-price interpretation of probability, which emphasizes the linearity properties of expectations. The interpretation is sometimes a useful guide to intuition.*

1. Why bother with measure theory?

Following the appearance of the little book by Kolmogorov (1933), which set forth a measure theoretic foundation for probability theory, it has been widely accepted that probabilities should be studied as special sorts of measures. (More or less true—see the Notes to the Chapter.) Anyone who wants to understand modern probability theory will have to learn something about measures and integrals, but it takes surprisingly little to get started.

For a rigorous treatment of probability, the measure theoretic approach is a vast improvement over the arguments usually presented in undergraduate courses. Let me remind you of some difficulties with the typical introduction to probability.

Independence

There are various elementary definitions of independence for random variables. For example, one can require factorization of distribution functions,

$$\mathbb{P}\{X \leq x, Y \leq y\} = \mathbb{P}\{X \leq x\}\,\mathbb{P}\{Y \leq y\} \qquad \text{for all real } x, y.$$

The problem with this definition is that one needs to be able to calculate distribution functions, which can make it impossible to establish rigorously some desirable properties of independence. For example, suppose $X_1, \ldots, X_4$ are independent random variables. How would you show that

$$Y = X_1 X_2 \left[\log\left(\frac{X_1^2 + X_2^2}{|X_1| + |X_2|} \right) + \frac{|X_1|^3 + X_2^3}{X_1^4 + X_2^4} \right]$$

is independent of

$$Z = \sin\left[X_3 + X_3^2 + X_3 X_4 + X_4^2 + \sqrt{X_3^4 + X_4^4} \right],$$

by means of distribution functions? Somehow you would need to express events $\{Y \le y, Z \le z\}$ in terms of the events $\{X_i \le x_i\}$, which is not an easy task. (If you did figure out how to do it, I could easily make up more taxing examples.)

You might also try to define independence via factorization of joint density functions, but I could invent further examples to make your life miserable, such as problems where the joint distribution of the random variables are not even given by densities. And if you could grind out the joint densities, probably by means of horrible calculations with Jacobians, you might end up with the mistaken impression that independence had something to do with the smoothness of the transformations.

The difficulty disappears in a measure theoretic treatment, as you will see in Chapter 4. Facts about independence correspond to facts about product measures.

Discrete versus continuous

Most introductory texts offer proofs of the Tchebychev inequality,

$$\mathbb{P}\{|X - \mu| \ge \epsilon\} \le \text{var}(X)/\epsilon^2,$$

where μ denotes the expected value of X. Many texts even offer two proofs, one for the discrete case and another for the continuous case. Indeed, introductory courses tend to split into at least two segments. First one establishes all manner of results for discrete random variables and then one reproves almost the same results for random variables with densities.

Unnecessary distinctions between discrete and continuous distributions disappear in a measure theoretic treatment, as you will see in Chapter 3.

Univariate versus multivariate

The unnecessary repetition does not stop with the discrete/continuous dichotomy. After one masters formulae for functions of a single random variable, the whole process starts over for several random variables. The univariate definitions acquire a prefix *joint*, leading to a whole host of new exercises in multivariate calculus: joint densities, Jacobians, multiple integrals, joint moment generating functions, and so on.

Again the distinctions largely disappear in a measure theoretic treatment. Distributions are just image measures; joint distributions are just image measures for maps into product spaces; the same definitions and theorems apply in both cases. One saves a huge amount of unnecessary repetition by recognizing the role of image measures (described in Chapter 2) and recognizing joint distributions as measures on product spaces (described in Chapter 4).

Approximation of distributions

Roughly speaking, the central limit theorem asserts:

If $\xi_1, \ldots, \xi_n$ are independent random variables with zero expected values and variances summing to one, and if none of the ξ_i makes too large a contribution to their sum, then $\xi_1 + \ldots + \xi_n$ is approximately $N(0, 1)$ distributed.

What exactly does that mean? How can something with a discrete distribution, such as a standardized Binomial, be approximated by a smooth normal distribution? The traditional answer (which is sometimes presented explicitly in introductory texts) involves pointwise convergence of distribution functions of random variables; but the central limit theorem is seldom established (even in introductory texts) by checking convergence of distribution functions. Instead, when proofs are given, they typically involve checking of pointwise convergence for some sort of generating function. The proof of the equivalence between convergence in distribution and pointwise convergence of generating functions is usually omitted. The treatment of convergence in distribution for random vectors is even murkier.

As you will see in Chapter 7, it is far cleaner to start from a definition involving convergence of expectations of "smooth functions" of the random variables, an approach that covers convergence in distribution for random variables, random vectors, and even random elements of metric spaces, all within a single framework.

***

In the long run the measure theoretic approach will save you much work and help you avoid wasted effort with unnecessary distinctions.

2. The cost and benefit of rigor

In traditional terminology, probabilities are numbers in the range [0, 1] attached to events, that is, to subsets of a sample space Ω. They satisfy the rules

(i) $\mathbb{P}\emptyset = 0$ *and* $\mathbb{P}\Omega = 1$

(ii) for disjoint events $A_1, A_2, \ldots$, the probability of their union, $\mathbb{P}(\cup_i A_i)$, is equal to $\sum_i \mathbb{P}A_i$, the sum of the probabilities of the individual events.

When teaching introductory courses, I find that it pays to be a little vague about the meaning of the dots in (ii), explaining only that it lets us calculate the probability of an event by breaking it into disjoint pieces whose probabilities are summed. Probabilities add up in the same way as lengths, areas, volumes, and masses. The fact that we sometimes need a countable infinity of pieces (as in calculations involving potentially infinite sequences of coin tosses, for example) is best passed off as an obvious extension of the method for an arbitrarily large, finite number of pieces.

In fact the extension is not at all obvious, mathematically speaking. As explained by Hawkins (1979), the possibility of having the additivity property (ii) hold for countable collections of disjoint events, a property known officially as ***countable additivity***, is one of the great discoveries of modern mathematics. In his 1902 doctoral dissertation, Henri Lebesgue invented a method for defining lengths of complicated subsets of the real line, in a countably additive way. The definition has the subtle feature that not every subset has a length. Indeed, under the usual axioms of set theory, it is impossible to extend the concept of length to *all* subsets of the real line while preserving countable additivity.

The same subtlety carries over to probability theory. In general, the collection of events to which countably additive probabilities are assigned cannot include all subsets of the sample space. The domain of the set function $\mathbb{P}$ (the ***probability measure***) is usually just a ***sigma-field***, a collection of subsets of Ω with properties that will be defined in Chapter 2.

Many probabilistic ideas are greatly simplified by reformulation as properties of sigma-fields. For example, the unhelpful multitude of possible definitions for independence coalesce nicely into a single concept of independence for sigma-fields.

The sigma-field limitation turns out to be less of a disadvantage than might be feared. In fact, it has positive advantages when we wish to prove some probabilistic fact about all events in some sigma-field, $\mathcal{A}$. The obvious line of attack—first find an explicit representation for the typical member of $\mathcal{A}$, then check the desired property directly—usually fails. Instead, as you will see in Chapter 2, an indirect approach often succeeds.

(a) Show directly that the desired property holds for all events in some subclass $\mathcal{E}$ of "simpler sets" from $\mathcal{A}$.

(b) Show that $\mathcal{A}$ is the smallest sigma-field for which $\mathcal{A} \supseteq \mathcal{E}$.

(c) Show that the desired property is preserved under various set theoretic operations. For example, it might be possible to show that if two events have the property then so does their union.

(d) Deduce from (c) that the collection $\mathcal{B}$ of all events with the property forms a sigma-field of subsets of Ω. That is, $\mathcal{B}$ is a sigma-field, which, by (a), has the property $\mathcal{B} \supseteq \mathcal{E}$.

(e) Conclude from (b) and (d) that $\mathcal{B} \supseteq \mathcal{A}$. That is, the property holds for all members of $\mathcal{A}$.

REMARK. Don't worry about the details for the moment. I include the outline in this Chapter just to give the flavor of a typical measure theoretic proof. I have found that some students have trouble adapting to this style of argument.

The indirect argument might seem complicated, but, with the help of a few key theorems, it actually becomes routine. In the literature, it is not unusual to see applications abbreviated to a remark like "a simple generating class argument shows ...," with the reader left to fill in the routine details.

Lebesgue applied his definition of length (now known as Lebesgue measure) to the construction of an integral, extending and improving on the Riemann integral. Subsequent generalizations of Lebesgue's concept of measure (as in the 1913 paper of Radon and other developments described in the Epilogue to

Hawkins 1979) eventually opened the way for Kolmogorov to identify probabilities with measures on sigma-fields of events on general sample spaces. From the Preface to Kolmogorov (1933), in the 1950 translation by Morrison:

> The purpose of this monograph is to give an axiomatic foundation for the theory of probability. The author set himself the task of putting in their natural place, among the general notions of modern mathematics, the basic concepts of probability theory—concepts which until recently were considered to be quite peculiar.
>
> This task would have been a rather hopeless one before the introduction of Lebesgue's theories of measure and integration. However, after Lebesgue's publication of his investigations, the analogies between measure of a set and probability of an event, and between integral of a function and mathematical expectation of a random variable, became apparent. These analogies allowed of further extensions; thus, for example, various properties of independent random variables were seen to be in complete analogy with the corresponding properties of orthogonal functions. But if probability theory was to be based on the above analogies, it still was necessary to make the theories of measure and integration independent of the geometric elements which were in the foreground with Lebesgue. This has been done by Fréchet.
>
> While a conception of probability theory based on the above general viewpoints has been current for some time among certain mathematicians, there was lacking a complete exposition of the whole system, free of extraneous complications. (Cf., however, the book by Fréchet ...)

Kolmogorov identified random variables with a class of real-valued functions (the ***measurable functions***) possessing properties allowing them to coexist comfortably with the sigma-field. Thereby he was also able to identify the expectation operation as a special case of integration with respect to a measure. For the newly restricted class of random variables, in addition to the traditional properties

(i) $\mathbb{E}(c_1X_1 + c_2X_2) = c_1\mathbb{E}(X_1) + c_2\mathbb{E}(X_2)$, for constants c_1 and c_2,
(ii) $\mathbb{E}(X) \geq \mathbb{E}(Y)$ if $X \geq Y$,

he could benefit from further properties implied by the countable additivity of the probability measure.

As with the sigma-field requirement for events, the measurability restriction on the random variables came with benefits. In modern terminology, no longer was $\mathbb{E}$ just an ***increasing linear functional*** on the space of real random variables (with some restrictions to avoid problems with infinities), but also it had acquired some continuity properties, making possible a rigorous treatment of limiting operations in probability theory.

3. Where to start: probabilities or expectations?

From the example set by Lebesgue and Kolmogorov, it would seem natural to start with probabilities of events, then extend, via the operation of integration, to the study of expectations of random variables. Indeed, in many parts of the mathematical world that is the way it goes: probabilities are the basic quantities, from which expectations of random variables are derived by various approximation arguments.

The apparently natural approach is by no means the only possibility, as anyone brought up on the works of the fictitious French author Bourbaki could affirm. (The treatment of measure theory, culminating with Bourbaki 1969, started from integrals defined as linear functionals on appropriate spaces of functions.) Moreover, historically speaking, expectation has a strong claim to being the preferred starting point for a theory of probability. For instance, in his discussion of the 1657 book *Calculating in Games of Chance* by Christian Huygens, Hacking (1978, page 97) commented:

> The fair prices worked out by Huygens are just what we would call the expectations of the corresponding gambles. His approach made expectation a more basic concept than probability, and this remained so for about a century.

The fair price interpretation is sketched in Section 5.

The measure theoretic history of integrals as linear functionals also extends back to the early years of the twentieth century, starting with Daniell (1918), who developed a general theory of integration via extension of linear functionals from small spaces of functions to larger spaces. It is also significant that, in one of the greatest triumphs of measure theory, Wiener (1923, Section 10) defined what is now known as Wiener measure (thereby providing a rigorous basis for the mathematical theory of Brownian motion) as an averaging operation for functionals defined on Brownian motion paths, citing Daniell (1919) for the basic extension theorem.

There are even better reasons than historical precedent for working with expectations as the basic concept. Whittle (1992), in the *Preface* to an elegant, intermediate level treatment of *Probability via Expectations*, presented some arguments:

(i) To begin with, people probably have a better intuition for what is meant by an 'average value' than for what is meant by a 'probability.'

(ii) Certain important topics, such as optimization and approximation problems, can be introduced and treated very quickly, just because they are phrased in terms of expectations.

(iii) Most elementary treatments are bedeviled by the apparent need to ring the changes of a particular proof or discussion for all the special cases of continuous or discrete distribution, scalar or vector variables, etc. In the expectations approach these are indeed seen as special cases, which can be treated with uniformity and economy.

His list continued. I would add that:

(a) It is often easier to work with the linearity properties of integrals than with the additivity properties of measures. For example, many useful probability inequalities are but thinly disguised consequences of pointwise inequalities, translated into probability form by the linearity and increasing properties of expectations.

(b) The linear functional approach, via expectations, can save needless repetition of arguments. Some theorems about probability measures, as set functions, are just special cases of more general results about expectations.

(c) When constructing new probability measures, we save work by defining the integral of measurable functions directly, rather than passing through the preliminary step of building the set function then establishing theorems about the corresponding integrals. As you will see repeatedly, definitions and theorems sometimes collapse into a single operation when expressed directly in terms of expectations, or integrals.

***

I will explain the essentials of measure theory in Chapter 2, starting from the traditional set-function approach but working as quickly as I can towards systematic use of expectations.

4. The de Finetti notation

The advantages of treating expectation as the basic concept are accentuated by the use of an elegant notation strongly advocated by de Finetti (1972, 1974). Knowing that many traditionally trained probabilists and statisticians find the notation shocking, I will introduce it slowly, in an effort to explain why it is worth at least a consideration. (Immediate enthusiastic acceptance is more than I could hope for.)

Ordinary algebra is easier than Boolean algebra. The correspondence $A \leftrightarrow \mathbb{I}_A$ between subsets A of a fixed set $\mathcal{X}$ and their indicator functions,

$$\mathbb{I}_A(x) = \begin{cases} 1 & \text{if } x \in A, \\ 0 & \text{if } x \in A^c, \end{cases}$$

transforms Boolean algebra into ordinary pointwise algebra with functions. I claim that probability theory becomes easier if one works systematically with expectations of indicator functions, $\mathbb{E}\mathbb{I}_A$, rather than with the corresponding probabilities of events.

Let me start with the assertions about algebra and Boolean algebra. The operations of union and intersection correspond to pointwise maxima (denoted by max or the symbol $\vee$) and pointwise minima (denoted by min or the symbol $\wedge$), or pointwise products:

$$\mathbb{I}_{\cup_i A_i}(x) = \bigvee_i \mathbb{I}_{A_i}(x) \qquad \text{and} \qquad \mathbb{I}_{\cap_i A_i}(x) = \bigwedge_i \mathbb{I}_{A_i}(x) = \prod_i \mathbb{I}_{A_i}(x).$$

Complements correspond to subtraction from one: $\mathbb{I}_{A^c}(x) = 1 - \mathbb{I}_A(x)$. Derived operations, such as the set theoretic difference $A \backslash B := A \cap B^c$ and the symmetric difference, $A \Delta B := (A \backslash B) \cup (B \backslash A)$, also have simple algebraic counterparts:

$$\mathbb{I}_{A \backslash B}(x) = (\mathbb{I}_A(x) - \mathbb{I}_B(x))^+ := \max\left(0, \mathbb{I}_A(x) - \mathbb{I}_B(x)\right),$$

$$\mathbb{I}_{A \Delta B}(x) = |\mathbb{I}_A(x) - \mathbb{I}_B(x)|.$$

To check these identities, just note that the functions take only the values 0 and 1, then determine which combinations of indicator values give a 1. For example, $|\mathbb{I}_A(x) - \mathbb{I}_B(x)|$ takes the value 1 when exactly one of $\mathbb{I}_A(x)$ and $\mathbb{I}_B(x)$ equals 1.

The algebra looks a little cleaner if we omit the argument x. For example, the horrendous set theoretic relationship

$$\left(\cap_{i=1}^n A_i\right) \Delta \left(\cap_{i=1}^n B_i\right) \subseteq \cup_{i=1}^n \left(A_i \Delta B_i\right)$$

corresponds to the pointwise inequality

$$\left|\prod_i \mathbb{I}_{A_i} - \prod_i \mathbb{I}_{B_i}\right| \le \max_i \left|\mathbb{I}_{A_i} - \mathbb{I}_{B_i}\right|,$$

whose verification is easy: when the right-hand side takes the value 1 the inequality is trivial, because the left-hand side can take only the values 0 or 1; and when right-hand side takes the value 0, we have $\mathbb{I}_{A_i} = \mathbb{I}_{B_i}$ for all i, which makes the left-hand side zero.

<1> **Example.** One could establish an identity such as

$$(A\Delta B)\Delta(C\Delta D) = A\Delta\left(B\Delta(C\Delta D)\right)$$

by expanding both sides into a union of many terms. It is easier to note the pattern for indicator functions. The set $A\Delta B$ is the region where $\mathbb{I}_A + \mathbb{I}_B$ takes an odd value (that is, the value 1); and $(A\Delta B)\Delta C$ is the region where $(\mathbb{I}_A + \mathbb{I}_B) + \mathbb{I}_C$ takes an odd value. And so on. In fact both sides of the set theoretic identity equal the region where $\mathbb{I}_A + \mathbb{I}_B + \mathbb{I}_C + \mathbb{I}_D$ takes an odd value. Associativity of set theoretic differences is a consequence of associativity of pointwise addition. □

<2> **Example.** The lim sup of a sequence of sets $\{A_n : n \in \mathbb{N}\}$ is defined as

$$\limsup_n A_n := \bigcap_{n=1}^{\infty} \bigcup_{i \ge n} A_i.$$

That is, the lim sup consists of those x for which, to each n there exists an $i \ge n$ such that $x \in A_i$. Equivalently, it consists of those x for which $x \in A_i$ for infinitely many i. In other words,

$$\mathbb{I}_{\limsup_n A_n} = \limsup_n \mathbb{I}_{A_n}.$$

Do you really need to learn the new concept of the lim sup of a sequence of sets? Theorems that work for lim sups of sequences of functions automatically carry over to theorems about sets. There is no need to prove everything twice. The correspondence between sets and their indicators saves us from unnecessary work. □

After some repetition, it becomes tiresome to have to keep writing the $\mathbb{I}$ for the indicator function. It would be much easier to write something like $\tilde{A}$ in place of $\mathbb{I}_A$. The indicator of the lim sup of a sequence of sets would then be written $\limsup_n \tilde{A}_n$, with only the tilde to remind us that we are referring to functions. But why do we need reminding? As the example showed, the concept for the lim sup of sets is really just a special case of the concept for sequences of functions. Why preserve a distinction that hardly matters?

There is a well established tradition in Mathematics for choosing notation that eliminates inessential distinctions. For example, we use the same symbol 3 for the natural number and the real number, writing $3 + 6 = 9$ as an assertion both about addition of natural numbers and about addition of real numbers.

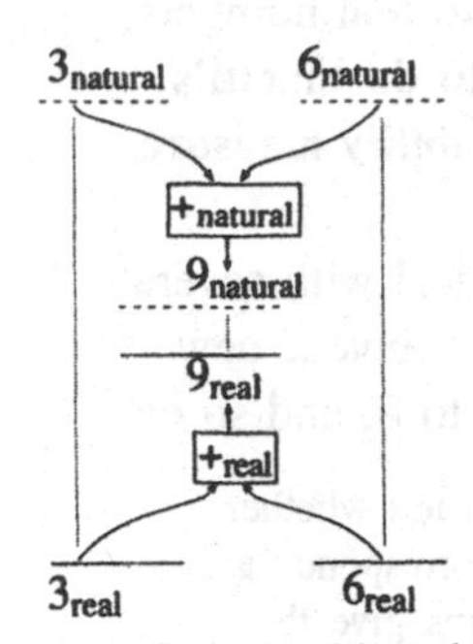

It does not matter if we cannot tell immediately which interpretation is intended, because we know there is a one-to-one correspondence between natural numbers and a subset of the real numbers, which preserves all the properties of interest. Formally, there is a map $\psi : \mathbb{N} \to \mathbb{R}$ for which

$$\psi(x +_{natural} y) = \psi(x) +_{real} \psi(y) \qquad \text{for all } x, y \text{ in } \mathbb{N},$$

with analogous equalities for other operations. (Notice that I even took care to distinguish between addition as a function from $\mathbb{N} \times \mathbb{N}$ to $\mathbb{N}$ and as a function from $\mathbb{R} \times \mathbb{R}$ to $\mathbb{R}$.) The map ψ is an isomorphism between $\mathbb{N}$ and a subset of $\mathbb{R}$.

REMARK. Of course there are some situations where we need to distinguish between a natural number and its real counterpart. For example, it would be highly confusing to use indistinguishable symbols when first developing the properties of the real number system from the properties of the natural numbers. Also, some computer languages get very upset when a function that expects a floating point argument is fed an integer variable; some languages even insist on an explicit conversion between types.

We are faced with a similar overabundance of notation in the correspondence between sets and their indicator functions. Formally, and traditionally, we have a map $A \mapsto \mathbb{I}_A$ from sets into a subset of the nonnegative real functions. The map preserves the important operations. It is firmly in the Mathematical tradition that we should follow de Finetti's suggestion and **use the same symbol for a set and its indicator function.**

REMARK. A very similar convention has been advocated by the renowned computer scientist, Donald Knuth, in an expository article (Knuth 1992). He attributed the idea to Kenneth Iversen, the inventor of the programming language APL.

In de Finetti's notation the assertion from Example <2> becomes

$$\limsup A_n = \limsup A_n,$$

a fact that is quite easy to remember. The theorem about lim sups of sequences of sets has become incorporated into the notation; we have one less theorem to remember.

The second piece of de Finetti notation is suggested by the same logic that encourages us to replace $+_{natural}$ and $+_{real}$ by the single addition symbol: use the same symbol when extending the domain of definition of a function. For example, the symbol "sin" denotes both the function defined on the real line and its extension to the complex domain. More generally, if we have a function g with domain G_0, which can be identified with a subset $\widetilde{G}_0$ of some $\widetilde{G}$ via a correspondence $x \leftrightarrow \tilde{x}$, and if $\tilde{g}$ is a function on $\widetilde{G}$ for which $\tilde{g}(\tilde{x}) = g(x)$ for x in G_0, then why not write g instead of $\tilde{g}$ for the function with the larger domain?

With probability theory we often use $\mathbb{P}$ to denote a probability measure, as a map from a class $\mathcal{A}$ (a sigma-field) of subsets of some Ω into the subinterval $[0, 1]$ of the real line. The correspondence $A \leftrightarrow \tilde{A} := \mathbb{I}_A$, between a set A and its indicator function $\tilde{A}$, establishes a correspondence between $\mathcal{A}$ and a subset of the collection of random variables on Ω. The expectation maps random variables into real numbers, in such a way that $\mathbb{E}(\tilde{A}) = \mathbb{P}(A)$. This line of thinking leads us to de Finetti's second suggestion: **use the same symbol for expectation and probability measure**, writing $\mathbb{P}X$ instead of $\mathbb{E}X$, and so on.

The de Finetti notation has an immediate advantage when we deal with several probability measures, $\mathbb{P}$, $\mathbb{Q}$, ... simultaneously. Instead of having to invent new symbols $\mathbb{E}_{\mathbb{P}}$, $\mathbb{E}_{\mathbb{Q}}$, ..., we reuse $\mathbb{P}$ for the expectation corresponding to $\mathbb{P}$, and so on.

REMARK. You might have the concern that you will not be able to tell whether $\mathbb{P}A$ refers to the probability of an event or the expected value of the corresponding indicator function. The ambiguity should not matter. Both interpretations give the same number; you will never be faced with a choice between two different values when choosing an interpretation. If this ambivalence worries you, I would suggest going systematically with the expectation/indicator function interpretation. It will never lead you astray.

<3> **Example.** For a finite collection of events $A_1, \ldots, A_n$, the so-called ***method of inclusion and exclusion*** asserts that the probability of the union $\cup_{i \le n} A_i$ equals

$$\sum_i \mathbb{P}A_i - \sum_{i \ne j} \mathbb{P}(A_i \cap A_j) + \sum_{i,j,k} \{i, j, k \text{ distinct}\} \mathbb{P}(A_i \cap A_j \cap A_k) - \ldots \pm \mathbb{P}(A_1 \cap A_2 \cap \ldots \cap A_n).$$

The equality comes by taking expectations on both sides of an identity for (indicator) functions,

$$\cup_{i \le n} A_i = \sum_i A_i - \sum_{i \ne j} A_i A_j + \sum_{i,j,k} \{i, j, k \text{ distinct}\} A_i A_j A_k - \ldots \pm A_1 A_2 \ldots A_n.$$

The right-hand side of this identity is just the expanded version of $1 - \prod_{i \le n} (1 - A_i)$. The identity is equivalent to

$$1 - \cup_{i \le n} A_i = \prod_{i \le n} (1 - A_i),$$

which presents two ways of expressing the indicator function of $\cap_{i \le n} A_i^c$. See Problem [1] for a generalization. □

<4> **Example.** Consider Tchebychev's inequality, $\mathbb{P}\{|X - \mu| \ge \epsilon\} \le \text{var}(X)/\epsilon^2$, for each $\epsilon > 0$, and each random variable X with expected value $\mu := \mathbb{P}X$ and finite variance, $\text{var}(X) := \mathbb{P}(X - \mu)^2$. On the left-hand side of the inequality we have the probability of an event. Or is it the expectation of an indicator function? Either interpretation is correct, but the second is more helpful. The inequality is a consequence of the increasing property for expectations invoked for a pair of functions, $\{|X - \mu| \ge \epsilon\} \le (X - \mu)^2/\epsilon^2$. The indicator function on the left-hand side takes only the values 0 and 1. The quadratic function on the right-hand side is nonnegative, and is ≥ 1 whenever the left-hand side equals 1. □

***

For the remainder of the book, I will be using the same symbol for a set and its indicator function, and writing $\mathbb{P}$ instead of $\mathbb{E}$ for expectation.

REMARK. For me, the most compelling reason to adopt the de Finetti notation, and work with $\mathbb{P}$ as a linear functional defined for random variables, was not that I would save on symbols, nor any of the other good reasons listed at the end of

Section 3. Instead, I favor the notation because, once the initial shock of seeing old symbols used in new ways wore off, it made probability theory easier. I can truly claim to have gained better insight into classical techniques through the mere fact of translating them into the new notation. I even find it easier to invent new arguments when working with a notation that encourages thinking in terms of linearity, and which does not overemphasize the special role for expectations of functions that take only the values 0 and 1 by according them a different symbol.

The hope that I might convince probability users of some of the advantages of de Finetti notation was, in fact, one of my motivations for originally deciding to write yet another book about an old subject.

*5. Fair prices

For the understanding of this book the interpretation of probability as a model for uncertainty is not essential. You could study it purely as a piece of mathematics, divorced from any interpretation but then you would forgo much of the intuition that accompanies the various interpretations.

The most widely accepted view interprets probabilities and expectations as long run averages, anticipating the formal laws of large numbers that make precise a sense in which averages should settle down to expectations over a long sequence of independent trials. As an aid to intuition I also like another interpretation, which does not depend on a preliminary concept of independence, and which concentrates attention on the linearity properties of expectations.

Consider a situation—a bet if you will–where you stand to receive an uncertain return X. You could think of X as a random variable, a real-valued function on a set Ω. For the moment forget about any probability measure on Ω. Suppose you consider $p(X)$ to be the fair price to pay now in order to receive X at some later time. (By *fair* I mean that you should be prepared to take either side of the bet. In particular, you should be prepared to accept a payment $p(X)$ from me now in return for giving me an amount X later.) What properties should $p(\cdot)$ have?

REMARK. As noted in Section 3, the value $p(X)$ corresponds to an expected value of the random variable X. If you already know about the possibility of infinite expectations, you will realize that I would have to impose some restrictions on the class of random variables for which fair prices are defined, if I were seriously trying to construct a rigorous system of axioms. It would suffice to restrict the argument to bounded random variables.

Your net return will be the random quantity $X'(\omega) := X(\omega) - p(X)$. Call the random variable X' a ***fair return***, the net return from a fair trade. Unless you start worrying about utilities—in which case you might consult Savage (1954) or Ferguson (1967, Section 1.4)—you should find the following properties reasonable.

(i) ***fair + fair = fair***. That is, if you consider $p(X)$ fair for X and $p(Y)$ fair for Y then you should be prepared to make both bets, paying $p(X) + p(Y)$ to receive $X + Y$.

(ii) ***constant × fair = fair***. That is, you shouldn't object if I suggest you pay $2p(X)$ to receive $2X$ (actually, that particular example is a special case of (i)) or $3.76p(X)$ to receive $3.76X$, or $-p(X)$ to receive $-X$. The last example corresponds to willingness to take either side of a fair bet. In general, to receive cX you should pay $cp(X)$, for constant c.

Properties (i) and (ii) imply that the collection of all fair returns is a vector space.

There is a third reasonable property that goes by several names: ***coherency*** or ***nonexistence of a Dutch book***, the ***no-arbitrage requirement***, or the ***no-free-lunch principle***:

(iii) There is no fair return X' for which $X'(\omega) \geq 0$ for all ω, with strict inequality for at least one ω.

(Students of decision theory might be reminded of the the concept of admissibility.) If you were to declare such an X' to be fair I would be delighted to offer you the opportunity to receive a net return of $-10^{100}X'$. I couldn't lose.

<5> **Lemma.** *Properties (i), (ii), and (iii) imply that $p(\cdot)$ is an increasing linear functional on random variables. The fair returns are those random variables for which $p(X) = 0$.*

Proof. For constants α and β, and random variables X and Y with fair prices $p(X)$ and $p(Y)$, consider the combined effect of the following fair bets:

you pay me $\alpha p(X)$ to receive αX

you pay me $\beta p(Y)$ to receive βY

I pay you $p(\alpha X + \beta Y)$ to receive $(\alpha X + \beta Y)$.

Your net return is a constant,

$$c = p(\alpha X + \beta Y) - \alpha p(X) - \beta p(Y).$$

If $c > 0$ you violate (iii); if $c < 0$ take the other side of the bet to violate (iii). That proves linearity.

To prove that $p(\cdot)$ is increasing, suppose $X(\omega) \geq Y(\omega)$ for all ω. If you claim that $p(X) < p(Y)$ then I would be happy for you to accept the bet that delivers

$$(Y - p(Y)) - (X - p(X)) = -(X - Y) - (p(Y) - p(X)),$$

which is always < 0.

If both X and $X - p(X)$ are considered fair, then the constant return $p(X) = X - (X - p(X))$ is fair, which would contradict (iii) unless $p(X) = 0$. □

As a special case, consider the bet that returns 1 if an event F occurs, and 0 otherwise. If you identify the event F with the random variable taking the value 1 on F and 0 on F^c (that is, the indicator of the event F), then it follows directly from Lemma <5> that $p(\cdot)$ is additive: $p(F_1 \cup F_2) = p(F_1) + p(F_2)$ for disjoint events F_1 and F_2. That is, p defines a finitely additive set-function on events. The set function $p(\cdot)$ has most of the properties required of a probability measure. As an exercise you might show that $p(\emptyset) = 0$ and $p(\Omega) = 1$.

Contingent bets

Things become much more interesting if you are prepared to make a bet to receive an amount X but only when some event F occurs. That is, the bet is made ***contingent*** on the occurrence of F. Typically, knowledge of the occurrence of F should change the fair price, which we could denote by $p(X \mid F)$. Expressed more compactly, the bet that returns $(X - p(X \mid F))F$ is fair. The indicator function F ensures that money changes hands only when F occurs.

<6> **Lemma.** *If Ω is partitioned into disjoint events $F_1, \ldots, F_k$, and X is a random variable, then $p(X) = \sum_{i=1}^{k} p(F_i)p(X \mid F_i)$.*

Proof. For a single F_i, argue by linearity that

$$0 = p(XF_i - p(X \mid F_i)F_i) = p(XF_i) - p(X \mid F_i)p(F_i).$$

Sum over i, using linearity again, together with the fact that $X = \sum_i XF_i$, to deduce that $p(X) = \sum_i p(XF_i) = \sum_i p(F_i)p(X \mid F_i)$, as asserted. □

Why should we restrict the Lemma to finite partitions? If we allowed countable partitions we would get the countable additivity property—the key requirement in the theory of measures. I would be suspicious of such an extension of the simple argument for finite partitions. It makes a tacit assumption that a combination of countably many fair bets is again fair. If we accept that assumption, then why not accept that arbitrary combinations of fair events are fair? For uncountably infinite collections we would run into awkward contradictions. For example, suppose ω is generated from a uniform distribution on $[0, 1]$. Let X_t be the random variable that returns 1 if $\omega = t$ and 0 otherwise. By symmetry one might expect $p(X_t) = c$ for some constant c that doesn't depend on t. But there can be no c for which

$$1 = p(1) = p\left(\sum_{0 \le t \le 1} X_t\right) \stackrel{?}{=} \sum_{0 \le t \le 1} p(X_t) = \begin{cases} 0 & \text{if } c = 0 \\ \pm\infty & \text{if } c \neq 0 \end{cases}$$

Perhaps our intuition about the infinite rests on shaky analogies with the finite.

REMARK. I do not insist that probabilities must be interpreted as fair prices, just as I do not accept that all probabilities must be interpreted as assertions about long run frequencies. It is convenient that both interpretations lead to almost the same mathematical formalism. You are free to join either camp, or both, and still play by the same probability rules.

6. Problems

[1] Let $A_1, \ldots, A_N$ be events in a probability space $(\Omega, \mathcal{F}, \mathbb{P})$. For each subset J of $\{1, 2, \ldots, N\}$ write A_J for $\cap_{i \in J} A_i$. Define $S_k := \sum_{|J|=k} \mathbb{P}A_J$, where $|J|$ denotes the number of indices in J. For $0 \le m \le N$ show that the probability $\mathbb{P}\{\text{exactly } m \text{ of the } A_i\text{'s occur}\}$ equals $\binom{m}{m}S_m - \binom{m+1}{m}S_{m+1} + \ldots \pm \binom{N}{m}S_N$. Hint: For a dummy variable z, show that $\prod_{i=1}^{N}(A_i^c + zA_i) = \sum_{k=0}^{n}\sum_{|J|=k}(z-1)^k A_J$. Expand the left-hand side, take expectations, then interpret the coefficient of z^m.

[2] Rederive the assertion of Lemma <6> by consideration of the net return from the following system of bets: (i) for each i, pay $c_i p(F_i)$ in order to receive c_i if F_i occurs, where $c_i := p(X \mid F_i)$; (ii) pay $-p(X)$ in order to receive $-X$; (iii) for each i, make a bet contingent on F_i, paying c_i (if F_i occurs) to receive X.

[3] For an increasing sequence of events $\{A_n : n \in \mathbb{N}\}$ with union A, show $\mathbb{P}A_n \uparrow \mathbb{P}A$.

7. Notes

See Dubins & Savage (1964) for an illustration of what is possible in a theory of probability without countable additivity.

The ideas leading up to Lebesgue's creation of his integral are described in fascinating detail in the excellent book of Hawkins (1979), which has been the starting point for most of my forays into the history of measure theory. Lebesgue first developed his new definition of the integral for his doctoral dissertation (Lebesgue 1902), then presented parts of his theory in the 1902–1903 Peccot course of lectures (Lebesgue 1904). The 1928 revision of the 1904 volume greatly expanded the coverage, including a treatment of the more general (Lebesgue-)Stieltjes integral. See also Lebesgue (1926), for a clear description of some of the ideas involved in the development of measure theory, and the *Note Historique* of Bourbaki (1969), for a discussion of later developments.

Of course it is a vast oversimplification to imagine that probability theory abruptly became a specialized branch of measure theory in 1933. As Kolmogorov himself made clear, the crucial idea was the measure theory of Lebesgue. Kolmogorov's little book was significant not just for "putting in their natural place, among the general notions of modern mathematics, the basic concepts of probability theory", but also for adding new ideas, such as probability distributions in infinite dimensional spaces (reinventing results of Daniell 1919) and a general theory of conditional probabilities and conditional expectations.

Measure theoretic ideas were used in probability theory well before 1933. For example, in the *Note* at the end of Lévy (1925) there was a clear statement of the countable additivity requirement for probabilities, but Lévy did not adopt the complete measure theoretic formalism; and Khinchin & Kolmogorov (1925) explicitly constructed their random variables as functions on [0, 1], in order to avail themselves of the properties of Lebesgue measure.

It is also not true that acceptance of the measure theoretic foundation was total and immediate. For example, eight years after Kolmogorov's book appeared, von Mises (1941, page 198) asserted (emphasis in the original):

> In recapitulating this paragraph I may say: First, the axioms of Kolmogorov are concerned with the distribution function within one kollektiv and are *supplementary to my theory, not a substitute for it.* Second, using the notion of measure zero in an absolute way without reference to the arbitrarily assumed measure system, *leads to essential inconsistencies.*

See also the argument for the measure theoretic framework in the accompanying paper by Doob (1941), and the comments by both authors that follow (von Mises & Doob 1941).

For more about Kolmogorov's pivotal role in the history of modern probability, see: Shiryaev (2000), and the other articles in the same collection; the memorial articles in the *Annals of Probability,* volume 17 (1989); and von Plato (1994), which also contains discussions of the work of von Mises and de Finetti.

References

Bourbaki, N. (1969), *Intégration sur les espaces topologiques séparés*, Éléments de mathématique, Hermann, Paris. Fascicule XXXV, Livre VI, Chapitre IX.

Daniell, P. J. (1918), 'A general form of integral', *Annals of Mathematics (series 2)* **19**, 279–294.

Daniell, P. J. (1919), 'Functions of limited variation in an infinite number of dimensions', *Annals of Mathematics (series 2)* **21**, 30–38.

de Finetti, B. (1972), *Probability, Induction, and Statistics*, Wiley, New York.

de Finetti, B. (1974), *Theory of Probability*, Wiley, New York. First of two volumes translated from *Teoria Delle probabilità*, published 1970. The second volume appeared under the same title in 1975.

Doob, J. L. (1941), 'Probability as measure', *Annals of Mathematical Statistics* **12**, 206–214.

Dubins, L. & Savage, L. (1964), *How to Gamble if You Must*, McGraw-Hill.

Ferguson, T. S. (1967), *Mathematical Statistics: A Decision Theoretic Approach*, Academic Press, Boston.

Fréchet, M. (1915), 'Sur l'intégrale d'une fonctionnelle étendue à un ensemble abstrait', *Bull. Soc. Math. France* **43**, 248–265.

Hacking, I. (1978), *The Emergence of Probability*, Cambridge University Press.

Hawkins, T. (1979), *Lebesgue's Theory of Integration: Its Origins and Development*, second edn, Chelsea, New York.

Khinchin, A. Y. & Kolmogorov, A. (1925), 'Über Konvergenz von Reihen, deren Glieder durch den Zufall bestimmt werden', *Mat. Sbornik* **32**, 668–677.

Knuth, D. E. (1992), 'Two notes on notation', *American Mathematical Monthly* **99**, 403–422.

Kolmogorov, A. N. (1933), *Grundbegriffe der Wahrscheinlichkeitsrechnung*, Springer-Verlag, Berlin. Second English Edition, *Foundations of Probability* 1950, published by Chelsea, New York.

Lebesgue, H. (1902), *Intégrale, longueur, aire*. Doctoral dissertation, submitted to Faculté des Sciences de Paris. Published separately in Ann. Mat. Pura Appl. 7. Included in the first volume of his *Œuvres Scientifiques*, published in 1972 by L'Enseignement Mathématique.

Lebesgue, H. (1904), *Leçons sur l'integration et la recherche des fonctions primitives*, first edn, Gauthier-Villars, Paris. Included in the second volume of his *Œuvres Scientifiques*, published in 1972 by L'Enseignement Mathématique. Second edition published in 1928. Third edition, 'an unabridged reprint of the second edition, with minor changes and corrections', published in 1973 by Chelsea, New York.

Lebesgue, H. (1926), 'Sur le développement de la notion d'intégrale', *Matematisk Tidsskrift B*. English version in the book *Measure and Integral*, edited and translated by Kenneth O. May.

Lévy, P. (1925), *Calcul des Probabilités*, Gauthier-Villars, Paris.

Radon, J. (1913), 'Theorie und Anwendungen der absolut additiven Mengenfunctionen', *Sitzungsberichten der Kaiserlichen Akademie der Wissenschaften in Wien. Mathematisch-naturwissenschaftliche Klasse* **122**, 1295–1438.

Savage, L. J. (1954), *The Foundations of Statistics*, Wiley, New York. Second edition, Dover, New York, 1972.

Shiryaev, A. N. (2000), Andrei Nikolaevich Kolmogorov: a biographical sketch of his life and creative paths, *in* 'Kolmogorov in Perspective', American Mathematical Society/London Mathematical Society.

von Mises, R. (1941), 'On the foundations of probability and statistics', *Annals of Mathematical Statistics* **12**, 191–205.

von Mises, R. & Doob, J. L. (1941), 'Discussion of papers on probability theory', *Annals of Mathematical Statistics* **12**, 215–217.

von Plato, J. (1994), *Creating Modern Probability: its Mathematics, Physics and Philosophy in Historical Perspective*, Cambridge University Press.

Whittle, P. (1992), *Probability via Expectation*, third edn, Springer-Verlag, New York. First edition 1970, under the title "Probability".

Wiener, N. (1923), 'Differential-space', *Journal of Mathematics and Physics* **2**, 131–174. Reprinted in *Selected papers of Norbert Wiener*, MIT Press, 1964.

Chapter 2

A modicum of measure theory

SECTION 1 defines measures and sigma-fields.
SECTION 2 defines measurable functions.
SECTION 3 defines the integral with respect to a measure as a linear functional on a cone of measurable functions. The definition sidesteps the details of the construction of integrals from measures.
*SECTION *4 constructs integrals of nonnegative measurable functions with respect to a countably additive measure.*
SECTION 5 establishes the Dominated Convergence theorem, the Swiss Army knife of measure theoretic probability.
SECTION 6 collects together a number of simple facts related to sets of measure zero.
*SECTION *7 presents a few facts about spaces of functions with integrable pth powers, with emphasis on the case p=2, which defines a Hilbert space.*
SECTION 8 defines uniform integrability, a condition slightly weaker than domination. Convergence in $\mathcal{L}^1$ is characterized as convergence in probability plus uniform integrability.
SECTION 9 defines the image measure, which includes the concept of the distribution of a random variable as a special case.
SECTION 10 explains how generating class arguments, for classes of sets, make measure theory easy.
*SECTION *11 extends generating class arguments to classes of functions.*

1. Measures and sigma-fields

As promised in Chapter 1, we begin with measures as set functions, then work quickly towards the interpretation of integrals as linear functionals. Once we are past the purely set-theoretic preliminaries, I will start using the de Finetti notation (Section 1.4) in earnest, writing the same symbol for a set and its indicator function.

Our starting point is a ***measure space***: a triple $(\mathcal{X}, \mathcal{A}, \mu)$, with $\mathcal{X}$ a set, $\mathcal{A}$ a class of subsets of $\mathcal{X}$, and μ a function that attaches a nonnegative number (possibly $+\infty$) to each set in $\mathcal{A}$. The class $\mathcal{A}$ and the set function μ are required to have properties that facilitate calculations involving limits along sequences.

<1> **Definition.** *Call a class $\mathcal{A}$ a sigma-field of subsets of $\mathcal{X}$ if:*

(i) the empty set $\emptyset$ and the whole space $\mathcal{X}$ both belong to $\mathcal{A}$;

(ii) if A belongs to $\mathcal{A}$ then so does its complement A^c;

(iii) if $A_1, A_2, \ldots$ is a countable collection of sets in $\mathcal{A}$ then both the union $\cup_i A_i$ and the intersection $\cap_i A_i$ are also in $\mathcal{A}$.

Some of the requirements are redundant as stated. For example, once we have $\emptyset \in \mathcal{A}$ then (ii) implies $\mathcal{X} \in \mathcal{A}$. When we come to establish properties about sigma-fields it will be convenient to have the list of defining properties pared down to a minimum, to reduce the amount of mechanical checking. The theorems will be as sparing as possible in the amount the work they require for establishing the sigma-field properties, but for now redundancy does not hurt.

The collection $\mathcal{A}$ need not contain every subset of $\mathcal{X}$, a fact forced upon us in general if we want μ to have the properties of a countably additive measure.

<2> **Definition.** *A function μ defined on the sigma-field $\mathcal{A}$ is called a (countably additive, nonnegative) measure if:*

(i) $0 \le \mu A \le \infty$ for each A in $\mathcal{A}$;

(ii) $\mu\emptyset = 0$;

(iii) if $A_1, A_2, \ldots$ is a countable collection of pairwise disjoint sets in $\mathcal{A}$ then $\mu(\cup_i A_i) = \sum_i \mu A_i$.

A measure μ for which $\mu\mathcal{X} = 1$ is called a ***probability measure***, and the corresponding $(\mathcal{X}, \mathcal{A}, \mu)$ is called a ***probability space***. For this special case it is traditional to use a symbol like $\mathbb{P}$ for the measure, a symbol like Ω for the set, and a symbol like $\mathcal{F}$ for the sigma-field. A triple $(\Omega, \mathcal{F}, \mathbb{P})$ will always denote a probability space.

Usually the qualifications "countably additive, nonnegative" are omitted, on the grounds that these properties are the most commonly assumed—the most common cases deserve the shortest names. Only when there is some doubt about whether the measures are assumed to have all the properties of Definition <2> should the qualifiers be attached. For example, one speaks of "finitely additive measures" when an analog of property (iii) is assumed only for finite disjoint collections, or "signed measures" when the value of μA is not necessarily nonnegative. When finitely additive or signed measures are under discussion it makes sense to mention explicitly when a particular measure is nonnegative or countably additive, but, in general, you should go with the shorter name.

Where do measures come from? The most basic constructions start from set functions μ defined on small collections of subsets $\mathcal{E}$, such as the collection of all subintervals of the real line. One checks that μ has properties consistent with the requirements of Definition <2>. One seeks to extend the domain of definition while preserving the countable additivity properties of the set function. As you saw in Chapter 1, Theorems guaranteeing existence of such extensions were the culmination of a long sequence of refinements in the concept of integration (Hawkins 1979). They represent one of the great achievements of modern mathematics, even though those theorems now occupy only a handful of pages in most measure theory texts.

Finite additivity has several appealing interpretations (such as the fair-prices of Section 1.5) that have given it ready acceptance as an axiom for a model of real-world uncertainty. Countable additivity is sometimes regarded with suspicion, or justified as a matter of mathematical convenience. (However, see Problem [6] for an equivalent form of countable additivity, which has some claim to intuitive appeal.) It is difficult to develop a simple probability theory without countable additivity, which gives one the licence (for only a small fee) to integrate series term-by-term, differentiate under integrals, and interchange other limiting operations.

The classical constructions are significant for my exposition mostly because they ensure existence of the measures needed to express the basic results of probability theory. I will relegate the details to the Problems and to Appendix A. If you crave a more systematic treatment you might consult one of the many excellent texts on measure theory, such as Royden (1968).

The constructions do not—indeed cannot, in general—lead to countably additive measures on the class of all subsets of a given $\mathcal{X}$. Typically, they extend a set function defined on a class of sets $\mathcal{E}$ to a measure defined on the ***sigma-field*** $\sigma(\mathcal{E})$ ***generated by*** $\mathcal{E}$, or to only slightly larger sigma-fields. By definition,

$$
\begin{aligned}
\sigma(\mathcal{E}) &:= \text{smallest sigma-field on } \mathcal{X} \text{ containing all sets from } \mathcal{E} \\
&= \{A \subseteq \mathcal{X} : A \in \mathcal{F} \text{ for every sigma-field } \mathcal{F} \text{ with } \mathcal{E} \subseteq \mathcal{F}\}.
\end{aligned}
$$

The representation given by the second line ensures existence of a smallest sigma-field containing $\mathcal{E}$. The method of definition is analogous to many definitions of "smallest... containing a fixed class" in mathematics—think of generated subgroups or linear subspaces spanned by a collection of vectors, for example. For the definition to work one needs to check that sigma-fields have two properties:

(i) If $\{\mathcal{F}_i : i \in \mathcal{I}\}$ is a nonempty collection of sigma-fields on $\mathcal{X}$ then $\cap_{i\in\mathcal{I}}\mathcal{F}_i$, the collection of all the subsets of $\mathcal{X}$ that belong to every $\mathcal{F}_i$, is also a sigma-field.

(ii) For each $\mathcal{E}$ there exists at least one sigma-field $\mathcal{F}$ containing all the sets in $\mathcal{E}$.

You should check property (i) as an exercise. Property (ii) is trivial, because the collection of all subsets of $\mathcal{X}$ is a sigma-field.

REMARK. Proofs of existence of nonmeasurable sets typically depend on some deep set-theoretic principle, such as the Axiom of Choice. Mathematicians who can live with different rules for set theory can have bigger sigma-fields. See Dudley (1989, Section 3.4) or Oxtoby (1971, Section 5) for details.

<3> **Exercise.** Suppose $\mathcal{X}$ consists of five points a, b, c, d, and e. Suppose $\mathcal{E}$ consists of two sets, $E_1 = \{a, b, c\}$ and $E_2 = \{c, d, e\}$. Find the sigma-field generated by $\mathcal{E}$.

SOLUTION: For this simple example we can proceed by mechanical application of the properties that a sigma-field $\sigma(\mathcal{E})$ must possess. In addition to the obvious $\emptyset$ and $\mathcal{X}$, it must contain each of the sets

$$
\begin{aligned}
F_1 &:= \{a, b\} = E_1 \cap E_2^c \qquad \text{and} \qquad F_2 := \{c\} = E_1 \cap E_2, \\
F_3 &:= \{d, e\} = E_1^c \cap E_2 \qquad \text{and} \qquad F_4 := \{a, b, d, e\} = F_1 \cup F_3.
\end{aligned}
$$

Further experimentation creates no new members of $\sigma(\mathcal{E})$; the sigma-field consists of the sets

$$\emptyset, \quad F_1, \quad F_2, \quad F_3, \quad F_1 \cup F_3, \quad F_1 \cup F_2 = E_1, \quad F_2 \cup F_3 = E_2, \quad \mathcal{X}.$$

The sets F_1, F_2, F_3 are the ***atoms*** of the sigma-field; every member of $\sigma(\mathcal{E})$ is a union of some collection (possibly empty) of F_i. The only measurable subsets of F_i are the empty set and F_i itself. There are no measurable protons or neutrons hiding inside these atoms. □

An unsystematic construction might work for finite sets, but it cannot generate all members of a sigma-field in general. Indeed, we cannot even hope to list all the members of an infinite sigma-field. Instead we must find a less explicit way to characterize its sets.

<4> **Example.** By definition, the Borel sigma-field on the real line, denoted by $\mathcal{B}(\mathbb{R})$, is the sigma-field generated by the open subsets. We could also denote it by $\sigma(\mathcal{G})$ where $\mathcal{G}$ stands for the class of all open subsets of $\mathbb{R}$. There are several other generating classes for $\mathcal{B}(\mathbb{R})$. For example, as you will soon see, the class $\mathcal{E}$ of all intervals $(-\infty, t]$, with $t \in \mathbb{R}$, is a generating class.

It might appear a hopeless task to prove that $\sigma(\mathcal{E}) = \mathcal{B}(\mathbb{R})$ if we cannot explicitly list the members of both sigma-fields, but actually the proof is quite routine. You should try to understand the style of argument because it is often used in probability theory.

The equality of sigma-fields is established by two inclusions, $\sigma(\mathcal{E}) \subseteq \sigma(\mathcal{G})$ and $\sigma(\mathcal{G}) \subseteq \sigma(\mathcal{E})$, both of which follow from more easily established results. First we must prove that $\mathcal{E} \subseteq \sigma(\mathcal{G})$, showing that $\sigma(\mathcal{G})$ is one of the sigma-fields $\mathcal{F}$ that enter into the intersection defining $\sigma(\mathcal{E})$, and hence $\sigma(\mathcal{E}) \subseteq \sigma(\mathcal{G})$. The other inclusion follows similarly if we show that $\mathcal{G} \subseteq \sigma(\mathcal{E})$.

Each interval $(-\infty, t]$ in $\mathcal{E}$ has a representation $\bigcap_{n=1}^{\infty}(-\infty, t+n^{-1})$, a countable intersection of open sets. The sigma-field $\sigma(\mathcal{G})$ contains all open sets, and it is stable under countable intersections. It therefore contains each $(-\infty, t]$. That is, $\mathcal{E} \subseteq \sigma(\mathcal{G})$.

The argument for $\mathcal{G} \subseteq \sigma(\mathcal{E})$ is only slightly harder. It depends on the fact that an open subset of the real line can be written as a countable union of open intervals. Such an interval has a representation $(a, b) = (-\infty, b) \cap (-\infty, a]^c$, and $(-\infty, b) = \bigcup_{n=1}^{\infty}(-\infty, b - n^{-1}]$. That is, every open set can be built up from sets in $\mathcal{E}$ using operations that are guaranteed not to take us outside the sigma-field $\sigma(\mathcal{E})$.

My explanation has been moderately detailed. In a published paper the reasoning would probably be abbreviated to something like "a generating class argument shows that . . . ," with the routine details left to the reader. □

REMARK. The generating class argument often reduces to an assertion like: $\mathcal{A}$ is a sigma-field and $\mathcal{A} \supseteq \mathcal{E}$, therefore $\mathcal{A} = \sigma(\mathcal{A}) \supseteq \sigma(\mathcal{E})$.

<5> **Example.** A class $\mathcal{E}$ of subsets of a set $\mathcal{X}$ is called a ***field*** if it contains the empty set and is stable under complements, finite unions, and finite intersections. For a field $\mathcal{E}$, write $\mathcal{E}_\delta$ for the class of all possible intersections of countable subclasses of $\mathcal{E}$, and $\mathcal{E}_\sigma$ for the class of all possible unions of countable subclasses of $\mathcal{E}$.

Of course if $\mathcal{E}$ is a sigma-field then $\mathcal{E} = \mathcal{E}_\delta = \mathcal{E}_\sigma$, but, in general, the inclusions $\sigma(\mathcal{E}) \supseteq \mathcal{E}_\delta \supseteq \mathcal{E}$ and $\sigma(\mathcal{E}) \supseteq \mathcal{E}_\sigma \supseteq \mathcal{E}$ will be proper. For example, if $\mathcal{X} = \mathbb{R}$ and $\mathcal{E}$ consists of all finite unions of half open intervals $(a, b]$, with possibly $a = -\infty$ or $b = +\infty$, then the set of rationals does not belong to $\mathcal{E}_\sigma$ and the complement of the same set does not belong to $\mathcal{E}_\delta$.

Let μ be a finite measure on $\sigma(\mathcal{E})$. Even though $\sigma(\mathcal{E})$ might be much larger than either $\mathcal{E}_\sigma$ or $\mathcal{E}_\delta$, a generating class argument will show that all sets in $\sigma(\mathcal{E})$ can be ***inner approximated by*** $\mathcal{E}_\delta$, in the sense that,

$$\mu A = \sup\{\mu F : A \supseteq F \in \mathcal{E}_\delta\} \qquad \text{for each } A \text{ in } \sigma(\mathcal{E}),$$

and ***outer approximated by*** $\mathcal{E}_\sigma$, in the sense that,

$$\mu A = \inf\{\mu G : A \subseteq G \in \mathcal{E}_\sigma\} \qquad \text{for each } A \text{ in } \sigma(\mathcal{E}).$$

REMARK. Incidentally, I chose the letters G and F to remind myself of open and closed sets, which have similar approximation properties for Borel measures on metric spaces—see Problem [12].

It helps to work on both approximation properties at the same time. Denote by $\mathcal{B}_0$ the class of all sets in $\sigma(\mathcal{E})$ that can be both innner and outer approximated. A set B belongs to $\mathcal{B}_0$ if and only if, to each $\epsilon > 0$ there exist $F \in \mathcal{E}_\delta$ and $G \in \mathcal{E}_\sigma$ such that $F \subseteq B \subseteq G$ and $\mu(G\backslash F) < \epsilon$. I'll call the sets F and G an ϵ-sandwich for B.

Trivially $\mathcal{B}_0 \supseteq \mathcal{E}$, because each member of $\mathcal{E}$ belongs to both $\mathcal{E}_\sigma$ and $\mathcal{E}_\delta$. The approximation result will follow if we show that $\mathcal{B}_0$ is a sigma-field, for then we will have $\mathcal{B}_0 = \sigma(\mathcal{B}_0) \supseteq \sigma(\mathcal{E})$.

Symmetry of the definition ensures that $\mathcal{B}_0$ is stable under complements: if $F \subseteq B \subseteq G$ is an ϵ-sandwich for B, then $G^c \subseteq B^c \subseteq F^c$ is an ϵ-sandwich for B^c. To show that $\mathcal{B}_0$ is stable under countable unions, consider a countable collection $\{B_n : n \in \mathbb{N}\}$ of sets from $\mathcal{B}_0$. We need to slice the bread thinner as n gets larger: choose $\epsilon/2^n$-sandwiches $F_n \subseteq B_n \subseteq G_n$ for each n. The union $\cup_n B_n$ is sandwiched between the sets $G := \cup_n G_n$ and $H = \cup_n F_n$; and the sets are close in μ measure because

$$\mu\left(\cup_n G_n \backslash \cup_n F_n\right) \le \sum_n \mu(G_n\backslash F_n) < \sum_n \epsilon/2^n = \epsilon.$$

REMARK. Can you prove this inequality? Do you see why $\cup_n G_n \backslash \cup_n F_n \subseteq \cup_n (G_n\backslash F_n)$ and why countable additivity implies that the measure of a countable union of (not necessarily disjoint) sets is smaller than the sum of their measures? If not, just wait until Section 3, after which you can argue that $\cup_n G_n \backslash \cup_n F_n \le \sum_n (G_n\backslash F_n)$, as an inequality between indicator functions, and $\mu\left(\sum_n (G_n\backslash F_n)\right) = \sum_n \mu(G_n\backslash F_n)$ by Monotone Convergence.

We have an ϵ-sandwich, but the bread might not be of the right type. It is certainly true that $G \in \mathcal{E}_\sigma$ (a countable union of countable unions is a countable union), but the set H need not belong to $\mathcal{E}_\delta$. However, the sets $H_N := \cup_{n \le N} F_n$ do belong to $\mathcal{E}_\delta$, and countable additivity implies that $\mu H_N \uparrow \mu H$.

REMARK. Do you see why? If not, wait for Monotone Convergence again.

□ If we choose a large enough N we have a 2ϵ-sandwich $H_N \subseteq \cup_n B_n \subseteq G$.

The measure $\mathfrak{m}$ on $\mathcal{B}(\mathbb{R})$ for which $\mathfrak{m}(a, b] = b - a$ is called ***Lebesgue measure***. Another sort of generating class argument (see Section 10) can be used to show that the values $\mathfrak{m}(B)$ for B in $\mathcal{B}(\mathbb{R})$ are uniquely determined by the values given to intervals; there can exist at most one measure on $\mathcal{B}(\mathbb{R})$ with the stated property. It is harder to show that at least one such measure exists. Despite any intuitions you might have about length, the construction of Lebesgue measure is not trivial—see Appendix A. Indeed, Henri Lebesgue became famous for proving existence of the measure and showing how much could be done with the new integration theory.

The name Lebesgue measure is also given to an extension of $\mathfrak{m}$ to a measure on a sigma-field, sometimes called the Lebesgue sigma-field, which is slightly larger than $\mathcal{B}(\mathbb{R})$. I will have more to say about the extension in Section 6.

Borel sigma-fields are defined in similar fashion for any topological space $\mathcal{X}$. That is, $\mathcal{B}(\mathcal{X})$ denotes the sigma-field generated by the open subsets of $\mathcal{X}$.

Sets in a sigma-field $\mathcal{A}$ are said to be ***measurable*** or $\mathcal{A}$-measurable. In probability theory they are also called ***events***. Good functions will also be given the title measurable. Try not to get confused when you really need to know whether an object is a set or a function.

2. Measurable functions

Let $\mathcal{X}$ be a set equipped with a sigma-field $\mathcal{A}$, and $\mathcal{Y}$ be a set equipped with a sigma-field $\mathcal{B}$, and T be a function (also called a map) from $\mathcal{X}$ to $\mathcal{Y}$. We say that T ***is*** $\mathcal{A}\backslash\mathcal{B}$***-measurable*** if the inverse image $\{x \in \mathcal{X} : Tx \in B\}$ belongs to $\mathcal{A}$ for each B in $\mathcal{B}$. Sometimes the inverse image is denoted by $\{T \in B\}$ or $T^{-1}B$. Don't be fooled by the T^{-1} notation into treating T^{-1} as a function from $\mathcal{Y}$ into $\mathcal{X}$: it's not, unless T is one-to-one (and onto, if you want to have domain $\mathcal{Y}$). Sometimes an $\mathcal{A}\backslash\mathcal{B}$-measurable map is referred to in abbreviated form as just $\mathcal{A}$-measurable, or just $\mathcal{B}$-measurable, or just measurable, if there is no ambiguity about the unspecified sigma-fields.

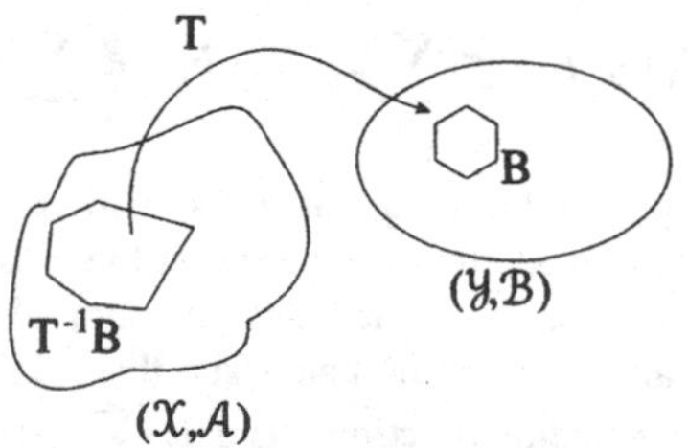

For example, if $\mathcal{Y} = \mathbb{R}$ and $\mathcal{B}$ equals the Borel sigma-field $\mathcal{B}(\mathbb{R})$, it is common to drop the $\mathcal{B}(\mathbb{R})$ specification and refer to the map as being $\mathcal{A}$-measurable, or as being Borel measurable if $\mathcal{A}$ is understood and there is any doubt about which sigma-field to use for the real line. *In this book, you may assume that any sigma-field on* $\mathbb{R}$ *is its Borel sigma-field, unless explicitly specified otherwise.* It can get confusing if you misinterpret where the unspecified sigma-fields live. My advice would be that you imagine a picture showing the two spaces involved, with any missing sigma-field labels filled in.

Sometimes the functions come first, and the sigma-fields are chosen specifically to make those functions measurable.

<6> **Definition.** *Let $\mathcal{H}$ be a class of functions on a set $\mathcal{X}$. Suppose the typical h in $\mathcal{H}$ maps $\mathcal{X}$ into a space $\mathcal{Y}_h$ equipped with a sigma-field $\mathcal{B}_h$. Then the sigma-field $\sigma(\mathcal{H})$ generated by $\mathcal{H}$ is defined as $\sigma\{h^{-1}(B) : B \in \mathcal{B}_h,\ h \in \mathcal{H}\}$. It is the smallest sigma-field $\mathcal{A}_0$ on $\mathcal{X}$ for which each h in $\mathcal{H}$ is $\mathcal{A}_0\backslash\mathcal{B}_h$-measurable.*

<7> **Example.** If $\mathcal{B} = \sigma(\mathcal{E})$ for some class $\mathcal{E}$ of subsets of $\mathcal{Y}$ then a map T is $\mathcal{A}\backslash\sigma(\mathcal{E})$-measurable if and only if $T^{-1}E \in \mathcal{A}$ for every E in $\mathcal{E}$. You should prove this assertion by checking that $\{B \in \mathcal{B} : T^{-1}B \in \mathcal{A}\}$ is a sigma-field, and then arguing from the definition of a generating class.

In particular, to establish $\mathcal{A}\backslash\mathcal{B}(\mathbb{R})$-measurability of a map into the real line it is enough to check the inverse images of intervals of the form (t, ∞), with t ranging over $\mathbb{R}$. (In fact, we could restrict t to a countable dense subset of $\mathbb{R}$, such as the set of rationals: How would you build an interval (t, ∞) from intervals (t_i, ∞) with rational t_i?) That is, a real-valued function f is Borel-measurable if $\{x \in \mathcal{X} : f(x) > t\} \in \mathcal{A}$ for each real t. There are many similar assertions obtained by using other generating classes for $\mathcal{B}(\mathbb{R})$. Some authors use particular generating classes for the definition of measurability, and then derive facts about inverse images of Borel sets as theorems. □

It will be convenient to consider not just real-valued functions on a set $\mathcal{X}$, but also functions from $\mathcal{X}$ into the extended real line $\overline{\mathbb{R}} := [-\infty, \infty]$. The Borel sigma-field $\mathcal{B}(\overline{\mathbb{R}})$ is generated by the class of open sets, or, more explicitly, by all sets in $\mathcal{B}(\mathbb{R})$ together with the two singletons $\{-\infty\}$ and $\{\infty\}$. It is an easy exercise to show that $\mathcal{B}(\overline{\mathbb{R}})$ is generated by the class of all sets of the form $(t, \infty]$, for t in $\mathbb{R}$, and by the class of all sets of the form $[-\infty, t)$, for t in $\mathbb{R}$. We could even restrict t to any countable dense subset of $\mathbb{R}$.

<8> **Example.** Let a set $\mathcal{X}$ be equipped with a sigma-field $\mathcal{A}$. Let $\{f_n : n \in \mathbb{N}\}$ be a sequence of $\mathcal{A}\backslash\mathcal{B}(\mathbb{R})$-measurable functions from $\mathcal{X}$ into $\mathbb{R}$. Define functions f and g by taking pointwise suprema and infima: $f(x) := \sup_n f_n(x)$ and $g(x) := \inf_n f_n(x)$. Notice that f might take the value $+\infty$, and g might take the value $-\infty$, at some points of $\mathcal{X}$. We may consider both as maps from $\mathcal{X}$ into $\overline{\mathbb{R}}$. (In fact, the whole argument is unchanged if the f_n functions themselves are also allowed to take infinite values.)

The function f is $\mathcal{A}\backslash\mathcal{B}(\overline{\mathbb{R}})$-measurable because

$$\{x : f(x) > t\} = \cup_n \{x : f_n(x) > t\} \in \mathcal{A} \qquad \text{for each real } t :$$

for each fixed x, the supremum of the real numbers $f_n(x)$ is strictly greater than t if and only if $f_n(x) > t$ for at least one n. Example <7> shows why we have only to check inverse images for such intervals.

The same generating class is not as convenient for proving measurability of g. It is not true that an infimum of a sequence of real numbers is strictly greater than t if and only if all of the numbers are strictly greater than t: think of the sequence $\{n^{-1} : n = 1, 2, 3, \ldots\}$, whose infimum is zero. Instead you should argue via the identity $\{x : g(x) < t\} = \cup_n \{x : f_n(x) < t\} \in \mathcal{A}$ for each real t. □

From Example <8> and the representations $\limsup f_n(x) = \inf_{n\in\mathbb{N}} \sup_{m\geq n} f_m(x)$ and $\liminf f_n(x) = \sup_{n\in\mathbb{N}} \inf_{m\geq n} f_m(x)$, it follows that the lim sup or lim inf of a sequence of measurable (real- or extended real-valued) functions is also measurable. In particular, if the limit exists it is measurable.

Measurability is also preserved by the usual algebraic operations—sums, differences, products, and so on—provided we take care to avoid illegal pointwise calculations such as $\infty - \infty$ or 0/0. There are several ways to establish these stability properties. One of the more direct methods depends on the fact that $\mathbb{R}$ has a countable dense subset, as illustrated by the following argument for sums.

<9> **Example.** Let f and g be $\mathcal{B}(\mathbb{R})$-measurable functions, with pointwise sum $h(x) = f(x) + g(x)$. (I exclude infinite values because I don't want to get caught up with inconclusive discussions of how we might proceed at points x where $f(x) = +\infty$ and $g(x) = -\infty$, or $f(x) = -\infty$ and $g(x) = +\infty$.) How can we prove that h is also a $\mathcal{B}(\mathbb{R})$-measurable function?

It is true that

$$\{x : h(x) > t\} = \cup_{s\in\mathbb{R}} \left(\{x : f(x) = s\} \cap \{x : g(x) > t - s\}\right),$$

and it is true that the set $\{x : f(x) = s\} \cap \{x : g(x) > t - s\}$ is measurable for each s and t, but sigma-fields are not required to have any particular stability properties for uncountable unions. Instead we should argue that at each x for which $f(x)+g(x) > t$ there exists a rational number r such that $f(x) > r > t - g(x)$. Conversely if there is an r lying strictly between $f(x)$ and $t - g(x)$ then $f(x) + g(x) > t$. Thus

$$\{x : h(x) > t\} = \cup_{r\in\mathbb{Q}} (\{x : f(x) > r\} \cap \{x : g(x) > t - r\}),$$

where $\mathbb{Q}$ denotes the countable set of rational numbers. A countable union of intersections of pairs of measurable sets is measurable. The sum is a measurable function. □

As a little exercise you might try to extend the argument from the last Example to the case where f and g are allowed to take the value $+\infty$ (but not the value $-\infty$). If you want practice at playing with rationals, try to prove measurability of products (be careful with inequalities if dividing by negative numbers) or try Problem [4], which shows why a direct attack on the lim sup requires careful handling of inequalities in the limit.

The real significance of measurability becomes apparent when one works through the construction of integrals with respect to measures, as in Section 4. For the moment it is important only that you understand that the family of all measurable functions is stable under most of the familiar operations of analysis.

<10> **Definition.** *The class* $\mathcal{M}(\mathcal{X}, \mathcal{A})$, *or* $\mathcal{M}(\mathcal{X})$ *or just* $\mathcal{M}$ *for short, consists of all* $\mathcal{A}\backslash\mathcal{B}(\overline{\mathbb{R}})$*-measurable functions from* $\mathcal{X}$ *into* $\overline{\mathbb{R}}$. *The class* $\mathcal{M}^+(\mathcal{X}, \mathcal{A})$, *or* $\mathcal{M}^+(\mathcal{X})$ *or just* $\mathcal{M}^+$ *for short, consists of the nonnegative functions in* $\mathcal{M}(\mathcal{X}, \mathcal{A})$.

If you desired exquisite precision you could write $\mathcal{M}(\mathcal{X}, \mathcal{A}, \overline{\mathbb{R}}, \mathcal{B}(\overline{\mathbb{R}}))$, to eliminate all ambiguity about domain, range, and sigma-fields.

The collection $\mathcal{M}^+$ is a cone (stable under sums and multiplication of functions by positive constants). It is also stable under products, pointwise limits of sequences, and suprema or infima of countable collections of functions. It is not a vector space, because it is not stable under subtraction; but it does have the property that if f and g belong to $\mathcal{M}^+$ and g takes only real values, then the positive part $(f-g)^+$, defined by taking the pointwise maximum of $f(x)-g(x)$ with 0, also belongs to $\mathcal{M}^+$. You could adapt the argument from Example <9> to establish the last fact.

It proves convenient to work with $\mathcal{M}^+$ rather than with the whole of $\mathcal{M}$, thereby eliminating many problems with $\infty - \infty$. As you will soon learn, integrals have some convenient properties when restricted to nonnegative functions.

For our purposes, one of the most important facts about $\mathcal{M}^+$ will be the possibility of approximation by ***simple functions*** that is by measurable functions of the form $s := \sum_i \alpha_i A_i$, for finite collections of real numbers α_i and events A_i from $\mathcal{A}$. If the A_i are disjoint, $s(x)$ equals α_i when $x \in A_i$, for some i, and is zero otherwise. If the A_i are not disjoint, the nonzero values taken by s are sums of various subsets of the $\{\alpha_i\}$. Don't forget: the symbol A_i gets interpreted as an indicator function when we start doing algebra. I will write $\mathcal{M}^+_{\text{simple}}$ for the cone of all simple functions in $\mathcal{M}^+$.

<11> **Lemma.** *For each f in $\mathcal{M}^+$ the sequence $\{f_n\} \subseteq \mathcal{M}^+_{\text{simple}}$, defined by*

$$f_n := 2^{-n} \sum_{i=1}^{4^n} \{f \geq i/2^n\},$$

has the property $0 \leq f_1(x) \leq f_2(x) \leq \ldots \leq f_n(x) \uparrow f(x)$ at every x.

REMARK. The definition of f_n involves algebra, so you must interpret $\{f \geq i/2^n\}$ as the indicator function of the set of all points x for which $f(x) \geq i/2^n$.

Proof. At each x, count the number of nonzero indicator values. If $f(x) \geq 2^n$, all 4^n summands contribute a 1, giving $f_n(x) = 2^n$. If $k2^{-n} \leq f(x) < (k+1)2^{-n}$, for some integer k from $\{0, 1, 2, \ldots, 4^n - 1\}$, then exactly k of the summands contribute a 1, giving $f_n(x) = k2^{-n}$. (Check that the last assertion makes sense when k equals 0.) That is, for $0 \leq f(x) < 2^n$, the function f_n rounds down to an integer multiple of 2^{-n}, from which the convergence and monotone increasing properties follow.

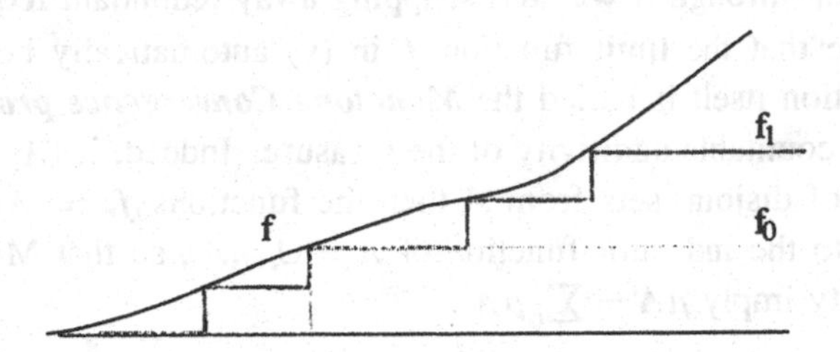

If you do not find the monotonicity assertion convincing, you could argue, more formally, that

$$f_n = \frac{1}{2^{n+1}} \sum_{i=1}^{4^n} 2\left\{f \geq \frac{2i}{2^{n+1}}\right\} \leq \frac{1}{2^{n+1}} \sum_{i=1}^{4\times 4^n} \left(\left\{f \geq \frac{2i}{2^{n+1}}\right\} + \left\{f \geq \frac{2i-1}{2^{n+1}}\right\}\right) = f_{n+1},$$

which reflects the effect of doubling the maximum value and halving the step size when going from the nth to the (n+1)st approximation. □

As an exercise you might prove that the product of functions in $\mathcal{M}^+$ also belongs to $\mathcal{M}^+$, by expressing the product as a pointwise limit of products of simple functions. Notice how the convention $0 \times \infty = 0$ is needed to ensure the correct limit behavior at points where one of the factors is zero.

3. Integrals

Just as $\int_a^b f(x)\,dx$ represents a sort of limiting sum of $f(x)$ values weighted by small lengths of intervals—the $\int$ sign is a long "S", for sum, and the dx is a sort of limiting increment—so can the general integral $\int f(x)\,\mu(dx)$ be defined as a limit of weighted sums but with weights provided by the measure μ. The formal definition involves limiting operations that depend on the assumed measurability of the function f. You can skip the details of the construction (Section 4) by taking the following result as an axiomatic property of the integral.

<12> **Theorem.** *For each measure μ on $(\mathcal{X}, \mathcal{A})$ there is a uniquely determined functional, a map $\tilde{\mu}$ from $\mathcal{M}^+(\mathcal{X}, \mathcal{A})$ into $[0, \infty]$, having the following properties:*

(i) $\tilde{\mu}(\mathbb{I}_A) = \mu A$ for each A in $\mathcal{A}$;

(ii) $\tilde{\mu}(0) = 0$, where the first zero stands for the zero function;

(iii) for nonnegative real numbers α, β and functions f, g in $\mathcal{M}^+$,

$$\tilde{\mu}(\alpha f + \beta g) = \alpha\tilde{\mu}(f) + \beta\tilde{\mu}(g);$$

(iv) if f, g are in $\mathcal{M}^+$ and $f \le g$ everywhere then $\tilde{\mu}(f) \le \tilde{\mu}(g)$;

(v) if $f_1, f_2, \ldots$ is a sequence in $\mathcal{M}^+$ with $0 \le f_1(x) \le f_2(x) \le \ldots \uparrow f(x)$ for each x in $\mathcal{X}$ then $\tilde{\mu}(f_n) \uparrow \tilde{\mu}(f)$.

I will refer to (iii) as ***linearity***, even though $\mathcal{M}^+$ is not a vector space. It will imply a linearity property when $\tilde{\mu}$ is extended to a vector subspace of $\mathcal{M}$. Property (iv) is redundant because it follows from (ii) and nonnegativity. Property (ii) is also redundant: put $A = \emptyset$ in (i); or, interpreting $0 \times \infty$ as 0, put $\alpha = \beta = 0$ and $f = g = 0$ in (iii). We need to make sure the bad case $\tilde{\mu} f = \infty$, for all f in $\mathcal{M}^+$, does not slip through if we start stripping away redundant requirements.

Notice that the limit function f in (v) automatically belongs to $\mathcal{M}^+$. The limit assertion itself is called the ***Monotone Convergence property***. It corresponds directly to countable additivity of the measure. Indeed, if $\{A_i : i \in \mathbb{N}\}$ is a countable collection of disjoint sets from $\mathcal{A}$ then the functions $f_n := A_1 + \ldots + A_n$ increase pointwise to the indicator function of $A = \cup_{i\in\mathbb{N}} A_i$, so that Monotone Convergence and linearity imply $\mu A = \sum_i \mu A_i$.

REMARK. You should ponder the role played by $+\infty$ in Theorem <12>. For example, what does $\alpha\tilde{\mu}(f)$ mean if $\alpha = 0$ and $\tilde{\mu}(f) = \infty$? The interpretation depends on the convention that $0 \times \infty = 0$.

In general you should be suspicious of any convention involving $\pm\infty$. Pay careful attention to cases where it operates. For example, how would the five assertions be affected if we adopted a new convention, whereby $0 \times \infty = 6$? Would the Theorem still hold? Where exactly would it fail? I feel uneasy if it is not clear how a convention is disposing of awkward cases. My advice: be very, very careful with any calculations involving infinity. Subtle errors are easy to miss when concealed within a convention.

There is a companion to Theorem <12> that shows why it is largely a matter of taste whether one starts from measures or integrals as the more primitive measure theoretic concept.

<13> **Theorem.** *Let $\tilde{\mu}$ be a map from $\mathcal{M}^+$ to $[0, \infty]$ that satisfies properties (ii) through (v) of Theorem <12>. Then the set function defined on the sigma-field $\mathcal{A}$ by (i) is a (countably additive, nonnegative) measure, with $\tilde{\mu}$ the functional that it generates.*

Lemma <11> provides the link between the measure μ and the functional $\tilde{\mu}$. For a given f in $\mathcal{M}^+$, let $\{f_n\}$ be the sequence defined by the Lemma. Then

$$\tilde{\mu} f = \lim_{n\to\infty} \tilde{\mu} f_n = \lim_{n\to\infty} 2^{-n} \sum_{i=1}^{4^n} \mu\{f \geq i/2^n\},$$

the first equality by Monotone Convergence, the second by linearity. The value of $\tilde{\mu} f$ is uniquely determined by μ, as a set function on $\mathcal{A}$. It is even possible to use the equality, or something very similar, as the basis for a direct construction of the integral, from which properties (i) through (v) are then derived, as you will see from Section 4.

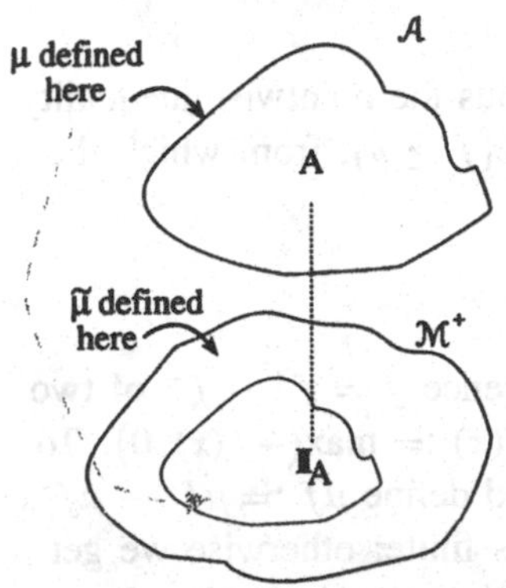

In summary: There is a one-to-one correspondence between measures on the sigma-field $\mathcal{A}$ and increasing linear functionals on $\mathcal{M}^+(\mathcal{A})$ with the Monotone Convergence property. To each measure μ there is a uniquely determined functional $\tilde{\mu}$ for which $\tilde{\mu}(\mathbb{I}_A) = \mu(A)$ for every A in $\mathcal{A}$. The functional $\tilde{\mu}$ is usually called an ***integral*** with respect to μ, and is variously denoted by $\int f\, d\mu$ or $\int f(x)\, \mu(dx)$ or $\int_{\mathcal{X}} f\, d\mu$ or $\int f(x)\, d\mu(x)$. With the de Finetti notation, where we identify a set A with its indicator function, the functional $\tilde{\mu}$ is just an extension of μ from a smaller domain (indicators of sets in $\mathcal{A}$) to a larger domain (all of $\mathcal{M}^+$). Accordingly, we should have no qualms about denoting it by the same symbol. I will write μf for the integral. With this notation, assertion (i) of Theorem <12> becomes: $\mu A = \mu A$ for all A in $\mathcal{A}$. You probably can't tell that the A on the left-hand side is an indicator function and the μ is an integral, but you don't need to be able to tell—that is precisely what (i) asserts.

REMARK. In elementary algebra we rely on parentheses, or precedence, to make our meaning clear. For example, both $(ax) + b$ and $ax + b$ have the same meaning, because multiplication has higher precedence than addition. With traditional notation, the $\int$ and the $d\mu$ act like parentheses, enclosing the integrand and separating it from following terms. With linear functional notation, we sometimes need explicit parentheses to make the meaning unambiguous. As a way of eliminating some parentheses, I often work with the convention that integration has lower precedence than exponentiation, multiplication, and division, but higher precedence than addition or subtraction. Thus I intend you to read $\mu fg + 6$ as $(\mu(fg)) + 6$. I would write $\mu(fg + 6)$ if the 6 were part of the integrand.

Some of the traditional notations also remove ambiguity when functions of several variables appear in the integrand. For example, in $\int f(x,y)\,\mu(dx)$ the y variable is held fixed while the μ operates on the first argument of the function. When a similar ambiguity might arise with linear functional notation, I will append a superscript, as in $\mu^x f(x,y)$, to make clear which variable is involved in the integration.

<14> **Example.** Suppose μ is a finite measure (that is, $\mu\mathcal{X} < \infty$) and f is a function in $\mathcal{M}^+$. Then $\mu f < \infty$ if and only if $\sum_{n=1}^{\infty}\mu\{f \geq n\} < \infty$.

The assertion is just a pointwise inequality in disguise. By considering separately values for which $k \leq f(x) < k+1$, for $k = 0, 1, 2, \ldots$, you can verify the pointwise inequality between functions,

$$\textstyle\sum_{n=1}^{\infty}\{f \geq n\} \leq f \leq 1 + \sum_{n=1}^{\infty}\{f \geq n\}.$$

In fact, the sum on the left-hand side defines $\lfloor f(x) \rfloor$, the largest integer $\leq f(x)$, and the right-hand side denotes the smallest integer $> f(x)$. From the leftmost inequality,

$$\begin{aligned} \mu f &\geq \mu\left(\textstyle\sum_{n=1}^{\infty}\{f \geq n\}\right) && \text{increasing} \\ &= \lim_{N\to\infty}\mu\left(\textstyle\sum_{n=1}^{N}\{f \geq n\}\right) && \text{Monotone Convergence} \\ &= \lim_{N\to\infty}\textstyle\sum_{n=1}^{N}\mu\{f \geq n\} && \text{linearity} \\ &= \textstyle\sum_{n=1}^{\infty}\mu\{f \geq n\}. \end{aligned}$$

A similar argument gives a companion upper bound. Thus the pointwise inequality integrates out to $\sum_{n=1}^{\infty}\mu\{f \geq n\} \leq \mu f \leq \mu\mathcal{X} + \sum_{n=1}^{\infty}\mu\{f \geq n\}$, from which the asserted equivalence follows. □

Extension of the integral to a larger class of functions

Every function f in $\mathcal{M}$ can be decomposed into a difference $f = f^+ - f^-$ of two functions in $\mathcal{M}^+$, where $f^+(x) := \max(f(x), 0)$ and $f^-(x) := \max(-f(x), 0)$. To extend μ from $\mathcal{M}^+$ to a linear functional on $\mathcal{M}$ we should define $\mu f := \mu f^+ - \mu f^-$. This definition works if at least one of μf^+ and μf^- is finite; otherwise we get the dreaded $\infty - \infty$. If both $\mu f^+ < \infty$ and $\mu f^- < \infty$ (or equivalently, f is measurable and $\mu|f| < \infty$) the function f is said to be ***integrable*** or μ-integrable. The linearity property (iii) of Theorem <12> carries over partially to $\mathcal{M}$ if $\infty - \infty$ problems are excluded, although it becomes tedious to handle all the awkward cases involving $\pm\infty$. The constants α and β need no longer be nonnegative. Also if both f and g are integrable and if $f \leq g$ then $\mu f \leq \mu g$, with obvious extensions to certain cases involving ∞.

<15> **Definition.** *The set of all real-valued, μ-integrable functions in $\mathcal{M}$ is denoted by $\mathcal{L}^1(\mu)$, or $\mathcal{L}^1(\mathcal{X}, \mathcal{A}, \mu)$.*

The set $\mathcal{L}^1(\mu)$ is a vector space (stable under pointwise addition and multiplication by real numbers). The integral μ defines an increasing linear functional on $\mathcal{L}^1(\mu)$, in the sense that $\mu f \geq \mu g$ if $f \geq g$ pointwise. The Monotone Convergence property implies other powerful limit results for functions in $\mathcal{L}^1(\mu)$, as described in Section 5. By restricting μ to $\mathcal{L}^1(\mu)$, we eliminate problems with $\infty - \infty$.

For each f in $\mathcal{L}^1(\mu)$, its $\mathcal{L}^1$ ***norm*** is defined as $\|f\|_1 := \mu|f|$. Strictly speaking, $\|\cdot\|_1$ is only a seminorm, because $\|f\|_1 = 0$ need not imply that f is the zero function—as you will see in Section 6, it implies only that $\mu\{f \neq 0\} = 0$. It is common practice to ignore the small distinction and refer to $\|\cdot\|_1$ as a norm on $\mathcal{L}^1(\mu)$.

<16> **Example.** Let Ψ be a convex, real-valued function on $\mathbb{R}$. The function Ψ is measurable (because $\{\Psi \le t\}$ is an interval for each real t), and for each x_0 in $\mathbb{R}$ there is a constant α such that $\Psi(x) \ge \Psi(x_0) + \alpha(x - x_0)$ for all x (Appendix C).

Let $\mathbb{P}$ be a probability measure, and X be an integrable random variable. Choose $x_0 := \mathbb{P}X$. From the inequality $\Psi(x) \ge -|\Psi(x_0)| - |\alpha|(|x| + |x_0|)$ we deduce that $\mathbb{P}\Psi(X)^- \le |\Psi(x_0)| + |\alpha|(\mathbb{P}|X| + |x_0|) < \infty$. Thus we should have no $\infty - \infty$ worries in taking expectations (that is, integrating with respect to $\mathbb{P}$) to deduce that $\mathbb{P}\Psi(X) \ge \Psi(\mathbb{P}X) + \alpha(\mathbb{P}X - x_0) = \Psi(\mathbb{P}X)$, a result known as ***Jensen's inequality***. One way to remember the direction of the inequality is to note that $0 \le \text{var}(X) = \mathbb{P}X^2 - (\mathbb{P}X)^2$, which corresponds to the case $\Psi(x) = x^2$. □

Integrals with respect to Lebesgue measure

Lebesgue measure $\mathfrak{m}$ on $\mathcal{B}(\mathbb{R})$ corresponds to length: $\mathfrak{m}[a, b] = b - a$ for each interval. I will occasionally revert to the traditional ways of writing such integrals,

$$\mathfrak{m}f = \int f(x)\,dx = \int_{-\infty}^{\infty} f(x)\,dx \quad \text{and} \quad \mathfrak{m}^x\big(f(x)\{a \le x \le b\}\big) = \int_a^b f(x)dx.$$

Don't worry about confusing the Lebesgue integral with the Riemann integral over finite intervals. Whenever the Riemann is well defined, so is the Lebesgue, and the two sorts of integral have the same value. The Lebesgue is a more general concept. Indeed, facts about the Riemann are often established by an appeal to theorems about the Lebesgue. You do not have to abandon what you already know about integration over finite intervals.

The improper Riemann integral, $\int_{-\infty}^{\infty} f(x)\,dx = \lim_{n\to\infty} \int_{-n}^{n} f(x)\,dx$, also agrees with the Lebesgue integral provided $\mathfrak{m}|f| < \infty$. If $\mathfrak{m}|f| = \infty$, as in the case of the function $f(x) := \sum_{n=1}^{\infty}\{n \le x < n+1\}(-1)^n/n$, the improper Riemann integral might exist as a finite limit, while the Lebesgue integral $\mathfrak{m}f$ does not exist.

*4. Construction of integrals from measures

To construct the integral $\tilde{\mu}$ as a functional on $\mathcal{M}^+(\mathcal{X}, \mathcal{A})$, starting from a measure μ on the sigma-field $\mathcal{A}$, we use approximation from below by means of simple functions.

First we must define $\tilde{\mu}$ on $\mathcal{M}^+_{\text{simple}}$. The representation of a simple function as a linear combination of indicator functions is not unique, but the additivity properties of the measure μ will let us use any representation to define the integral. For example, if $s := 3A_1 + 7A_2 = 3A_1A_2^c + 10A_1A_2 + 7A_1^cA_2$, then

$$3\mu(A_1) + 7\mu(A_2) = 3\mu(A_1A_2^c) + 10\mu(A_1A_2) + 7\mu(A_1^cA_2).$$

More generally, if $s := \sum_i \alpha_i A_i$ has another representation $s = \sum_j \beta_j B_j$, then $\sum_i \alpha_i \mu A_i = \sum_j \beta_j \mu B_j$. Proof? Thus we can uniquely define $\tilde{\mu}(s)$ for a simple function $s := \sum_i \alpha_i A_i$ by $\tilde{\mu}(s) := \sum_i \alpha_i \mu A_i$.

Define the increasing functional $\tilde{\mu}$ on $\mathcal{M}^+$ by

$$\tilde{\mu}(f) := \sup\{\tilde{\mu}(s) : f \geq s \in \mathcal{M}^+_{\text{simple}}\}.$$

That is, the integral of f is a supremum of integrals of nonnegative simple functions less than f.

From the representation of simple functions as linear combinations of disjoint sets in $\mathcal{A}$, it is easy to show that $\tilde{\mu}(\mathbb{I}_A) = \mu A$ for every A in $\mathcal{A}$. It is also easy to show that $\tilde{\mu}(0) = 0$, and $\tilde{\mu}(\alpha f) = \alpha\tilde{\mu}(f)$ for nonnegative real α, and

$$\tilde{\mu}(f + g) \geq \tilde{\mu}(f) + \tilde{\mu}(g). \qquad \text{<17>}$$

The last inequality, which is usually referred to as the superadditivity property, follows from the fact that if $f \geq u$ and $g \geq v$, and both u and v are simple, then $f + g \geq u + v$ with $u + v$ simple.

Only the Monotone Convergence property and the companion to <17>,

$$\tilde{\mu}(f + g) \leq \tilde{\mu}(f) + \tilde{\mu}(g), \qquad \text{<18>}$$

require real work. Here you will see why measurability is needed.

Proof of inequality <18>. Let s be a simple function $\leq f + g$, and let ϵ be a small positive number. It is enough to construct simple functions u, v with $u \leq f$ and $v \leq g$ such that $u + v \geq (1 - \epsilon)s$. For then $\tilde{\mu}f + \tilde{\mu}g \geq \tilde{\mu}u + \tilde{\mu}v \geq (1 - \epsilon)\tilde{\mu}s$, from which the subadditivity inequality <18> follows by taking a supremum over simple functions then letting ϵ tend to zero.

For simplicity of notation I will assume s to be very simple: $s := A$. You can repeat the argument for each A_i in a representation $\sum_i \alpha_i A_i$ with disjoint A_i to get the general result. Suppose $\epsilon = 1/m$ for some positive integer m. Write ℓ_j for j/m. Define simple functions

$$u := A\{f \geq 1\} + \sum_{j=1}^m A\left\{\ell_{j-1} \leq f < \ell_j\right\}\ell_{j-1},$$
$$v := \sum_{j=1}^m A\left\{\ell_{j-1} \leq f < \ell_j\right\}\left(1 - \ell_j\right).$$

The measurability of f ensures $\mathcal{A}$-measurability of all the sets entering into the definitions of u and v. For the inequality $v \leq g$, notice that $f + g \geq 1$ on A, so $g > 1 - \ell_j = v$ when $\ell_{j-1} \leq f < \ell_j$ on A. Finally, note that the simple functions were chosen so that

$$u + v = A\{f \geq 1\} + \sum_{j=1}^m A\left\{\ell_{j-1} \leq f < \ell_j\right\}(1 - \epsilon) \geq (1 - \epsilon)A,$$

as desired. □

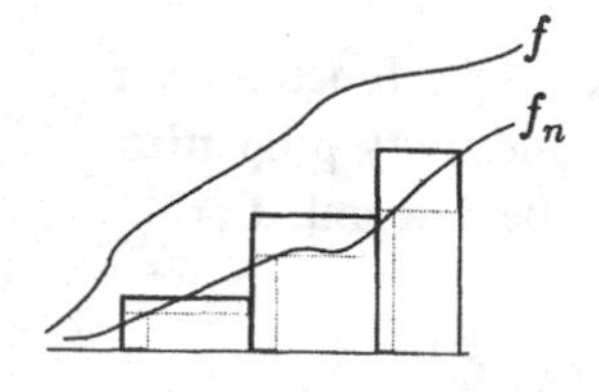

Proof of the Monotone Convergence property. Suppose $f_n \in \mathcal{M}^+$ and $f_n \uparrow f$. Suppose $f \geq s := \sum \alpha_i A_i$, with the A_i disjoint sets in $\mathcal{A}$ and $\alpha_i > 0$. Define approximating simple functions $s_n := \sum_i (1 - \epsilon)\alpha_i A_i\{f_n \geq (1 - \epsilon)\alpha_i\}$. Clearly $s_n \leq f_n$. The simple function s_n is one of those that enters into the supremum defining $\tilde{\mu} f_n$. It follows that

$$\tilde{\mu} f_n \geq \tilde{\mu}(s_n) = (1-\epsilon) \sum_i \alpha_i \mu \left(A_i \{ f_n \geq (1-\epsilon)\alpha_i \} \right).$$

On the set A_i the functions f_n increase monotonely to f, which is $\geq \alpha_i$. The sets $A_i\{f_n \geq (1-\epsilon)\alpha_i\}$ expand up to the whole of A_i. Countable additivity implies that the μ measures of those sets increase to μA_i. It follows that

$$\lim \tilde{\mu} f_n \geq \limsup \tilde{\mu} s_n \geq (1-\epsilon)\tilde{\mu} s.$$

□ Take a supremum over simple $s \leq f$ then let ϵ tend to zero to complete the proof.

5. Limit theorems

Theorem <13> identified an integral on $\mathcal{M}^+$ as an increasing linear functional with the Monotone Convergence property :

<19> $$\text{if } 0 \leq f_n \uparrow \text{ then } \mu\left(\lim_{n\to\infty} f_n\right) = \lim_{n\to\infty} \mu f_n.$$

Two direct consequences of this limit property have important applications throughout probability theory. The first, ***Fatou's Lemma***, asserts a weaker limit property for nonnegative functions when the convergence and monotonicity assumptions are dropped. The second, ***Dominated Convergence***, drops the monotonicity and nonnegativity but imposes an extra domination condition on the convergent sequence $\{f_n\}$. I have slowly realized over the years that many simple probabilistic results can be established by Dominated Convergence arguments. The Dominated Convergence Theorem is the Swiss Army Knife of probability theory.

It is important that you understand why some conditions are needed before we can interchange integration (which is a limiting operation) with an explicit limit, as in <19>. Variations on the following example form the basis for many counterexamples.

<20> **Example.** Let μ be Lebesgue measure on $\mathcal{B}[0,1]$ and let $\{\alpha_n\}$ be a sequence of positive numbers. The function $f_n(x) := \alpha_n\{0 < x < 1/n\}$ converges to zero, pointwise, but its integral $\mu(f_n) = \alpha_n/n$ need not converge to zero. For example, $\alpha_n = n^2$ gives $\mu f_n \to \infty$; the integrals diverge. And

$$\alpha_n = \begin{cases} 6n & \text{for } n \text{ even} \\ 3n & \text{for } n \text{ odd} \end{cases} \qquad \text{gives} \qquad \mu f_n = \begin{cases} 6 & \text{for } n \text{ even} \\ 3 & \text{for } n \text{ odd.} \end{cases}$$

□ The integrals oscillate.

<21> **Fatou's Lemma.** *For every sequence $\{f_n\}$ in $\mathcal{M}^+$ (not necessarily convergent), $\mu(\liminf_{n\to\infty} f_n) \leq \liminf_{n\to\infty} \mu(f_n)$.*

Proof. Write f for $\liminf f_n$. Remember what a lim inf means. Define $g_n := \inf_{m\geq n} f_m$. Then $g_n \leq f_n$ for every n and the $\{g_n\}$ sequence increases monotonely to the function f. By Monotone Convergence, $\mu f = \lim_{n\to\infty} \mu g_n$. By the increasing property, $\mu g_n \leq \mu f_n$ for each n, and hence $\lim_{n\to\infty} \mu g_n \leq \liminf_{n\to\infty} \mu f_n$. □

For dominated sequences of functions, a splicing together of two Fatou Lemma assertions gives two lim inf consequences that combine to produce a limit result. (See Problem [10] for a generalization.)

<22> **Dominated Convergence.** *Let $\{f_n\}$ be a sequence of μ-integrable functions for which $\lim_n f_n(x)$ exists for all x. Suppose there exists a μ-integrable function F, which does not depend on n, such that $|f_n(x)| \le F(x)$ for all x and all n. Then the limit function is integrable and $\mu(\lim_{n\to\infty} f_n) = \lim_{n\to\infty} \mu f_n$.*

Proof. The limit function is also bounded in absolute value by F, and hence it is integrable.

Apply Fatou's Lemma to the two sequences $\{F + f_n\}$ and $\{F - f_n\}$ in $\mathcal{M}^+$, to get

$$\mu(\liminf(F + f_n)) \le \liminf \mu(F + f_n) = \liminf (\mu F + \mu f_n),$$
$$\mu(\liminf(F - f_n)) \le \liminf \mu(F - f_n) = \liminf (\mu F - \mu f_n).$$

Simplify, using the fact that a lim inf is the same as a lim for convergent sequences.

$$\mu(F \pm \lim f_n)) \le \mu F + \liminf (\pm \mu f_n).$$

Notice that we cannot yet assert that the lim inf on the right-hand side is actually a limit. The negative sign turns a lim inf into a lim sup.

$$\mu F \pm \mu(\lim f_n) \le \begin{cases} \mu F + \liminf \mu f_n \\ \mu F - \limsup \mu f_n \end{cases}$$

Cancel out the finite number μF then rearrange, leaving

$$\limsup \mu f_n \le \mu(\lim f_n) \le \liminf \mu f_n.$$

□ The convergence assertion follows.

REMARK. You might well object to some of the steps in the proof on $\infty - \infty$ grounds. For example, what does $F(x) + f_n(x)$ mean at a point where $F(x) = \infty$ and $f_n(x) = -\infty$? To eliminate such problems, replace F by $F\{F < \infty\}$ and f_n by $f_n\{F < \infty\}$, then appeal to Lemma <26> in the next Section to ensure that the integrals are not affected.

The function F is said to ***dominate*** the sequence $\{f_n\}$. The assumption in Theorem <22> could also be written as $\mu\left(\sup_n |f_n|\right) < \infty$, with $F := \sup_n |f_n|$ as the dominating function. It is a common mistake amongst students new to the result to allow F to depend on n.

Dominated Convergence turns up in many situations that you might not at first recognize as examples of an interchange in the order of two limit procedures.

<23> **Example.** Do you know why

$$\frac{d}{dt}\int_0^1 e^{xt}x^{5/2}(1-x)^{3/2}\,dx = \int_0^1 e^{xt}x^{7/2}(1-x)^{3/2}\,dx\,?$$

Of course I just differentiated under the integral sign, but why is that allowed? The neatest justification uses a Dominated Convergence argument.

More generally, for each t in an interval $(-\delta, \delta)$ about the origin let $f(\cdot, t)$ be a μ-integrable function on $\mathcal{X}$, such that the function $f(x, \cdot)$ is differentiable in $(-\delta, \delta)$

for each x. We need to justify taking the derivative at $t = 0$ inside the integral, to conclude that

$$\frac{d}{dt}\left(\mu^x f(x,t)\right)\Big|_{t=0} = \mu^x\left(\frac{\partial}{\partial t} f(x,t)\Big|_{t=0}\right). \tag{<24>}$$

Domination of the partial derivative will suffice.

Write $g(t)$ for $\mu^x f(x,t)$ and $\Delta(x,t)$ for the partial derivative $\frac{\partial}{\partial t} f(x,t)$. Suppose there exists a μ-integrable function M such that

$$|\Delta(x,t)| \le M(x) \qquad \text{for all } x, \text{ all } t \in (-\delta, \delta).$$

To establish <24>, it is enough to show that

$$\frac{g(h_n) - g(0)}{h_n} \to \mu^x \Delta(x, 0) \tag{<25>}$$

for every sequence $\{h_n\}$ of nonzero real numbers tending to zero. (Please make sure that you understand why continuous limits can be replaced by sequential limits in this way. It is a common simplification.) With no loss of generality, suppose $\delta > h_n > 0$ for all n. The ratio on the left-hand side of <25> equals the μ integral of the function $f_n(x) := (f(x,h_n) - f(x,0))/h_n$. By assumption, $f_n(x) \to \Delta(x,0)$ for every x. The sequence $\{f_n\}$ is dominated by M: by the mean-value theorem, for each x there exists a t_x in $(-h_n, h_n) \subseteq (-\delta, \delta)$ for which $|f_n(x)| = |\Delta(x, t_x)| \le M(x)$. An appeal to Dominated Convergence completes the argument. □

6. Negligible sets

A set N in $\mathcal{A}$ for which $\mu N = 0$ is said to be ***μ-negligible***. (Some authors use the term μ-null, but I find it causes confusion with null as a name for the empty set.) As the name suggests, we can usually ignore bad things that happen only on a negligible set. A property that holds everywhere except possibly for those x in a μ-negligible set of points is said to hold ***μ-almost everywhere*** or ***μ-almost surely***, abbreviated to ***a.e.*** $[\mu]$ or ***a.s.*** $[\mu]$, with the $[\mu]$ omitted when understood.

There are several useful facts about negligible sets that are easy to prove and exceedingly useful to have formally stated. They depend on countable additivity, via its Monotone Convergence generalization. I state them only for nonnegative functions, leaving the obvious extensions for $\mathcal{L}^1(\mu)$ to you.

<26> **Lemma.** *For every measure μ:*

(i) *if $g \in \mathcal{M}^+$ and $\mu g < \infty$ then $g < \infty$* a.e. $[\mu]$;

(ii) *if $g, h \in \mathcal{M}^+$ and $g = h$* a.e. $[\mu]$ *then $\mu g = \mu h$;*

(iii) *if $N_1, N_2, \ldots$ is a sequence of negligible sets then $\bigcup_i N_i$ is also negligible;*

(iv) *if $g \in \mathcal{M}^+$ and $\mu g = 0$ then $g = 0$* a.e. $[\mu]$.

Proof. For (i): Integrate out the inequality $g \ge n\{g = \infty\}$ for each positive integer n to get $\infty > \mu g \ge n\mu\{g = \infty\}$. Let n tend to infinity to deduce that $\mu\{g = \infty\} = 0$.

For (ii): Invoke the increasing and Monotone Convergence properties of integrals, starting from the pointwise bound $h \le g + \infty\{h \ne g\} = \lim_n (g + n\{h \ne g\})$

to deduce that $\mu h \le \lim_n (\mu g + n\mu\{h \ne g\}) = \mu g$. Reverse the roles of g and h to get the reverse inequality.

For (iii): Invoke Monotone Convergence for the right-hand side of the pointwise inequality $\cup_i N_i \le \sum_i N_i$ to get $\mu(\cup_i N_i) \le \mu\left(\sum_i N_i\right) = \sum_i \mu N_i = 0$.

For (iv): Put $N_n := \{g \ge 1/n\}$ for $n = 1, 2, \ldots$. Then $\mu N_n \le n\mu g = 0$, from which it follows that $\{g > 0\} = \bigcup_n N_n$ is negligible. □

REMARK. Notice the appeals to countable additivity, via the Monotone Convergence property, in the proofs. Results such as (iv) fail without countable additivity, which might trouble those brave souls who would want to develop a probability theory using only finite additivity.

Property (iii) can be restated as: if $A \in \mathcal{A}$ and A is covered by a countable family of negligible sets then A is negligible. Actually we can drop the assumption that $A \in \mathcal{A}$ if we enlarge the sigma-field slightly.

<27> **Definition.** *The μ-completion of the sigma-field $\mathcal{A}$ is the class $\mathcal{A}_\mu$ of all those sets B for which there exist sets A_0, A_1 in $\mathcal{A}$ with $A_0 \subseteq B \subseteq A_1$ and $\mu(A_1 \backslash A_0) = 0$.*

You should check that $\mathcal{A}_\mu$ is a sigma-field and that μ has a unique extension to a measure on $\mathcal{A}_\mu$ defined by $\mu B := \mu A_0 = \mu A_1$, with A_0 and A_1 as in the Definition. More generally, for each f in $\mathcal{M}^+(\mathcal{X}, \mathcal{A}_\mu)$, you should show that there exist functions f_0, g_0 in $\mathcal{M}^+(\mathcal{X}, \mathcal{A})$ for which $f_0 \le f \le f_0 + g_0$ and $\mu g_0 = 0$. Of course, we then have $\mu f := \mu f_0$.

The Lebesgue sigma-field on the real line is the completion of the Borel sigma-field with respect to Lebesgue measure.

<28> **Example.** Here is one of the standard methods for proving that some measurable set A has zero μ measure. Find a measurable function f for which $f(x) > 0$, for all x in A, and $\mu(fA) = 0$. From part (iv) of Lemma <26> deduce that $fA = 0$ a.e. $[\mu]$. That is, $f(x) = 0$ for almost all x in A. The set $A = \{x \in A : f(x) > 0\}$ must be negligible. □

Many limit theorems in probability theory assert facts about sequences that hold only almost everywhere.

<29> **Example.** (Generalized Borel-Cantelli lemma) Suppose $\{f_n\}$ is a sequence in $\mathcal{M}^+$ for which $\sum_n \mu f_n < \infty$. By Monotone Convergence, $\mu \sum_n f_n = \sum_n \mu f_n < \infty$. Part (i) of Lemma <26> then gives $\sum_n f_n(x) < \infty$ for μ almost all x.

For the special case of probability measure with each f_n an indicator function of a set in $\mathcal{A}$, the convergence property is called the ***Borel-Cantelli lemma***: If $\sum_n \mathbb{P}A_n < \infty$ then $\sum_n A_n < \infty$ almost surely. That is,

$$\mathbb{P}\{\omega \in \Omega : \omega \in A_n \text{ for infinitely many } n\} = 0,$$

a trivial result that, nevertheless, is the basis for much probabilistic limit theory. The event in the last display is often written in abbreviated form, $\{A_n \text{ i. o.}\}$.

REMARK. For sequences of independent events, there is a second part to the Borel-Cantelli lemma (Problem [1]), which asserts that if $\sum_n \mathbb{P}A_n = \infty$ then $\mathbb{P}\{A_n \text{ i. o.}\} = 1$. Problem [2] establishes an even stronger converse, replacing independence by a weaker limit property.

The Borel-Cantelli argument often takes the following form when invoked to establish almost sure convergence. You should make sure you understand the method, because the details are usually omitted in the literature.

Suppose $\{X_n\}$ is a sequence of random variables (all defined on the same Ω) for which $\sum_n \mathbb{P}\{|X_n| > \epsilon\} < \infty$ for each $\epsilon > 0$. By Borel-Cantelli, to each $\epsilon > 0$ there is a $\mathbb{P}$-negligible set $N(\epsilon)$ for which $\sum_n \{|X_n(\omega)| > \epsilon\} < \infty$ if $\omega \in N(\epsilon)^c$. A sum of integers converges if and only if the summands are eventually zero. Thus to each ω in $N(\epsilon)^c$ there exists a finite $n(\epsilon, \omega)$ such that $|X_n(\omega)| \le \epsilon$ when $n \ge n(\epsilon, \omega)$.

We have an uncountable family of negligible sets $\{N(\epsilon) : \epsilon > 0\}$. We are allowed to neglect only countable unions of negligible sets. Replace ϵ by a sequence of values such as 1, 1/2, 1/3, 1/4, ..., tending to zero. Define $N := \bigcup_{k=1}^{\infty} N(1/k)$, which, by part (iii) of Lemma <26>, is negligible. For each ω in N^c we have $|X_n(\omega)| \le 1/k$ when $n \ge n(1/k, \omega)$. Consequently, $X_n(\omega) \to 0$ as $n \to \infty$ for each ω in N^c; the sequence $\{X_n\}$ converges to zero almost surely. □

For measure theoretic arguments with a fixed μ, it is natural to treat as identical those functions that are equal almost everywhere. Many theorems have trivial modifications with equalities replaced by almost sure equalities, and convergence replaced by almost sure convergence, and so on. For example, Dominated Convergence holds in a slightly strengthened form:

> Let $\{f_n\}$ be a sequence of measurable functions for which $f_n(x) \to f(x)$ at μ *almost* all x. Suppose there exists a μ-integrable function F, which does not depend on n, such that $|f_n(x)| \le F(x)$ for μ *almost* all x and all n. Then $\mu f_n \to \mu f$.

Most practitioners of probability learn to ignore negligible sets (and then suffer slightly when they come to some stochastic process arguments where the handling of uncountable families of negligible sets requires more delicacy). For example, if I could show that a sequence $\{f_n\}$ converges almost everywhere I would hardly hesitate to write: Define $f := \lim_n f_n$. What happens at those x where $f_n(x)$ does not converge? If hard pressed I might write:

$$\text{Define } f(x) := \begin{cases} \lim_n f_n(x) & \text{on the set where the limit exists,} \\ 0 & \text{otherwise.} \end{cases}$$

You might then wonder if the function so-defined were measurable (it is), or if the set where the limit exists is measurable (it is). A sneakier solution would be to write: Define $f(x) := \limsup_n f_n(x)$. It doesn't much matter what happens on the negligible set where the limsup is not equal to the liminf, which happens only when the limit does not exist.

A more formal way to equate functions equal almost everywhere is to work with equivalence classes, $[f] := \{g \in \mathcal{M} : f = g \text{ a.e. } [\mu]\}$. The almost sure equivalence also partitions $\mathcal{L}^1(\mu)$ into equivalence classes, for which we can define $\mu[f] := \mu g$ for an arbitrary choice of g from $[f]$. The collection of all these equivalence classes is denoted by $L^1(\mu)$. The L^1 norm, $\|[f]\|_1 := \|f\|_1$, is a true norm on L^1, because $[f]$ equals the equivalence class of the zero function when $\|[f]\|_1 = 0$. Few authors are careful about maintaining the distinction between f and $[f]$, or between $L^1(\mu)$ and $\mathcal{L}^1(\mu)$.

*7. L^p spaces

For each real number p with $p \geq 1$ the ***$\mathcal{L}^p$-norm*** is defined on $\mathcal{M}(\mathcal{X}, \mathcal{A}, \mu)$ by $\|f\|_p := (\mu|f|^p)^{1/p}$. Problem [17] shows that the map $f \mapsto \|f\|_p$ satisfies the triangle inequality, $\|f + g\|_p \leq \|f\|_p + \|g\|_p$, at least when restricted to real-valued functions in $\mathcal{M}$.

As with the $\mathcal{L}^1$-norm, it is not quite correct to call $\|\cdot\|_p$ a norm, for two reasons: there are measurable functions for which $\|f\|_p = \infty$, and there are nonzero measurable functions for which $\|f\|_p = 0$. We avoid the first complication by restricting attention to the vector space $\mathcal{L}^p := \mathcal{L}^p(\mathcal{X}, \mathcal{A}, \mu)$ of all real-valued, $\mathcal{A}$-measurable functions for which $\|f\|_p < \infty$. We could avoid the second complication by working with the vector space $L^p := L^p(\mathcal{X}, \mathcal{A}, \mu)$ of μ-equivalence classes of functions in $\mathcal{L}^p(\mathcal{X}, \mathcal{A}, \mu)$. That is, the members of L^p are the μ-equivalence classes, $[f] := \{g \in \mathcal{L}^p : g = f \text{ a.e. } [\mu]\}$, with f in $\mathcal{L}^p$. (See Problem [20] for the limiting case, $p = \infty$.)

REMARK. The correct term for $\|\cdot\|_p$ on $\mathcal{L}^p$ is ***pseudonorm***, meaning that it has all the properties of a norm (triangle inequality, and $\|cf\| = |c|\,\|f\|$ for real constants c) except that it might be zero for nonzero functions. Again, few authors are careful about maintaining the distinction between $\mathcal{L}^p$ and L^p.

Problem [19] shows that the norm defines a complete pseudometric on $\mathcal{L}^p$ (and a complete metric on L^p). That is, if $\{f_n\}$ is a Cauchy sequence of functions in $\mathcal{L}^p$ (meaning that $\|f_n - f_m\|_p \to 0$ as $\min(m, n) \to \infty$) then there exists a function f in $\mathcal{L}^p$ for which $\|f_n - f\|_n \to 0$. The limit function f is unique up to a μ-equivalence.

For our purposes, the case where p equals 2 will be the most important. The pseudonorm is then generated by an inner product (or, more correctly, a "pseudo" inner product), $\langle f, g\rangle := \mu(fg)$. That is, $\|f\|_2^2 := \langle f, f\rangle$. The inner product has the properties:

(a) $\langle \alpha f + \beta g, h\rangle = \alpha\langle f, h\rangle + \beta\langle g, h\rangle$ for all real α, β all f, g, h in $\mathcal{L}^2$;

(b) $\langle f, g\rangle = \langle g, f\rangle$ for all f, g in $\mathcal{L}^2$;

(c) $\langle f, f\rangle \geq 0$ with equality if and only if $f = 0$ a.e. $[\mu]$.

If we work with the equivalence classes of L^2 then (c) is replaced by the assertion that $\langle [f], [f]\rangle$ equals zero if and only if $[f]$ is zero, as required for a true inner product.

A vector space equipped with an inner product whose corresponding norm defines a complete metric is called a ***Hilbert space***, a generalization of ordinary Euclidean space. Arguments involving Hilbert spaces look similar to their analogs for Euclidean space, with an occasional precaution against possible difficulties with infinite dimensionality. Many results in Probability and Statistics rely on Hilbert space methods: information inequalities; the Blackwell-Rao theorem; the construction of densities and abstract conditional expectations; Hellinger differentiability; prediction in time series; Gaussian process theory; martingale theory; stochastic integration; and much more.

Some of the basic theory for Hilbert space is established in Appendix B. For the next several Chapters, the following two Hilbert space results, specialized to L^2 spaces, will suffice.

(1) **Cauchy-Schwarz inequality:** $|\mu(fg)| \le \|f\|_2 \|g\|_2$ for all f, g in $L^2(\mu)$, which follows from the Hölder inequality (Problem [15]).

(2) **Orthogonal projections:** Let $\mathcal{H}_0$ be a closed subspace of $\mathcal{L}^2(\mu)$. For each f in $\mathcal{L}^2$ there is a f_0 in $\mathcal{H}_0$, the (orthogonal) projection of f onto $\mathcal{H}_0$, for which $f - f_0$ is orthogonal to $\mathcal{H}_0$, that is, $\langle f - f_0, g\rangle = 0$ for all g in $\mathcal{H}_0$. The point f_0 minimizes $\|f - h\|$ over all h in $\mathcal{H}_0$. The projection f_0 is unique up to a μ-almost sure equivalence.

REMARK. A closed subspace $\mathcal{H}_0$ of $\mathcal{L}^2$ must contain all f in $\mathcal{L}^2$ for which there exist $f_n \in \mathcal{H}_0$ with $\|f_n - f\|_2 \to 0$. In particular, if f belongs to $\mathcal{H}_0$ and $g = f$ a.e. $[\mu]$ then g must also belong to $\mathcal{H}_0$. If $\mathcal{H}_0$ is closed, the set of equivalence classes $\widetilde{\mathcal{H}}_0 = \{[f] : f \in \mathcal{H}_0\}$ must be a closed subspace of $L^2(\mu)$, and $\mathcal{H}_0$ must equal the union of all equivalence classes in $\widetilde{\mathcal{H}}_0$.

For us the most important subspaces of $\mathcal{L}^2(\mathcal{X}, \mathcal{A}, \mu)$ will be defined by the sub-sigma-fields $\mathcal{A}_0$ of $\mathcal{A}$. Let $\mathcal{H}_0 = \mathcal{L}^2(\mathcal{X}, \mathcal{A}_0, \mu)$. The corresponding $L^2(\mathcal{X}, \mathcal{A}_0, \mu)$ is a Hilbert space in its own right, and therefore it is a closed subspace of $L^2(\mathcal{X}, \mathcal{A}, \mu)$. Consequently $\mathcal{H}_0$ is a complete subspace of $\mathcal{L}^2$: if $\{f_n\}$ is a Cauchy sequence in $\mathcal{H}_0$ then there exists an $f_0 \in \mathcal{H}_0$ such that $\|f_n - f_0\|_2 \to 0$. However, $\{f_n\}$ also converges to every other $\mathcal{A}$-measurable f for which $f = f_0$ a.e. $[\mu]$. Unless $\mathcal{A}_0$ contains all μ-negligible sets from $\mathcal{A}$, the limit f need not be $\mathcal{A}_0$-measurable; the subspace $\mathcal{H}_0$ need not be closed. If we work instead with the corresponding $L^2(\mathcal{X}, \mathcal{A}, \mu)$ and $L^2(\mathcal{X}, \mathcal{A}_0, \mu)$ we do get a closed subspace, because the equivalence class of the limit function is uniquely determined.

*8. Uniform integrability

Suppose $\{f_n\}$ is a sequence of measurable functions converging almost surely to a limit f. If the sequence is dominated by some μ-integrable function F, then $2F \ge |f_n - f| \to 0$ almost surely, from which it follows, via Dominated Convergence, that $\mu|f_n - f| \to 0$. That is, domination plus almost sure convergence imply convergence in $\mathcal{L}^1(\mu)$ norm. The converse is not true: μ equal to Lebesgue measure and $f_n(x) := n\{(n+1)^{-1} < x \le n^{-1}\}$ provides an instance of $\mathcal{L}^1$ convergence without domination.

At least when we deal with finite measures, there is an elegant circle of equivalences, involving a concept (convergence in measure) slightly weaker than almost sure convergence and a concept (uniform integrability) slightly weaker than domination. With no loss of generality, I will explain the connections for a sequence of random variables $\{X_n\}$ on a probability space $(\Omega, \mathcal{F}, \mathbb{P})$.

The sequence is said to ***converge in probability*** to a random variable X, sometimes written $X_n \xrightarrow{\mathbb{P}} X$, if $\mathbb{P}\{|X_n - X| > \epsilon\} \to 0$ for each $\epsilon > 0$. Problem [14] guides you through the proofs of the following facts.

(a) If $\{X_n\}$ converges to X almost surely then $X_n \to X$ in probability, but the converse is false: there exist sequences that converge in probability but not almost surely.

(b) If $\{X_n\}$ converges in probability to X, there is an increasing sequence of positive integers $\{n(k)\}$ for which $\lim_{k\to\infty} X_{n(k)} = X$ almost surely.

If a random variable Z is integrable then a Dominated Convergence argument shows that $\mathbb{P}|Z|\{|Z| > M\} \to 0$ as $M \to \infty$. Uniform integrability requires that the convergence holds uniformly over a class of random variables. Very roughly speaking, it lets us act almost as if all the random variables were bounded by a constant M, at least as far as $\mathcal{L}^1$ arguments are concerned.

<30> **Definition.** *A family of random variables $\{Z_t : t \in T\}$ is said to be uniformly integrable if $\sup_{t\in T} \mathbb{P}|Z_t|\{|Z_t| > M\} \to 0$ as $M \to \infty$.*

It is sometimes slightly more convenient to check for uniform integrability by means of an ϵ-δ characterization.

<31> **Lemma.** *A family of random variables $\{Z_t : t \in T\}$ is uniformly integrable if and only if both the following conditions hold:*

(i) $\sup_{t\in T} \mathbb{P}|Z_t| < \infty$

(ii) for each $\epsilon > 0$ there exists a $\delta > 0$ such that $\sup_{t\in T} \mathbb{P}|Z_t|F \le \epsilon$ for every event F with $\mathbb{P}F < \delta$.

REMARK. Requirement (i) is superfluous if, for each $\delta > 0$, the space Ω can be partitioned into finitely many pieces each with measure less than δ.

Proof. Given uniform integrability, (i) follows from $\mathbb{P}|Z_t| \le M + \mathbb{P}|Z_t|\{|Z_t| > M\}$, and (ii) follows from $\mathbb{P}|Z_t|F \le M\mathbb{P}F + \mathbb{P}|Z_t|\{|Z_t| > M\}$.

Conversely, if (i) and (ii) hold then the event $\{|Z_t| > M\}$ is a candidate for the F in (ii) when M is so large that $\mathbb{P}\{|Z_t| > M\} \le \sup_{t\in T} \mathbb{P}|Z_t|/M < \delta$. It follows that $\sup_{t\in T} \mathbb{P}|Z_t|\{|Z_t| > M\} \le \epsilon$ if M is large enough. □

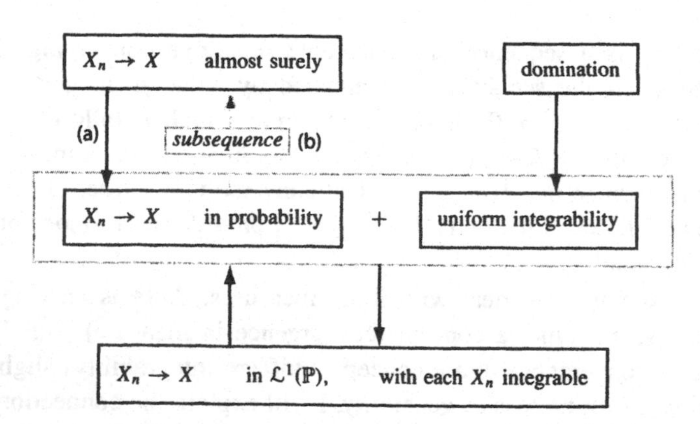

The diagram summarizes the interconnections between the various convergence concepts, with each arrow denoting an implication. The relationship between almost sure convergence and convergence in probability corresponds to results (a) and (b) noted above. A family $\{Z_t : t \in T\}$ dominated by an integrable random variable Y

is also uniformly integrable, because $Y\{Y \geq M\} \geq |Z_t|\{|Z_t| \geq M\}$ for every t. Only the implications leading to and from the box for the $\mathcal{L}^1$ convergence remain to be proved.

<32> **Theorem.** *Let $\{X_n : n \in \mathbb{N}\}$ be a sequence of integrable random variables. The following two conditions are equivalent.*

(i) *The sequence is uniformly integrable and it converges in probability to a random variable X_∞, which is necessarily integrable.*

(ii) *The sequence converges in $\mathcal{L}^1$ norm, $\mathbb{P}|X_n - X_\infty| \to 0$, with a limit X_∞ that is necessarily integrable.*

Proof. Suppose (i) holds. The assertion about integrability of X_∞ follows from Fatou's lemma, because $|X_{n'}| \to |X_\infty|$ almost surely along some subsequence, so that $\mathbb{P}|X_\infty| \leq \liminf_{n'} \mathbb{P}|X_{n'}| \leq \sup_n \mathbb{P}|X_n| < \infty$. To prove $\mathcal{L}^1$ convergence, first split according to whether $|X_n - X_\infty|$ is less than ϵ or not, and then split according to whether $\max(|X_n|, |X_\infty|)$ is less than some large constant M or not.

$$\begin{aligned}\mathbb{P}|X_n - X_\infty| &\leq \epsilon + \mathbb{P}\left(|X_n| + |X_\infty|\right)\{|X_n - X_\infty| > \epsilon\}\\ &\leq \epsilon + 2M\mathbb{P}\{|X_n - X_\infty| > \epsilon\} + \mathbb{P}\left(|X_n| + |X_\infty|\right)\{|X_n| \vee |X_\infty| > M\}.\end{aligned}$$

Split the event $\{|X_n| \vee |X_\infty| > M\}$ according to which of the two random variables is larger, to bound the last term by $2\mathbb{P}|X_n|\{|X_n| > M\} + 2\mathbb{P}|X_\infty|\{|X_\infty| > M\}$. Invoke uniform integrability of $\{X_n\}$ and integrability of X_∞ to find an M that makes this bound small, uniformly in n. With M fixed, the convergence in probability sends $M\mathbb{P}\{|X_n - X_\infty| > \epsilon\}$ to zero as $n \to \infty$.

Conversely, if the sequence converges in $\mathcal{L}^1$, then X_∞ must be integrable, because $\mathbb{P}|X_\infty| \leq \mathbb{P}|X_n| + \mathbb{P}|X_n - X_\infty|$ for each n. When $|X_n| \leq M$ or $|X_\infty| > M/2$, the inequality

$$|X_n|\{|X_n| > M\} \leq |X_\infty|\{|X_\infty| > M/2\} + 2|X_n - X_\infty|,$$

is easy to check; and when $|X_\infty| \leq M/2$ and $|X_n| > M$, it follows from the inequality $|X_n - X_\infty| \geq |X_n| - |X_\infty| \geq |X_n|/2$. Take expectations, choose M large enough to make the contribution from X_∞ small, then let n tend to infinity to find an n_0 such that $\mathbb{P}|X_n|\{|X_n| > M\} < \epsilon$ for $n > n_0$. Increase M if necessary to handle the corresponding tail contributions for $n \leq n_0$. □

9. Image measures and distributions

Suppose μ is a measure on a sigma-field $\mathcal{A}$ of subsets of $\mathcal{X}$ and T is a map from $\mathcal{X}$ into a set $\mathcal{Y}$, equipped with a sigma-field $\mathcal{B}$. If T is $\mathcal{A}\backslash\mathcal{B}$-measurable we can carry μ over to $\mathcal{Y}$, by defining

<33> $$\nu B := \mu(T^{-1}B) \qquad \text{for each } B \text{ in } \mathcal{B}.$$

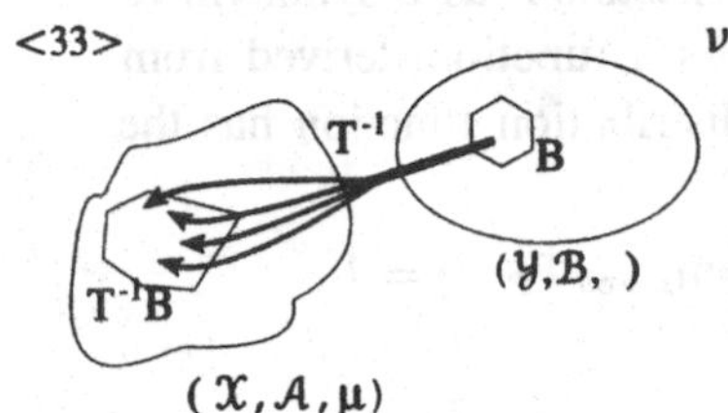

Actually the operation is more one of carrying the sets back to the measure rather than carrying the measure over to the sets, but the net result is the definition of a new set function on $\mathcal{B}$.

It is easy to check that ν is a measure on $\mathcal{B}$, using facts such as $T^{-1}(B^c) = (T^{-1}B)^c$ and $T^{-1}(\cup_i B_i) = \cup_i T^{-1}B_i$. It is called the ***image measure of μ under T***, or just the image measure, and is denoted by μT^{-1} or μ_T or $T(\mu)$, or even just $T\mu$. The third and fourth forms, which I prefer to use, have the nice property that if μ is a point mass concentrated at x then $T(\mu)$ denotes a point mass concentrated at $T(x)$.

Starting from definition <33> we could prove facts about integrals with respect to the image measure ν. For example, we could show

<34> $$\nu g = \mu(g \circ T) \qquad \text{for all } g \in \mathcal{M}^+(\mathcal{Y}, \mathcal{B}).$$

The small circle symbol $\circ$ denotes the composition of functions: $(g \circ T)(x) := g(Tx)$.

The proof of <34> could follow the traditional path: first argue by linearity from <33> to establish the result for simple functions; then take monotone limits of simple functions to extend to $\mathcal{M}^+(\mathcal{Y}, \mathcal{B})$.

There is another method for constructing image measures that gets <34> all in one step. *Define* an increasing linear functional ν on $\mathcal{M}^+(\mathcal{Y}, \mathcal{B})$ by $\nu g := \mu(g \circ T)$. It inherits the Monotone Convergence property directly from μ, because, if $0 \le g_n \uparrow g$ then $0 \le g_n \circ T \uparrow g \circ T$. By Theorem <13> it corresponds to a uniquely determined measure on $\mathcal{B}$. When restricted to indicator functions of measurable sets the new measure coincides with the measure defined by <33>, because if g is the indicator function of B then $g \circ T$ is the indicator function of $T^{-1}B$. (Why?) We have gained a theorem with almost no extra work, by starting with the linear functional as the definition of the image measure.

Using the notation $T\mu$ for image measure, we could rewrite the defining equality as $(T\mu)(g) := \mu(g \circ T)$ at least for all $g \in \mathcal{M}^+(\mathcal{Y}, \mathcal{B})$, a relationship that I find easier to remember.

REMARK. In the last sentence I used the qualifier *at least*, as a reminder that the equality could easily be extended to other cases. For example, by splitting into positive and negative parts then subtracting, we could extend the equality to functions in $\mathcal{L}^1(\mathcal{Y}, \mathcal{B}, \nu)$. And so on.

Several familiar probabilistic objects are just image measures. If X is a random variable, the image measure $X(\mathbb{P})$ on $\mathcal{B}(\mathbb{R})$ is often written $\mathbb{P}_X$, and is called the ***distribution of* X**. More generally, if X and Y are random variables defined on the same probability space, they together define a ***random vector***, a (measurable—see Chapter 4) map $T(\omega) = (X(\omega), Y(\omega))$ from Ω into $\mathbb{R}^2$. The image measure $T(\mathbb{P})$ on $\mathcal{B}(\mathbb{R}^2)$ is called the ***joint distribution of* X *and* Y**, and is often denoted by $\mathbb{P}_{X,Y}$. Similar terminology applies for larger collections of random variables.

Image measures also figure in a construction that is discussed nonrigorously in many introductory textbooks. Let P be a probability measure on $\mathcal{B}(\mathbb{R})$. Its ***distribution function*** (also known as a cumulative distribution function) is defined by $F_P(x) := P(-\infty, x]$ for $x \in \mathbb{R}$. Don't confuse *distribution*, as a synonym for probability measure, with *distribution function*, which is a function derived from the measures of a particular collection of sets. The distribution function has the following properties.

(a) It is increasing, with $\lim_{x \to -\infty} F_P(x) = 0$ and $\lim_{x \to \infty} F_P(x) = 1$.

(b) It is continuous from the right: to each $\epsilon > 0$ and $x \in \mathbb{R}$, there exists a $\delta > 0$ such that $F_P(x) \leq F_P(y) \leq F_P(x) + \epsilon$ for $x \leq y \leq x + \delta$.

Property (a) follows from that fact that the integral is an increasing functional, and from Dominated Convergence applied to the sequences $(-\infty, -n] \downarrow \emptyset$ and $(-\infty, n] \uparrow \mathbb{R}$ as $n \to \infty$. Property (b) also follows from Dominated Convergence, applied to the sequence $(-\infty, x + 1/n] \downarrow (-\infty, x]$.

Except in introductory textbooks, and in works dealing with the order properties of the real line (such as the study of ranks and order statistics), distribution functions have a reduced role to play in modern probability theory, mostly in connection with the following method for building measures on $\mathcal{B}(\mathbb{R})$ as images of Lebesgue measure. In probability theory the construction often goes by the name of ***quantile transformation***.

<35> **Example.** There is a converse to the assertions (a) and (b) about distribution functions. Suppose F is a right-continuous, increasing function on $\mathbb{R}$ for which $\lim_{x\to-\infty} F(x) = 0$ and $\lim_{x\to\infty} F(x) = 1$. Then there exists a probability measure P such that $P(-\infty, x] = F(x)$ for all real x. To construct such a P, consider the ***quantile function*** q, defined by $q(t) := \inf\{x : F(x) \geq t\}$ for $0 < t < 1$.

By right continuity of the increasing function F, the set $\{x \in \mathbb{R} : F(x) \geq t\}$ is a closed interval of the form $[\alpha, \infty)$, with $\alpha = q(t)$. That is, for all $x \in \mathbb{R}$ and all $t \in (0, 1)$,

<36>
$$F(x) \geq t \qquad \text{if and only if} \qquad x \geq q(t).$$

In general there are many plausible, but false, equalities related to <36>. For example, it is not true in general that $F(q(t)) = t$. However, if F is continuous and strictly increasing, then q is just the inverse function of F, and the plausible equalities hold.

Let $\mathfrak{m}$ denote Lebesgue measure restricted to the Borel sigma-field on $(0, 1)$. The image measure $P := q(\mathfrak{m})$ has the desired property,

$$P(-\infty, x] = \mathfrak{m}\{t : q(t) \leq x\} = \mathfrak{m}\{t : t \leq F(x)\} = F(x),$$

the first equality by definition of the image measure, and the second by equality <36>. The result is often restated as: *if ξ has a Uniform*(0, 1) *distribution then $q(\xi)$ has distribution function F.* □

10. Generating classes of sets

To prove that all sets in a sigma-field $\mathcal{A}$ have some property one often resorts to a generating-class argument. The simplest form of such an argument has three steps:

(i) Show that all members of a subclass $\mathcal{E}$ have the property.

(ii) Show that $\mathcal{A} \subseteq \sigma(\mathcal{E})$.

(iii) Show that $\mathcal{A}_0 := \{A \in \mathcal{A} : A \text{ has the property }\}$ is a sigma-field.

Then one deduces that $\mathcal{A}_0 = \sigma(\mathcal{A}_0) \supseteq \sigma(\mathcal{E}) \supseteq \mathcal{A}$, whence $\mathcal{A}_0 = \mathcal{A}$. That is, the property holds for all sets in $\mathcal{A}$.

For some properties, direct verification of all the sigma-field requirements for $\mathcal{A}_0$ proves too difficult. In such situations an indirect argument sometimes succeeds if $\mathcal{E}$ has some extra structure. For example, if is possible to establish that $\mathcal{A}_0$ is a ***λ-system of sets***, then one needs only check one extra requirement for $\mathcal{E}$ in order to produce a successful generating-class argument.

<37> **Definition.** *A class $\mathcal{D}$ of subsets of $\mathcal{X}$ is called a λ-system if*

(i) $\mathcal{X} \in \mathcal{D}$,

(ii) if $D_1, D_2 \in \mathcal{D}$ and $D_1 \supseteq D_2$ then $D_1 \backslash D_2 \in \mathcal{D}$,

(iii) if $\{D_n\}$ is an increasing sequence of sets in $\mathcal{D}$ then $\cup_1^\infty D_n \in \mathcal{D}$.

REMARK. Some authors start from a slightly different definition, replacing requirement (iii) by

(iii)′ if $\{D_n\}$ is a sequence of disjoint sets in $\mathcal{D}$ then $\cup_1^\infty D_n \in \mathcal{D}$.

The change in definition would have little effect on the role played by λ-systems.

Many authors (including me, until recently) use the name ***Dynkin class*** instead of λ-system, but the name ***Sierpiński class*** would be more appropriate. See the Notes at the end of this Chapter.

Notice that a λ-system is also a sigma-field if and only if it is stable under finite intersections. This stability property can be inherited from a subclass $\mathcal{E}$, as in the next Theorem, which is sometimes referred to as the π–λ theorem. The π stands for *product*, an indirect reference to the stability of the subclass $\mathcal{E}$ under finite intersections (products). I think that the letter λ stands for *limit*, an indirect reference to property (iii).

<38> **Theorem.** *If $\mathcal{E}$ is stable under finite intersections, and if $\mathcal{D}$ is a λ-system with $\mathcal{D} \supseteq \mathcal{E}$, then $\mathcal{D} \supseteq \sigma(\mathcal{E})$.*

Proof. It would be enough to show that $\mathcal{D}$ is a sigma-field, by establishing that it is stable under finite intersections, but that is a little more than I know how to do. Instead we need to work with a (possibly) smaller λ-system $\mathcal{D}_0$, with $\mathcal{D} \supseteq \mathcal{D}_0 \supseteq \mathcal{E}$, for which generating class arguments can extend the assumption

<39>
$$E_1 E_2 \in \mathcal{E} \qquad \text{for all } E_1, E_2 \text{ in } \mathcal{E}$$

to an assertion that

<40>
$$D_1 D_2 \in \mathcal{D}_0 \qquad \text{for all } D_1, D_2 \text{ in } \mathcal{D}_0.$$

It will then follow that $\mathcal{D}_0$ is a sigma-field, which contains $\mathcal{E}$, and hence $\mathcal{D}_0 \supseteq \sigma(\mathcal{E})$.

The choice of $\mathcal{D}_0$ is easy. Let $\{\mathcal{D}_\alpha : \alpha \in A\}$ be the collection of all λ-systems with $\mathcal{D}_\alpha \supseteq \mathcal{E}$, one of them being the $\mathcal{D}$ we started with. Let $\mathcal{D}_0$ equal the intersection of all these $\mathcal{D}_\alpha$. That is, let $\mathcal{D}_0$ consist of all sets D for which $D \in \mathcal{D}_\alpha$ for each α. I leave it to you to check the easy details that prove $\mathcal{D}_0$ to be a λ-system. In other words, $\mathcal{D}_0$ is the smallest λ-system containg $\mathcal{E}$; it is the λ-system generated by $\mathcal{E}$.

To upgrade <39> to <40> we have to replace each E_i on the left-hand side by a D_i in $\mathcal{D}_0$, without going outside the class $\mathcal{D}_0$. The trick is to work one component at a time. Start with the E_1. Define $\mathcal{D}_1 := \{A : AE \in \mathcal{D}_0 \text{ for each } E \in \mathcal{E}\}$. From <39>, we have $\mathcal{D}_1 \supseteq \mathcal{E}$. If we show that $\mathcal{D}_1$ is a λ-system then it will follow that $\mathcal{D}_1 \supseteq \mathcal{D}_0$, because $\mathcal{D}_0$ is the smallest λ-system containing $\mathcal{E}$. Actually, the assertion that $\mathcal{D}_1$ is

λ-system is trivial; it follows immediately from the λ-system properties for $\mathcal{D}_0$ and identities like $(A_1\backslash A_2)E = (A_1E)\backslash(A_2E)$ and $(\cup_i A_i)\, E = \cup(A_iE)$.

The inclusion $\mathcal{D}_1 \supseteq \mathcal{D}_0$ implies that $D_1E_2 \in \mathcal{D}_0$ for all $D_1 \in \mathcal{D}_0$ and all $E_2 \in \mathcal{E}$. Put another way—this step is the only subtlety in the proof—we can assert that the class $\mathcal{D}_2 := \{B : BD \in \mathcal{D}_0 \text{ for each } D \in \mathcal{D}_0\}$ contains $\mathcal{E}$. Just write D_1 instead of D, and E_1 instead of B, in the definition to see that it is only a matter of switching the order of the sets.

Argue in the same way as for $\mathcal{D}_1$ to show that $\mathcal{D}_2$ is also a λ-system. It then follows that $\mathcal{D}_2 \supseteq \mathcal{D}_0$, which is another way of expressing assertion <40>. □

The proof of the last Theorem is typical of many generating class arguments, in that it is trivial once one knows what one has to check. The Theorem, or its analog for classes of functions (see the next Section), will be my main method for establishing sigma-field properties. You will be getting plenty of practice at filling in the details behind frequent assertions of "a generating class argument shows that ...." Here is a typical example to get you started.

<41> **Exercise.** Let μ and ν be finite measures on $\mathcal{B}(\mathbb{R})$ with the same distribution function. That is, $\mu(-\infty, t] = \nu(-\infty, t]$ for all real t. Show that $\mu B = \nu B$ for all $B \in \mathcal{B}(\mathbb{R})$, that is, $\mu = \nu$ as Borel measures.

SOLUTION: Write $\mathcal{E}$ for the class of all intervals $(-\infty, t]$, with $t \in \mathbb{R}$. Clearly $\mathcal{E}$ is stable under finite intersections. From Example <4>, we know that $\sigma(\mathcal{E}) = \mathcal{B}(\mathbb{R})$. It is easy to check that the class $\mathcal{D} := \{B \in \mathcal{B}(\mathbb{R}) : \mu B = \nu B\}$ is a λ-system. For example, if $B_n \in \mathcal{D}$ and $B_n \uparrow B$ then $\mu B = \lim_n \mu B_n = \lim_n \nu B_n = \nu B$, by Monotone Convergence. It follows from Theorem <38> that $\mathcal{D} \supseteq \sigma(\mathcal{E}) = \mathcal{B}(\mathbb{R})$, and the equality of the two Borel measures is established. □

When you employ a λ-system argument be sure to verify the properties required of $\mathcal{E}$. The next Example shows what can happen if you forget about the stability under finite intersections.

<42> **Example.** Consider a set $\mathcal{X}$ consisting of four points, labelled nw, ne, sw, and se. Let $\mathcal{E}$ consist of $\mathcal{X}$ and the subsets $N = \{\text{nw}, \text{ne}\}$, $S = \{\text{sw}, \text{se}\}$, $E = \{\text{ne}, \text{se}\}$, and $W = \{\text{nw}, \text{sw}\}$. Notice that $\mathcal{E}$ generates the sigma-field of all subsets of $\mathcal{X}$, but it is not stable under finite intersections. Let μ and ν be probability measures for which

$$\mu(\text{nw}) = 1/2 \quad \mu(\text{ne}) = 0 \quad \nu(\text{nw}) = 0 \quad \nu(\text{ne}) = 1/2$$
$$\mu(\text{sw}) = 0 \quad \mu(\text{se}) = 1/2 \quad \nu(\text{sw}) = 1/2 \quad \nu(\text{se}) = 0$$

Both measures give the the value 1/2 to each of N, S, E, and W, but they differ in the values they give to the four singletons. □

*11. Generating classes of functions

Theorem <38> is often used as the starting point for proving facts about measurable functions. One first invokes the Theorem to establish a property for sets in a sigma-field, then one extends by taking limits of simple functions to $\mathcal{M}^+$ and beyond, using Monotone Convergence and linearity arguments. Sometimes it is simpler to invoke an analog of the λ-system property for classes of functions.

<43> **Definition.** *Call a class $\mathcal{H}^+$ of bounded, nonnegative functions on a set $\mathcal{X}$ a* ***λ-cone*** *if:*

(i) *$\mathcal{H}^+$ is a cone, that is, if $h_1, h_2 \in \mathcal{H}^+$ and α_1 and α_2 are nonnegative constants then $\alpha_1 h_1 + \alpha_2 h_2 \in \mathcal{H}^+$;*

(ii) *each nonnegative constant function belongs to $\mathcal{H}^+$;*

(iii) *if $h_1, h_2 \in \mathcal{H}^+$ and $h_1 \geq h_2$ then $h_1 - h_2 \in \mathcal{H}^+$;*

(iv) *if $\{h_n\}$ is an increasing sequence of functions in $\mathcal{H}^+$ whose pointwise limit h is bounded then $h \in \mathcal{H}^+$.*

Typically $\mathcal{H}^+$ consists of the nonnegative functions in a vector space of bounded functions that is stable under pairwise maxima and minima.

REMARK. The name λ-cone is not standard. I found it hard to come up with a name that was both suggestive of the defining properties and analogous to the name for the corresponding classes of sets. For a while I used the term Dynkin-cones but abandoned it for historical reasons. (See the Notes.) I also toyed with the name cdl-cone, as a reminder that the cone contains the (positive) constant functions and that it is stable under (proper) differences and (monotone increasing) limits of uniformly bounded sequences.

The sigma-field properties of λ-cones are slightly harder to establish than their λ-system analogs, but the reward of more streamlined proofs will make the extra, one-time effort worthwhile. First we need an analog of the fact that a λ-system that is stable under finite intersections is also a sigma-field.

<44> **Lemma.** *If a λ-cone $\mathcal{H}^+$ is stable under the formation of pointwise products of pairs of functions then it consists of all bounded, nonnegative, $\sigma(\mathcal{H}^+)$-measurable functions, where $\sigma(\mathcal{H}^+)$ denotes the sigma-field generated by $\mathcal{H}^+$.*

Proof. First note that $\mathcal{H}^+$ must be stable under uniform limits. For suppose $h_n \to h$ uniformly, with $h_n \in \mathcal{H}^+$. Write δ_n for 2^{-n}. With no loss of generality we may suppose $h_n + \delta_n \geq h \geq h_n - \delta_n$ for all n. Notice that

$$h_n + 3\delta_n = h_n + \delta_n + \delta_{n-1} \geq h + \delta_{n-1} \geq h_{n-1}.$$

From the monotone convergence, $0 \leq h_n + 3(\delta_1 + \ldots + \delta_n) \uparrow h + 3$, deduce that $h + 3 \in \mathcal{H}^+$, and hence, via the proper difference property (iii), $h \in \mathcal{H}^+$.

Via uniform limits we can now show that $\mathcal{H}^+$ is stable under composition with any continuous nonnegative function f. Let h be a member of $\mathcal{H}^+$, bounded above by a constant D. By a trivial generalization of Problem [25], there exists a sequence of polynomials $p_n(\cdot)$ such that $\sup_{0 \leq t \leq D} |p_n(t) - f(t)| < 1/n$. The function $f_n(h) := p_n(h) + 1/n$ takes only nonegative values, and it converges uniformly to $f(h)$. Suppose $f_n(t) = a_0 + a_1 t + \ldots + a_k t^k$. Then

$$f_n(h) = (a_0^+ + a_1^+ h + \ldots + a_k^+ h^k) - (a_0^- + a_1^- h + \ldots + a_k^- h^k) \geq 0.$$

By virtue of properties (i) and (ii) of λ-cones, and the assumed stability under products, both terms on the right-hand side belong to $\mathcal{H}^+$. The proper differencing property then gives $f_n(h) \in \mathcal{H}^+$. Pass uniformly to the limit to get $f(h) \in \mathcal{H}^+$.

Write $\mathcal{E}$ for the class of all sets of the form $\{h < C\}$, with $h \in \mathcal{H}^+$ and C a positive constant. From Example <7>, every h in $\mathcal{H}^+$ is $\sigma(\mathcal{E})$-measurable, and hence $\sigma(\mathcal{E}) = \sigma(\mathcal{H}^+)$. For a fixed h and C, the continuous function $(1 - (h/C)^n)^+$ of h belongs to $\mathcal{H}^+$, and it increases monotonely to the indicator of $\{h < C\}$. Thus the indicators of all sets in $\mathcal{E}$ belong to $\mathcal{H}^+$. The assumptions about $\mathcal{H}^+$ ensure that the class $\mathcal{B}$ of all sets whose indicator functions belong to $\mathcal{H}^+$ is stable under finite intersections (products), complements (subtract from 1), and increasing countable unions (montone increasing limits). That is, $\mathcal{B}$ is a λ-system, stable under finite intersections, and containing $\mathcal{E}$. It is a sigma-field containing $\mathcal{E}$. Thus $\mathcal{B} \supseteq \sigma(\mathcal{E}) = \sigma(\mathcal{H}^+)$. That is, $\mathcal{H}^+$ contains all indicators of sets in $\sigma(\mathcal{H}^+)$.

Finally, let k be a bounded, nonnegative, $\sigma(\mathcal{H}^+)$-measurable function. From the fact that each of the sets $\{k \ge i/2^n\}$, for $i = 1, \ldots, 4^n$, belongs to the cone $\mathcal{H}^+$, we have $k_n := 2^{-n} \sum_{i=1}^{4^n} \{k \ge i/2^n\} \in \mathcal{H}^+$. The functions k_n increase monotonely to k, which consequently also belongs to $\mathcal{H}^+$. □

<45> **Theorem.** *Let $\mathcal{H}^+$ be a λ-cone of bounded, nonnegative functions, and $\mathcal{G}$ be a subclass of $\mathcal{H}^+$ that is stable under the formation of pointwise products of pairs of functions. Then $\mathcal{H}^+$ contains all bounded, nonnegative, $\sigma(\mathcal{G})$-measurable functions.*

Proof. Let $\mathcal{H}_0^+$ be the smallest λ-cone containing $\mathcal{G}$. From the previous Lemma, it is enough to show that $\mathcal{H}_0^+$ is stable under pairwise products.

Argue as in Theorem <38> for λ-systems of sets. A routine calculation shows that $\mathcal{H}_1^+ := \{h \in \mathcal{H}_0^+ : hg \in \mathcal{H}_0^+ \text{ for all } g \text{ in } \mathcal{G}\}$ is a λ-cone containing $\mathcal{G}$, and hence $\mathcal{H}_1^+ = \mathcal{H}_0^+$. That is, $h_0 g \in \mathcal{H}_0^+$ for all $h_0 \in \mathcal{H}_0^+$ and $g \in \mathcal{G}$. Similarly, the class $\mathcal{H}_2^+ := \{h \in \mathcal{H}_0^+ : h_0 h \in \mathcal{H}_0^+ \text{ for all } h_0 \text{ in } \mathcal{H}_0^+\}$ is a λ-cone. By the result for $\mathcal{H}_1^+$ we have $\mathcal{H}_2^+ \supseteq \mathcal{G}$, and hence $\mathcal{H}_2^+ = \mathcal{H}_0^+$. That is, $\mathcal{H}_0^+$ is stable under products. □

<46> **Exercise.** Let μ be a finite measure on $\mathcal{B}(\mathbb{R}^k)$. Write $\mathbb{C}_0$ for the vector space of all continuous real functions on $\mathbb{R}^k$ with compact support. Suppose f belongs to $\mathcal{L}^1(\mu)$. Show that for each $\epsilon > 0$ there exists a g in $\mathbb{C}_0$ such that $\mu|f - g| < \epsilon$. That is, show that $\mathbb{C}_0$ is dense in $\mathcal{L}^1(\mu)$ under its $\mathcal{L}^1$ norm. □

SOLUTION: Define $\mathcal{H}$ as the collection of all bounded functions in $\mathcal{L}^1(\mu)$ that can be approximated arbitrarily closely by functions from $\mathbb{C}_0$. Check that the class $\mathcal{H}^+$ of nonnegative functions in $\mathcal{H}$ is a λ-cone. Trivially it contains $\mathbb{C}_0^+$, the class of nonnegative members of $\mathbb{C}_0$. The sigma-field $\sigma(\mathbb{C}_0^+)$ coincides with the Borel sigma-field. Why? The class $\mathcal{H}^+$ consists of all bounded, nonnegative Borel measurable functions.

To approximate a general f in $\mathcal{L}^1(\mu)$, first reduce to the case of nonnegative functions by splitting into positive and negative parts. Then invoke Dominated Convergence to find a finite n for which $\mu|f^+ - f^+ \wedge n| < \epsilon$, then approximate $f^+ \wedge n$ by a member of $\mathbb{C}_0^+$. See Problem [26] for the extension of the approximation result to infinite measures. □

12. Problems

[1] Suppose events $A_1, A_2, \ldots$, in a probability space $(\Omega, \mathcal{F}, \mathbb{P})$, are independent: meaning that $\mathbb{P}(A_{i_1} A_{i_2} \ldots A_{i_k}) = \mathbb{P}A_{i_1} \mathbb{P}A_{i_2} \ldots \mathbb{P}A_{i_k}$ for all choices of distinct subscripts $i_1, i_2, \ldots, i_k$, all k. Suppose $\sum_{i=1}^{\infty} \mathbb{P}A_i = \infty$.

(i) Using the inequality $e^{-x} \geq 1 - x$, show that

$$\mathbb{P}\max_{n \leq i \leq m} A_i = 1 - \prod_{n \leq i \leq m}(1 - \mathbb{P}A_i) \geq 1 - \exp\left(-\sum_{n \leq i \leq m} \mathbb{P}A_i\right)$$

(ii) Let m then n tend to infinity, to deduce (via Dominated Convergence) that $\mathbb{P}\limsup_i A_i = 1$. That is, $\mathbb{P}\{A_i \text{ i. o.}\} = 1$.

REMARK. The result gives a converse for the Borel-Cantelli lemma from Example <29>. The next Problem establishes a similar result under weaker assumptions.

[2] Let $A_1, A_2, \ldots$ be events in a probability space $(\Omega, \mathcal{F}, \mathbb{P})$. Define $X_n = A_1 + \ldots + A_n$ and $\sigma_n = \mathbb{P}X_n$. Suppose $\sigma_n \to \infty$ and $\|X_n/\sigma_n\|_2 \to 1$. *(Compare with the inequality $\|X_n/\sigma_n\|_2 \geq 1$, which follows from Jensen's inequality.)*

(i) Show that

$$\{X_n = 0\} \leq \frac{(k - X_n)(k + 1 - X_n)}{k(k+1)}$$

for each positive integer k.

(ii) By an appropriate choice of k (depending on n) in (i), deduce that $\sum_1^\infty A_i \geq 1$ almost surely.

(iii) Prove that $\sum_m^\infty A_i \geq 1$ almost surely, for each fixed m. Hint: Show that the two convergence assumptions also hold for the sequence $A_m, A_{m+1}, \ldots$.

(iv) Deduce that $\mathbb{P}\{\omega \in A_i \text{ i. o. }\} = 1$.

(v) If $\{B_i\}$ is a sequence of events for which $\sum_i \mathbb{P}B_i = \infty$ and $\mathbb{P}B_iB_j = \mathbb{P}B_i\mathbb{P}B_j$ for $i \neq j$, show that $\mathbb{P}\{\omega \in B_i \text{ i. o. }\} = 1$.

[3] Suppose T is a function from a set $\mathcal{X}$ into a set $\mathcal{Y}$, and suppose that $\mathcal{Y}$ is equipped with a σ-field $\mathcal{B}$. Define $\mathcal{A}$ as the sigma-field of sets of the form $T^{-1}B$, with B in $\mathcal{B}$. Suppose $f \in \mathcal{M}^+(\mathcal{X}, \mathcal{A})$. Show that there exists a $\mathcal{B}\backslash\mathcal{B}[0,\infty]$-measurable function g from $\mathcal{Y}$ into $[0,\infty]$ such that $f(x) = g(T(x))$, for all x in $\mathcal{X}$, by following these steps.

(i) Show that $\mathcal{A}$ is a σ-field on $\mathcal{X}$. (It is called the σ-field generated by the map T. It is often denoted by $\sigma(T)$.)

(ii) Show that $\{f \geq i/2^n\} = T^{-1}B_{i,n}$ for some $B_{i,n}$ in $\mathcal{B}$. Define

$$f_n = 2^{-n}\sum_{i=1}^{4^n}\{f \geq i/2^n\} \qquad \text{and} \qquad g_n = 2^{-n}\sum_{i=1}^{4^n} B_{i,n}.$$

Show that $f_n(x) = g_n(T(x))$ for all x.

(iii) Define $g(y) = \limsup g_n(y)$ for each y in $\mathcal{Y}$. Show that g has the desired property. (Question: Why can't we define $g(y) = \lim g_n(y)$?)

[4] Let $g_1, g_2, \ldots$ be $\mathcal{A}\backslash\mathcal{B}(\mathbb{R})$-measurable functions from $\mathcal{X}$ into $\mathbb{R}$. Show that $\{\limsup_n g_n > t\} = \bigcup_{r \in \mathbb{Q}, r > t} \bigcap_{m=1}^\infty \bigcup_{i \geq m}\{g_i > r\}$. Deduce, without any appeal to Example <8>, that $\limsup g_n$ is $\mathcal{A}\backslash\mathcal{B}(\overline{\mathbb{R}})$-measurable. Warning: Be careful about

strict inequalities that turn into nonstrict inequalities in the limit—it is possible to have $x_n > x$ for all n and still have $\limsup_n x_n = x$.

[5] Suppose a class of sets $\mathcal{E}$ cannot separate a particular pair of points x, y: for every E in $\mathcal{E}$, either $\{x, y\} \subseteq E$ or $\{x, y\} \subseteq E^c$. Show that $\sigma(\mathcal{E})$ also cannot separate the pair.

[6] A collection of sets $\mathcal{F}_0$ that is stable under finite unions, finite intersections, and complements is called a field. A nonnegative set function μ defined on $\mathcal{F}_0$ is called a finitely additive measure if $\mu\left(\cup_{i\le n} F_i\right) = \sum_{i\le n} \mu F_i$ for every finite collection of disjoint sets in $\mathcal{F}_0$. The set function is said to be countably additive on $\mathcal{F}_0$ if $\mu\left(\cup_{i\in\mathbb{N}} F_i\right) = \sum_{i\in\mathbb{N}} \mu F_i$ for every countable collection of disjoint sets in $\mathcal{F}_0$ *whose union belongs to* $\mathcal{F}$. Suppose $\mu\mathcal{X} < \infty$. Show that μ is countably additive on $\mathcal{F}_0$ if and only if $\mu A_n \downarrow 0$ for every decreasing sequence in $\mathcal{F}_0$ with empty intersection. Hint: For the argument in one direction, consider the union of differences $A_i \backslash A_{i+1}$.

[7] Let $f_1, \ldots, f_n$ be functions in $\mathcal{M}^+(\mathcal{X}, \mathcal{A})$, and let μ be a measure on $\mathcal{A}$. Show that $\mu\left(\vee_i f_i\right) \le \sum_i \mu f_i \le \mu\left(\vee_i f_i\right) + \sum_{i<j} \mu\left(f_i \wedge f_j\right)$ where $\vee$ denotes pointwise maxima of functions and $\wedge$ denotes pointwise minima.

[8] Let μ be a finite measure and f be a measurable function. For each positive integer k, show that $\mu|f|^k < \infty$ if and only if $\sum_{n=1}^{\infty} n^{k-1} \mu\{|f| \ge n\} < \infty$.

[9] Suppose $\nu := T\mu$, the image of the measure μ under the measurable map T. Show that $f \in \mathcal{L}^1(\nu)$ if and only if $f \circ T \in \mathcal{L}^1(\mu)$, in which case $\nu f = \mu(f \circ T)$.

[10] Let $\{h_n\}$, $\{f_n\}$, and $\{g_n\}$ be sequences of μ-integrable functions that converge μ almost everywhere to limits h, f and g. Suppose $h_n(x) \le f_n(x) \le g_n(x)$ for all x. Suppose also that $\mu h_n \to \mu h$ and $\mu g_n \to \mu g$. Adapt the proof of Dominated Convergence to prove that $\mu f_n \to \mu f$.

[11] A collection of sets is called a monotone class if it is stable under unions of increasing sequences and intersections of decreasing sequences. Adapt the argument from Theorem <38> to prove: if a class $\mathcal{E}$ is stable under finite unions and complements then $\sigma(\mathcal{E})$ equals the smallest monotone class containing $\mathcal{E}$.

[12] Let μ be a finite measure on the Borel sigma-field $\mathcal{B}(\mathcal{X})$ of a metric space $\mathcal{X}$. Call a set B ***inner regular*** if $\mu B = \sup\{\mu F : B \supseteq F \text{ closed}\}$ and ***outer regular*** if $\mu B = \inf\{\mu F : B \subseteq G \text{ open}\}$

(i) Prove that the class $\mathcal{B}_0$ of all Borel sets that are both inner and outer regular is a sigma-field. Deduce that every Borel set is inner regular.

(ii) Suppose μ is tight: for each $\epsilon > 0$ there exists a compact K_ϵ such that $\mu K_\epsilon^c < \epsilon$. Show that the F in the definition of inner regularity can then be assumed compact.

(iii) When μ is tight, show that there exists a sequence of disjoint compacts subsets $\{K_i : i \in \mathbb{N}\}$ of $\mathcal{X}$ such that $\mu\left(\cup_i K_i\right)^c = 0$.

[13] Let μ be a finite measure on the Borel sigma-field of a complete, separable metric space $\mathcal{X}$. Show that μ is tight: for each $\epsilon > 0$ there exists a compact K_ϵ such that $\mu K_\epsilon^c < \epsilon$. Hint: For each positive integer n, show that the space $\mathcal{X}$ is a countable

union of closed balls with radius $1/n$. Find a finite family of such balls whose union B_n has μ measure greater than $\mu\mathcal{X} - \epsilon/2^n$. Show that $\cap_n B_n$ is compact, using the total-boundedness characterization of compact subsets of complete metric spaces.

[14] A sequence of random variables $\{X_n\}$ is said to ***converge in probability*** to a random variable X, written $X_n \xrightarrow{\mathbb{P}} X$, if $\mathbb{P}\{|X_n - X| > \epsilon\} \to 0$ for each $\epsilon > 0$.

(i) If $X_n \to X$ almost surely, show that $1 \geq \{|X_n - X| > \epsilon\} \to 0$ almost surely. Deduce via Dominated Convergence that X_n converges in probability to X.

(ii) Give an example of a sequence $\{X_n\}$ that converges to X in probability but not almost surely.

(iii) Suppose $X_n \to X$ in probability. Show that there is an increasing sequence of positive integers $\{n(k)\}$ for which $\sum_k \mathbb{P}\{|X_{n(k)} - X| > 1/k\} < \infty$. Deduce that $X_{n(k)} \to X$ almost surely.

[15] Let f and g be measurable functions on $(\mathcal{X}, \mathcal{A}, \mu)$, and r and s be positive real numbers for which $r^{-1} + s^{-1} = 1$. Show that $\mu|fg| \leq (\mu|f|^r)^{1/r}(\mu|g|^s)^{1/s}$ by arguing as follows. First dispose of the trivial case where one of the factors on the righthand side is 0 or ∞. Then, without loss of generality (why?), assume that $\mu|f|^r = 1 = \mu|g|^s$. Use concavity of the logarithm function to show that $|fg| \leq |f|^r/r + |g|^s/s$, and then integrate with respect to μ. *This result is called the* ***Hölder inequality***.

[16] Generalize the Hölder inequality (Problem [15]) to more than two measurable functions $f_1, \ldots, f_k$, and positive real numbers $r_1, \ldots, r_k$ for which $\sum_i r_i^{-1} = 1$. Show that $\mu|f_1 \ldots f_k| \leq \prod_i (\mu|f_i|^{r_i})^{1/r_i}$.

[17] Let $(\mathcal{X}, \mathcal{A}, \mu)$ be a measure space, f and g be measurable functions, and r be a real number with $r \geq 1$. Define $\|f\|_r = (\mu|f|^r)^{1/r}$. Follow these steps to prove ***Minkowski's inequality***: $\|f + g\|_r \leq \|f\|_r + \|g\|_r$.

(i) From the inequality $|x + y|^r \leq |2x|^r + |2y|^r$ deduce that $\|f + g\|_r < \infty$ if $\|f\|_r < \infty$ and $\|g\|_r < \infty$.

(ii) Dispose of trivial cases, such as $\|f\|_r = 0$ or $\|f\|_r = \infty$.

(iii) For arbitrary positive constants c and d argue by convexity that

$$\left(\frac{|f| + |g|}{c + d}\right)^r \leq \frac{c}{c + d}\left(\frac{|f|}{c}\right)^r + \frac{d}{c + d}\left(\frac{|g|}{d}\right)^r$$

(iv) Integrate, then choose $c = \|f\|_r$ and $d = \|g\|_r$ to complete the proof.

[18] For f in $\mathcal{L}^1(\mu)$ define $\|f\|_1 = \mu|f|$. Let $\{f_n\}$ be a Cauchy sequence in $\mathcal{L}^1(\mu)$, that is, $\|f_n - f_m\|_1 \to 0$ as $\min(m, n) \to \infty$. Show that there exists an f in $\mathcal{L}^1(\mu)$ for which $\|f_n - f\|_1 \to 0$, by following these steps.

(i) Find an increasing sequence $\{n(k)\}$ such that $\sum_{k=1}^{\infty} \|f_{n(k)} - f_{n(k+1)}\|_1 < \infty$. Deduce that the function $H := \sum_{k=1}^{\infty} |f_{n(k)} - f_{n(k+1)}|$ is integrable.

(ii) Show that there exists a real-valued, measurable function f for which

$$H \geq |f_{n(k)}(x) - f(x)| \to 0 \qquad \text{as } k \to \infty, \text{ for } \mu \text{ almost all } x.$$

Deduce that $\|f_{n(k)} - f\|_1 \to 0$ as $k \to \infty$.

(iii) Show that f belongs to $\mathcal{L}^1(\mu)$ and $\|f_n - f\|_1 \to 0$ as $n \to \infty$.

[19] Let $\{f_n\}$ be a Cauchy sequence in $\mathcal{L}^p(\mathcal{X}, \mathcal{A}, \mu)$, that is, $\|f_n - f_m\|_p \to 0$ as $\min(m, n) \to \infty$. Show that there exists a function f in $\mathcal{L}^p(\mathcal{X}, \mathcal{A}, \mu)$ for which $\|f_n - f\|_p \to 0$, by following these steps.

(i) Find an increasing sequence $\{n(k)\}$ such that $C := \sum_{k=1}^{\infty} \|f_{n(k)} - f_{n(k+1)}\|_p < \infty$. Define $H_\infty = \lim_{N\to\infty} H_N$, where $H_N = \sum_{k=1}^{N} |f_{n(k)} - f_{n(k+1)}|$ for $1 \le N < \infty$. Use the triangle inequality to show that $\mu H_N^p \le C^p$ for all finite N. Then use Monotone Convergence to deduce that $\mu H_\infty^p \le C^p$.

(ii) Show that there exists a real-valued, measurable function f for which $f_{n(k)}(x) \to f(x)$ as $k \to \infty$, a.e. $[\mu]$.

(iii) Show that $|f_{n(k)} - f| \le \sum_{i=k}^{\infty} |f_{n(i)} - f_{n(i+1)}| \le H_\infty$ a.e. $[\mu]$. Use Dominated Convergence to deduce that $\|f_{n(k)} - f\|_p \to 0$ as $k \to \infty$.

(iv) Deduce from (iii) that f belongs to $\mathcal{L}^p(\mathcal{X}, \mathcal{A}, \mu)$ and $\|f_n - f\|_p \to 0$ as $n \to \infty$.

[20] For each random variable on a probability space $(\Omega, \mathcal{F}, \mathbb{P})$ define

$$\|X\|_\infty := \inf\{c \in [0, \infty] : |X| \le c \text{ almost surely}\}.$$

Let $L^\infty := L^\infty(\Omega, \mathcal{F}, \mathbb{P})$ denote the set of equivalence classes of real-valued random variables with $\|X\|_\infty < \infty$. Show that $\|\cdot\|_\infty$ is a norm on L^∞, which is a vector space, complete under the metric defined by $\|X\|_\infty$.

[21] Let $\{X_t : t \in T\}$ be a collection of $\overline{\mathbb{R}}$-valued random variables with possibly uncountable index set T. Complete the following argument to show that there exists a countable subset T_0 of T such that the random variable $X = \sup_{t\in T_0} X_t$ has the properties

(a) $X \ge X_t$ almost surely, for each $t \in T$

(b) if $Y \ge X_t$ almost surely, for each $t \in T$, then $Y \ge X$ almost surely

(The random variable X is called the ***essential supremum*** of the family. It is denoted by $\text{ess sup}_{t\in T} X_t$. Part (b) shows that it is, unique up to an almost sure equivalence.)

(i) Show that properties (a) and (b) are unaffected by a monotone, one-to-one transformation such as $x \mapsto x/(1 + |x|)$. Deduce that there is no loss of generality in assuming $|X_t| \le 1$ for all t.

(ii) Let $\delta = \sup\{\mathbb{P}\sup_{t\in S} X_t : \text{ countable } S \subseteq T\}$. Choose countable T_n such that $\mathbb{P}\sup_{t\in T_n} X_t \ge \delta - 1/n$. Let $T_0 = \cup_n T_n$. Show that $\mathbb{P}\sup_{t\in T_0} X_t = \delta$.

(iii) Suppose $t \notin T_0$. From the inequality $\delta \ge \mathbb{P}(X_t \vee X) \ge \mathbb{P}X = \delta$ deduce that $X \ge X_t$ almost surely.

(iv) For a Y as in assertion (b), show that $Y \ge \sup_{t\in T_0} X_t = X$ almost surely.

[22] Let Ψ be a convex, increasing function for which $\Psi(0) = 0$ and $\Psi(x) \to \infty$ as $x \to \infty$. (For example, $\Psi(x)$ could equal x^p for some fixed $p \ge 1$, or $\exp(x) - 1$ or $\exp(x^2) - 1$.) Define $\mathcal{L}^\Psi(\mathcal{X}, \mathcal{A}, \mu)$ to be the set of all real-valued measurable functions on $\mathcal{X}$ for which $\mu\Psi(|f|/c_0) < \infty$ for some positive real c_0. Define

$\|f\|_\Psi := \inf\{c > 0 : \mu\Psi(|f|/c) \le 1\}$, with the convention that the infimum of an empty set equals $+\infty$. For each f, g in $\mathcal{L}^\Psi(\mathcal{X}, \mathcal{A}, \mu)$ and each real t prove the following assertions.

(i) $\|f\|_\Psi < \infty$. Hint: Apply Dominated Convergence to $\mu\Psi(|f|/c)$.

(ii) $f+g \in \mathcal{L}^\Psi(\mathcal{X}, \mathcal{A}, \mu)$ and the triangle inequality holds: $\|f+g\|_\Psi \le \|f\|_\Psi + \|g\|_\Psi$. Hint: If $c > \|f\|_\Psi$ and $d > \|g\|_\Psi$, deduce that

$$\Psi\left(\frac{|f+g|}{c+d}\right) \le \frac{c}{c+d}\Psi\left(\frac{|f|}{c}\right) + \frac{d}{c+d}\Psi\left(\frac{|g|}{d}\right),$$

by convexity of Ψ.

(iii) $tf \in \mathcal{L}^\Psi(\mathcal{X}, \mathcal{A}, \mu)$ and $\|tf\|_\Psi = |t|\,\|f\|_\Psi$.

REMARK. $\|\cdot\|_\Psi$ is called an Orlicz "norm"—to make it a true norm one should work with equivalence classes of functions equal μ almost everywhere. The L^p norms correspond to the special case $\Psi(x) = x^p$, for some $p \ge 1$.

[23] Define $\|f\|_\Psi$ and $\mathcal{L}^\Psi$ as in Problem [22]. Let $\{f_n\}$ be a Cauchy sequence in $\mathcal{L}^\Psi(\mu)$, that is, $\|f_n - f_m\|_\Psi \to 0$ as $\min(m, n) \to \infty$. Show that there exists an f in $\mathcal{L}^\Psi(\mu)$ for which $\|f_n - f\|_\Psi \to 0$, by following these steps.

(i) Let $\{g_i\}$ be a nonnegative sequence in $\mathcal{L}^\Psi(\mu)$ for which $C := \sum_i \|g_i\|_\Psi < \infty$. Show that the function $G := \sum_i g_i$ is finite almost everywhere and $\|G\|_\Psi \le \sum_i \|g_i\|_\Psi < \infty$. Hint: Use Problem [22] to show that $\mathbb{P}\Psi\left(\sum_{i\le n} g_i/C\right) \le 1$ for each n, then justify a passage to the limit.

(ii) Find an increasing sequence $\{n(k)\}$ such that $\sum_{k=1}^{\infty} \|f_{n(k)} - f_{n(k+1)}\|_\Psi < \infty$. Deduce that the functions $H_L := \sum_{k=L}^{\infty} |f_{n(k)} - f_{n(k+1)}|$ satisfy

$$\infty > \|H_1\|_\Psi \ge \|H_2\|_\Psi \ge \ldots \to 0.$$

(iii) Show that there exists a real-valued, measurable function f for which

$$|f_{n(k)}(x) - f(x)| \to 0 \qquad \text{as } k \to \infty, \text{ for } \mu \text{ almost all } x.$$

(iv) Given $\epsilon > 0$, choose L so that $\|H_L\|_\Psi < \epsilon$. For $i > L$, show that

$$\Psi\left(H_L/\epsilon\right) \ge \Psi\left(|f_{n(L)} - f_{n(i)}|/\epsilon\right) \to \Psi\left(|f_{n(L)} - f|/\epsilon\right).$$

Deduce that $\|f_{n(L)} - f\|_\Psi \le \epsilon$.

(v) Show that f belongs to $\mathcal{L}^\Psi(\mu)$ and $\|f_n - f\|_\Psi \to 0$ as $n \to \infty$.

[24] Let Ψ be a convex increasing function with $\Psi(0) = 0$, as in Problem [22]. Let Ψ^{-1} denote its inverse function. If $X_1, \ldots, X_N \in \mathcal{L}^\Psi(\mathcal{X}, \mathcal{A}, \mu)$, show that

$$\mathbb{P}\max_{i\le N}|X_i| \le \Psi^{-1}(N)\max_{i\le N}\|X_i\|_\Psi.$$

Hint: Consider $\Psi(\mathbb{P}\max|X_I|/C)$ with $C > \max_{i\le N}\|X_i\|_\Psi$.

REMARK. Compare with van der Vaart & Wellner (1996, page 96): if also $\limsup_{x,y\to\infty}\Psi(x)\Psi(y)/\Psi(cxy) < \infty$ for some constant $c > 0$ then $\|\max_{i\le N}|X_i|\|_\Psi \le K\Psi^{-1}(N)\max_{i\le N}\|X_i\|_\Psi$ for a constant K depending only on Ψ. See page 105 of their Problems and Complements for related counterexamples.

[25] For each θ in $[0, 1]$ let $X_{n,\theta}$ be a random variable with a Binomial(n, θ) distribution. That is, $\mathbb{P}\{X_{n,\theta} = k\} = \binom{n}{k}\theta^k(1-\theta)^{n-k}$ for $k = 0, 1, \dots, n$. You may assume these elementary facts: $\mathbb{P}X_{n,\theta} = n\theta$ and $\mathbb{P}(X_{n,\theta} - n\theta)^2 = n\theta(1-\theta)$. Let f be a continuous function defined on $[0, 1]$.

(i) Show that $p_n(\theta) = \mathbb{P}f(X_{n,\theta}/n)$ is a polynomial in θ.

(ii) Suppose $|f| \le M$, for a constant M. For a fixed ϵ, invoke (uniform) continuity to find a $\delta > 0$ such that $|f(s) - f(t)| \le \epsilon$ whenever $|s - t| \le \delta$, for all s, t in $[0, 1]$. Show that

$$|f(x/n) - f(\theta)| \le \epsilon + 2M\{|(x/n) - \theta| > \delta\} \le \epsilon + \frac{2M|(x/n) - \theta|^2}{\delta^2}.$$

(iii) Deduce that $\sup_{0\le\theta\le1} |p_n(\theta) - f(\theta)| < 2\epsilon$ for n large enough. That is, deduce that $f(\cdot)$ can be uniformly approximated by polynomials over the range $[0, 1]$, a result known as the ***Weierstrass approximation theorem***.

[26] Extend the approximation result from Example <46> to the case of an infinite measure μ on $\mathcal{B}(\mathbb{R}^k)$ that gives finite measure to each compact set. Hint: Let B be a closed ball of radius large enough to ensure $\mu|f|B < \epsilon$. Write μ_B for the restriction of μ to B. Invoke the result from the Example to find a g in $\mathbb{C}_0$ such that $\mu_B|f - g| < \epsilon$. Find $\mathbb{C}_0$ functions $1 \ge h_i \downarrow B$. Consider approximations gh_i for i large enough.

13. Notes

I recommend Royden (1968) as a good source for measure theory. The books of Ash (1972) and Dudley (1989) are also excellent references, for both measure theory and probability. Dudley's book contains particularly interesting historical notes.

See Hawkins (1979, Chapter 4) to appreciate the subtlety of the idea of a negligible set.

The result from Problem [10] is often attributed to (Pratt 1960), but, as he noted (in his 1966 Acknowledgment of Priority), it is actually much older.

Theorem <38> (the π–λ theorem for generating classes of sets) is often attributed to Dynkin (1960, Section 1.1), although Sierpiński (1928) had earlier proved a slightly stronger result (covering generation of sigma-rings, not just sigma-fields). I adapted the analogous result for classes of functions, Theorem <45>, from Protter (1990, page 7) and Dellacherie & Meyer (1978, page 14). Compare with the "Sierpiński Stability Lemma" for sets, and the "Functional Sierpiński Lemma" presented by Hoffmann-Jørgensen (1994, pages 8, 54, 60).

References

Ash, R. B. (1972), *Real Analysis and Probability*, Academic Press, New York.

Dellacherie, C. & Meyer, P. A. (1978), *Probabilities and Potential*, North-Holland, Amsterdam.

Dudley, R. M. (1989), *Real Analysis and Probability*, Wadsworth, Belmont, Calif.

Dynkin, E. B. (1960), *Theory of Markov Processes*, Pergamon.

Hawkins, T. (1979), *Lebesgue's Theory of Integration: Its Origins and Development*, second edn, Chelsea, New York.

Hoffmann-Jørgensen, J. (1994), *Probability with a View toward Statistics*, Vol. 1, Chapman and Hall, New York.

Oxtoby, J. (1971), *Measure and Category*, Springer-Verlag.

Pratt, J. W. (1960), 'On interchanging limits and integrals', *Annals of Mathematical Statistics* **31**, 74–77. Acknowledgement of priority, *same journal*, vol 37 (1966), page 1407.

Protter, P. (1990), *Stochastic Integration and Differential Equations*, Springer, New York.

Royden, H. L. (1968), *Real Analysis*, second edn, Macmillan, New York.

Sierpiński, W. (1928), 'Un théorème général sur les familles d'ensembles', *Fundamenta Mathematicae* **12**, 206–210.

van der Vaart, A. W. & Wellner, J. A. (1996), *Weak Convergence and Empirical Process: With Applications to Statistics*, Springer-Verlag.

Chapter 3

Densities and derivatives

SECTION 1 explains why the traditional split of introductory probability courses into two segments—the study of discrete distributions, and the study of "continuous" distributions—is unnecessary in a measure theoretic treatment. Absolute continuity of one measure with respect to another measure is defined. A simple case of the Radon-Nikodym theorem is proved.

*SECTION *2 establishes the Lebesgue decomposition of a measure into parts absolutely continuous and singular with respect to another measure, a result that includes the Radon-Nikodym theorem as a particular case.*

SECTION 3 shows how densities enter into the definitions of various distances between measures.

SECTION 4 explains the connection between the classical concept of absolute continuity and its measure theoretic generalization. Part of the Fundamental Theorem of Calculus is deduced from the Radon-Nikodym theorem.

*SECTION *5 establishes the Vitali covering lemma, the key to the identification of derivatives as densities.*

*SECTION *6 presents the proof of the other part of the Fundamental Theorem of Calculus, showing that absolutely continuous functions (on the real line) are Lebesgue integrals of their derivatives, which exist almost everywhere.*

1. Densities and absolute continuity

Nonnegative measurable functions create new measures from old.

Let $(\mathcal{X}, \mathcal{A}, \mu)$ be a measure space, and let $\Delta(\cdot)$ be a function in $\mathcal{M}^+(\mathcal{X}, \mathcal{A})$. The increasing, linear functional defined on $\mathcal{M}^+(\mathcal{X}, \mathcal{A})$ by $\nu f := \mu(f\Delta)$ inherits from μ the Monotone Convergence property, which identifies it as an integral with respect to a measure on $\mathcal{A}$.

The measure μ is said to ***dominate*** ν; the measure ν is said to have ***density*** Δ with respect to μ. This relationship is often indicated symbolically as $\Delta = d\nu/d\mu$, which fits well with the traditional notation,

$$\int f(x)\, d\nu(x) = \int f(x) \frac{d\nu}{d\mu}\, d\mu(x).$$

The $d\mu$ symbols "cancel out," as in the change of variable formula for Lebesgue integrals.

REMARK. The density $d\nu/d\mu$ is often called the Radon-Nikodym derivative of ν with respect to μ, a reference to the result described in Theorem <4> below. The word *derivative* suggests a limit of a ratio of ν and μ measures of "small" sets. For μ equal to Lebesgue measure on a Euclidean space, $d\nu/d\mu$ can indeed be recovered as such a limit. Section 4 explains the one-dimensional case. Chapter 6 will give another interpretation via martingales.

For example, if μ is Lebesgue measure on $\mathcal{B}(\mathbb{R})$, the probability measure defined by the density $\Delta(x) = (2\pi)^{-1/2}\exp(-x^2/2)$ with respect to μ is called the standard normal distribution, usually denoted by $N(0,1)$. If μ is counting measure on $\mathbb{N}_0$ (that is, mass 1 at each nonnegative integer), the probability measure defined by the density $\Delta(x) = e^{-\theta}\theta^x/x!$ is called the Poisson(θ) distribution, for each positive constant θ. If μ is Lebesgue measure on $\mathcal{B}(\mathbb{R}^2)$, the probability measure defined by the density $\Delta(x,y) = (2\pi)^{-1}\exp\left(-(x^2+y^2)/2\right)$ with respect to μ is called the standard bivariate normal distribution. The qualifier *joint* sometimes creeps into the description of densities with respect to Lebesgue measure on $\mathcal{B}(\mathbb{R}^2)$ or $\mathcal{B}(\mathbb{R}^k)$. From a measure theoretic point of view the qualifier is superfluous, but it is a comforting probabilistic tradition perhaps worthy of preservation.

Under mild assumptions, which rule out the sort of pathological behavior involving sets with infinite measure described by Problems [1] and [2], it is not hard to show (Problem [3]) that the density is unique up to a μ-equivalence. The simplest way to avoid the pathologies is an assumption that the dominating measure μ is ***sigma-finite***, meaning that there exists a partition of $\mathcal{X}$ into countably many disjoint measurable sets $\mathcal{X}_1, \mathcal{X}_2, \dots$ with $\mu\mathcal{X}_i < \infty$ for each i.

Existence of a density is a property that depends on two measures. Even measures that don't fit the traditional idea of a continuous distribution can be specified by densities, as in the case of measures dominated by a counting measure. Some introductory texts use the technically correct term *density* in that case, much to the confusion of students who have come to think that densities have something to do with continuous functions. More generally, every measure could be thought of as having a density, because $d\mu/d\mu = 1$, which is a perfectly useless fact. Densities are useful because they allow integrals with respect to one measure to be reexpressed as integrals with respect to a different measure.

The distribution function of a probability measure dominated by Lebesgue measure on $\mathcal{B}(\mathbb{R})$ is continuous, as a map from $\mathbb{R}$ into $\mathbb{R}$; but not every probability measure with a continuous distribution function has a density with respect to Lebesgue measure

<1> **Example.** Let $\mathfrak{m}$ denote Lebesgue measure on $[0,1)$. Each point x in $[0,1)$ has a binary expansion $x = \sum_{n=1}^{\infty} x_n 2^{-n}$ with each x_i either 0 or 1. To ensure uniqueness of the $\{x_n\}$, choose the expansion that ends in a string of zeros when x is a dyadic rational. The set $\{x_n = 1\}$ is then a finite union of intervals of the form $[a,b)$, with both endpoints dyadic rationals. The map $T(x) := \sum_{n=1}^{\infty} 2x_n 3^{-n}$ from $[0,1)$ back into itself is measurable. The image measure $\nu := T(\mathfrak{m})$ concentrates on the

compact subset $C = \cap_n C_n$ of [0, 1] obtained by successive removal of "middle thirds" of subintervals:

$$C_1 = [0, 1/3] \cup [2/3, 1]$$
$$C_2 = [0, 1/9] \cup [2/9, 3/9] \cup [6/9, 7/9] \cup [8/9, 1]$$

and so on. The set C, which is called the ***Cantor set***, has Lebesgue measure less than $mC_n = (2/3)^n$ for every n. That is, $mC = 0$.

The distribution function $F(x) := \nu[0, x]$, for $0 \le x \le 1$ has the strange property that it is constant on each of the open intervals that make up each C_n^c, because ν puts zero mass in those intervals. Thus F has a zero derivative at each point of $C^c = \cup_n C_n^c$, a set with Lebesgue measure one. □

The distinction between a continuous function and a function expressible as an integral was recognized early in the history of measure theory, with the name ***absolute continuity*** being used to denote the stronger property. The original definition of absolute continuity (Section 4) is now a special case of a streamlined characterization that applies not just to measures on the real line.

> REMARK. Probability measures dominated by Lebesgue measure correspond to the *continuous distributions* of introductory courses, although the correct term is *distributions absolutely continuous with respect to Lebesgue measure*. By extension, random variables whose distributions are dominated by Lebesgue measure are sometimes called "continuous random variables," which I regard as a harmful abuse of terminology. There need be nothing continuous about a "continuous random variable" as a function from a set Ω into $\mathbb{R}$. Indeed, there need be no topology on Ω; the very concept of continuity for the function might be void. Many a student of probability has been misled into assuming topological properties for "continuous random variables." I try to avoid using the term.

<2> **Definition.** *A measure ν is said to be absolutely continuous with respect to a measure μ, written $\nu \ll \mu$, if every μ-negligible set is also ν-negligible.*

Clearly, a measure ν given by a density with respect to a dominating measure μ is also absolutely continuous with respect to μ.

<3> **Example.** If the measure ν is finite, there is an equivalent formulation of absolute continuity that looks more like a continuity property.

Let ν and μ be measures on $(\mathcal{X}, \mathcal{A})$, with $\nu\mathcal{X} < \infty$. Suppose that to each $\epsilon > 0$ there exists a $\delta > 0$ such that $\nu A < \epsilon$ for each $A \in \mathcal{A}$ with $\mu A < \delta$. Then clearly ν must be absolutely continuous with respect to μ: for each measurable set A with $\mu A = 0$ we must have $\nu A < \epsilon$ for every positive ϵ.

Conversely, if ν fails to have the ϵ–δ property then there exists some $\epsilon > 0$ and a sequence of sets $\{A_n\}$ with $\mu A_n \to 0$ but $\nu A_n \ge \epsilon$ infinitely often. With no loss of generality (or by working with a subsequence) we may assume that $\mu A_n < 2^{-n}$ and $\nu A_n \ge \epsilon$ for all n. Define $A := \{A_n \text{ i.o.}\} = \limsup_n A_n$. Finiteness of $\sum_n \mu A_n$ implies $\sum_n A_n < \infty$ a.e. $[\mu]$, and hence $\mu A = 0$; but Dominated Convergence, using the assumption $\nu\mathcal{X} < \infty$, gives $\nu A = \lim_n \nu\left(\sup_{i \ge n} A_i\right) \ge \epsilon$. Thus ν is not absolutely continuous with respect to μ.

In other words, the ϵ–δ property is equivalent to absolute continuity, at least when ν is a finite measure. The equivalence can fail if ν is not a finite measure.

For example, if μ denotes Lebesgue measure on $\mathcal{B}(\mathbb{R})$ and ν is the measure defined by the density $|x|$, then the interval $(n, n+n^{-1})$ has μ measure n^{-1} but ν measure greater than 1. □

REMARK. For finite measures it might seem that absolute continuity should also have an equivalent formulation in terms of functionals on $\mathcal{M}^+$. However, even if $\nu \ll \mu$, it need not be true that to each $\epsilon > 0$ there exists a $\delta > 0$ such that: $f \in \mathcal{M}^+$ and $\mu f < \delta$ imply $\nu f < \epsilon$. For example, let μ be Lebesgue measure on $\mathcal{B}(0,1)$ and ν be the finite measure with density $\Delta(x) := x^{-1/2}$ with respect to μ. The functions $f_n(x) := \Delta(x)\{n^{-2} \le x \le n^{-1}\}$ have the property that $\mu f_n \le \int_0^{1/n} x^{-1/2} dx \to 0$ as $n \to \infty$, but $\nu f_n = \int_{1/n^2}^{1/n} x^{-1} dx = \log n \to \infty$, even though $\nu \ll \mu$.

Existence of a density and absolute continuity are equivalent properties if we exclude some pathological examples, such as those presented in Problems [1] and [2].

<4> **Radon-Nikodym Theorem.** *Let μ be a sigma-finite measure on a space $(\mathcal{X}, \mathcal{A})$. Then every sigma-finite measure that is absolutely continuous with respect to μ has a density, which is unique up to μ-equivalence.*

The Theorem is a special case of the slightly more general result known as the ***Lebesgue decomposition***, which is proved in Section 2 using projections in Hilbert spaces. Most of the ideas needed to prove the general version of the Theorem appear in simpler form in the following proof of a special case.

<5> **Lemma.** *Suppose ν and μ are finite measures on $(\mathcal{X}, \mathcal{A})$, with $\nu \le \mu$, that is, $\nu f \le \mu f$ for all f in $\mathcal{M}^+(\mathcal{X}, \mathcal{A})$. Then ν has a density Δ with respect to μ for which $0 \le \Delta \le 1$ everywhere.*

Proof. The linear subspace $\mathcal{H}_0 := \{f \in \mathcal{L}^2(\mu) : \nu f = 0\}$ of $\mathcal{L}^2(\mu)$ is closed for convergence in the $\mathcal{L}^2(\mu)$ norm: if $\mu|f_n - f|^2 \to 0$ then, by the Cauchy-Schwarz inequality,

$$|\nu f_n - \nu f|^2 \le \left(\nu 1^2\right)\left(\nu |f_n - f|^2\right) \le \nu(\mathcal{X})\mu|f_n - f|^2 \to 0.$$

Except when ν is the zero measure (in which case the result is trivial), the constant function 1 does not belong to $\mathcal{H}_0$. From Section 2.7, there exist functions $g_0 \in \mathcal{H}_0$ and g_1 orthogonal to $\mathcal{H}_0$ for which $1 = g_0 + g_1$. Notice that $\nu g_1 = \nu 1 \ne 0$. The desired density will be a constant multiple of g_1.

Consider any f in $\mathcal{L}^2(\mu)$. With $C := \nu f / \nu g_1$ the function $f - C g_1$ belongs to $\mathcal{H}_0$, because $\nu(f - C g_1) = \nu f - C \nu g_1 = 0$. Orthogonality of $f - C g_1$ and g_1 gives

$$0 = \langle f - C g_1, g_1 \rangle = \mu(f g_1) - \frac{\nu f}{\nu g_1}\mu(g_1^2),$$

which rearranges to $\nu f = \mu(f\Delta)$ where $\Delta := (\nu g_1 / \mu g_1^2) g_1$.

The inequality $0 \le \nu\{\Delta < 0\} = \mu\Delta\{\Delta < 0\}$ ensures that $\Delta \ge 0$ a.e. $[\mu]$; and the inequalities $\nu\{\Delta > 1\} = \mu\Delta\{\Delta > 1\} \ge \mu\{\Delta > 1\} \ge \nu\{\Delta > 1\}$ force $\Delta \le 1$ a.e. $[\mu]$, for otherwise the middle inequality would be strict. Replacement of Δ by $\Delta\{0 \le \Delta \le 1\}$ therefore has no effect on the representation $\nu f = \mu(f\Delta)$ for $f \in \mathcal{L}^2(\mu)$. To extend the equality to f in $\mathcal{M}^+$, first invoke it for the $\mathcal{L}^2(\mu)$ function $n \wedge f$, then invoke Monotone Convergence twice as n tends to infinity. □

Not all measures on $(\mathcal{X}, \mathcal{A})$ need be dominated by a given μ. The extreme example is a measure ν that is ***singular*** with respect to μ, meaning that there exists a measurable subset S for which $\mu S^c = 0 = \nu S$. That is, the two measures concentrate on disjoint parts of $\mathcal{X}$, a situation denoted by writing $\nu \perp \mu$. Perhaps it would be better to say that the two measures are mutually singular, to emphasize the symmetry of the relationship. For example, discrete measures—those that concentrate on countable subsets—are singular with respect to Lebesgue measure on the real line. There also exist singular measures (with respect to Lebesgue measure) that give zero mass to each countable set, such as the probability measure ν from Example <1>.

Avoidance of all probability measures except those dominated by a counting measure or a Lebesgue measure—as in introductory probability courses—imposes awkward constraints on what one can achieve with probability theory. The restriction becomes particularly tedious for functions of more than a single random variable, for then one is limited to smooth transformations for which image measures and densities (with respect to Lebesgue measure) can be calculated by means of Jacobians. The unfortunate effects of an artificially restricted theory permeate much of the statistical literature, where sometimes inappropriate and unnecessary requirements are imposed merely to accommmodate a lack of an appropriate measure theoretic foundation.

REMARK. Why should absolute continuity with respect to Lebesgue measure or counting measure play such a central role in introductory probability theory? I believe the answer is just a matter of definition, or rather, a lack of definition. For a probability measure $\mathbb{P}$ concentrated on a countable set of points, expectations $\mathbb{P}g(X)$ become countable sums, which can be handled by elementary methods. For general probability measures the definition of $\mathbb{P}g(X)$ is typically not a matter of elementary calculation. However, if X has a distribution P with density Δ with respect to Lebesgue measure, then $\mathbb{P}g(X) = Pg = \int \Delta(x)g(x)\,dx$. The last integral has the familiar look of a Riemann integral, which is the subject of elementary Calculus courses. Seldom would Δ or g be complicated enough to require the interpretation as a Lebesgue integral—one stays away from such functions when teaching an introductory course.

From the measure theoretic viewpoint, densities are not just a crutch for support of an inadequate integration theory; they become a useful tool for exploiting absolute continuity for pairs of measures.

In much statistical theory, the actual choice of dominating measure matters little. The following result, which is often called Scheffé's lemma, is typical.

<6> **Exercise.** Suppose $\{P_n\}$ is a sequence of probability measures with densities $\{\Delta_n\}$ with respect to a measure μ. Suppose Δ_n converges almost everywhere $[\mu]$ to the density $\Delta := dP/d\mu$ of a *probability measure* P. Show that $\mu|\Delta_n - \Delta| \to 0$.

SOLUTION: Write $\mu|\Delta - \Delta_n|$ as

$$\mu(\Delta - \Delta_n)^+ + \mu(\Delta - \Delta_n)^- = 2\mu(\Delta - \Delta_n)^+ - (\mu\Delta - \mu\Delta_n).$$

On the right-hand side, the second term equals zero, because both densities integrate to 1, and the first term tends to zero by Dominated Convergence, because $\Delta \geq (\Delta - \Delta_n)^+ \to 0$ a.e. $[\mu]$. □

Convergence in $\mathcal{L}^1(\mu)$ of densities is equivalent to convergence of the probability measures in ***total variation***, a topic discussed further in Section 3.

*2. The Lebesgue decomposition

Absolute continuity and singularity represent the two extremes for the relationship between two measures on the same space.

<7> **Lebesgue decomposition.** *Let μ be a sigma-finite measure on a space $(\mathcal{X}, \mathcal{A})$. To each sigma-finite measure ν on $\mathcal{A}$ there exists a μ-negligible set $\mathcal{N}$ and a real-valued Δ in $\mathcal{M}^+$ such that*

$$\nu f = \nu(f\mathcal{N}) + \mu(f\Delta) \qquad \text{for all } f \in \mathcal{M}^+(\mathcal{X}, \mathcal{A}). \tag{8}$$

The set $\mathcal{N}$ and the function $\Delta\mathcal{N}^c$ are unique up to a $\nu + \mu$ almost sure equivalence.

REMARK. Of course the value of Δ on $\mathcal{N}$ has no effect on $\mu(f\Delta)$ or $\nu(f\mathcal{N})$. Some authors adopt the convention that $\Delta = \infty$ on the set $\mathcal{N}$. The convention has no effect on the equality <8>, but it does make Δ unique $\nu + \mu$ almost surely.

The restriction $\nu_\perp$ of ν to $\mathcal{N}$ is called the part of ν that is ***singular*** with respect to μ. The restriction ν_{abs} of ν to $\mathcal{N}^c$ is called the part of ν that is ***absolutely continuous*** with respect to μ. Problem [11] shows that the decomposition $\nu = \nu_\perp + \nu_{\text{abs}}$, into singular and dominated components, is unique.

There is a corresponding decomposition $\mu = \mu_\perp + \mu_{\text{abs}}$ into parts singular and absolutely continuous with respect to ν. Together the two decompositions partition the underlying space into four measurable sets: a set that is negligible for both measures; a set where they are mutually absolutely continuous; and two sets where the singular components $\mu_\perp$ and $\nu_\perp$ concentrate.

$\mathcal{N}^c$	$\mathcal{N}$
$\Delta = 0$ $\nu = 0$ $\mu_\perp$ concentrates here	$\Delta = \infty$? $\mu = \nu = 0$ neither measure puts mass here
$0 < \Delta < \infty$ $\dfrac{d\nu_{\text{abs}}}{d\mu_{\text{abs}}} = \Delta, \quad \dfrac{d\mu_{\text{abs}}}{d\nu_{\text{abs}}} = 1/\Delta$	$1/\Delta = 0$? $\mu = 0$ $\nu_\perp$ concentrates here

Proof. Consider first the question of existence. With no loss of generality we may assume that both ν and μ are finite measures. The general decomposition would follow by piecing together the results for countably many disjoint subsets of $\mathcal{X}$.

Define $\lambda = \nu + \mu$. Note that $\nu \le \lambda$. From Lemma <5>, there is an $\mathcal{A}$-measurable function Δ_0, taking values in $[0, 1]$, for which $\nu f = \lambda(f\Delta_0)$ for all f in $\mathcal{M}^+$. Define $\mathcal{N} := \{\Delta_0 = 1\}$, a set that has zero μ measure because

$$\nu\{\Delta_0 = 1\} = \nu\Delta_0\{\Delta_0 = 1\} + \mu\Delta_0\{\Delta_0 = 1\} = \nu\{\Delta_0 = 1\} + \mu\{\Delta_0 = 1\}.$$

Define

$$\Delta := \frac{\Delta_0}{1-\Delta_0}\{\Delta_0 < 1\}.$$

We have to show that $\nu(f\mathcal{N}^c) = \mu(f\Delta)$ for all f in $\mathcal{M}^+$. For such an f, and each positive integer n, the defining property of Δ_0 gives

$$\nu(f \wedge n)\{\Delta_0 < 1\} = \nu(f \wedge n)\Delta_0\{\Delta_0 < 1\} + \mu(f \wedge n)\Delta_0\{\Delta_0 < 1\},$$

which rearranges (no problems with $\infty - \infty$) to

$$\nu(f \wedge n)(1-\Delta_0)\{\Delta_0 < 1\} = \mu(f \wedge n)\Delta_0\{\Delta_0 < 1\}.$$

Appeal twice to Monotone Convergence as $n \to \infty$ to deduce

$$\nu f(1-\Delta_0)\{\Delta_0 < 1\} = \mu f \Delta_0\{\Delta_0 < 1\} \qquad \text{for all } f \in \mathcal{M}^+.$$

Replace f by $f\{\Delta_0 < 1\}/(1-\Delta_0)$, which also belongs to $\mathcal{M}^+$, to complete the proof of existence.

The proof of uniqueness of the representation, up to various almost sure equivalences, follows a similar style of argument. Problem [9] will step you through the details. □

3. Distances and affinities between measures

Let μ_1 and μ_2 be two finite measures on $(\mathcal{X}, \mathcal{A})$. We may suppose both μ_1 and μ_2 are absolutely continuous with respect to some dominating (nonnegative) measure λ, with densities m_1 and m_2. For example, we could choose $\lambda = \mu_1 + \mu_2$.

For the purposes of this subsection, a ***finite signed measure*** will be any set function expressible as a difference $\mu_1 - \mu_2$ of finite, nonnegative measures. In fact every real valued, countably additive set function defined on a sigma field can be represented in that way, but we shall not be needing the more basic characterization. If μ_i has density m_i with respect to λ then $\mu_1 - \mu_2$ has density $m_1 - m_2$.

Throughout the section, $\mathcal{M}_{\text{bdd}}$ will denote the space of all bounded, real-valued, $\mathcal{A}$-measurable functions on $\mathcal{X}$, and $\mathcal{M}^+_{\text{bdd}}$ will denote the cone of nonnegative functions in $\mathcal{M}_{\text{bdd}}$.

There are a number of closely related distances between the measures, all of which involve calculations with densities. Several easily proved facts about these distances have important application in mathematical statistics.

Total variation distance

The total variation norm $\|\mu\|_1$ of the signed measure μ is defined as $\sup_\pi \sum_{A\in\pi} |\mu A|$, where the supremum ranges over all partitions π of $\mathcal{X}$ into finitely many measurable sets. The total variation distance between two signed measures is the norm of their difference. In terms of the density m for μ with respect to λ, we have

$$\sum_{A\in\pi} |\mu A| = \sum_{A\in\pi} |\lambda(mA)| \le \sum_{A\in\pi} \lambda|m|A = \lambda|m|.$$

Equality is achieved for a partition with two sets: $A_1 := \{m \geq 0\}$ and $A_2 := \{m < 0\}$. In particular, the total variation distance between two measures equals the $\mathcal{L}^1$ distance between their densities; in fact, total variation distance is often referred to as $\mathcal{L}^1$ distance. The initial definition of $\|\mu\|_1$, in which the dominating measure does not appear, shows that the $\mathcal{L}^1$ norm does not depend on the particular choice of dominating measure. The $\mathcal{L}^1$ norm also equals $\sup_{|f|\leq 1} |\mu f|$, because $|\mu f| = |\lambda(mf)| \leq \lambda(|m|\,|f|) \leq \lambda|m|$ for $|f| \leq 1$, with equality for $f := \{m \geq 0\} - \{m < 0\}$.

REMARK. Some authors also refer to $v(\mu_1, \mu_2) := \sup_{A\in\mathcal{A}} |\mu_1 A - \mu_2 A|$ as the total variation distance, which can be confusing. In the special case when $\mu_1\mathcal{X} = \mu_2\mathcal{X}$ we have $\mu\{m_1 \geq m_2\} = -\mu\{m_1 < m_2\}$, whence $\|\mu\| = 2v(\mu_1, \mu_2)$.

The affinity between two finite signed measures

The affinity between μ_1 and μ_2 is defined as

$$\begin{aligned}\alpha_1(\mu_1, \mu_2) &:= \inf\{\mu_1 f_1 + \mu_2 f_2 : f_1, f_2 \in \mathcal{M}^+_{\text{bdd}},\ f_1 + f_2 \geq 1\} \\ &= \inf\{\lambda(m_1 f_1 + m_2 f_2) : f_1, f_2 \in \mathcal{M}^+_{\text{bdd}},\ f_1 + f_2 \geq 1\} \\ &= \lambda(m_1 \wedge m_2),\end{aligned}$$

the infimum being achieved by $f_1 = \{m_1 \leq m_2\}$ and $f_2 = \{m_1 > m_2\}$. That is, the affinity equals the minimum of $\mu_1 A + \mu_2 A^c$ over all A in $\mathcal{A}$.

REMARK. For probability measures, the minimizing set has the statistical interpretation of the (nonrandomized) test between the two hypotheses μ_1 and μ_2 that minimizes the sum of the type one and type two errors.

The pointwise equality $2(m_1 \wedge m_2) = m_1 + m_2 - |m_1 - m_2|$ gives the connection between affinity and $\mathcal{L}^1$ distance,

$$2\alpha_1(\mu_1, \mu_2) = \lambda(m_1 + m_2 - |m_1 - m_2|) = \mu_1\mathcal{X} + \mu_2\mathcal{X} - \|\mu_1 - \mu_2\|_1.$$

Both the affinity and the $\mathcal{L}^1$ distance are related to a natural ordering on the space of all finite signed measures on $\mathcal{A}$, defined by:

$$\nu \leq \nu' \text{ means } \nu f \leq \nu' f \text{ for each } f \text{ in } \mathcal{M}^+_{\text{bdd}}.$$

<9> **Example.** To each pair of finite, signed measures μ_1 and μ_2 there exists a largest measure ν for which $\nu \leq \mu_1$ and $\nu \leq \mu_2$. It is determined by its density $m_1 \wedge m_2$ with respect to the dominating λ. It is easy to see that the ν so defined is smaller than both μ_1 and μ_2. To prove that it is the largest such measure, let μ be any other signed measure with the same property. Without loss of generality, we may assume μ has a density m with respect to λ. For each bounded, nonnegative f, we then have

$$\begin{aligned}\lambda(mf) = \mu f &= \mu f\{m_1 \geq m_2\} + \mu f\{m_1 < m_2\} \\ &\leq \mu_2 f\{m_1 \geq m_2\} + \mu_1 f\{m_1 < m_2\} \\ &= \lambda(m_1 \wedge m_2)f.\end{aligned}$$

The particular choice $f = \{m > m_1 \wedge m_2\}$ then leads to the conclusion that $m \leq m_1 \wedge m_2$ a.e. $[\lambda]$, and hence $\mu \leq \nu$ as measures.

The measure ν is also denoted by $\mu_1 \wedge \mu_2$ and is called the measure theoretic minimum of μ_1 and μ_2. For nonnegative measures, the affinity $\alpha_1(\mu_1, \mu_2)$ equals the $\mathcal{L}^1$ norm of $\mu_1 \wedge \mu_2$. □

By a similar argument, it is easy to show that the density $|m_1 - m_2|$ defines the smallest measure ν (necessarily nonnegative) for which $\nu \geq \mu_1 - \mu_2$ and $\nu \geq \mu_2 - \mu_1$.

Hellinger distance between probability measures

Let P and Q be probability measures with densities p and q with respect to a dominating measure λ. The square roots of the densities, $\sqrt{p}$ and $\sqrt{q}$ are both square integrable; they both belong to $\mathcal{L}^2(\lambda)$. The Hellinger distance between the two measures is defined as the $\mathcal{L}^2$ distance between the square roots of their densities,

$$H(P, Q)^2 := \lambda\left(\sqrt{p} - \sqrt{q}\right)^2 = \lambda\left(p + q - 2\sqrt{pq}\right) = 2 - 2\lambda\sqrt{pq}.$$

Again the distance does not depend on the choice of dominating measure (Problem [13]). The quantity $\lambda\sqrt{pq}$ is called the ***Hellinger affinity*** between the two probabilities, and is denoted by $\alpha_2(P, Q)$. Clearly $\sqrt{pq} \geq p \wedge q$, from which it follows that $\alpha_2(P, Q) \geq \alpha_1(P, Q)$ and $H(P, Q)^2 \leq \|P - Q\|_1$. The Cauchy-Schwarz inequality gives a useful lower bound:

$$\begin{aligned} \|P - Q\|_1^2 &= \left(\lambda|\sqrt{p} - \sqrt{q}||\sqrt{p} + \sqrt{q}|\right)^2 \\ &\leq \left(\lambda|\sqrt{p} - \sqrt{q}|^2\right)\left(\lambda|\sqrt{p} + \sqrt{q}|^2\right) \\ &= H(P, Q)^2\,(2 + 2\alpha_2(P, Q)). \end{aligned}$$

Substituting for the Hellinger affinity, we get $\|P - Q\|_1 \leq H(P, Q)\left(4 - H(P, Q)^2\right)^{1/2}$, which is smaller than $2H(P, Q)$. The Hellinger distance defines a bounded metric on the space of all probability measures on $\mathcal{A}$. Convergence in that metric is equivalent to convergence in $\mathcal{L}^1$ norm, because

$$H(P, Q)^2 \leq \|P - Q\|_1 \leq H(P, Q). \qquad \text{<10>}$$

The Hellinger distance satisfies the inequality $0 \leq H(P, Q) \leq \sqrt{2}$. The equality at 0 occurs when $\sqrt{p} = \sqrt{q}$ almost surely $[\lambda]$, that is, when $P = Q$ as measures on $\mathcal{A}$. Equality at $\sqrt{2}$ occurs when the Hellinger affinity is zero, that is, when $pq = 0$ almost surely $[\lambda]$, which is the condition that P and Q be mutually singular. For example, discrete distributions (concentrated on a countable set) are always at the maximum Hellinger distance from nonatomic distributions (zero mass at each point).

REMARK. Some authors prefer to have an upper bound of 1 for the Hellinger distance; they include an extra factor of a half in the definition of $H(P, Q)^2$.

Relative entropy

Let P and Q be two probability measures with densities p and q with respect to some dominating measure λ. The ***relative entropy*** (also known as the Kullback-Leibler "distance," even though it is not a metric) between P and Q is defined as $D(P\|Q) = \lambda(p\log(p/q))$.

At first sight, it is not obvious that the definition cannot suffer from the $\infty - \infty$ problem. A Taylor expansion comes to the rescue: for $x > -1$,

$$(1 + x)\log(1 + x) = x + R(x), \tag{<11>}$$

with $R(x) = \frac{1}{2}x^2/(1 + x^*)$ for some x^* between 0 and x. When $p > 0$ and $q > 0$, put $x = (p - q)/q$, discard the nonnegative remainder term, then multiply through by q to get $p\log(p/q) \geq p - q$. The same inequality also holds at points where $p > 0$ and $q = 0$, with the left-hand side interpreted as ∞; and at points where $p = 0$ we get no contribution to the defining integral. It follows not only that $\lambda p\,(\log(p/q))^- < \infty$ but also that $D(P\|Q) \geq \lambda(p - q) = 0$. The relative entropy is well defined and nonnegative. (Nonnegativity would also follow via Jensen's inequality.) Moreover, if $\lambda\{p > 0 = q\} > 0$ then $p\log(p/q)$ is infinite on a set of positive measure, which forces $D(P\|Q) = \infty$. That is, the relative entropy is infinite unless P is absolutely continuous with respect to Q. It can also be infinite even if P and Q are mutually absolutely continuous (Problem [15]).

As with the $\mathcal{L}^1$ and Hellinger distances, the relative entropy does not depend on the choice of the dominating measure λ (Problem [14]).

It is easy to deduce from the conditions for equality in Jensen's inequality that $D(P\|Q) = 0$ if and only if $P = Q$. An even stronger assertion follows from the inequality

$$D(P\|Q) \geq H^2(P, Q). \tag{<12>}$$

This inequality is trivially true unless P is absolutely continuous with respect to Q, in which case we can take λ equal to Q. For that case, define $\eta = \sqrt{p} - 1$. Note that $Q\eta^2 = H^2(P, Q)$ and

$$1 = Qp = Q(1 + \eta)^2 = 1 + 2Q\eta + Q\eta^2,$$

which implies that $2Q\eta = -H^2(P, Q)$. Hence

$$\begin{aligned} D(P\|Q) &= 2Q\left((1 + \eta)^2\log(1 + \eta)\right) \\ &\geq 2Q\left((1 + \eta)^2\frac{\eta}{1 + \eta}\right) \\ &= 2Q\eta + 2Q\eta^2 \\ &= H^2(P, Q), \end{aligned}$$

as asserted.

In a similar vein, there is a lower bound for the relative entropy involving the $\mathcal{L}^1$-distance. Inequalities <10> and <12> together imply $D(P\|Q) \geq \frac{1}{4}\|P - Q\|_1^2$. A more direct argument will give a slightly better bound,

$$D(P\|Q) \geq \tfrac{1}{2}\|P - Q\|_1^2. \tag{<13>}$$

The improvement comes from a refinement (Problem [19]) of the error term in equality <11>, namely, $R(x) \geq \frac{1}{2}x^2/(1 + x/3)$ for $x \geq -1$.

To establish <13> we may once more assume, with no loss of generality, that P is absolutely continuous with respect to Q. Write $1+\delta$ for the density. Notice that $Q\delta = 0$. Then deduce that

$$D(P\|Q) = Q\left((1+\delta)\log(1+\delta) - \delta\right) \geq \tfrac{1}{2}Q\left(\frac{\delta^2}{1+\delta/3}\right).$$

Multiply the right-hand side by $1 = Q(1+\delta/3)$, then invoke the Cauchy-Schwarz inequality to bound the product from below by half the square of

$$Q\left(\frac{|\delta|}{\sqrt{1+\delta/3}}\sqrt{1+\delta/3}\right) = Q|\delta| = \|P - Q\|_1.$$

The asserted inequality <13> follows.

<14> **Example.** Let P_θ denote the $N(\theta, 1)$ distribution on the real line with density $\phi(x-\theta)$ with respect to Lebesgue measure, where $\phi(x) = \exp(-x^2/2)/\sqrt{2\pi}$. Each of the three distances between P_0 and P_θ can be calculated in closed form:

$$D(P_0\|P_\theta) = P_0^x\left(-\tfrac{1}{2}x^2 + \tfrac{1}{2}(x-\theta)^2\right) = \tfrac{1}{2}\theta^2,$$

and

$$H^2(P_0, P_\theta) = 2 - \frac{2}{\sqrt{2\pi}}\int_{-\infty}^{\infty} \exp\left(-\tfrac{1}{4}x^2 + \tfrac{1}{4}(x-\theta)^2\right) = 2\left(1 - \exp(-\theta^2/8)\right),$$

and

$$\begin{aligned}
\|P_0 - P_\theta\|_1 &= \int_{-\infty}^{\infty} |\phi(x) - \phi(x-\theta)|\,dx \\
&= 2\int_{-\infty}^{\infty} (\phi(x) - \phi(x-\theta))^+\,dx \\
&= 2\int_{-\infty}^{\theta/2} (\phi(x) - \phi(x-\theta))\,dx = 2\left(\Phi\left(\tfrac{1}{2}\theta\right) - \Phi\left(-\tfrac{1}{2}\theta\right)\right),
\end{aligned}$$

where Φ denotes the $N(0,1)$ distribution function.

For θ near zero, Taylor expansion gives

$$H^2(P_0, P_\theta) = \tfrac{1}{4}\theta^2 + O(\theta^4) \qquad \text{and} \qquad \|P_0 - P_\theta\|_1 = \sqrt{\frac{2}{\pi}}|\theta| + O(|\theta|^3).$$

Inequalities <10>, <12>, and <13> then become, for θ near zero,

$$\begin{aligned}
\tfrac{1}{4}\theta^2 + O(\theta^4) &\leq \sqrt{\frac{2}{\pi}}|\theta| + O(|\theta|^3) \leq |\theta| + O(|\theta|^3) \\
\tfrac{1}{2}\theta^2 &\geq \tfrac{1}{4}\theta^2 + O(\theta^4) \\
\tfrac{1}{2}\theta^2 &\geq \frac{1}{\pi}\theta^2 + O(\theta^4)
\end{aligned}$$

The comparisons show that there is little room for improvement, except possibly for the lower bound in <10>. □

<15> **Example.** The method from the previous Example can be extended to other families of densities, providing a better indication of how much improvement might be possible in the three inequalities. I will proceed heuristically, ignoring questions

of convergence and integrability, and assuming existence of well behaved Taylor expansions.

Suppose P_θ has a density $\exp(g(x-\theta))$ with respect to Lebesgue measure on the real line. Write $\dot{g}$ and $\ddot{g}$ for the first and second derivatives of the function $g(x)$, so that

$$g(x-\theta) - g(x) = -\theta\dot{g}(x) + \tfrac{1}{2}\theta^2\ddot{g}(x) + \text{ terms of order } |\theta|^3$$

and

$$\begin{aligned}\exp(g(x-\theta) - g(x)) &= 1 - \theta\dot{g}(x) + \tfrac{1}{2}\theta^2\ddot{g}(x) + \tfrac{1}{2}\left(-\theta\dot{g}(x) + \tfrac{1}{2}\theta^2\ddot{g}(x)\right)^2 + \dots \\ &= 1 - \theta\dot{g}(x) + \tfrac{1}{2}\theta^2\left(\ddot{g}(x) + \dot{g}(x)^2\right) + \text{ terms of order } |\theta|^3\end{aligned}$$

From the equalities

$$\begin{aligned}1 &= \int_{-\infty}^{\infty} \exp(g(x-\theta)) \\ &= P_0^x \exp(g(x-\theta) - g(x)) \\ &= 1 - \theta P_0\dot{g}(x) + \tfrac{1}{2}\theta^2 P_0\left(\ddot{g}(x) + \dot{g}(x)^2\right) + \text{ terms of order } |\theta|^3\end{aligned}$$

we can conclude first that $P_0\dot{g} = 0$, and then $P_0\left(\ddot{g}(x) + \dot{g}(x)^2\right) = 0$.

REMARK. The last two assertions can be made rigorous by domination assumptions, allowing derivatives to be taken inside integral signs. Readers familiar with the information inequality from theoretical statistics will recognize the two ways of representing the information function for the family of densities, together with the zero-mean property of the score function.

Continuing the heuristic, we will get approximations for the three distances for values of θ near zero.

$$D(P_0\|P_\theta) = P_0^x\left(g(x) - g(x-\theta)\right) = \tfrac{1}{2}\theta^2 P\dot{g}^2 + \dots$$

From the Taylor approximation,

$$\exp\left(\tfrac{1}{2}g(x-\theta) - \tfrac{1}{2}g(x)\right) = 1 - \tfrac{1}{2}\theta\dot{g} + \tfrac{1}{8}\theta^2\left(2\ddot{g} + \dot{g}^2\right) + \dots$$

we get

$$H^2(P_0, P_\theta) = 2 - 2P_0^x \exp\left(\tfrac{1}{2}g(x-\theta) - \tfrac{1}{2}g(x)\right) = \tfrac{1}{4}\theta^2 P\dot{g}^2 + \dots$$

And finally,

$$\|P_0 - P_\theta\|_1 = P_0\left|g(x-\theta) - g(x)\right| = |\theta|\, P_0|\dot{g}| + \dots .$$

If we could find a g for which $(P_0|\dot{g}|)^2 = P_0\dot{g}^2$, the two inequalities

$$2H(P_0, P_\theta) \geq \|P_0 - P_\theta\|_1 \qquad \text{and} \qquad D(P_0\|P_\theta) \geq \tfrac{1}{2}\|P_0 - P_\theta\|_1^2$$

would become sharp up to terms of order θ^2. The desired equality for g would force $|\dot{g}|$ to be a constant, which would lead us to the density $f(x) = \frac{1}{2}e^{-|x|}$. Of course $\log f$ is not differentiable at the origin, an oversight we could remedy by a slight smoothing of $|x|$ near the origin. In that way we could construct arbitrarily smooth densities with $P_0\dot{g}^2$ as close as we please to $(P_0|\dot{g}|)^2$. There can be no improvement in the constants in the last two displayed inequalities. □

4. The classical concept of absolute continuity

The fundamental problem of Calculus concerns the interpretation of differentiation and integration, for functions of a real variable, as inverse operations: Which functions on the real line can be expressed as integrals of their derivatives? Clearly, if $H(x) = \int_a^x h(t)\,dt$, with h Lebesgue integrable, then H is a continuous function, but it also possesses a stronger property.

<16> **Definition.** *A real valued function H defined on an interval $[a, b]$ of the real line is said to be* ***absolutely continuous*** *if to each $\epsilon > 0$ there exists a $\delta > 0$ such that $\sum_i |H(b_i) - H(a_i)| < \epsilon$ for all finite collections of nonoverlapping subintervals $[a_i, b_i]$ of $[a, b]$ for which $\sum_i (b_i - a_i) < \delta$.*

REMARK. In the Definition, and subsequently, ***nonoverlapping*** means that the interiors of the intervals are disjoint. For example, $[0, 1]$ and $[1, 2]$ are nonoverlapping. Their intersection has zero Lebesgue measure.

Note the strong similarity between Definition <16> and the reformulation of the measure theoretic definition of absolute continuity as an ϵ-δ property, in Example <3>.

The connection between absolute continuity of functions and integration of derivatives was established by Lebesgue (1904). It is one of the most celebrated results of classical analysis.

<17> **Fundamental Theorem of Calculus.** *A real valued function H defined on an interval $[a, b]$ is absolutely continuous if and only if the following three conditions hold*

(i) the derivative $H'(x)$ exists at Lebesgue almost all points of $[a, b]$

(ii) the derivative H' is Lebesgue integrable

(iii) $H(x) - H(a) = \int_a^x H'(t)\,dt$ for each x in $[a, b]$

REMARK. Of course it is actually immaterial for (ii) and (iii) how H' is defined on the Lebesgue negligible set of points at which the derivative does not exist. For example, we could take H' as the measurable function $\limsup_{n\to\infty} n\big(H(x + n^{-1}) - H(x)\big)$. The proof in Section 6 provides two other natural choices.

We may fruitfully think of the Fundamental Theorem as making two separate assertions:

(FT_1) *A real-valued function H on $[a, b]$ is absolutely continuous if and only if it has the representation*

<18>
$$H(x) = H(a) + \int_a^x h(t)\,dt \qquad \text{for all } x \text{ in } [a, b]$$

for some Lebesgue integrable h defined on $[a, b]$. The function h is unique up to Lebesgue almost sure equivalence.

(FT_2) *If H is absolutely continuous then it is differentiable almost everywhere, with derivative equal Lebesgue almost everywhere to the function h from <18>.*

As shown at the end of this Section, Assertion FT_1 can be recovered from the Radon-Nikodym theorem for measures. The proof of Assertion FT_2 (Section 6), which identifies the density h with a derivative defined almost everywhere, requires an auxiliary result known as the Vitali Covering Lemma (Section 5).

Notice what the Fundamental Theorem does not assert: that differentiability almost everywhere should allow the function to be recovered as an integral of that derivative. As a pointwise limit of continuous functions, $(H(x+\delta) - H(x))/\delta$, the derivative $H'(x)$ is measurable, when it exists. However, it need not be Lebesgue integrable, as noted by Lebesgue in his doctoral thesis (Lebesgue 1902): the function $F(x) := x^2 \sin(1/x^2)$, with $F(0) = 0$, has a derivative at all points of the real line, but $F'(x)$ behaves like $-2x^{-1}\cos(1/x^2)$ for x near 0, which prevents integrability on any interval that contains 0 (Problem [20]). The function F is not absolutely continuous.

Neither does the Fundamental Theorem assert that absolute continuity follows from almost sure existence of H' with $\int_a^b |H'(x)|\,dx$ finite. For example, the function F constructed in Example <1>, whose derivative exists and is equal to zero outside a set of zero Lebesgue measure, cannot be recovered by integration of the derivative, even though that derivative is integrable. It is therefore quite surprising that existence of even a one sided, integrable derivative *everywhere* is enough to ensure absolute continuity.

<19> **Theorem.** *Let H be a continuous function defined on an interval $[a, b]$, with a (real-valued) right-hand derivative $h(x) := \lim_{\delta \downarrow 0} (H(x+\delta) - H(x))/\delta$ existing at each point of $[a, b)$. If h is Lebesgue integrable then $H(x) = H(a) + \int_a^x h(t)\,dt$ for each x in $[a, b]$, and hence H is an absolutely continuous function on $[a, b]$.*

See Problem [21] for an outline of the proof. The Theorem justifies the usual treatment in introductory Calculus courses of integration as an inverse operation to differentiation.

Proof of FT_1. If H is given by representation <18> then its increments are controlled by the measure ν defined on $[a, b]$ by its density $|h|$ with respect to Lebesgue measure $\mathfrak{m}$ on $\mathcal{B}[a, b]$. If $\{[a_i, b_i] : i = 1, \ldots, k\}$ is a family of nonoverlapping subintervals of $[a, b]$ then $\sum_{i=1}^k |H(b_i) - H(a_i)| \le \sum_{i=1}^k \nu[a_i, b_i]$. From Example <3>, when the Lebesgue measure $\sum_i (b_i - a_i)$ of the set $A = \cup_i [a_i, b_i]$ is small enough, the ν measure is also small, because ν is absolutely continuous (in the sense of measures) with respect to $\mathfrak{m}$. It follows that H is absolutely continuous as a function on $[a, b]$.

Conversely, an absolutely continuous function H defines two nonnegative functions F^+ and F^- on $[a, b]$ by

<20>
$$F^{\pm}(x) = \sup_{\pi(x)} \sum_i (H(x_i) - H(x_{i-1}))^{\pm}$$

where the suprema run over the collection $\pi(x)$ of all finite partitions $a = x_0 \le x_1 \le \ldots \le x_k = x$ of $[a, x]$, for each x in $[a, b]$. At $x = a$ all partitions are degenerate and $F^{\pm}(a) = 0$. By splitting $[a, x]$ into a finite union of intervals of length less than δ, you will see that both functions are real valued. More precisely, $F^{\pm}(x) \le \epsilon(b-a)/\delta$, where ϵ and δ come from Definition <16>.

Both $F^{\pm}$ are increasing functions, for the following reason. First note that insertion of an extra point into a partition $\{x_i\}$ increases the defining sums. In particular, if $a < y < x$ then we may arrange that y is a point in the $\pi(x)$ partition, so that the sums on the right-hand side of <20> are larger than the corresponding sums for $\pi(y)$. More precisely, $F^{\pm}(x) = F^{\pm}(y) + \sup \sum_i (H(y_i) - H(y_{i-1}))^{\pm}$, where the supremum runs over all finite partitions $y = y_0 < y_1 < \ldots < y_k = x$ of $[y, x]$. When $|x - y| < \delta$, with $\delta > 0$ as in Definition <16>, each sum for the supremum is less than ϵ, and hence $|F^{\pm}(x) - F^{\pm}(y)| \le \epsilon$. That is, both $F^{\pm}$ are continuous functions on $[a, b]$. A similar argument applied to finite nonoverlapping collections of intervals $\{[y_i, x_i]\}$ leads to the stronger conclusion that both $F^{\pm}$ are absolutely continuous functions.

REMARK. We didn't really need absolute continuity to break H into a difference of two increasing functions. It would suffice to have $\sup \sum_i |H(y_i) - H(y_{i-1})| < \infty$, where the supremum runs over all finite partitions $y = y_0 < y_1 < \ldots < y_k = x$ of $[y, x]$. Such a function is said to have ***bounded variation*** on the interval $[a, b]$. Close inspection of the arguments in Section 6 would reveal that functions of bounded variation have a derivative almost everywhere. Without absolute continuity, we cannot recover the function by integrating that derivative, as shown by Example <1>.

As shown in Section 2.9 using quantile transformations, increasing functions such as $F^{\pm}$ correspond to distribution functions of measures: there exists two finite measures $\nu^{\pm}$ on $\mathcal{B}[a, b]$ such that $\nu^{\pm}(a, x] = F^{\pm}(x)$ for x in $[a, b]$. Notice that $\nu^{\pm}$ put zero mass at each point, by continuity of $F^{\pm}$. The absolute continuity of the $F^{\pm}$ functions translates into an assertion regarding the class $\mathcal{E}$ of all subsets of $[a, b]$ expressible as a finite union of intervals $(y_i, x_i]$: for each $\epsilon > 0$ there exists a $\delta > 0$ such that $\nu E \le \epsilon$ for every set E in $\mathcal{E}$ with $\mathfrak{m}E < \delta$.

A simple λ-class generating argument shows that $\mathcal{E}$ is a dense subclass of $\mathcal{B}[a, b]$ in the $\mathcal{L}^1(\mu)$ sense, where $\mu = \mathfrak{m} + \nu^+ + \nu^-$. That is, for each set B in $\mathcal{B}[a, b]$ and each $\epsilon' > 0$ there exists an E in $\mathcal{E}$ for which $\mu|B - E| < \epsilon'$. In particular, if we put $\epsilon' = \min(\delta/2, \epsilon)$ then $\mathfrak{m}B < \delta/2$ implies $\mathfrak{m}E < \delta$ from which we get $\nu^{\pm}E \le \epsilon$ and $\nu^{\pm}B \le 2\epsilon$. That is, both $\nu^{\pm}$ are absolutely continuous (as measures) with respect to $\mathfrak{m}$.

By the Radon-Nikodym Theorem, there exist $\mathfrak{m}$-integrable functions $h^{\pm}$ for which $\nu^{\pm} f = \mathfrak{m}(h^{\pm} f)$ for all f in $\mathcal{M}^+$. In particular, $F^{\pm}(x) = \nu^{\pm}(a, x] = \int_a^x h^{\pm}(t)\, dt$ for $a \le x \le b$.

Finally, note that for each partition $\{x_i\}$ in $\pi(x)$ we have

$$\sum_i (H(x_i) - H(x_{i-1}))^+ - \sum_i (H(x_i) - H(x_{i-1}))^- = \sum_i (H(x_i) - H(x_{i-1})),$$

which reduces to $H(x) - H(a)$ after cancellations. We can choose the partition to give simultaneously values as close as we please to both suprema in <20>. In the limit we get

$$H(x) - H(a) = F^+(x) - F^-(x) = \int_a^x (h^+(t) - h^-(t))\, dt.$$

That is, representation <18> holds with $h = h^+ - h^-$. The uniqueness of the representing h can be established by another λ-class generating argument. □

*5. Vitali covering lemma

Suppose D is a Borel subset of $\mathbb{R}^d$ with finite Lebesgue measure $\mathfrak{m}D$. There are various ways in which we may approximate D by simpler sets. For example, from Section 2.1 and related problems, to each $\epsilon > 0$ there exists an open set $G \supset D$ and a compact set $K \subseteq D$ for which $\mathfrak{m}(G \backslash K) < \epsilon$. In particular, we could take G as a countable union of open cubes. In general, we cannot hope to represent K as union of closed cubes, because those sets would all have to lie in the interior of D, a set whose measure might be strictly smaller than $\mathfrak{m}D$. However, if we allow the cubes to poke slightly outside D we can even approximate by a finite union of *disjoint* closed cubes, as a consequence of a Vitali covering theorem.

There are many different results presented in texts as the Vitali theorem, each involving different levels of generality and different degrees of subtlety in the proof. For the application to FT_2, a simple version for approximation by intervals on the real line would suffice, but I will present a version for $\mathbb{R}^d$, in which the sets are not assumed to be cubes, but are instead required to be "not too skinny". The extra generality is useful; and it causes few extra difficulties in the proof, indeed, it helps to highlight the beautiful underlying idea; and the pictures look nicer in more than one dimension.

Of course *skinny* is not a technical term. Instead, for a fixed constant $\gamma > 0$, let us say that a measurable set F is ***γ-regular*** if there exists an open ball B_F with $B_F \supseteq F$ and $\mathfrak{m}F/\mathfrak{m}B_F \geq \gamma$. We will sometimes need to write B_F as $B(x, r)$, the open ball with center x and radius r.

<21> **Definition.** *Call a collection $\mathcal{V}$ of closed subsets of $\mathbb{R}^d$ a γ-regular* ***Vitali covering*** *of a set E if each member of $\mathcal{V}$ is γ-regular and if, to each $\epsilon > 0$, each point of E belongs to a set F (depending on the point and ϵ) from $\mathcal{V}$ with diameter less than ϵ.*

Put another way, if we write $\mathcal{V}_\epsilon$ for $\{F \in \mathcal{V} : \text{diam}(F) < \epsilon\}$, then the Vitali covering property is equivalent to $E \subseteq \cup\{F : F \in \mathcal{V}_\epsilon\}$ for every $\epsilon > 0$. Notice that if G is an open set with $G \supseteq E$, then $\{F \in \mathcal{V} : F \subseteq G\}$ is also a Vitali covering for E: for each x in E, all points close enough to x must lie within G.

<22> **Vitali Covering Lemma.** *For some fixed $\gamma > 0$, let $\mathcal{V}$ be a Vitali covering for a set D in $\mathcal{B}(\mathbb{R}^d)$ with finite Lebesgue measure. Then there exists a countable family $\{F_i\}$ of disjoint sets from $\mathcal{V}$ for which the set $D \backslash (\cup_i F_i)$ has zero Lebesgue measure.*

REMARK. More refined versions of the Theorem (such as the results presented by Saks 1937, Section IV.3, or Wheeden & Zygmund 1977, Section 7.3) allow γ to vary from point to point within D, and relax assumptions of measurablity or finiteness of $\mathfrak{m}D$. We will have no need for the more general versions.

Proof. The result is trivial if $\mathfrak{m}D$ is zero, so we may assume that $\mathfrak{m}D > 0$. The proof works via repeated reduction of $\mathfrak{m}D$ by removal of unions of finite families of disjoint sets from $\mathcal{V}$. The method for the first step sets the pattern.

Fix an $\epsilon > 0$. As you will see later, we need ϵ small enough that the constant $\rho := 3^{-d}(1 - \epsilon)\gamma - \epsilon$ is strictly positive.

Find an open set G and a compact set K for which $G \supseteq D \supseteq K$ and $m(G\backslash K) < \epsilon m D$. Discard all those members of $\mathcal{V}$ that are not subsets of G. What remains is still a Vitali covering of D.

REMARK. As the proof proceeds, various sets will be discarded from $\mathcal{V}$. Rather than invent a new symbol for the class of sets that remains after each discard, I will reuse the symbol $\mathcal{V}$. That is, $\mathcal{V}$ will denote different things at different stages of the proof, but in every case it will be a Vitali covering for a subset of interest.

The family of open balls $\{B_F : F \in \mathcal{V}\}$ covers the compact set K. It has a finite subcover, corresponding to sets $F_1, \ldots, F_m$. The ball B_{F_i} is of the form $B(x_i, r_i)$. We may assume that the sets are numbered so that $r_1 \geq r_2 \geq \ldots \geq r_m$.

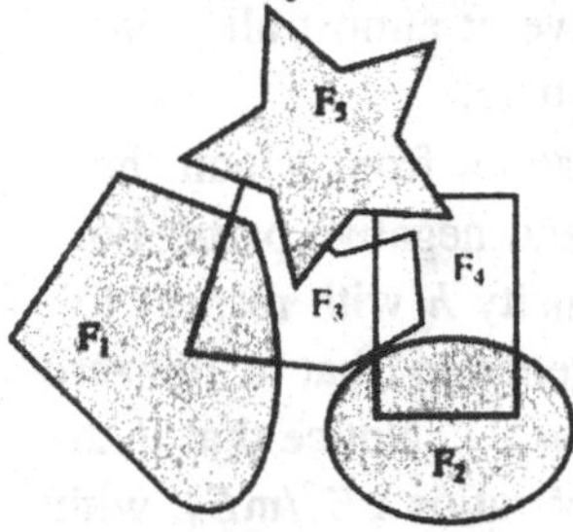

Define a subset J of $\{1, 2, \ldots, m\}$ by successively considering each j, in the order 1, 2, ..., m, for inclusion in J, rejecting j if F_j intersects an F_i for an i already accepted into J. For example, with the five sets shown in the picture, we would have $J = \{1, 2, 5\}$: we include 1, then 2 (because $F_2 \cap F_1 = \emptyset$), reject 3 (because $F_3 \cap F_1 \neq \emptyset$), reject 4 (because $F_4 \cap F_2 \neq \emptyset$), then accept 5 (because $F_5 \cap F_1 = \emptyset$ and $F_5 \cap F_2 = \emptyset$).

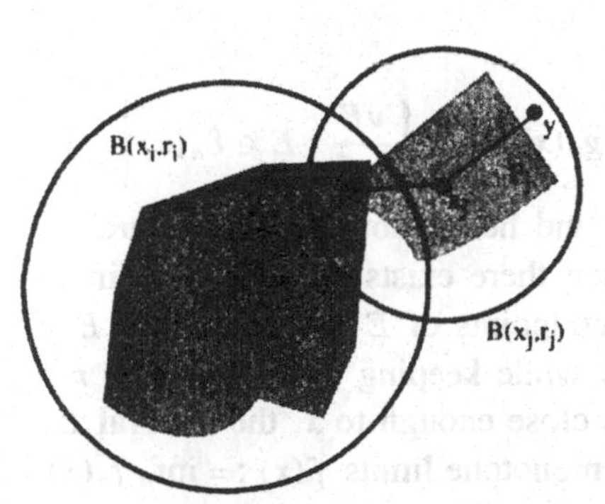

For each excluded j there is an i in J for which $i < j$ and $F_i \cap F_j \neq \emptyset$. The ordering of the radii ensures that $B(x_i, 3r_i) \supseteq B(x_j, r_j)$: if $z \in F_i \cap F_j$ and $y \in B(x_j, r_j)$, we have

$$|x_i - y| \leq |x_i - z| + |z - y| \leq r_i + 2r_j \leq 3r_i.$$

Thus $\cup_{i \in J} B(x_i, 3r_i) \supseteq \cup_{j=1}^{m} B(x_j, r_i) \supseteq K$. The open balls $B(x_i, 3r_i)$ might poke outside G, but the corresponding sets F_i from $\mathcal{V}$, and their union, $E_1 := \cup_{i \in J} F_i$, are closed subsets of G.

Pairwise disjointness of the sets $\{F_i : i \in J\}$ allows us to calculate the measure of their union by adding, which, together with the regularity property

$$mF_i \geq \gamma m B(x_i, r_i) = (\gamma 3^{-d})\, m B(x_i, 3r_i),$$

gives us a lower bound for the measure of E_1,

$$\frac{3^d}{\gamma} m E_1 = \frac{3^d}{\gamma} \sum_{i \in J} mF_i \geq \sum_{i \in J} m B(x_i, 3r_i) \geq m\left(\cup_{i \in J} B(x_i, 3r_i)\right) \geq mK \geq (1 - \epsilon) m D.$$

It follows that

$$m(D\backslash E_1) \leq mG - mE_1 \leq (1 + \epsilon) m D - \left(3^{-d}(1 - \epsilon)\gamma\right) m D = (1 - \rho) m D.$$

That is, E_1 carves away at least a fraction ρ of the Lebesgue measure of D, thereby completing the first step in the construction.

The second step is analogous. Discard all those members of $\mathcal{V}$ that intersect E_1. What remains is a new Vitali covering $\mathcal{V}$ for $D\backslash E_1$, because each point of $D\backslash E_1$ is at a strictly positive distance from the closed set E_1. Repeat the argument from the previous paragraphs to find disjoint sets from $\mathcal{V}$ whose union, E_2, removes at least a fraction ρ of the Lebesgue measure of $D\backslash E_1$. Then $\mathfrak{m}(D\backslash(E_1 \cup E_2)) \le (1-\rho)^2 \mathfrak{m} D$, and the sets that make up $E_1 \cup E_2$ are all disjoint. And so on. In the limit we have the countable disjoint family whose existence is asserted by the Lemma. □

*6. Densities as almost sure derivatives

With Lemma <22> we have the means to prove FT$_2$: if

$$H(x) - H(a) = \int_a^x h(t)\, dt \qquad \text{for } a \le x \le b$$

for a Lebesgue integrable function h then H has a derivative at almost all points of $[a, b]$ and that derivative coincides with h almost everywhere.

It suffices if we consider the case where h is nonnegative, from which the general case would follow by breaking h into its positive and negative parts. For nonnegative h, write ν for the measure on $\mathcal{B}[a, b]$ with density h with respect to Lebesgue measure $\mathfrak{m}$. Define $\mathcal{E}_n(x)$ as the set of all nondegenerate (that is, nonzero length) closed intervals E for which $x \in E \subseteq (x - n^{-1}, x + n^{-1})$. Notice that if the derivative $H'(x)$ exists then it can be recovered as a limit of ratios $\nu E_n / \mathfrak{m} E_n$, with $E_n \in \mathcal{E}_n(x)$.

Define functions

$$f_n(x) = \sup\left\{ \frac{\nu E}{\mathfrak{m} E} : E \in \mathcal{E}_n(x) \right\} \qquad \text{and} \qquad g_n(x) = \inf\left\{ \frac{\nu E}{\mathfrak{m} E} : E \in \mathcal{E}_n(x) \right\}$$

Both the sets $\{f_n > r\}$ and $\{g_n < r\}$ are open, and hence both f_n and g_n are Borel measurable. For example, if $f_n(x) > r$ then there exists an interval E in $\mathcal{E}_n(x)$ with $\nu E > r \mathfrak{m} E$. Continuity of H at the endpoints of E lets us expand E slightly, ensuring that x is an interior point of E while keeping the ratio above r and keeping E within $(x - n^{-1}, x + n^{-1})$. If y is close enough to x, the interval E also belongs to $\mathcal{E}_n(y)$, and hence $f_n(y) > r$. The monotone limits $f(x) := \inf_n f_n(x)$ and $g(x) := \sup_n g_n(x)$ are also Borel measurable.

Clearly $f(x) \ge g(x)$ everywhere. For $0 < \delta < 1/n$, both $[x, x+\delta]$ and $[x-\delta, x]$ belong to $\mathcal{E}_n(x)$, and hence both

$$\frac{H(x+\delta) - H(x)}{\delta} \qquad \text{and} \qquad \frac{H(x) - H(x-\delta)}{\delta}$$

lie between $g_n(x)$ and $f_n(x)$. In the limit as first δ tends to zero then n tends to infinity we have

$$g(x) \le \liminf_{\delta \to 0} \frac{H(x+\delta) - H(x)}{\delta} \le \limsup_{\delta \to 0} \frac{H(x+\delta) - H(x)}{\delta} \le f(x),$$

with an analogous pair of inequalities for $[x - \delta, x]$. At points where $g(x) = h(x) = f(x)$ the derivative $H'(x)$ must exist and be equal to $h(x)$. The next Lemma will help us prove the almost sure equality of f, g, and h.

<23> **Lemma.** *Let A be a Borel set with finite Lebesgue measure and r be a positive constant.*

(i) If $f(x) > r$ for all x in A then $\nu A \ge r \mathfrak{m} A$.

(ii) If $g(x) < r$ for all x in A then $\nu A \le r\mathbb{m}A$.

Proof. Let G be an open set and K be a compact set, with $G \supseteq A \supseteq K$.

For (i), note that the collection $\mathcal{V}$ of all nondegenerate closed subintervals E of G that satisfy the inequality $\nu E \ge r\mathbb{m}E$ is a Vitali covering of A (because $f_n > r$ on A for every n). By Lemma <22> there is a countable family of disjoint intervals $\{E_i\}$ from $\mathcal{V}$ whose union L covers A up to a m-negligible set. We then have

$$\begin{aligned} \nu G &\ge \nu L && \text{because } G \text{ contains all intervals from } \mathcal{V} \\ &= \textstyle\sum_i \nu E_i && \text{disjoint intervals} \\ &\ge \textstyle\sum_i r\mathbb{m}E_i && \text{definition of } \mathcal{V} \\ &= r\mathbb{m}L && \text{disjoint intervals} \\ &\ge r\mathbb{m}A && \text{covering property.} \end{aligned}$$

Take the infimum over G to obtain the assertion from (i).

The argument for (ii) is similar: Reverse the direction of the inequality in the definition of $\mathcal{V}$, then interchange the roles of ν and $r\mathbb{m}$ in the string of inequalities for the covering. □

To complete the proof of FT$_2$, for any pair of rational numbers with $r > s > 0$ let $A = \{x \in [a, b] : h(x) < s < r < f(x)\}$. From Lemma <23> and the fact that ν has density h with respect to m we get $s\mathbb{m}A \ge \nu A \ge r\mathbb{m}A$. Conclude that $\mathbb{m}A = 0$. Cast out countably many negligible sets, one for each pair of rationals, to deduce that $h \ge f$ a.e. $[\lambda]$ on $[a, b]$. Argue similarly to show that $h \le g$ a.e. $[\lambda]$ on $[a, b]$. Together with the fact that $f \ge g$ these two inequalities imply that $f = g = h$ at almost all points of $[a, b]$.

7. Problems

[1] Let μ be the measure defined for all subsets of $\mathcal{X}$ by $\mu A = +\infty$ if $A \ne \emptyset$, and $\mu(\emptyset) = 0$. Show that $\mu f = \mu(2f)$ for all nonnegative functions f.

[2] Let λ denote the measure on $\mathbb{R}$ with λA equal to the number of points in A (possibly infinite), and let $\mu A = \infty$ for all nonempty A. Show that λ has no density with respect to μ, even though both measures have the same negligible sets.

[3] Suppose Δ_1 and Δ_2 are functions in $\mathcal{M}^+(\mathcal{X}, \mathcal{A})$ for which $\mu(f\Delta_1) = \mu(f\Delta_2)$ for all f in $\mathcal{M}^+(\mathcal{X}, \mathcal{A})$, for some measure μ.

(i) Show that $\Delta_1 = \Delta_2$ a.e. $[\mu]$ if $\mu\Delta_1 < \infty$. Hint: Consider the equality $\mu\Delta_1\{\Delta_1 > \Delta_2\} = \mu\Delta_2\{\Delta_1 > \Delta_2\}$.

(ii) Show that the assumption of finiteness of $\mu\Delta_1$ is not required for the conclusion $\Delta_1 = \Delta_2$ a.e. $[\mu]$ if μ is a sigma-finite measure. Hint: For each positive rational number r show that $\mu(\Delta_1 - \Delta_2)\{\Delta_1 > r \ge \Delta_2\}A = 0$ for each set with $\mu A < \infty$.

[4] Let P and Q be probability measures on $(\Omega, \mathcal{F})$, and let $\mathcal{E}$ be a generating class for $\mathcal{F}$ that is stable under finite intersections and contains Ω. Suppose there exists a

p in $\mathcal{L}^1(Q)$ such that $PE = Q(pE)$ for every E in $\mathcal{E}$. Show that P has density p with respect to Q.

[5] Does Example <3> have an analog for $\mathcal{L}^p$ convergence, for some $p > 1$: if ν is a finite measure dominated by a measure λ, and if $\{f_n\}$ is a sequence of measurable functions for which $\lambda|f_n|^p \to 0$, does it follow that $\nu|f_n|^p \to 0$?

[6] Let ν and μ be finite measures on the sigma-field $\sigma(\mathcal{E})$ generated by a countable field $\mathcal{E}$. Suppose that for each $\epsilon > 0$ there exists a $\delta > 0$ such that $\nu E < \epsilon$ for each E in $\mathcal{E}$ with $\mu E < \delta$. Show that ν is absolutely continuous with respect to μ, as measures on $\sigma(\mathcal{E})$. Hint: Suppose $\mu A = 0$ for an A in $\sigma(\mathcal{E})$. Show that there exists a countable subclass $\{E_n\}$ of $\mathcal{E}$ such that $\cup_n E_n \supseteq A$ and $\mu(\cup_n E_n) < \delta$. Use the properties of fields to argue that the E_n sets can be chosen disjoint. Deduce that $\mu\left(\cup_{n\le N} E_n\right) < \delta$ for each finite N. Conclude that $\nu A \le \lim \nu\left((\cup_n E_n\right) \le \epsilon$.

[7] Let ν and μ be finite measures on $(\mathcal{X}, \mathcal{A})$. Suppose ν has Lebesgue decomposition $(\Delta, \mathcal{N})$ with respect to μ. Find the Lebesgue decomposition of μ with respect to ν. Hint: Consider $1/\Delta$ on $\{0 < \Delta < \infty\}$.

[8] Show that a measure μ is sigma-finite if and only if there exists a strictly positive, measurable function Ψ for which $\mu\Psi < \infty$. Hint: If μ is sigma-finite, consider functions of the form $\Phi(x) = \sum_i \alpha_i\{x \in A_i\}$, for appropriate partitions.

[9] Suppose $(\mathcal{N}_1, \Delta_1)$ and $(\mathcal{N}_2, \Delta_2)$ are both Lebesgue decompositions for a finite measure ν with respect to a finite measure μ. Prove $\mathcal{N}_1 = \mathcal{N}_2$ a.e. $[\nu + \mu]$ and $\Delta_1\mathcal{N}_1^c = \Delta_2\mathcal{N}_2^c$ a.e. $[\nu + \mu]$ by the following steps.

(i) Show that $\mathcal{N}_1 = \mathcal{N}_2$ a.e. $[\nu]$ by means of the equality $\nu(\mathcal{N}_2\mathcal{N}_1^c) = \nu(\mathcal{N}_2\mathcal{N}_1^c\mathcal{N}_1) + \mu(\mathcal{N}_2\mathcal{N}_1^c\Delta_1) = 0$, and a companion equality obtained by interchanging the roles of the two decompositions. Using the trivial fact that $\mathcal{N}_1 = \mathcal{N}_2$ a.e. $[\mu]$ (because both sets are μ-negligible), deduce that $\mathcal{N}_1 = \mathcal{N}_2$ a.e. $[\nu + \mu]$.

(ii) Use the result from part (i) to show that

$$\mu(f\Delta_1\mathcal{N}_1^c) = \nu(f\mathcal{N}_1^c) = \nu(f\mathcal{N}_2^c) = \mu(f\Delta_2\mathcal{N}_2^c) \qquad \text{for all } f \in \mathcal{M}^+.$$

Deduce that $\Delta_1\mathcal{N}_1^c = \Delta_2\mathcal{N}_2^c$ a.e. $[\mu]$ by arguing as in Problem [3].

(iii) Use part (ii) to show that

$$\nu\{\Delta_1\mathcal{N}_1^c > \Delta_2\mathcal{N}_2^c\} = \nu\{\Delta_1\mathcal{N}_1^c > \Delta_2\mathcal{N}_2^c\}\mathcal{N}_1 + \mu\{\Delta_1\mathcal{N}_1^c > \Delta_2\mathcal{N}_2^c\}\Delta_1\mathcal{N}_1^c = 0$$

Hint: What value does $\Delta_1\mathcal{N}_1^c$ take on $\mathcal{N}_1$? Argue similarly for the companion equality. Deduce that $\Delta_1\mathcal{N}_1^c = \Delta_2\mathcal{N}_2^c$ a.e. $[\nu + \mu]$.

(iv) Extend the argument to prove the analogous uniqueness assertion for the Lebesgue decomposition when ν and μ are sigma-finite measures. Hint: Decompose $\mathcal{X}$ into countably many sets $\{\mathcal{X}_i : i \in \mathbb{N}\}$ with $(\nu + \mu)\mathcal{X}_i < \infty$ for each i.

[10] Suppose ν and μ are both sigma-finite measures on $(\mathcal{X}, \mathcal{A})$.

(i) Show that $\nu + \mu$ is also sigma-finite. Deduce via Problem [8] that there exists a strictly positive, measurable function Φ_0 for which $\nu\Phi_0 < \infty$ and $\mu\Phi_0 < \infty$.

(ii) Define $\nu_0 f = \nu(f\Phi_0)$ and $\mu_0 f = \mu(f\Phi_0)$ for each f in $\mathcal{M}^+$. Show that ν_0 and μ_0 are both finite measures.

(iii) From the Lebesgue decomposition $\nu_0 f = \nu_0(f\mathcal{N}) + \mu_0(f\Delta)$ for all $f \in \mathcal{M}^+$, where $\mu_0\mathcal{N} = 0$ and $\Delta \in \mathcal{M}^+$, derive the corresponding Lebesgue decomposition for λ with respect to μ.

(iv) From the uniqueness result of Problem [9], applied to λ_0 and μ_0, deduce that the Lebesgue decomposition for λ is unique up to the appropriate almost sure equivalences.

[11] Let ν be a finite measure for which $\nu = \nu_1 + \lambda_1 = \nu_2 + \lambda_2$, where each ν_i is dominated by a fixed sigma-finite measure μ and each λ_i is singular with respect to μ. Show that $\nu_1 = \nu_2$ and $\lambda_1 = \lambda_2$. Hint: If $\mu S_i^c = 0 = \lambda_i S_i$ and $d\nu_i/d\mu = \Delta_i$, for $i = 1, 2$, show that $\mu(f\Delta_1 f S_1 S_2) = \mu(f\Delta_2 f S_1 S_2)$ for all f in $\mathcal{M}^+$, by arguing as in Problem [3].

[12] Let μ_1 and μ_2 be finite measures with densities m_1 and m_2 with respect to a sigma-finite measure λ. Define $(\mu_1 - \mu_2)^+$ as the measure with density $(m_1 - m_2)^+$ with respect to λ.

(i) Show that $(m_1 - m_2)^+ B = \sup_\pi \sum_{A\in\pi}(m_1 - m_2)AB$, the supremum running over all finite partitions of $\mathcal{X}$.

(ii) Show that $|\mu_1 - \mu_2| = (\mu_1 - \mu_2)^+ + (\mu_2 - \mu_1)^+$.

[13] Let P and Q be probability measures with densities p and q with respect to a *sigma-finite measure* λ. *For fixed* $\alpha \geq 1$, *show that* $\Delta_\alpha(P, Q) := \lambda|p^{1/\alpha} - q^{1/\alpha}|^\alpha$ does not depend on the choice of dominating measure. Hint: Let μ be another sigma-finite dominating measure. Write ψ for the density of λ with respect to $\lambda + \mu$. Show that $dP/d(\lambda + \mu) = \psi p$ and $dQ/d(\lambda + \mu) = \psi q$. Express $\Delta_\alpha(P, Q)$ as an integral with respect to $\lambda + \mu$. Argue similarly for μ.

[14] Adapt the argument from the previous Problem to show that the relative entropy $D(P\|Q)$ does not depend on the choice of dominating measure.

[15] Let P be the standard Cauchy distribution on the real line, and let Q be the standard normal distribution. Show that $D(P\|Q) = \infty$, even though P and Q are mutually absolutely continuous.

[16] Let P and Q be probability measures defined on the same sigma-field $\mathcal{F}$. A *randomized test* is a measurable function f with $0 \leq f \leq 1$. (For observation ω, the value $f(\omega)$ gives the probability of rejecting the "hypothesis" P.) Find the test that minimizes $Pf + Q(1 - f)$.

[17] Let P and Q be finite measures defined on the same sigma-field $\mathcal{F}$, with densities p and q with respect to a measure μ. Suppose $\mathcal{X}_0$ is a measurable set with the property that there exists a nonnegative constant K such that $q \geq Kp$ on $\mathcal{X}_0$ and $q \leq Kp$ on $\mathcal{X}_0^c$. For each $\mathcal{F}$-measurable function with $0 \leq f \leq 1$ and $Pf \leq P\mathcal{X}_0$, prove that $Qf \leq Q\mathcal{X}_0$. Hint: Prove that $(q - Kp)(2\mathcal{X}_0 - 1) \geq (q - Kp)(2f - 1)$, then integrate. *To statisticians this result is known as the Neyman-Pearson Lemma.*

[18] Let $\mu = \mu_1 - \mu_2$ and $\mu' = \mu'_1 - \mu'_2$ be signed measures, in the sense of Section 3. Show that $\nu := (\mu_1 + \mu'_2) \wedge (\mu'_1 + \mu_2) - (\mu_2 + \mu'_2)$ is the largest signed measure for which $\nu \le \mu$ and $\nu \le \mu'$.

[19] Let $f(x) = (1+x)\log(1+x) - x$, for $x \ge -1$. Use the representations

$$f(x) = x \int\{0 \le s \le 1\} f'(xs)\, ds = x^2 \iint\{0 \le t \le s \le 1\} f''(xt)\, dt\, ds$$

to show that $f(x) \ge \frac{1}{2}x^2(1 + x/3)^{-1}$.

[20] Show that the function $x^{-1}\cos\left(x^{-2}\right)$ is not integrable on [0, 1]. Hint: Consider contributions from intervals where x^{-2} lies in a range $n\pi \pm \pi/4$. The intervals have lengths of order $n^{-3/2}$, but the x^{-1} contributes a factor of order $n^{1/2}$.

[21] Prove Theorem <19> by establishing the following assertions. With no loss of generality, assume $a = 0$ and $b = 1$ and $H(0) = 0$. Write $\mathfrak{m}$ for Lebesgue measure on $\mathcal{B}[0,1]$. Fix $\epsilon > 0$. Define $A_n := \{n\epsilon \le h < (n+1)\epsilon\}$ for each integer n, both positive and negative. Note that

$$h^+(x) = \textstyle\sum_{n \ge 0} h(x)\{x \in A_n\} \qquad \text{and} \qquad h^-(x) = -\textstyle\sum_{n<0} h(x)\{x \in A_n\}.$$

(i) There exist open sets G_n and compact sets K_n with $K_n \subseteq A_n \subseteq G_n$ and $\mathfrak{m}(G_n \backslash K_n) < \epsilon_n$, with ϵ_n so small that the functions

$$f(x) := \textstyle\sum_{n \ge 0} (n+1)\epsilon\{x \in G_n\} \qquad \text{and} \qquad g(x) := \textstyle\sum_{n<0} |n+1|\epsilon\{x \in K_n\}$$

satisfy the inequalities $f \ge h^+$ and $g \le h^-$ and $\mathfrak{m}\left(f - h^+\right) + \mathfrak{m}\left(h^- - g\right) < \epsilon$.

(ii) The function $h_\epsilon := \epsilon + f - \sum_{n=-N}^{-1} |n+1|\epsilon\{x \in K_n\}$, for a large enough positive integer N, has the properties $h_\epsilon \ge h + \epsilon$ and $\mathfrak{m}(h_\epsilon - h) < 3\epsilon$.

(iii) The function h_ϵ is lower semicontinuous, that is, for each real number r the set $\{h_\epsilon > r\}$ is open.

(iv) For $0 \le x \le 1$, define a continuous function $G(x) := \int_0^x h_\epsilon(t)\, dt - H(x)$. It achieves its maximum at some point y in [0, 1]. We cannot have $y < 1$, for otherwise there would be a small $\delta > 0$ for which $h_\epsilon(t) > h(y) + \epsilon$ for $y \le t \le y + \delta < 1$, and $(H(y+\delta) - H(y))/\delta < h(y) + \epsilon/2$, from which we could conclude that $(G(y+\delta) - G(y))/\delta \ge h(y) + \epsilon - (h(y) + \epsilon/2) = \epsilon/2$, a contradiction. Thus $G(1) > G(0) = 0$. That is, $H(1) < \int_0^1 h_\epsilon(t)\, dt \le \int_0^1 h(t)\, dt + 4\epsilon$, for each $\epsilon > 0$.

(v) Similarly $H(x) \le \int_0^x h(t)\, dt$ for $0 \le x \le 1$.

(vi) An analogous argument applied to $-H$ gives the reverse inequality, letting us conclude that $H(x) = \int_0^x h(t)\, dt$ for $0 \le x \le 1$. The function H is absolutely continuous on [0, 1].

[22] Let H be a continuous real function defined on [0, 1). Suppose the right-hand derivative h exists and is finite everywhere. If h is an increasing function, show that H is convex. Hint: For fixed $0 \le x_0 < x_1 < 1$ and $0 < \alpha < 1$ define $x_\alpha = (1-\alpha)x_0 + \alpha x_1$. Show that

$$\begin{aligned}(1-\alpha)H(x_0) &+ \alpha H(x_1) - H(x_\alpha) \\ &= \int_0^1 \left(\alpha\{x_\alpha < t \le x_1\} - (1-\alpha)\{x_0 < t \le x_\alpha\}\right) h(t)\, dt \ge 0.\end{aligned}$$

[23] Let h be integrable with respect to Lebesgue measure $\mathfrak{m}$ on $\mathbb{R}^d$. Show that

$$\lim_{r\to 0} \frac{1}{\mathfrak{m}B(z,r)} \lambda^x \left(\{|x-z|<r\}h(x)\right) = h(z) \qquad \text{a.e. } [\mathfrak{m}].$$

Generalize by replacing open balls by decreasing sequences of closed sets $F_n \downarrow \{z\}$ that are regular in the sense of Section 5.

8. Notes

The Fundamental Theorem <17> is due to Lebesgue, proved in part in his doctoral dissertation (Lebesgue 1902), and completed in his Peccot lectures (Lebesgue 1904). Read Hawkins (1979) if you want to appreciate the subtlety and great significance of Lebesgue's contributions. Note, in particular, Hawkins's comments on page 145 regarding introduction of the term *absolute continuity*. Compare the footnotes on page 129 (as reprinted in his collected works) of the 1904 edition of the Peccot lectures,

> Pour qu'une fonction soit intégrale indéfinie, il faut de plus que sa variation totale dans une infinité dénombrable d'intervalles de longeur totale ℓ, tende vers zéro avec ℓ.
>
> Si, dans l'énoncé de la page 94, on n'assujettit pas $f(x)$ à être bornée, ni $F(x)$ à être à nombres dérivés bornés, mais seulement à la condition précédente, on a une définition de l'intégrale équivalente à celle développée dans ce Chapitre et applicable à toutes les fonctions sommables, bornées ou non.

and on page 188 of the reprinted 1928 edition,

> Dans la première edition de ce livre, j'avais signalé cet énoncé, en note de la page 128, de façon tout à fait incidente et sans démonstration. M. Vitali a retrouvé ce théorème et en a publié la première démonstration (*Acc. Reale delle Sc. di Torino*, 1904–1905). C'est à l'occasion de ce théorème que M. Vitali a introduit, pour les fonctions d'une variable, la dénomination de fonction absolument continue et qu'il a montré la simplicité et la clarté que prend toute la théorie quand on met cette notion à sa base.

The essay by Lebesgue (1926) contains a clear account of the transformation of absolute continuity from a property of functions to a property of measures. The book of Benedetto (1976), which is particularly interesting for its discussion of the role played by Vitali, and the Notes to Chapters 5 and 7 of Dudley (1989), provide more historical background.

For the discussion in Sections 4 through 6, I borrowed ideas from Saks (1937, Chapters 4 and 7), Royden (1968, Chapter 8), Benedetto (1976, Chapter 4), and Wheeden & Zygmund (1977, Chapter 7). The methods of Section 6 extend easily to higher dimensional Euclidean spaces.

See Royden (1968, Chapter 6) and Dunford & Schwartz (1958, Chapter 3) for more about the Radon-Nikodym theorem. According to Dudley (1989, page 141), the method of proof used in Section 2 is due to von Neumann (1940), but I have not seen the original paper.

Inequality <13> is due to Kemperman (1969), Csiszar (1967), and Kullback (1967). (In fact, Kullback established a slightly better inequality.) My proof is based on Kemperman's argument. See Devroye (1987, Chapter 1) for other inequalities involving distances between probability densities.

References

Benedetto, J. J. (1976), *Real Variable and Integration*, Mathematische Leitfäden, Teubner, Stuttgart. Subtitled "with Historical Notes".

Csiszar, I. (1967), 'Information-type measures of difference of probability distributions and indirect observations', *Studia Scientarium Mathematicarum Hungarica* **2**, 299–318.

Devroye, L. (1987), *A Course in Density Estimation*, Birkhäuser, Boston.

Dudley, R. M. (1989), *Real Analysis and Probability*, Wadsworth, Belmont, Calif.

Dunford, N. & Schwartz, J. T. (1958), *Linear Operators, Part I: General Theory*, Wiley.

Hawkins, T. (1979), *Lebesgue's Theory of Integration: Its Origins and Development*, second edn, Chelsea, New York.

Kemperman, J. H. B. (1969), On the optimum rate of transmitting information, *in* 'Probability and Information Theory', Springer-Verlag. Lecture Notes in Mathematics, 89, pages 126–169.

Kullback, S. (1967), 'A lower bound for discrimination information in terms of variation', *IEEE Transactions on Information Theory* **13**, 126–127.

Lebesgue, H. (1902), *Intégrale, longueur, aire*. Doctoral dissertation, submitted to Faculté des Sciences de Paris. Published separately in Ann. Mat. Pura Appl. 7. Included in the first volume of his *Œuvres Scientifiques*, published in 1972 by L'Enseignement Mathématique.

Lebesgue, H. (1904), *Leçons sur l'integration et la recherche des fonctions primitives*, first edn, Gauthier-Villars, Paris. Included in the second volume of his *Œuvres Scientifiques*, published in 1972 by L'Enseignement Mathématique. Second edition published in 1928. Third edition, 'an unabridged reprint of the second edition, with minor changes and corrections', published in 1973 by Chelsea, New York.

Lebesgue, H. (1926), 'Sur le développement de la notion d'intégrale', *Matematisk Tidsskrift B*. English version in the book *Measure and Integral*, edited and translated by Kenneth O. May.

Royden, H. L. (1968), *Real Analysis*, second edn, Macmillan, New York.

Saks, S. (1937), *Theory of the Integral*, second edn, Dover. English translation of the second edition of a volume first published in French in 1933. Page references are to the 1964 Dover edition.

von Neumann, J. (1940), 'On rings of operators, III', *Annals of Mathematics* **41**, 94–161.

Wheeden, R. & Zygmund, A. (1977), *Measure and Integral: An Introduction to Real Analysis*, Marcel Dekker.

Chapter 4

Product spaces and independence

SECTION 1 introduces independence as a property that justifies some sort of factorization of probabilities or expectations. A key factorization Theorem is stated, with proof deferred to the next Section, as motivation for the measure theoretic approach. The Theorem is illustrated by a derivation of a simple form of the strong law of large numbers, under an assumption of bounded fourth moments.

SECTION 2 formally defines independence as a property of sigma-fields. The key Theorem from Section 1 is used as motivation for the introduction of a few standard techniques for dealing with independence. Product sigma-fields are defined.

SECTION 3 describes a method for constructing measures on product spaces, starting from a family of kernels.

SECTION 4 specializes the results from Section 3 to define product measures. The Tonelli and Fubini theorems are deduced. Several important applications are presented.

*SECTION *5 discusses some difficulties encountered in extending the results of Sections 3 and 4 when the measures are not sigma-finite.*

SECTION 6 introduces a blocking technique to refine the proof of the strong law of large numbers from Section 1, to get a version that requires only a second moment condition.

*SECTION *7 introduces a truncation technique to further refine the proof of the strong law of large numbers, to get a version that requires only a first moment condition for identically distributed summands.*

*SECTION *8 discusses the construction of probability measures on products of countably many spaces.*

1. Independence

Much classical probability theory, such as the laws of large numbers and central limit theorems, rests on assumptions of independence, which justify factorizations for probabilities of intersections of events or expectations for products of random variables.

An elementary treatment usually starts from the definition of independence for events. Two events A and B are said to be independent if $\mathbb{P}(AB) = (\mathbb{P}A)(\mathbb{P}B)$; three events A, B, and C, are said to be independent if not only $\mathbb{P}(ABC) = (\mathbb{P}A)(\mathbb{P}B)(\mathbb{P}C)$ but also $\mathbb{P}(AB) = (\mathbb{P}A)(\mathbb{P}B)$ and $\mathbb{P}(AC) = (\mathbb{P}A)(\mathbb{P}C)$ and $\mathbb{P}(BC) = (\mathbb{P}B)(\mathbb{P}C)$. And so on. There are similar definitions for independence of random variables, in terms of joint distribution functions or joint densities. The definitions have two things in common: they all assert some type of factorization; and they do not lend themselves to elementary derivation of desirable facts about independence. The measure theoretic approach, by contrast, simplifies the study of independence by eliminating unnecessary duplications of definitions, replacing them by a single concept of independence for sigma-fields, from which useful consequences are easily deduced. For example, the following key assertion is impossible to derive by elementary means, but requires only routine effort (see Section 2) to establish by measure theoretic arguments.

<1> **Theorem.** *Let $Z_1, \ldots, Z_n$ be independent random variables on a probability space $(\Omega, \mathcal{F}, \mathbb{P})$. If $f \in \mathcal{M}^+(\mathbb{R}^k, \mathcal{B}(\mathbb{R}^k))$ and $g \in \mathcal{M}^+(\mathbb{R}^{n-k}, \mathcal{B}(\mathbb{R}^{n-k}))$ then $f(Z_1, \ldots, Z_k)$ and $g(Z_{k+1}, \ldots, Z_n)$ are independent random variables, and*

$$\mathbb{P}f(Z_1, \ldots, Z_k)g(Z_{k+1}, \ldots, Z_n) = \mathbb{P}f(Z_1, \ldots, Z_k)\mathbb{P}g(Z_{k+1}, \ldots, Z_n).$$

<2> **Corollary.** *The same conclusion (independendence and factorization) holds for Borel measurable functions f and g taking both positive and negative values if both $f(Z_1, \ldots, Z_k)$ and $g(Z_{k+1}, \ldots, Z_n)$ are integrable.*

As you will see at the end of Section 2, the Corollary follows easily from addition and subtraction of analogous results for the functions $f^{\pm}$ and $g^{\pm}$. Problem [10] shows that the result also extends to cases where some of the integrals are infinite, provided $\infty - \infty$ problems are ruled out.

The best way for you to understand the worth of Theorem <1> and its Corollary is to see it used. At the risk of interrupting the flow of ideas, I will digress slightly to present an instructive application.

The proof of the ***strong law of large numbers*** (often referrred to by means of the acronym SLLN) illustrates well the use of Corollary <2>. Actually, several slightly different results answer to the name SLLN. A law of large numbers asserts convergence of averages to expectations, in some sense. The word "strong" specifies almost sure convergence. The various SLLN's differ in the assumptions made about the individual summands. The most common form invoked in statistical applications goes as follows.

<3> **Theorem.** *(Kolmogorov) Let $X_1, X_2, \ldots$ be independent, integrable random variables, each with the same distribution and common expectation μ. Then the average $(X_1 + \ldots + X_n)/n$ converges almost surely to μ.*

REMARK. If $\mathbb{P}|X_1| = \infty$ then $(X_1 + \ldots + X_n)/n$ cannot converge almost surely to a finite limit (Problem [21]). Moreover Kolmogorov's zero-one law (Example <12>) implies that it cannot even converge to a finite limit at each point of a set with strictly positive probability. If only one of $\mathbb{P}X_1^{\pm}$ is infinite, the average still converges almost surely to $\mathbb{P}X_1$ (Problem [20]).

A complete proof of this form of the SLLN is quite a challenge. The classical proof (a modified version of which appears in Sections 6 and 7) combines a number of tricks that are more easily understood if introduced as separate ideas and not just rolled into one monolithic argument. The basic idea is not too hard to grasp when we have bounded fourth moments; it involves little more than an application of Corollary <2> and an appeal to the Borel-Cantelli lemma from Section 2.6.

For theoretical purposes, for summands that need not all have the same distribution, it is cleaner to work with the centered variables $X_i - \mathbb{P}X_i$, which is equivalent to an assumption that all variables have zero expected values.

<4> **Theorem.** *Let $X_1, X_2, \ldots$ be independent random variables with $\mathbb{P}X_i = 0$ for every i and $\sup_i \mathbb{P}X_i^4 < \infty$. Then $(X_1 + \ldots + X_n)/n \to 0$ almost surely.*

Proof. Define $S_n = X_1 + \ldots + X_n$. It is good enough to show, for each $\epsilon > 0$, that

<5>
$$\sum_{n=1}^{\infty} \mathbb{P}\left\{\frac{|S_n|}{n} > \epsilon\right\} < \infty.$$

Do you remember why? If not, you should refer to Section 2.6 for a detailed explanation of the Borel-Cantelli argument: the series $\sum_n \{|S_n|/n > \epsilon\}$ must converge almost surely, which implies that $\limsup |S_n/n| \le \epsilon$ almost surely, from which the conclusion $\limsup |S_n/n| = 0$ follows after a casting out of a sequence of negligible sets.

Bound the nth term of the sum in <5> by $(n\epsilon)^{-4}\mathbb{P}(X_1 + \ldots + X_n)^4$. Expand the fourth power.

$$\begin{aligned}(X_1 + \ldots + X_n)^4 = \; & X_1^4 + \ldots + X_n^4 & \boxed{1}\\ & + \text{(lots of terms like } X_1^3 X_2) & \boxed{2}\\ & + \binom{n}{2} \text{ terms like } 6X_1^2 X_2^2 & \boxed{3}\\ & + \text{(lots of terms like } X_1^2 X_2 X_3) & \boxed{4}\\ & + \text{(lots of terms like } X_1 X_2 X_3 X_4) & \boxed{5}\end{aligned}$$

The contributions to $\mathbb{P}(X_1 + \ldots + X_n)^4$ from the five groups of terms are:

$\boxed{1}$ $\sum_{i \le n} \mathbb{P}X_i^4 \le nM$, where $M = \sup_i \mathbb{P}X_i^4$;
$\boxed{2}$ zero, because $\mathbb{P}(X_1^3 X_2) = (\mathbb{P}X_1^3)(\mathbb{P}X_2) = 0$;
$\boxed{3}$ less than $12\binom{n}{2}M$, because $\mathbb{P}(X_1^2 X_2^2) \le \mathbb{P}X_1^4 + \mathbb{P}X_2^4 \le 2M$;
$\boxed{4}$ zero, because $\mathbb{P}(X_1^2 X_2 X_3) = (\mathbb{P}X_1^2 X_2)(\mathbb{P}X_3) = 0$;
$\boxed{5}$ zero, because $\mathbb{P}(X_1 X_2 X_3 X_4) = (\mathbb{P}X_1 X_2 X_3)(\mathbb{P}X_4) = 0$.

Notice all the factorizations due to independence. Combining these bounds and equalities we get $\mathbb{P}\{|S_n|/n > \epsilon\} = O(n^{-2})$, from which <5> follows. □

If you feel that Theorem <4> is good enough for 'practical purposes,' and that all the extra work to whittle a fourth moment assumption down to a first moment assumption is hardly worth the gain in generality, you might like to contemplate the following example. How natural, or restrictive, would it be if we were to assume finite fourth moments?

<6> **Example.** Let $\{P_\theta : \theta = 0, 1, \ldots, N\}$ be a finite family of distinct probability measures, defined by densities $\{p_\theta\}$ with respect to a measure μ. Suppose observations $X_1, X_2, \ldots$ are generated independently from P_0. The maximum likelihood estimator $\widehat{\theta}_n(\omega)$ is defined as the value that maximizes $L_n(\theta, \omega) := \prod_{i \le n} p_\theta(X_i(\omega))$. The SLLN will show that $\mathbb{P}\{\widehat{\theta}_n = 0 \text{ eventually}\} = 1$. That is, the maximum likelihood estimator eventually picks the true value of θ.

It will be enough to show, for each $\theta \neq 0$, that with probability one, $\log(L_n(\theta)/L_n(0)) < 0$ eventually. For fixed $\theta \neq 0$ define $\ell_i = \log(p_\theta(X_i)/p_0(X_i))$. By Jensen's inequality, with a strict inequality because $P_\theta \neq P_0$,

$$\mathbb{P}\ell_i = P_0^x \log\left(\frac{p_\theta(x)}{p_0(x)}\right) < \log \mu^x \left(p_0(x)\frac{p_\theta(x)}{p_0(x)}\right)$$
$$= \log \mu^x p_\theta(x)\{p_0(x) \neq 0\} \leq 0.$$

By the SLLN (or its extension from Problem [20] if $\mathbb{P}\ell_i = -\infty$), for almost all ω there exists a finite $n_0(\omega, \theta)$ for which $0 > n^{-1}\sum_{i\leq n} \ell_i := n^{-1}\log(L_n(\theta)/L_n(0))$ when $n \geq n_0(\omega,\theta)$. When $n \geq \max_{\theta=1}^N n_0(\omega,\theta)$, we have $\max_{\theta=1}^N L_n(\theta) < L_n(0)$, in which case the maximizing $\widehat{\theta}_n$ prefers 0 to each $\theta \geq 1$. □

REMARK. Notice that the argument would not work if the index set were infinite. To handle such sets, one typically imposes compactness assumptions to reduce to the finite case, by means of a much-imitated method originally due to Wald (1949).

2. Independence of sigma-fields

Technically speaking, the best treatment of independence starts with the concept of independent sub-sigma-fields of $\mathcal{F}$, for a fixed probability space $(\Omega, \mathcal{F}, \mathbb{P})$. This Section will develop the appropriate definitions and techniques for dealing with independence of sigma-fields, using the ideas needed for the proof of Theorem <2> as motivation.

<7> **Definition.** *Let $(\Omega, \mathcal{F}, \mathbb{P})$ be a probability space. Sub-sigma-fields $\mathcal{G}_1, \ldots, \mathcal{G}_n$ of $\mathcal{F}$ are said to be independent if*

$$\mathbb{P}(G_1 \ldots G_n) = (\mathbb{P}G_1)\ldots(\mathbb{P}G_n) \qquad \text{for all } G_i \in \mathcal{G}_i, \text{ for } i = 1, \ldots n.$$

An infinite collection of sub-sigma-fields $\{\mathcal{G}_i : i \in I\}$ is said to be independent if each finite subcollection is independent, that is, if $\mathbb{P}(\cap_{i\in S} G_i) = \prod_{i\in S}\mathbb{P}G_i$ for all finite subsets S of I, and all choices $G_i \in \mathcal{G}_i$ for each i in S.

The definition neatly captures all the factorizations involved in the elementary definitions of independence for more than two events.

<8> **Example.** Let A, B, and C be events. They generate sigma-fields $\mathcal{A} = \{\emptyset, A, A^c, \Omega\}$, and $\mathcal{B} = \{\emptyset, B, B^c, \Omega\}$, and $\mathcal{C} = \{\emptyset, C, C^c, \Omega\}$. Independence of the three sigma-fields requires factorization for $4^3 = 64$ triples of events, amongst which are the four factorizations stated at the start of Section 1 as the elementary definition of independence for the three events A, B, and C. In fact, all 64 factorizations are consequences of those four. For example, any factorization where one of the factors is the empty set will reduce to the identity $0 = 0$. The factorization $\mathbb{P}(AB^cC) = (\mathbb{P}A)(\mathbb{P}B^c)(\mathbb{P}C)$ follows from $\mathbb{P}(AC) = (\mathbb{P}A)(\mathbb{P}C)$ and $\mathbb{P}(ABC) = (\mathbb{P}A)(\mathbb{P}B)(\mathbb{P}C)$, by subtraction. And so on. □

Generating class arguments, such as the π–λ Theorem from Section 2.10, make it easy to derive facts about independent sigma-fields. For example, Problem [8] uses such arguments in a routine way to establish the following result.

<9> **Theorem.** *Let $\mathcal{E}_1, \ldots, \mathcal{E}_n$ be classes of measurable sets, each class stable under finite intersections and containing the whole space Ω. If*

$$\mathbb{P}(E_1E_2\ldots E_n) = (\mathbb{P}E_1)(\mathbb{P}E_2)\ldots(\mathbb{P}E_n) \qquad \text{for all } E_i \in \mathcal{E}_i, \text{ for } i = 1, 2, \ldots, n,$$

then the sigma-fields $\sigma(\mathcal{E}_1), \sigma(\mathcal{E}_2), \ldots, \sigma(\mathcal{E}_n)$ are independent.

REMARK. The requirement that $\Omega \in \mathcal{E}_i$ for each i is just a sneaky way of getting factorizations for intersections of fewer than n sets.

<10> **Corollary.** *Let $\{\mathcal{E}_i : i \in I\}$ be classes of measurable sets, each stable under finite intersections. If $\mathbb{P}(\cap_{i\in S}E_i) = \prod_{i\in S}\mathbb{P}E_i$ for all finite subsets S of I, and all choices $E_i \in \mathcal{E}_i$ for each i in S, then the sigma-fields $\sigma(\mathcal{E}_i)$, for $i \in I$, are independent.*

Proof. Notice the alternative to requiring $\Omega \in \mathcal{E}_i$ for every i. Theorem <9> establishes independence for each finite subcollection. □

<11> **Corollary.** *Let $\{\mathcal{G}_i : i \in I\}$ be independent sigma-fields. If $\{I_j : j \in J\}$ are disjoint subsets of I, then the sigma-fields $\sigma\left(\cup_{i\in I_j}\mathcal{G}_i\right)$, for $j \in J$, are independent.*

Proof. Invoke Corollary <10> with $\mathcal{E}_j$ consisting of the collection of all finite intersections of sets chosen from $\cup_{i\in I_j}\mathcal{G}_i$. □

<12> **Example.** Let $\{\mathcal{G}_i : i \in \mathbb{N}\}$ be a sequence of independent sigma-fields. For each n let $\mathcal{H}_n$ denote the sigma-field generated by $\cup_{i>n}\mathcal{G}_i$. The ***tail sigma-field*** is defined as $\mathcal{H}_\infty := \cap_n\mathcal{H}_n$. Kolmogorov's ***zero-one law*** asserts that, for each H in $\mathcal{H}_\infty$, either $\mathbb{P}H = 0$ or $\mathbb{P}H = 1$. Equivalently, the sigma-field $\mathcal{H}_\infty$ is independent of itself, so that $\mathbb{P}(HH) = (\mathbb{P}H)(\mathbb{P}H)$ for every H in $\mathcal{H}_\infty$.

For each finite n, Corollary <11> implies independence of $\mathcal{H}_n, \mathcal{G}_1, \ldots, \mathcal{G}_n$. From the fact that $\mathcal{H}_\infty \subseteq \mathcal{H}_n$ for every n, it then follows that each finite subcollection of $\{\mathcal{H}_\infty, \mathcal{G}_i : i \in \mathbb{N}\}$ is independent, and hence the whole collection of sigma-fields is independent. From Corollary <11> again, $\mathcal{H}_\infty$ and $\mathcal{F}_\infty := \sigma(\cup_{i\in\mathbb{N}}\mathcal{G}_i)$ are independent. To complete the argument, note that $\mathcal{F}_\infty \supseteq \mathcal{H}_\infty$. □

Random variables (or random vectors, or random elements of more general spaces) inherit their definition of independence from the sigma-fields they generate. Recall that if X is a map from Ω into a set $\mathcal{X}$, equipped with a sigma-field $\mathcal{A}$, then the sigma-field $\sigma(X)$ on Ω generated by X is defined as the smallest sigma-field $\mathcal{G}$ for which X is $\mathcal{G}\backslash\mathcal{A}$-measurable. It consists of all sets of the form $\{\omega \in \Omega : X(\omega) \in A\}$, with $A \in \mathcal{A}$.

REMARK. The extra generality gained by allowing maps into arbitrary measurable spaces will not be wasted; but in the first instance you could safely imagine each space to be the real line, ignoring the fact that the definition also covers independence of random vectors and independence of stochastic processes.

<13> **Definition.** *Measurable maps X_i, for $i \in I$, from Ω into measurable spaces $(\mathcal{X}_i, \mathcal{A}_i)$ are said to be independent if the sigma-fields that they generate are independent, that is, if*

$$\mathbb{P}\left(\bigcap_{i\in S}\{X_i \in A_i\}\right) = \prod_{i\in S}\mathbb{P}\{X_i \in A_i\}, \tag{14}$$

for all finite subsets S of the index set I, and all choices of $A_i \in \mathcal{A}_i$ for $i \in S$.

Results about independent random variables are usually easy to deduce from the corresponding results about independent sigma-fields.

<15> **Example.** Real random variables X_1 and X_2 for which

$$\mathbb{P}\{X_1 \le x_1, X_2 \le x_2\} = \mathbb{P}\{X_1 \le x_1\}\mathbb{P}\{X_2 \le x_2\} \qquad \text{for all } x_1, x_2 \text{ in } \mathbb{R}$$

are independent, because the collections of sets $\mathcal{E}_i = \{\{X_i \le x\} : x \in \mathbb{R}\}$ are both stable under finite intersections, and $\sigma(X_i) = \sigma(\mathcal{E}_i)$. □

We now have the tools needed to establish Theorem <2>. Write X_1 for $f(Z_1, \dots, Z_k)$ and X_2 for $g(Z_{k+1}, \dots, Z_n)$. Write $\mathcal{G}_i$ for $\sigma(Z_i)$, the sigma-field generated by the random variable Z_i. From Corollary <11>, the sigma-fields $\mathcal{F}_1 := \sigma(\mathcal{G}_1 \cup \dots \cup \mathcal{G}_k)$ and $\mathcal{F}_2 := \sigma(\mathcal{G}_{k+1} \cup \dots \cup \mathcal{G}_n)$ are independent. If we can show that X_1 is $\mathcal{F}_1\backslash\mathcal{B}(\mathbb{R})$-measurable and X_2 is $\mathcal{F}_2\backslash\mathcal{B}(\mathbb{R})$-measurable, then their independence will follow: we will have the desired factorization for all sets of the form $\{X_1 \in A_1\}$ and $\{X_2 \in A_2\}$, for Borel sets A_1 and A_2.

Consider first the measurability property for X_1. Temporarily write $\mathbf{Z}$ for $(Z_1, \dots, Z_k)$, a map from Ω into $\mathbb{R}^k$. We need to show that the set

$$\{X_1 \in A\} = \{\mathbf{Z} \in f^{-1}(A)\}$$

belongs to $\mathcal{F}_1$ for every A in $\mathcal{B}(\mathbb{R})$. The $\mathcal{B}(\mathbb{R}^k)\backslash\mathcal{B}(\mathbb{R})$-measurability of f ensures that $f^{-1}(A) \in \mathcal{B}(\mathbb{R}^k)$. We therefore need only show that $\{\mathbf{Z} \in B\} \in \mathcal{F}_1$ for every B in $\mathcal{B}(\mathbb{R}^k)$, that is, that the map $\mathbf{Z}$ is $\mathcal{F}_1\backslash\mathcal{B}(\mathbb{R}^k)$-measurable.

As with many measure theoretic problems, it is better to turn the question around and ask: For how extensive a class of sets B does $\{\mathbf{Z} \in B\}$ belong to $\mathcal{F}_1$? It is very easy to show that the class $\mathcal{B}_0$ of all such B is a sigma-field; so $\mathbf{Z}$ is an $\mathcal{F}_1\backslash\mathcal{B}_0$-measurable function. Moreover, for all choices of $D_i \in \mathcal{B}(\mathbb{R})$, the set

$$D := \{(z_1, \dots, z_k) \in \mathbb{R}^k : z_i \in D_i \text{ for } i = 1, \dots, k\}$$

belongs to $\mathcal{B}_0$ because $\{\mathbf{Z} \in D\} = \cap_i\{Z_i \in D_i\} \in \mathcal{F}_1$. As shown in Problem [6], the collection of all such D sets generates the Borel sigma-field $\mathcal{B}(\mathbb{R}^k)$. Thus $\mathcal{B}(\mathbb{R}^k) \subseteq \mathcal{B}_0$, and $\{\mathbf{Z} \in B\} \in \mathcal{F}_1$ for all $B \in \mathcal{B}(\mathbb{R}^k)$. It follows that X_1 is $\mathcal{F}_1\backslash\mathcal{B}(\mathbb{R})$-measurable. Similarly, X_2 is $\mathcal{F}_2\backslash\mathcal{B}(\mathbb{R})$-measurable. The random variables X_1 and X_2 are independent, as asserted by Theorem <2>.

The whole argument can be carried over to random elements of more general spaces if we work with the right sigma-fields.

<16> **Definition.** *Let $\mathcal{X}_1, \dots, \mathcal{X}_n$ be sets equipped with sigma-fields $\mathcal{A}_1, \dots, \mathcal{A}_n$. The set of all ordered n-tuples $(x_1, \dots, x_n)$, with $x_i \in \mathcal{X}_i$ for each i is denoted by $\mathcal{X}_1 \times \dots \times \mathcal{X}_n$ or $\mathsf{X}_{i \le n} \mathcal{X}_i$. It is called the **product** of the $\{\mathcal{X}_i\}$. A set of the form*

$$A_1 \times \dots \times A_n = \{(x_1, \dots, x_n) \in \mathcal{X}_1 \times \dots \times \mathcal{X}_n : x_i \in A_i \text{ for each } i\},$$

*with $A_i \in \mathcal{A}_i$ for each i, is called a **measurable rectangle**. The product sigma-field $\mathcal{A}_1 \otimes \dots \otimes \mathcal{A}_n$ on $\mathcal{X}_1 \times \dots \times \mathcal{X}_n$ is defined to be the sigma-field generated by all measurable rectangles.*

REMARK. Even if n equals 2 and $\mathcal{X}_1 = \mathcal{X}_2 = \mathbb{R}$, there is is no presumption that either A_1 or A_2 is an interval—a measurable rectangle might be composed of many disjoint pieces. The symbol $\otimes$ in place of $\times$ is intended as a reminder that $\mathcal{A}_1 \otimes \mathcal{A}_2$ consists of more than the set of all measurable rectangles $A_1 \times A_2$.

If Z_i is an $\mathcal{F}\backslash\mathcal{A}_i$-measurable map from Ω into $\mathcal{X}_i$, for $i = 1, \ldots, n$, then the map $\omega \mapsto \mathbf{Z}(\omega) = (Z_1(\omega), \ldots, Z_n(\omega))$ from Ω into $\mathcal{X} = \mathcal{X}_1 \times \ldots \times \mathcal{X}_n$ is $\mathcal{F}\backslash\mathcal{A}$-measurable, where $\mathcal{A}$ denotes the product sigma-field $\mathcal{A}_1 \otimes \ldots \otimes \mathcal{A}_n$. If f is an $\mathcal{A}\backslash\mathcal{B}(\mathbb{R})$-measurable real-valued function on $\mathcal{X}$ then $f(\mathbf{Z})$ is $\mathcal{F}\backslash\mathcal{B}(\mathbb{R})$-measurable.

The second assertion of Theorem <1> is now reduced to a factorization property for products of independent random variables, a result easily deduced from the defining factorization for independence of sigma-fields by means of the usual approximation arguments.

<17> **Lemma.** *Let X and Y be independent random variables. If either $X \geq 0$ and $Y \geq 0$, or both X and Y are integrable, then $\mathbb{P}(XY) = (\mathbb{P}X)(\mathbb{P}Y)$. The product XY is integrable if both X and Y are integrable.*

Proof. Consider first the case of nonnegative variables. Express X and Y as monotone increasing limits of simple random variables (as in Section 2.2), $X_n := 2^{-n}\sum_{1\leq i\leq 4^n}\{X \geq i/2^n\}$ and $Y_n := 2^{-n}\sum_{1\leq i\leq 4^n}\{Y \geq i/2^n\}$. Then, for each n,

$$\begin{aligned}\mathbb{P}(X_nY_n) &= 4^{-n}\textstyle\sum_{i,j}\mathbb{P}(\{X \geq i/2^n\}\{Y \geq j/2^n\})\\ &= 4^{-n}\textstyle\sum_{i,j}(\mathbb{P}\{X \geq i/2^n\})(\mathbb{P}\{Y \geq j/2^n\}) \qquad \text{by independence}\\ &= \left(2^{-n}\textstyle\sum_i \mathbb{P}\{X \geq i/2^n\}\right)\left(2^{-n}\textstyle\sum_j \mathbb{P}\{Y \geq j/2^n\}\right)\\ &= (\mathbb{P}X_n)(\mathbb{P}Y_n).\end{aligned}$$

Invoke Monotone Convergence twice in the passage to the limit to deduce $\mathbb{P}(XY) = (\mathbb{P}X)(\mathbb{P}Y)$.

For the case of integrable random variables, factorize expectations for products of positive and negative parts, $\mathbb{P}(X^{\pm}Y^{\pm}) = (\mathbb{P}X^{\pm})(\mathbb{P}Y^{\pm})$. Each of the four products represented by the right-hand side is finite. Complete the argument by splitting each term on the right-hand side of the decomposition

$$\mathbb{P}(XY) = \mathbb{P}(X^+Y^+) - \mathbb{P}(X^+Y^-) - \mathbb{P}(X^-Y^+) + \mathbb{P}(X^-Y^-)$$

into a product of expectations, then refactorize as $(\mathbb{P}X^+ - \mathbb{P}X^-)(\mathbb{P}Y^+ - \mathbb{P}Y^-)$. Integrability of XY follows from a similar decomposition for $\mathbb{P}|XY|$. □

3. Construction of measures on a product space

The probabilistic concepts of independence and conditioning are both closely related to the measure theoretic constructions for measures on product spaces. As you will see in Chapter 5, conditioning may be thought of as a inverse operation to a general construction whereby a measure on a product space is built from families of measures on the component spaces. For probability measures the components have the interpretation of distributions involved in a two-stage experiment. Product measures, and independence, correspond to the special case where the second stage of the experiment does not depend on the first stage. Many traditional facts about independence, such as the assertion of Theorem <2>, have interpretations as facts about product measures.

If you want to understand independence then you should learn about product measures. If you want to understand conditioning you should learn about the more general construction. We kill two birds with one stone by starting with the general case. The full generality will be needed in Chapter 5.

To keep the notation simple, I will mostly consider only measures on a product of two spaces, $(\mathcal{X}, \mathcal{A})$ and $(\mathcal{Y}, \mathcal{B})$. Sometimes I will abbreviate symbols like $\mathcal{M}^+(\mathcal{X} \times \mathcal{Y}, \mathcal{A} \otimes \mathcal{B})$ to $\mathcal{M}^+(\mathcal{X} \times \mathcal{Y})$, with the product sigma-field assumed, or to $\mathcal{M}^+(\mathcal{A} \otimes \mathcal{B})$, with the product space assumed.

To each measure Γ on $\mathcal{A} \otimes \mathcal{B}$ there correspond two ***marginal measures*** μ and λ, defined by

$$\mu A := \Gamma(A \times \mathcal{Y}) \text{ for } A \in \mathcal{A} \qquad \text{and} \qquad \lambda B := \Gamma(\mathcal{X} \times B) \text{ for } B \in \mathcal{B}.$$

Equivalently, μ is the image of Γ under the coordinate projection X, which takes (x, y) to x, and λ is the image under the projection onto the other coordinate space. In particular, if $\Gamma(\mathcal{X} \times \mathcal{Y}) = 1$ then the marginals give the distributions of the coordinate projections as $\mathcal{X}$- or $\mathcal{Y}$-valued random variables on the probability space $(\mathcal{X} \times \mathcal{Y}, \mathcal{A} \otimes \mathcal{B}, \Gamma)$.

In general, each marginal has the same total mass as Γ, which can lead to bizarre behavior if $\Gamma(\mathcal{X} \times \mathcal{Y}) = \infty$. For example, if $\mathcal{X} = \mathcal{Y} = \mathbb{R}$ and Γ is Lebesgue measure on $\mathcal{B}(\mathbb{R}^2)$ then each marginal assigns infinite mass to all except the Lebesgue negligible subsets of $\mathbb{R}$.

As you will see in Section 4, if Γ is a probability measure under which the coordinate projections define independent random variables then Γ can be reconstructed as a product of its marginal distributions. Without independence, Γ is not completely determined by its marginals. Instead we need a whole family $\Lambda = \{\lambda_x : x \in \mathcal{X}\}$ of measures on $\mathcal{B}$, together with the marginal μ. The construction will make sense—and be useful—for more than just probability measures, provided the members of Λ are tied together by a measurability assumption.

<18> **Definition.** *Call a family of measures $\Lambda = \{\lambda_x : x \in \mathcal{X}\}$ on $\mathcal{B}$ a **kernel from** $(\mathcal{X}, \mathcal{A})$ **to** $(\mathcal{Y}, \mathcal{B})$ if the map $x \mapsto \lambda_x B$ is $\mathcal{A}$-measurable for each B in $\mathcal{B}$. In addition, call Λ a **probability kernel** if $\lambda_x(\mathcal{Y}) = 1$ for each x.*

When there is no ambiguity regarding the sigma-fields, I will also speak of kernels from $\mathcal{X}$ to $\mathcal{Y}$.

REMARK. Probability kernels are also known by several other names: Markov kernels, randomizations, conditional distributions, and (particularly when $\mathcal{X}$ is interpreted as a parameter space) statistical models.

Suppose μ is a measure on $\mathcal{A}$ and $\Lambda = \{\lambda_x : x \in \mathcal{X}\}$ is a kernel from $(\mathcal{X}, \mathcal{A})$ to $(\mathcal{Y}, \mathcal{B})$. The main idea behind the construction is definition of a measure on $\mathcal{A} \otimes \mathcal{B}$ via an iterated integral, $\mu^x \left(\lambda_x^y f(x, y)\right)$.

REMARK. Remember the notation for distinguishing between arguments. The superscript y means that the λ_x measure integrates out $f(x, y)$ with x held fixed. The measure μ then integrates out over the resulting function of x. Notice that superscripts denote dummy variables of integration, and subscripts denotes variables that are held fixed. In traditional notation the iterated integral would be written $\iint f(x,y)\,\lambda_x(dy)\,\mu(dx)$, or $\int \left(\int f(x,y)\,\lambda_x(dy)\right)\,\mu(dx)$.

For the iterated integral to make sense, we need to establish two key measurability properties: for each product measurable function f and each fixed x, the map $y \mapsto f(x,y)$ should be $\mathcal{B}$-measurable; and the map $x \mapsto \lambda_x^y f(x,y)$ should be $\mathcal{A}$-measurable. In order to establish conditions under which these two measurability properties hold, I will make use of the generating class method for λ-cones, as developed in Section 2.11.

REMARK. It is unfortunate that the letter λ should have two distinct meanings in this Chapter: as a measure or member of a family of measures on $\mathcal{Y}$, and as a prefix suggesting the idea of stability under bounded limits. Sometimes there are not enough good symbols to go around.

Recall that a λ-cone on a set Ω is a family $\mathcal{H}^+$ of bounded, nonnegative functions on $\mathcal{X}$ with the properties:

(i) $\mathcal{H}^+$ is a cone, that is, if $h_1, h_2 \in \mathcal{H}^+$ and α_1 and α_2 are nonnegative constants then $\alpha_1 h_1 + \alpha_2 h_2 \in \mathcal{H}^+$;

(ii) each nonnegative constant function belongs to $\mathcal{H}^+$;

(iii) if $h_1, h_2 \in \mathcal{H}^+$ and $h_1 \geq h_2$ then $h_1 - h_2 \in \mathcal{H}^+$;

(iv) if $\{h_n\}$ is an increasing sequence of functions in $\mathcal{H}^+$ whose pointwise limit h is bounded then $h \in \mathcal{H}^+$.

Recall also that: *if a sigma-field $\mathcal{F}$ on Ω is generated by a subclass $\mathcal{G}$ (of a λ-cone $\mathcal{H}^+$) that is stable under the formation of pointwise products of pairs of functions, then every bounded, nonnegative, $\mathcal{F}$-measurable function belongs to $\mathcal{H}^+$.*

Let me illustrate the application of λ-cone methods, by establishing the first of the desired measurability properties. It would be more elegant to combine the proofs for both properties into a single generating class argument, but I feel it is good to see the method first in a simple setting.

<19> **Lemma.** *For each f in $\mathcal{M}(\mathcal{X} \times \mathcal{Y}, \mathcal{A} \otimes \mathcal{B})$, and each fixed x in $\mathcal{X}$, the function $y \mapsto f(x,y)$ is $\mathcal{B}$-measurable.*

Proof. It is enough to consider the case of bounded nonnegative f. The general case would then follow by splitting f into positive and negative parts, and representing each part as a pointwise limit of bounded functions, $f^{\pm} = \lim_n \left(f^{\pm} \wedge n\right)$.

Write $\mathcal{H}^+$ for the collection of all bounded, nonnegative, $\mathcal{A} \otimes \mathcal{B}$-measurable functions on $\mathcal{X} \times \mathcal{Y}$ for which the stated measurablity property holds. It is routine to check the four properties identifying $\mathcal{H}^+$ as a λ-cone. It contains the class $\mathcal{G}$ of all indicator functions $g(x,y) := \{x \in A, y \in B\}$ of measurable rectangles, because $g(x,\cdot)$ is either the zero function or the indicator of the set B. The class $\mathcal{G}$ is stable under pointwise products, and it generates $\mathcal{A} \otimes \mathcal{B}$. □

The application of the generating argument to establish the second desired measurablity property can be surprisingly delicate for kernels assigning infinite measure to $\mathcal{Y}$. A finiteness assumption eliminates all difficulties.

<20> **Theorem.** *Let $\Lambda = \{\lambda_x : x \in \mathcal{X}\}$ be a kernel from $(\mathcal{X}, \mathcal{A})$ to $(\mathcal{Y}, \mathcal{B})$ with $\lambda_x \mathcal{Y} < \infty$ for all x, and let μ be a measure on $\mathcal{A}$. Then, for each function f in $\mathcal{M}^+(\mathcal{X} \times \mathcal{Y}, \mathcal{A} \otimes \mathcal{B})$,*

(i) $y \mapsto f(x, y)$ is $\mathcal{B}$-measurable for each fixed x;

(ii) $x \mapsto \lambda_x^y f(x, y)$ is $\mathcal{A}$-measurable;

(iii) the iterated integral $(\mu \otimes \Lambda)(f) := \mu^x \left(\lambda_x^y f(x, y)\right)$, for f in $\mathcal{M}^+(\mathcal{X} \times \mathcal{Y}, \mathcal{A} \otimes \mathcal{B})$, defines a measure on $\mathcal{A} \otimes \mathcal{B}$.

Proof. Property (i) merely restates the assertion of Lemma <19>. For (ii), consider the class $\mathcal{H}^+ := \{f \in \mathcal{M}^+(\mathcal{X} \times \mathcal{Y}) : f$ bounded and satisfying (ii)$\}$. Note that $\lambda_x^y f(x, y) < \infty$ for each x if $f \in \mathcal{H}^+$, so there is no problem with infinite values when subtracting to show that $\lambda^y f_1(x, y) - \lambda_x^y f_2(x, y)$ is $\mathcal{A}$-measurable if $f_1 \geq f_2$ and both functions belong to $\mathcal{H}^+$. It is just as easy to check the other three properties needed to show that $\mathcal{H}^+$ is a λ-cone. All indicator functions of measurable rectangles belong to $\mathcal{H}^+$, because $\lambda_x^y\{x \in A,\ y \in B\} = \{x \in A\}(\lambda_x B)$. Property (ii) therefore holds for all bounded functions in $\mathcal{M}^+(\mathcal{X} \times \mathcal{Y})$. A Monotone Convergence argument,

$$\lambda_x^y f(x, y) = \lim_{n \to \infty} \lambda_x^y (n \wedge f(x, y)),$$

extends the property to all of $\mathcal{M}^+(\mathcal{X} \times \mathcal{Y})$.

It follows immediately that the iterated integral is well defined, thereby defining an increasing linear functional $\mu \otimes \Lambda$ on $\mathcal{M}^+(\mathcal{X} \times \mathcal{Y})$. The functional has the Monotone Convergence property: if $0 \leq f_n \uparrow f$ then

$$\begin{aligned}(\mu \otimes \Lambda) f &= \mu^x \left(\lambda_x^y \lim_n f_n(x, y)\right) \\ &= \mu^x \left(\lim_n \lambda_x^y f_n(x, y)\right) && \text{Monotone Convergence for } \lambda_x \\ &= \lim_n \mu^x \left(\lambda_x^y f_n(x, y)\right) && \text{Monotone Convergence for } \mu.\end{aligned}$$

Thus $\mu \otimes \Lambda$ has all the properties required of a functional corresponding to a measure on $\mathcal{A} \otimes \mathcal{B}$. □

If μ is a probability measure and Λ is a probabilty kernel, then $\mu \otimes \Lambda$ defines a probability measure on the product sigma-field of $\mathcal{X} \times \mathcal{Y}$. It defines a joint distribution for the coordinate projections, X and Y. In Chapter 5, the probability measure λ_x will be identified as the conditional distribution for Y, given that $X = x$. The construction can be extended further by means of a second probability kernel, $N = \{\nu_{x,y} : (x, y) \in \mathcal{X} \times \mathcal{Y}\}$, from $(\mathcal{X} \times \mathcal{Y}, \mathcal{A} \otimes \mathcal{B})$ to another space, $(\mathcal{Z}, \mathcal{C})$. For f in $\mathcal{M}^+(\mathcal{X} \times \mathcal{Y} \times \mathcal{Z}, \mathcal{A} \otimes \mathcal{B} \otimes \mathcal{C})$, the iterated integral

$$((\mu \otimes \Lambda) \otimes N) f = (\mu \otimes \Lambda)^{x,y} \left(\nu_{x,y}^z f(x, y, z)\right) = \mu^x \left(\lambda_x^y \left(\nu_{x,y}^z f(x, y, z)\right)\right)$$

is well defined. It corresponds to a probability measure on $\mathcal{A} \otimes \mathcal{B} \otimes \mathcal{C}$, which defines a joint distribution for the three coordinate projections, X, Y, Z. We can also define a probability kernel $\Lambda \otimes N$ from $\mathcal{X}$ to $\mathcal{Y} \times \mathcal{Z}$ by means of the map $x \mapsto \lambda_x^y \left(\nu_{x,y}^z g(y, z)\right)$, for g in $\mathcal{M}^+(\mathcal{Y} \times \mathcal{Z}, \mathcal{B} \otimes \mathcal{C})$, thereby identifying the joint distribution as $\mu \otimes (\Lambda \otimes N)$. It makes no difference which way we interpret the probability on $\mathcal{A} \otimes \mathcal{B} \otimes \mathcal{C}$, which from now on I will write as $\mu \otimes \Lambda \otimes N$. A similar construction works for any finite number of probability kernels. The extension to a countably infinite sequence involves a more delicate argument, which will be presented in Section 8.

Infinite kernels

The construction of $\mu \otimes \Lambda$, for general measures and kernels, extends easily to product measurable sets that can be covered by a countable collection of measurable rectangles, in each of which an analog of the conditions of Theorem <20> hold.

<21> **Definition.** *Say that a kernel is* ***sigma-finite*** *if there exist countably many measurable rectangles $A_i \times B_i$ for which $\mathcal{X} \times \mathcal{Y} = \cup_{i \in \mathbb{N}} A_i \times B_i$ and for which $\lambda_x B_i < \infty$ for all $x \in A_i$, for each i.*

<22> **Corollary.** *If Λ is a sigma-finite kernel, then the three assertions of Theorem <20> still hold. The countably additive measure $\mu \otimes \Lambda$ on $\mathcal{A} \otimes \mathcal{B}$, defined via the iterated integral, is sigma-finite if μ is a sigma-finite measure on $\mathcal{A}$.*

Proof. Only property (ii), the measurablity of $x \mapsto \lambda_x^y f(x, y)$ for each f in $\mathcal{M}^+(\mathcal{X} \times \mathcal{Y})$, requires any new argument, because the finiteness of $\lambda_x \mathcal{Y}$ was used only in the proof of the corresponding assertion in the Theorem.

Temporarily, I will use the word *rectangle* as an abbreviation for *measurable rectangle*. The difference of two rectangles can be written as a disjoint union of two other rectangles: $(A \times B) \setminus (C \times D) = (AC^c \times B) \cup (AC \times BD^c)$. In particular, we can write $A_2 \times B_2$ as a disjoint union of two rectangles, each disjoint from $A_1 \times B_1$; then write $A_3 \times B_3$ as a disjoint union of at most four rectangles, each disjoint from both $A_1 \times B_1$ and $A_2 \times B_2$; and so on. In other words, with no loss of generality, we may assume the rectangles $A_i \times B_i$ are pairwise disjoint, thereby ensuring that $\sum_{i \in \mathbb{N}} A_i \times B_i = 1$.

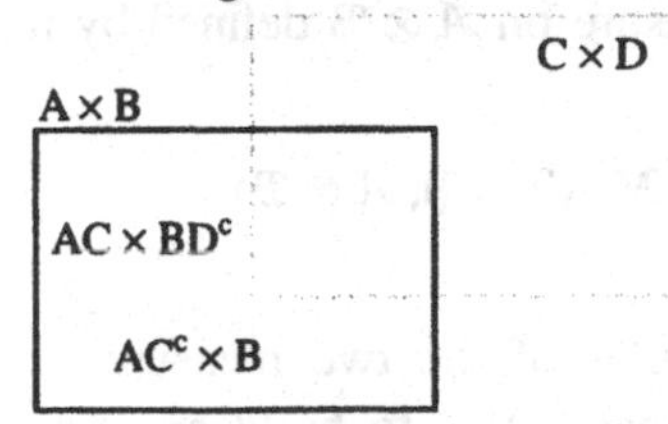

For f in $\mathcal{M}^+(\mathcal{X} \times \mathcal{Y})$, the integral $\lambda_x^y f(x, y)$ breaks into a countable sum,

$$\lambda_x^y \textstyle\sum_{i \in \mathbb{N}} \left(\{x \in A_i,\ y \in B_i\} f(x, y)\right) = \sum_{i \in \mathbb{N}} \{x \in A_i\} \lambda_x^y \left(\{y \in B_i\} f(x, y)\right).$$

For the ith summand, we may regard λ_x as a kernel from A_i to B_i, which lets us invoke Theorem <20> to establish measurability of that summand as a function of x. Thus $\lambda_x^y f(x, y)$ is a countable sum of $\mathcal{A}$-measurable functions, which establishes (ii).

If μ is sigma-finite, we may assume, with no loss of generality, that $\mu A_i < \infty$ for every i. Define $A_{i,n} = \{x \in A_i : \lambda_x B_i \le n\}$. Then $\mu \otimes \Lambda \left(A_{i,n} \times B_i\right) \le n \mu A_{i,n} < \infty$, and $\mathcal{X} \times \mathcal{Y} = \cup_{i \in \mathbb{N}, n \in \mathbb{N}} \left(A_{i,n} \times B_i\right)$. □

<23> **Example.** If $\mathcal{X} = \mathcal{Y} = \mathbb{R}$ and each λ_x and μ equals (one-dimensional) Lebesgue measure λ on $\mathcal{B}(\mathbb{R})$, then the resulting $\mu \otimes \Lambda$ may be taken as the definition of two-dimensional Lebesgue measure $\mathfrak{m}_2$ on $\mathcal{B}(\mathbb{R}^2)$. (For a more direct construction, see Section A.5 of Appendix A.) □

<24> **Example.** Let μ denote the $N(0,1)$ distribution on $\mathcal{B}(\mathbb{R})$, and λ_x denote the $N(\rho x, 1 - \rho^2)$ distribution, also on $\mathcal{B}(\mathbb{R})$, for a constant ρ with $|\rho| < 1$. For $f \in \mathcal{M}^+(\mathbb{R}^2)$,

$$(\mu \otimes \Lambda) f = \mathfrak{m}^x \left(\frac{\exp(-x^2/2)}{\sqrt{2\pi}} \mathfrak{m}^y \left(\frac{\exp(-(y - \rho x)^2/2(1 - \rho^2)}{\sqrt{2\pi(1 - \rho^2)}} f(x, y) \right) \right)$$

$$= \mathfrak{m}^x \mathfrak{m}^y \left(\frac{\exp\left(-x^2/2 - (y - \rho x)^2/2(1 - \rho^2)\right)}{2\pi\sqrt{(1 - \rho^2)}} f(x, y) \right).$$

That is, $\mu \otimes \Lambda$ is absolutely continuous with respect to two-dimensional Lebesgue measure, with density

$$\frac{1}{2\pi\sqrt{1-\rho^2}} \exp\left(-\frac{(x^2 - 2\rho xy + y^2)}{2(1-\rho^2)}\right).$$

The probability measure $\mu \otimes \Lambda$ is called the ***standard bivariate normal distribution with correlation*** ρ.
□

4. Product measures

A particularly important special case of the constructions from the previous Section arises when $\Lambda = \{\lambda\}$ for a fixed measure λ on $\mathcal{B}$, that is, when $\lambda_x = \lambda$ for all x. Sigma-finiteness of the kernel Λ is then equivalent to sigma-finiteness of λ in the usual sense: the space $\mathcal{Y}$ should be a countable union of sets each with finite λ measure. I will abbreviate $\mu \otimes \{\lambda\}$ to $\mu \otimes \lambda$, the conventional notation for the ***product of the measures*** μ ***and*** λ. That is, $\mu \otimes \lambda$ is the measure on $\mathcal{A} \otimes \mathcal{B}$ defined by the linear functional

$$(\mu \otimes \lambda)(f) := \mu^x \left(\lambda^y f(x,y)\right) \qquad \text{for } f \in \mathcal{M}^+(\mathcal{X} \times \mathcal{Y}, \mathcal{A} \otimes \mathcal{B}).$$

In particular, $(\mu \otimes \lambda)(A \times B) = (\mu A)(\lambda B)$.

If μ is also sigma-finite, we can reverse the roles of the two measures, integrating first with respect to μ and then with respect to λ, to build another measure on $\mathcal{A} \otimes \mathcal{B}$, which again would give the value $(\mu A)(\lambda B)$ to the measurable rectangle $A \times B$. As shown in Problem [11], the equality for the generating class of all measurable rectangles ensures that the new linear functional defines the same measure on $\mathcal{A} \otimes \mathcal{B}$.

<25> **Tonelli Theorem.** *If μ is a sigma-finite measure on $(\mathcal{X}, \mathcal{A})$, and λ is a sigma-finite measure on $(\mathcal{Y}, \mathcal{B})$, then, for each f in $\mathcal{M}^+(\mathcal{X} \times \mathcal{Y}, \mathcal{A} \otimes \mathcal{B})$,*

(i) $y \mapsto f(x,y)$ is $\mathcal{B}$-measurable for each fixed x, and $x \mapsto f(x,y)$ is $\mathcal{A}$-measurable for each fixed y;

(ii) $x \mapsto \lambda^y f(x,y)$ is $\mathcal{A}$-measurable, and $y \mapsto \mu^x f(x,y)$ is $\mathcal{B}$-measurable;

(iii) $(\mu \otimes \lambda) f = \mu^x (\lambda^y f(x,y)) = \lambda^y (\mu^x f(x,y))$.

Proof. Assertions (i) and (ii) follow immediately from Corollary <22>. The third assertion merely restates the fact that the linear functionals defined by both iterated integrals correspond to the same measure on $\mathcal{A} \otimes \mathcal{B}$.
□

See Problem [12] for an example emphasizing the need for sigma-finiteness.

<26> **Example.** Let μ be a sigma-finite measure on $\mathcal{A}$. For f in $\mathcal{M}^+(\mathcal{X}, \mathcal{A})$ and each constant $p \geq 1$, we can express $\mu(f^p)$ as an iterated integral,

$$\mu\left(f^p\right) = \mu^x\left(\mathfrak{m}^y\left(py^{p-1}\{y : f(x) > y > 0\}\right)\right),$$

where $\mathfrak{m}$ denotes Lebesgue measure on $\mathcal{B}(\mathbb{R})$. It is not hard—although a little messy, as you will see from Problem [2]—to show that the function $g(x,y) := py^{p-1}\{f(x) > y > 0\}$, on $\mathcal{X} \times \mathbb{R}$, is product measurable. Tonelli lets us reverse the order of integration. Abbreviating $\mu^x\{y : f(x) > y > 0\}$ to $\mu\{f > y\}$ and writing the Lebesgue integral in traditional notational, we then conclude that

$$\mu(f^p) = p\int_0^\infty y^{p-1}\mu\{f > y\}\,dy.$$

In particular, if $\mu(f^p) < \infty$ then $\mu\{f > y\}$ must decrease to zero faster than y^{-p} as $y \to \infty$. □

The definition of product measures, and the Tonelli Theorem, can be extended to collections $\mu_1, \ldots, \mu_n$ of more than two sigma-finite measures, as for kernels.

<27> **Example.** Apparently every mathematician is supposed to know the value of the constant $C := \int_{-\infty}^{\infty} \exp(-x^2)\,dx$. With the help of Tonelli, you too will discover that $C = \sqrt{\pi}$. Let $\mathfrak{m}$ denote Lebesgue measure on $\mathcal{B}(\mathbb{R})$ and $\mathfrak{m}_2 = \mathfrak{m} \otimes \mathfrak{m}$ denote Lebesgue measure on $\mathcal{B}(\mathbb{R}^2)$. Then

$$C^2 = \mathfrak{m}^x\mathfrak{m}^y \exp(-x^2 - y^2) = \mathfrak{m}_2^{x,y}\left(\mathfrak{m}^z\{x^2 + y^2 \le z\}e^{-z}\right).$$

The $\mathfrak{m}_2$ measure of the ball $\{x^2 + y^2 \le z\}$, for fixed positive z, equals πz. A change in the order of integration leaves $\mathfrak{m}^z\left(\pi\{0 \le z\}ze^{-z}\right) = \pi$ as the value for C^2. □

The Tonelli Theorem is often invoked to establish integrability of a product measurable (extended-) real-valued function f, by showing that at least one of the iterated integrals $\mu^x(\lambda^y|f(x,y)|)$ or $\lambda^y(\mu^x|f(x,y)|)$ is finite. In that case, the Theorem also asserts equality for pairs of iterated integrals for the positive and negative parts of the function:

$$\mu^x\lambda^y f^+(x,y) = \lambda^y\mu^x f^+(x,y) < \infty,$$

with a similar assertion for f^-. As a consequence, the $\mathcal{A}$-measurable set

$$N_\mu := \{x : \lambda^y f^+(x,y) = \infty \text{ or } \lambda^y f^-(x,y) = \infty\}$$

has zero μ-measure, and the analogously defined $\mathcal{B}$-measurable set N_λ has zero λ measure. For $x \notin N_\mu$, the integral $\lambda^y f(x,y) := \lambda^y f^+(x,y) - \lambda^y f^-(x,y)$ is well defined and finite. If we replace f by the product measurable function $\widetilde{f}(x,y) := f(x,y)\{x \notin N_\mu, y \notin N_\lambda\}$, the negligible sets of bad behavior disappear, leaving an assertion similar to the Tonelli Theorem but for integrable functions taking both positive and negative values. Less formally, we can rely on the convention that a function can be left undefined on a negligible set without affecting its integrability properties.

<28> **Corollary (Fubini Theorem).** *For sigma-finite measures μ and λ, and a product measurable function f with $(\mu \otimes \lambda)|f| < \infty$,*

(i) $y \mapsto f(x,y)$ is $\mathcal{B}$-measurable for each fixed x; and $x \mapsto f(x,y)$ is $\mathcal{A}$-measurable for each fixed y;

(ii) the integral $\lambda^y f(x,y)$ is well defined and finite μ almost everywhere, and $x \mapsto \lambda^y f(x,y)$ is μ-integrable; the integral $\mu^x f(x,y)$ is well defined and finite λ almost everywhere, and $y \mapsto \mu^x f(x,y)$ is λ-integrable;

(iii) $(\mu \otimes \lambda) f = \mu^x(\lambda^y f(x,y)) = \lambda^y(\mu^x f(x,y))$.

REMARKS. If we add similar almost sure qualifiers to assertion (i), then the Fubini Theorem also works for functions that are measurable with respect to $\mathcal{F}$, the $\mu \otimes \lambda$ completion of the product sigma-field. The result is easy to deduce from the Theorem as stated, because each $\mathcal{F}$-measurable function f can be sandwiched between two product measurable functions, $f_0 \le f \le f_1$, with $f_0 = f_1$, a.e. $[\mu \otimes \lambda]$. Many authors work with the slightly more general version, stated for the completion, but then the Tonelli Theorem also needs almost sure qualifiers.

Without integrability of the function f, the Fubini Theorem can fail, as shown by Problem [13]. Strictly speaking, the sigma-finiteness of the measures is not essential, but little is gained by eliminating it from the assumptions of the Theorem. As explained in the next Section, under the traditional definition of products for general measures, integrable functions must almost concentrate on a countable union of measurable rectangles each with finite product measure.

Product measures correspond to joint distributions for independent random variables or random elements. For example, suppose X is a random element of a space $(\mathcal{X}, \mathcal{A})$ (that is, X is an $\mathcal{F}\backslash\mathcal{A}$-measurable map from Ω into $\mathcal{X}$), and Y is a random element of a space $(\mathcal{Y}, \mathcal{B})$. Let X have (marginal) distribution μ and Y have (marginal) distribution λ. That is, $\mathbb{P}\{X \in A\} = \mu A$ for each A in $\mathcal{A}$, and $\mathbb{P}\{Y \in B\} = \lambda B$ for each B in $\mathcal{B}$. The joint distribution of X and Y is the image measure of $\mathbb{P}$ under the map $\omega \mapsto (X(\omega), Y(\omega))$. It is the probability measure $\mathbb{Q}$ on $\mathcal{A} \otimes \mathcal{B}$ defined by $\mathbb{Q}D = \mathbb{P}\{(X, Y) \in D\}$. If X and Y are independent, and if D is a measurable rectangle, then $\mathbb{Q}D$ factorizes:

$$\mathbb{Q}(A \times B) = \mathbb{P}\{X \in A,\ Y \in B\} = \mathbb{P}\{X \in A\}\mathbb{P}\{Y \in B\} = (\mu A)(\lambda B).$$

That is, $\mathbb{Q}D = (\mu \otimes \lambda)\, D$ for each D in the generating class of measurable rectangles. It follows that $\mathbb{Q} = \mu \otimes \lambda$, as measures on the product sigma-field.

Conversely, if $\mathbb{Q}$ is a product measure then $\mathbb{P}\{X \in A,\ Y \in B\}$ factorizes, implying that X and Y are independent.

In short: random variables (or random elements of general spaces) are independent if and only if their joint distribution is the product of their marginal distributions.

Facts about independence can often be deduced from analogous facts about product measures. In effect, the proofs of the Tonelli/Fubini Theorems are equivalent to several of the standard generating class and Monotone Convergence tricks used for independence arguments. Moreover, the results for product measures are stated for functions, which eliminates a layer of argument needed to extend independence factorizations from sets to functions.

<29> **Example.** The factorization asserted by Theorem <2> follows from Tonelli's Theorem. Let $X = (Z_1, \ldots, Z_k)$ and $Y = (Z_{k+1}, \ldots, Z_n)$. The function $(x, y) \mapsto f(x)g(y)$ is product measurable. Why? With $\mathbb{Q}$, μ, and ν as above,

$$
\begin{aligned}
\mathbb{P}(f(X)g(Y)) &= \mathbb{Q}^{x,y}(f(x)g(y)) && \text{image measure}\\
&= (\mu \otimes \nu)^{x,y}(f(x)g(y)) && \text{independence}\\
&= \nu^y\left(\mu^x f(x)g(y)\right) && \text{Tonelli}\\
&= \left(\nu^y g(y)\right)\left(\mu^x f(x)\right) = \\
&= (\mathbb{P}f(X))(\mathbb{P}g(Y)) && \text{images measures.}
\end{aligned}
$$

In the third line on the right-hand side the factor $g(y)$ behaves like a constant for the μ integral. □

<30> **Example.** The image of $\mu \otimes \nu$, a product of finite measures on $\mathbb{R}^k \times \mathbb{R}^k$, under the map $T(x, y) = x + y$ is called the ***convolution*** of the two measures, and is denoted by $\mu \star \nu$ (or $\nu \star \mu$, the order of the factors being irrelevant). If μ and ν are probability measures, and X and Y are independent random vectors with distributions μ and ν, then the product measure $\mu \otimes \nu$ gives their joint distribution, and the convolution $\mu \star \nu$ gives the distribution of the sum $X + Y$.

By the Tonelli Theorem and the definition of image measure,

$$(\mu \star \nu)(f) = \mu^x \nu^y f(x + y) \qquad \text{for } f \in \mathcal{M}^+(\mathbb{R}^k).$$

When ν has a density $\delta(\cdot)$ with respect to k-dimensional Lebesgue measure $\mathfrak{m}$, the innermost integral on the right-hand side can be written, in traditional notation, as $\int \delta(y) f(x + y)\, dy$, which invites us to make a change of variable and rewrite it as $\int \delta(y - x) f(y)\, dy$. More formally, we could invoke the invariance of Lebesgue measure under translations (Problem [15]) to justify the reexpression. Whichever way we justify the change, the convolution becomes

$$\begin{aligned}(\mu \star \nu) f &= \mu^x \mathfrak{m}^y \delta(y - x) f(y) \\ &= \mathfrak{m}^y \mu^x \delta(y - x) f(y) \qquad \text{by Tonelli.}\end{aligned}$$

Writing $g(y) := \mu^x \delta(y - x)$, we have $(\mu \star \nu) f = \mathfrak{m}^y (g(y) f(y))$. That is, $\mu \star \nu$ has density g with respect to $\mathfrak{m}$. □

REMARK. The statistical techniques known as *density estimation* and *nonparametric smoothing* rely heavily on convolutions.

<31> **Exercise.** The $N(\theta, \sigma^2)$ distribution on the real line is defined by the density

$$f(x, \theta, \sigma) = \frac{1}{\sigma\sqrt{2\pi}} \exp\left(\frac{-(x - \theta)^2}{2\sigma^2}\right)$$

with respect to Lebesgue measure. Show that the convolution of two normal distributions is normal: $N(\theta_1, \sigma_1^2) \star N(\theta_2, \sigma_2^2) = N(\theta_1 + \theta_2, \sigma_1^2 + \sigma_2^2)$.

SOLUTION: The convolution formula from Example <30> (with μ as the $N(\theta_1, \sigma_1^2)$ distribution and δ as the $N(\theta_2, \sigma_2^2)$ density) becomes

$$g(y) = \frac{1}{2\pi\sigma_1\sigma_2} \int_{-\infty}^{\infty} \exp\left(-\frac{(x - \theta_1)^2}{2\sigma_1^2} - \frac{(y - x - \theta_2)^2}{2\sigma_2^2}\right) dx.$$

Make the change of variable $z = x - \theta_1$, and replace y by $\theta_1 + \theta_2 + y$ (to get a neater expression).

$$g(\theta_1 + \theta_2 + y) = \frac{1}{2\pi\sigma_1\sigma_2} \int_{-\infty}^{\infty} \exp\left(-\frac{z^2}{2\sigma_1^2} - \frac{(y - z)^2}{2\sigma_2^2}\right) dz.$$

Complete the square, to rewrite the exponent as

$$-\frac{1}{2\tau^2}\left(z - \frac{\tau^2 y}{\sigma_2^2}\right)^2 + \left(\frac{\tau^2}{2\sigma_2^4} - \frac{1}{2\sigma_2^2}\right) y^2 \qquad \text{where } \tau^{-2} = \sigma_1^{-2} + \sigma_2^{-2}.$$

The coefficient of y^2 simplifies to $-(2\sigma^2)^{-1}$, where $\sigma^2 = \sigma_1^2 + \sigma_2^2$. When integrated out, the quadratic in z contributes a factor $\sqrt{2\pi\tau}$, leaving the appropriate multiple of $\exp(-y^2/2\sigma^2)$. The convolution result follows. □

REMARK. We could avoid most of the algebra in the Example, by noting that $g(y + \theta_1 + \theta_2)$ has the form $C_1 \exp(-C_2 y^2)$, for constants C_1 and C_2. Nonnegativity and integrability of the density would force both constants to be strictly positive, which would show that g must be some $N(\mu, \sigma^2)$ density. Calculation of means and variances would then identify μ and σ^2.

There is also a quicker derivation of the result, based on Fourier transforms (see Chapter 8), but a distaste for circular reasoning has compelled me to inflict on you the grubby Calculus details of the direct convolution argument: I will derive the form of the $N(\theta, \sigma^2)$ Fourier transform by means of a central limit theorem, for the proof of which I will use the convolution property of normals.

<32> **Example.** Recall from Section 2.2 the definition of the distribution function F_X and its corresponding quantile function for a random variable X:

$$F_X(x) = \mathbb{P}\{X \le x\} \qquad \text{for } x \in \mathbb{R},$$
$$q_X(u) = \inf\{t : F_X(t) \ge u\} \qquad \text{for } 0 < u < 1.$$

The quantile function is almost an inverse to the distribution function, in the sense that $F_X(x) \ge u$ if and only if $q_X(u) \le x$. As a random variable on $(0, 1)$ equipped with its Borel sigma-field and Lebesgue measure $\widetilde{\mathbb{P}}$, the function $\widetilde{X} := q_X(u)$ has the same distribution as X. Similarly, if Y has distribution function F_Y and quantile function q_Y, the random variable $\widetilde{Y} := q_Y(u)$ has the same distribution as Y.

Notice that $\widetilde{X}$ and $\widetilde{Y}$ are both defined on the same $(0, 1)$, even though the original variables need not be defined on the same space. If X and Y do happen to be defined on the same Ω their joint distribution need not be the same as the joint distribution for $\widetilde{X}$ and $\widetilde{Y}$. In fact, the new variables are closer to each other, in various senses. For example, several applications of Tonelli will show that $\mathbb{P}|X - Y|^p \ge \widetilde{\mathbb{P}}|\widetilde{X} - \widetilde{Y}|^p$ for each $p \ge 1$.

As a first step, calculate an inequality for tail probabilities.

$$\begin{aligned}
\mathbb{P}\{X > x,\ Y > y\} &\le \min\left(\mathbb{P}\{X > x\},\ \mathbb{P}\{Y > y\}\right) \\
&= \min\left(1 - F_X(x),\ 1 - F_Y(y)\right) \\
&= 1 - F_X(x) \vee F_Y(y) \\
&= \int_0^1 \{u > F_X(x) \vee F_Y(y)\}\, du \\
&= \int_0^1 \{x < q_X(u),\ y < q_Y(u)\}\, du \\
&= \widetilde{\mathbb{P}}\{\widetilde{X} > x,\ \widetilde{Y} > y\}
\end{aligned} \qquad \text{<33>}$$

We also have $\mathbb{P}\{X > x\} = \widetilde{\mathbb{P}}\{\widetilde{X} > x\}$ and $\mathbb{P}\{Y > y\} = \widetilde{\mathbb{P}}\{\widetilde{Y} > y\}$, from equality of the marginal distributions. By subtraction,

$$\begin{aligned}
&\mathbb{P}\big(\{X > x\} + \{Y > y\} - 2\{X > x,\ Y > y\}\big) \\
&\qquad \ge \widetilde{\mathbb{P}}\big(\{\widetilde{X} > x\} + \{\widetilde{Y} > y\} - 2\{\widetilde{X} > x,\ \widetilde{Y} > y\}\big) \qquad \text{for all } x \text{ and } y.
\end{aligned} \qquad \text{<34>}$$

The left-hand side can be rewritten as

$$\mathbb{P}^{\omega}\left(\{X(\omega) > x,\ y \geq Y(\omega)\} + \{X(\omega) \leq x, y < Y(\omega)\}\right),$$

a nonnegative function just begging for an application of Tonelli. For each real constant s, put $y = x + s$ then integrate over x with respect to Lebesgue measure $\mathfrak{m}$ on $\mathcal{B}(\mathbb{R})$. Tonelli lets us interchange the order of integration, leaving

$$\begin{aligned}
&\mathbb{P}^{\omega}\left(\mathfrak{m}^{x}\{X(\omega) > x \geq Y(\omega) - s\} + \mathfrak{m}^{x}\{X(\omega) \leq x < Y(\omega) - s\}\right)\\
&= \mathbb{P}^{\omega}\left((X(\omega) - Y(\omega) + s)^{+} + (Y(\omega) - s - X(\omega))^{+}\right)\\
&= \mathbb{P}|X - Y + s|.
\end{aligned}$$

Argue similarly for the right-hand side of <34>, to deduce that

$$\mathbb{P}|X - Y + s| \geq \widetilde{\mathbb{P}}|\widetilde{X} - \widetilde{Y} + s| \qquad \text{for all real } s.$$

For each nonnegative t, invoke the inequality for $s = t$ then $s = -t$, then add.

$$\mathbb{P}\left(|X - Y + t| + |X - Y - t|\right) \geq \widetilde{\mathbb{P}}\left(|\widetilde{X} - \widetilde{Y} + t| + |\widetilde{X} - \widetilde{Y} - t|\right) \qquad \text{for all } t \geq 0.$$

An appeal to the identity $|z + t| + |z - t| = 2t + 2\left(|z| - t\right)^{+}$, for $z \in \mathbb{R}$ and $t \geq 0$, followed by a cancellation of common terms, then leaves us with a useful relationship, which neatly captures the idea that $\widetilde{X}$ and $\widetilde{Y}$ are more tightly coupled than X and Y.

<35> $$\mathbb{P}\left(|X - Y| - t\right)^{+} \geq \widetilde{\mathbb{P}}\left(|\widetilde{X} - \widetilde{Y}| - t\right)^{+} \qquad \text{for all } t \geq 0.$$

Various interesting inequalities follow from <35>. Putting t equal to zero we get $\mathbb{P}|X - Y| \geq \widetilde{\mathbb{P}}|\widetilde{X} - \widetilde{Y}|$. For $p > 1$, note the identity

$$D^{p} = p(p-1)\int_{0}^{D}(D - t)t^{p-2}\,dt = p(p-1)\mathfrak{m}_0^{t}\left(t^{p-2}(D - t)^{+}\right) \qquad \text{for } D \geq 0,$$

where $\mathfrak{m}_0$ denotes Lebesgue measure on $\mathcal{B}\left(\mathbb{R}^{+}\right)$. Temporarily write Δ for $|X - Y|$ and $\widetilde{\Delta}$ for $|\widetilde{X} - \widetilde{Y}|$. Two more appeals to Tonelli then give

$$\begin{aligned}
\mathbb{P}|X - Y|^{p} &= p(p-1)\mathfrak{m}_0^{t}\left(t^{p-2}\mathbb{P}^{\omega}\left(\Delta(\omega) - t\right)^{+}\right)\\
&\geq p(p-1)\mathfrak{m}_0^{t}\left(t^{p-2}\widetilde{\mathbb{P}}^{u}\left(\widetilde{\Delta}(u) - t\right)^{+}\right) = \widetilde{\mathbb{P}}|\widetilde{X} - \widetilde{Y}|^{p}.
\end{aligned}$$

See Problem [17] for the analogous inequality, $\mathbb{P}\Psi(|X - Y|) \geq \widetilde{\mathbb{P}}\Psi(|\widetilde{X} - \widetilde{Y}|)$, for every convex, increasing function Ψ on $\mathbb{R}^{+}$. □

*5. Beyond sigma-finiteness

Even when the kernel Λ is not sigma-finite, it is possible to define $\mu \otimes \Lambda$ as a measure on the whole of $\mathcal{A} \otimes \mathcal{B}$. Unfortunately, the obvious method—by means of an iterated integral—can fail for measurability reasons. Even when λ_x does not depend on x, the map $x \mapsto \lambda_x^{y} f(x, y)$ need not be measurable for all f in $\mathcal{M}^{+}(\mathcal{X} \times \mathcal{Y}, \mathcal{A} \otimes \mathcal{B})$, as shown by the following counterexample.

<36> **Example.** Let $\mathcal{X}$ equal $[0,1]$, equipped with Borel sigma-field $\mathcal{A}$, and $\mathcal{Y}$ also equal $[0,1]$, but equipped with the sigma-field $\mathcal{B}$ of all subsets of $[0,1]$. Let E be a subset of $[0,1]$, not $\mathcal{A}$-measurable. Let λ be the counting measure on E, interpreted as a measure on $\mathcal{B}$. That is, λB equals the cardinality of $E \cap B$, for every subset B of $[0,1]$. The Borel measurable subset $D = \{x = y\}$ of $[0,1]^2$ also belongs to the larger sigma-field $\mathcal{A} \otimes \mathcal{B}$. The function $x \mapsto \lambda^y\{(x,y) \in D\}$ equals 1 when $x \in E$, and is 0 otherwise; it is not $\mathcal{A}$-measurable. □

In general, the difficulty lies in the possible nonmeasurability of the function $x \mapsto \lambda_x^y f(x,y)$. When f equals the indicator function of a measurable rectangle the difficulty disappears, because $x \mapsto \lambda_x B$ is assumed to be $\mathcal{A}$-measurable for every B in $\mathcal{B}$. If $\mu^x\left(\lambda_x^y\{x \in A, y \in B\}\right) < \infty$ then $A_0 := \{x \in A : \lambda_x B = \infty\}$ has zero μ measure. The method of Theorem <20> defines a measure on the product sigma-field of $(A \backslash A_0) \times B$, by means of an iterated integral,

$$\Gamma f := \mu^x\left(\lambda_x^y f(x,y)\right) \qquad \text{if } f \in \mathcal{M}^+ \text{ and } f = 0 \text{ outside } (A \backslash A_0) \times B.$$

For the remainder of the Section, I will use the letter Γ, instead of the symbol $\mu \otimes \Lambda$, to avoid any inadvertent suggestion of an iterated integral.

The same definition also works if f is required to be zero only outside $A \times B$, provided we ignore possible nonmeasurability of $\lambda_x^y f(x,y)$ on a μ-negligible set. More formally, we could assume μ to be complete, so that the behavior for x in A_0 has no effect on the measurability. The method of Corollary <22> can then be applied to extend Γ to a measure on a larger collection of subsets.

<37> **Definition.** *Write $\mathcal{R}$ for the collection of all $\mathcal{A} \otimes \mathcal{B}$-measurable sets D for which there exist measurable rectangles with $D \subseteq \cup_{i \in \mathbb{N}} A_i \times B_i$ and $\mu^x\left(\{x \in A_i\}\lambda_x B_i\right) < \infty$ for each i. Denote by $\mathcal{M}^+(\mathcal{X} \times \mathcal{Y}, \mathcal{R})$, or just $\mathcal{M}^+(\mathcal{R})$, the collection of functions f in $\mathcal{M}^+(\mathcal{X} \times \mathcal{Y}, \mathcal{A} \otimes \mathcal{B})$ for which $\{f \neq 0\} \in \mathcal{R}$.*

The collection $\mathcal{R}$ is stable under countable unions and countable intersections, and differences (that is, if $R_i \in \mathcal{R}$ then $R_1 \backslash R_2 \in \mathcal{R}$)—that is, $\mathcal{R}$ is a ***sigma-ring***. It need not be stable under complements, because the whole product space $\mathcal{X} \times \mathcal{Y}$ might not have a covering by a sequence of measurable rectangles with the desired property—that is, $\mathcal{R}$ need not be a sigma-field.

The definition of a countably additive measure on a sigma-ring is essentially the same as the definition for a sigma-field. There is a one-to-one correspondence between measures defined on $\mathcal{R}$ and increasing linear functionals on $\mathcal{M}^+(\mathcal{R})$ with the Monotone Convergence property. In particular, the iterated integral $\mu^x\left(\lambda_x^y f(x,y)\right)$ defines an increasing linear functional on $f \in \mathcal{M}^+(\mathcal{R})$, corresponding to a measure on $\mathcal{R}$.

For product measurable sets not in $\mathcal{R}$, there is no unique way to proceed.

A minor extension of the classical approach for product measures (using approximation from above, as in Section 12.4 of Royden 1968) suggests we should define $\Gamma(D)$ as the infimum of $\sum_{i \in \mathbb{N}} \mu^x\left(\{x \in A_i\}\lambda_x B_i\right)$, taken over all countable coverings of D by measurable rectangles, $\cup_{i \in \mathbb{N}} A_i \times B_i$. This definition is equivalent to putting $\Gamma(D) = \infty$ when $D \in (\mathcal{A} \otimes \mathcal{B}) \backslash \mathcal{R}$. Consequently, we would have $\Gamma(f) = \infty$ for each nonnegative, product measurable f not in $\mathcal{M}^+(\mathcal{R})$. For the particular case where $\lambda \equiv \lambda$, if f is product measurable and $\Gamma(|f|) < \infty$, then $\{f \neq 0\} \subseteq \cup_{i\in\mathbb{N}} A_i \times B_i$, for some countable collection of measurable rectangles with $\sum_{i\in\mathbb{N}} (\mu A_i)(\lambda B_i) < \infty$. For each i in the set $J_\mu := \{i : \mu A_i = \infty\}$ we must have $\lambda B_i = 0$, which implies that the set $N_\lambda := \cup_{i\in J_\mu} B_i$ is λ-negligible. Similarly, the set $N_\mu := \cup_{i\in J_\lambda} A_i$, where $J_\lambda := \{i : \lambda B_i = \infty\}$, is μ-negligible. When restricted to $\mathcal{X}_0 := \cup_{i\notin J_\mu} A_i$, the measure μ is sigma-finite; and when restricted to $\mathcal{Y}_0 := \cup_{i\notin J_\lambda} B_i$, the measure λ is sigma-finite. Corollary <28> gives the assertions of the Fubini Theorem for the restriction of f to $\mathcal{X}_0 \times \mathcal{Y}_0$. For trivial reasons, the Fubini Theorem also holds for the restriction of f to $N_\mu \times \mathcal{Y}$ or $\mathcal{X} \times N_\lambda$.

REMARK. In effect, with the classical approach, the general Fubini Theorem is made to hold by forcing integrable functions to concentrate on regions where both measures are sigma-finite. The general form of the Fubini Theorem is really just a minor extension of the theorem for sigma-finite measures. I see some virtue in being content with the definition of the general $\mu \otimes \Lambda$ as a measure on the sigma-ring $\mathcal{R}$.

6. SLLN via blocking

The fourth moment bound in Theorem <4> is an unnecessarily strong requirement. It reflects the crudity of the Borel-Cantelli argument based on <5> as a method for proving almost sure convergence. Successive terms contribute almost the same tail probability to the sum, because the averages do not change very rapidly with n. It is possible to do better by breaking the sequence of averages into blocks of similar terms. We need to choose the blocks so that the maximum of the terms over each block behaves almost like a single average, for then the Borel-Cantelli argument need be applied only to a subsequence of the averages.

Probabilistic bounds involving maxima of collections of random variables are often called ***maximal inequalities***. For the SLLN problem there are several types of maximal inequality that could be used. I am fond of the following result because it also proves useful in other contexts. The proof introduces the important idea of a first passage time.

<38> **Maximal Inequality.** *Let $Z_1, \ldots, Z_N$ be independent random variables, and ϵ_1, ϵ_2, and β be positive constants for which $\mathbb{P}\{|Z_i + Z_{i+1} + \ldots + Z_N| \le \epsilon_2\} \ge 1/\beta$ for each i. Then*

$$\mathbb{P}\left\{\max_{i\le N} |Z_1 + \ldots + Z_i| > \epsilon_1 + \epsilon_2\right\} \le \beta\mathbb{P}\{|Z_1 + \ldots + Z_N| > \epsilon_1\}.$$

REMARK. The same inequality holds if the $\{Z_i\}$ are independent random vectors, and $|\cdot|$ is interpreted as length. The inequality is but one example of a large class of results based on a simple principle. Suppose we wish to bound the probability that a sequence $\{S_i : i = 0, 1, \ldots, N\}$ ever enters some region $\mathcal{R}$. If it does enter $\mathcal{R}$, there will be a first time, τ, at which it does so. If we can show that the process has a (conditional) probability of at least $1/\beta$ of 'staying near the scene of the crime' (that is, of remaining within a distance ϵ of $\mathcal{R}$ at time N), then the probability of the event $\{Z_N$ within ϵ of $\mathcal{R}\}$ should be at least $1/\beta$ times the probability of the event {the process hits $\mathcal{R}$}. Of course the time τ will be a random variable, which means that $S_N - S_\tau$ need not behave like a typical increment $S_N - S_i$. We avoid that complication by arguing separately for each event $\{\tau = i\}$.

There are also conditional probability versions of the inequality, relaxing the independence assumption.

Proof. Define

$$S_i := Z_1 + \ldots + Z_i,$$
$$T_i := \qquad Z_{i+1} + \ldots + Z_N = S_N - S_i.$$

Define a random variable τ (a first passage time) by

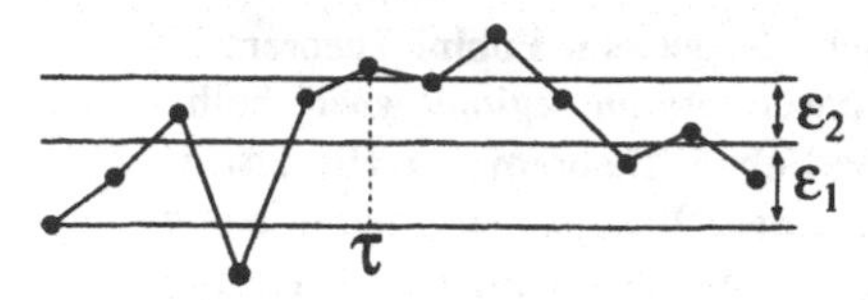

$$\tau := \begin{cases} \text{first } i \text{ for which } |S_i| > \epsilon_1 + \epsilon_2, \\ N \text{ if } |S_i| \le \epsilon_1 + \epsilon_2 \text{ for all } i. \end{cases}$$

Notice that the events $\{\tau = i\}$ for $i = 1, \ldots, N$ are disjoint. The probability on the left-hand side of the asserted inequality equals

$$\begin{aligned} &\mathbb{P}\{\tau = i \text{ and } |S_i| > \epsilon_1 + \epsilon_2 \text{ for some } i\} \\ &\quad = \textstyle\sum_{i=1}^N \mathbb{P}\{\tau = i,\ |S_i| > \epsilon_1 + \epsilon_2\} \qquad \text{disjoint events} \\ &\quad \le \textstyle\sum_{i=1}^N \mathbb{P}\{\tau = i,\ |S_i| > \epsilon_1 + \epsilon_2\}\beta\mathbb{P}\{|T_i| \le \epsilon_2\} \qquad \text{definition of } \beta. \end{aligned}$$

The event $\{\tau = i,\ |S_i| > \epsilon_1 + \epsilon_2\}$ is a function of $Z_1, \ldots, Z_i$; the event $\{|T_i| \le \epsilon_2\}$ is a function of $Z_{i+1}, \ldots, Z_N$. By independence, the product of the probabilities in the last displayed line is equal to the probability of the intersection. The sum is less than

$$\beta \textstyle\sum_{i=1}^N \mathbb{P}\{\tau = i,\ |S_i| > \epsilon_1 + \epsilon_2,\ |T_i| \le \epsilon_2\}.$$

If $|S_i| > \epsilon_1 + \epsilon_2$ and $|T_i| \le \epsilon_2$ then certainly $|S_N| > \epsilon_1$. The sum is less than $\beta \sum_{i=1}^N \mathbb{P}\{\tau = i, |S_N| > \epsilon_1\} = \beta\mathbb{P}\{|S_N| > \epsilon_1\}$, as asserted. □

REMARK. Notice how the disjointness of the events $\{\tau = i\}$ for $i = 1, \ldots, N$ was used twice: first to break the maximal event into pieces and then to reassemble the bounds on the pieces. Also, you should ponder the choice of τ value for the case where $|S_i| \le \epsilon_1 + \epsilon_2$ for all i. Where would the proof fail if I had chosen $\tau = 1$ instead of $\tau = N$ for that case?

The Maximal Inequality will control the behavior of the averages over blocks of successive partial sums, chosen short enough that the β constants stay bounded. The Borel-Cantelli argument can then be applied along the subsequence corresponding to the endpoints of the blocks. The longer the blocks, the sparser the subsequence of endpoints, and the easier it becomes to establish convergence for the sum of tail probabilities along the subsequence. With block lengths increasing geometrically fast, under mild second moment conditions we get both acceptable β values and tail probabilities decreasing rapidly enough for the Borel-Cantelli argument.

<39> **Theorem.** *(Kolmogorov) Let $X_1, X_2, \ldots$ be independent random variables with $\mathbb{P}X_i = 0$ for every i. If $\sum_i \mathbb{P}X_i^2/i^2 < \infty$ then $(X_1 + \ldots + X_n)/n \to 0$ almost surely.*

Proof. Define $S_i := X_1 + \ldots + X_i$ and

$$V(i) := \sigma_1^2 + \ldots + \sigma_i^2 = \mathbb{P}S_i^2 \qquad \text{where } \sigma_i^2 := \mathbb{P}X_i^2,$$
$$B_k := \{n : n_k < n \le n_{k+1}\} \qquad \text{where } n_k := 2^k, \text{ for } k = 1, 2, \ldots.$$

REMARK. The n_k define the blocks of terms to which the Maximal Inequality will be applied. You might try experimenting with other block sizes, to get a feel for how much latitude we have here. For example, what would happen if we took $n_k = 3^k$, or k^k?

The series $\sum_k V(n_k)/n_k^2$ is convergent, because

$$\sum_{k=1}^{\infty} \frac{V(n_k)}{n_k^2} = \sum_{k=1}^{\infty}\sum_{i=1}^{\infty}\{i \le 2^k\}\sigma_i^2 4^{-k} = \sum_{i=1}^{\infty}\sigma_i^2 \sum_{k=1}^{\infty}\{i \le 2^k\}4^{-k}.$$

The innermost sum on the right-hand side is just the tail of a geometric series, started at $k(i)$, the smallest k for which $i \le 2^k$. The sum equals

$$\sum_i \sigma_i^2 4^{-k(i)}/\left(1 - \tfrac{1}{4}\right) \le \tfrac{4}{3}\sum_i \sigma_i^2/i^2 < \infty.$$

The almost sure convergence assertion of the Theorem is equivalent to

$$\max_{n \in B_k} \frac{|S_n|}{n} \to 0 \qquad \text{as } k \to \infty. \tag{<40>}$$

It therefore suffices if we show, for each $\epsilon > 0$, that

$$\sum_k \mathbb{P}\left\{\max_{n \in B_k} \frac{|S_n|}{n} > 2\epsilon\right\} < \infty. \tag{<41>}$$

Replace the n in the denominator for the kth summand by the smaller number n_k then expand the range from B_k to $\{n \le n_{k+1}\}$ to bound the probability by

$$\mathbb{P}\left\{\max_{n \le n_{k+1}} |S_n| > 2\epsilon n_k\right\}.$$

By the Maximal Inequality, this probability is less than

$$\beta_k \mathbb{P}\{|S_{n_{k+1}}| > \epsilon n_k\} \le \beta_k V(n_{k+1})/(\epsilon n_k)^2,$$

where

$$\begin{aligned}\beta_k^{-1} &= \min_{n \le n_{k+1}} \mathbb{P}\{|S_{n_{k+1}} - S_n| \le \epsilon n_k\} \\ &\ge 1 - \max_{n \le n_{k+1}} \frac{\mathbb{P}(S_{n_{k+1}} - S_n)^2}{\epsilon^2 n_k^2} \ge 1 - \frac{4V(n_{k+1})}{\epsilon^2 n_{k+1}^2} \to 1 \qquad \text{as } k \to \infty.\end{aligned}$$

The kth term in <41> is eventually smaller than a constant multiple of $V(n_{k+1})/n_{k+1}^2$, which establishes the desired convergence. □

*7. SLLN for identically distributed summands

Theorem <39> is not only useful in its own right, but also it acts as the stepping stone towards the version of the SLLN stated as Theorem <3>, where the moment assumptions are relaxed even further, at the cost of a requirement that the variables have identical distributions. The proof of Theorem <3> requires a truncation argument and several appeals to the assumption of identical distributions in order to reduce the calculation of bounds involving many X_i to pointwise inequalities involving only a typical X_1. Actually slightly less would suffice. We only need a way to bound various integrals by quantities that do not change with n.

Proof of Theorem <3>. With no loss of generality, suppose $\mu = 0$ (or replace X_i by the centered variable $X_i - \mu$), so that the Theorem becomes: for independent, identically distributed, integrable random variables $X_1, X_2, \ldots$ with $\mathbb{P}X_i = 0$,

$$\frac{X_1 + \ldots + X_n}{n} \to 0 \qquad \text{almost surely.}$$

Break each X_i into two contributions: a central part, $Y_i = X_i\{|X_i| \le i\}$, which, when recentered at its expectation $\mu_i = \mathbb{P}X_i\{|X_i| \le i\}$, becomes a candidate for Theorem <39>; and a tail part, $X_i - Y_i = X_i\{|X_i| > i\}$, which can be handled by relatively crude summability arguments. Notice that, after truncation, the summands no longer have identical distributions.

REMARK. The choice of i as the ith truncation level is determined by the available moment information. More generally, finiteness of a pth moment would allow us to truncate at $i^{1/p}$ for almost sure convergence arguments.

The truncation constants increase fast enough to allow us to dispose of various fragments by means of the first moment assumption. For example, using the identity $0 = \mathbb{P}X_i = \mu_i + \mathbb{P}X_i\{|X_i| > i\}$, and the identical distribution assumption, we have

$$\text{<42>} \qquad \left|\frac{1}{n}\sum_{i\le n}\mu_i\right| \le \mathbb{P}\frac{1}{n}\sum_{i\le n}|X_1|\{|X_1| > i\} \le \mathbb{P}|X_1|\min\left(1, \frac{|X_1|}{n}\right) \to 0,$$

the final convergence to zero following by Dominated Convergence. Notice how the contribution from $X_i\{|X_i| \le i\}$ was related to a contribution from $-X_i\{|X_i| > i\}$. Without that trick, the analog of <42> would have given the useless upper bound $\mathbb{P}|X_1|$.

The rate of growth of the truncation constants is also fast enough to ensure that the truncation has little effect on the summands, as far as the almost sure convergence of the averages is concerned:

$$\sum_{i=1}^{\infty}\mathbb{P}\{X_i \ne Y_i\} \le \sum_{i=1}^{\infty}\mathbb{P}\{|X_1| > i\} \le \mathbb{P}|X_1| < \infty.$$

Again the identical distributions have reduced the argument to calculation of a pointwise bound involving the single X_1. It follows that, with probability one, there exists a positive integer $i_0(\omega)$ such that, for $n > i_0(\omega)$,

$$\text{<43>} \qquad \left|\frac{1}{n}\sum_{i\le n}X_i(\omega) - \frac{1}{n}\sum_{i\le n}Y_i(\omega)\right| \le \frac{1}{n}\sum_{i\le i_0(\omega)}|X_i(\omega)| \to 0.$$

Notice that the last sum does not change as n increases.

The truncation constants increase slowly enough to allow us to bound second moment quantities for the truncated variables by first moments of the original variables. Here I leave some minor calculus details to you (see Problem [25]). First you should establish a deterministic inequality: there exists a finite constant C such that

$$\text{<44>} \qquad |x|^2\sum_{i=1}^{\infty}\{|x| \le i\}\frac{1}{i^2} \le C|x| \qquad \text{for each real } x.$$

From this bound and the inequality $\mathbb{P}(Y_i - \mu_i)^2 \le \mathbb{P}Y_i^2 = \mathbb{P}X_1^2\{|X_1| \le i\}$, deduce that

$$\sum_{i=1}^{\infty} \frac{\mathbb{P}(Y_i - \mu_i)^2}{i^2} \le C\mathbb{P}|X_1| < \infty.$$

It follows by Theorem <39> that

<45> $$\frac{1}{n}\sum_{i \le n}(Y_i - \mu_i) \to 0 \quad \text{almost surely.}$$

The asserted SLLN for the identically distributed $\{X_i\}$ then follows from the results <42> and <43> and <45>.

*8. Infinite product spaces

The SLLN makes an assertion about an infinite sequence $\{X_n : n \in \mathbb{N}\}$ of independent random variables. How do we know such sequences exist? More generally, for an assertion about any sequence of random variables $\{X_n\}$, with prescribed behavior for the joint distributions of finite subcollections of the variables, how do we know there exists a probability space $(\Omega, \mathcal{F}, \mathbb{P})$ on which the $\{X_n\}$ can be defined? Depending on one's attitude towards rigor, the question of existence is either a technical detail or a vital preliminary to any serious study of limit theorems. If you prefer, at least initially, to take the existence as a matter of faith, then you will probably want to skip this Section.

To accommodate more general random objects, suppose X_i takes values in a measurable space $(\mathcal{X}_i, \mathcal{A}_i)$. For finite collections of random elements $X_1, \ldots, X_n$, we can take Ω as a product space $\times_{i \le n} \mathcal{X}_i$ equipped with its product sigma-field $\bigotimes_{i \le n} \mathcal{A}_i$, with the random elements X_i as the coordinate projections. For infinite collections $\{X_n : n \in \mathbb{N}\}$ of random elements, we need a way of building a probability measure on $\Omega = \times_{i \in \mathbb{N}} \mathcal{X}_i$, the set of all infinite sequences $\omega := (x_1, x_2, \ldots)$ with $x_i \in \mathcal{X}_i$ for each i. The measure should live on the product sigma field $\mathcal{F} := \bigotimes_{i \in \mathbb{N}} \mathcal{A}_i$, which is generated by the measurable hyperrectangles $\times_{i \in \mathbb{N}} A_i$ with $A_i \in \mathcal{A}_i$ for each i.

It is natural to start from a prescribed family $\{\mathbb{P}_n : n \in \mathbb{N}\}$ of desired ***finite dimensional distributions*** for the random elements. That is, $\mathbb{P}_n$ is a probability measure defined on the product sigma-field $\mathcal{F}_n := \mathcal{A}_1 \otimes \ldots \otimes \mathcal{A}_n$ of the product space $\Omega_n := \mathcal{X}_1 \times \ldots \times \mathcal{X}_n$, for each n. The X_i's correspond to the coordinate projections on these product spaces. Of course we need the $\mathbb{P}_n$'s to be consistent in the distributions they give to the variables, in the sense that

<46> $$\mathbb{P}_{n+1}(F \times \mathcal{X}_{n+1}) = \mathbb{P}_{n+1}\{(X_1, \ldots, X_n) \in F,\ X_{n+1} \in \mathcal{X}_{n+1}\} = \mathbb{P}_n F \qquad \text{all } F \text{ in } \mathcal{F}_n,$$

or, equivalently,

$$\mathbb{P}_{n+1} g(x_1, \ldots x_n) = \mathbb{P}_n g(x_1, \ldots x_n) \qquad \text{for all } g \in \mathcal{M}^+(\Omega_n, \mathcal{F}_n).$$

Such a family of probability measures is said to be a ***consistent family*** of finite dimensional distributions. Within this framework, the existence problem becomes:

For a consistent family of finite dimensional distributions $\{\mathbb{P}_n : n \in \mathbb{N}\}$, when does there exist a probability measure $\mathbb{P}$ on $\mathcal{F}$ for which the joint distribution of the coordinate projections $(X_1, \ldots, X_n)$ equals $\mathbb{P}_n$, for each finite n?

Roughly speaking, when such a $\mathbb{P}$ exists (as a countably additive probability measure on $\mathcal{F}$), we are justified in speaking of the joint distribution of the whole sequence $\{X_n : n \in \mathbb{N}\}$. If such a $\mathbb{P}$ does not exist, assertions such as almost sure convergence require delicate interpretation.

The sigma-field $\mathcal{F}_n$ is also generated by the cone $\mathcal{G}_n^+$ of all bounded, nonnegative, $\mathcal{F}_n$-measurable real functions on Ω_n. Write $\mathcal{H}_n^+$ for the corresponding cone of all functions on Ω of the form $h(\omega) = g_n(\omega|n)$, where $\omega|n$ denotes the initial segment $(x_1, \ldots, x_n)$ of ω and $g_n \in \mathcal{G}_n^+$. A routine argument shows that the cone $\mathcal{H}^+ := \cup_{n\in\mathbb{N}}\mathcal{H}_n^+$ generates the product sigma-field $\mathcal{F}$ on Ω. Consistency of the family $\{\mathbb{P}_n : n \in \mathbb{N}\}$ lets us define unambiguously an increasing linear functional $\mathbb{P}$ on $\mathcal{H}^+$ by

<47>
$$\mathbb{P}h := \mathbb{P}_n g_n \qquad \text{if } h(\omega) = g_n(\omega|n).$$

The functional is well defined because condition <46> ensures that $\mathbb{P}_n g_n = \mathbb{P}_{n+1} g_{n+1}$ when $h(\omega) = g_{n+1}(\omega|n+1) = g_n(\omega|n)$, that is, when h depends on only the first n coordinates of ω. Some authors would call $\mathbb{P}$ a finitely additive probability.

From Appendix A.6, the functional $\mathbb{P}$ has an extension to a (countably additive) probability measure on $\mathcal{F}$ if and only if it is ***σ-smooth at* 0**, meaning that $\mathbb{P}h_i \downarrow 0$ for every sequence $\{h_i\}$ in $\mathcal{H}^+$ for which $1 \geq h_i \downarrow 0$ pointwise.

REMARK. The σ-smoothness property requires more than countable additivity of each individual $\mathbb{P}_n$. Indeed, Andersen & Jessen (1948) gave an example of weird spaces $\{\mathcal{X}_i\}$ for which $\mathbb{P}$ is not countably additive (see Problem [29] for details). The failure occurs because a decreasing sequence $\{h_i\}$ in $\mathcal{H}^+$ need not depend on only a fixed, finite set of coordinates. If there were such a finite set, that is, if $h_i(\omega) = g_i(\omega_n)$ for a fixed n, then we would have $\mathbb{P}h_i = \mathbb{P}_n h_i \to 0$ if $h_i \downarrow 0$. In general, h_i might depend on the first n_i coordinates, with $n_i \to \infty$, precluding the reduction to a fixed $\mathbb{P}_n$.

In the literature, sufficient conditions for σ-smoothness generally take one of two forms. The first, due to Daniell (1919), and later rediscovered by Kolmogorov (1933, Section III.4) in slightly greater generality, imposes topological assumptions on the $\mathcal{X}_i$ coordinate spaces and a tightness assumption on the $\mathbb{P}_n$'s. Probabilists who have a distaste for topological solutions to probability puzzles sometimes find the Daniell/Kolmogorov conditions unpalatable, in general, even though those conditions hold automatically for products of real lines.

REMARK. Remember that a probability measure $\mathbb{Q}$ on a topological space $\mathcal{X}$ is called tight if, to each $\epsilon > 0$ there is a compact subset K_ϵ of $\mathcal{X}$ such that $\mathbb{Q}K_\epsilon^c < \epsilon$. If $\mathcal{X} = \mathcal{X}_1 \times \ldots \times \mathcal{X}_n$, then $\mathbb{Q}$ is tight if and only if each of its marginals is tight.

The other sufficient condition, due to Ionescu Tulcea (1949), makes no assumptions about existence of a topology, but instead requires a connection between the $\{\mathbb{P}_n\}$ by means of a family of probability kernels, $\Lambda_{n+1} = \{\lambda_{\omega_n} : \omega_n \in \Omega_n\}$ from Ω_n to $\mathcal{X}_{n+1}$, for each $n \in \mathbb{N}$. Indeed, a sequence of probability measures can be constructed from such a family, starting from an arbitrary probability measure $\mathbb{P}_1$ on $(\mathcal{X}_1, \mathcal{A}_1)$,

<48>
$$\mathbb{P}_n := \mathbb{P}_1 \otimes \Lambda_2 \otimes \Lambda_2 \otimes \ldots \otimes \Lambda_n.$$

More succinctly, $\mathbb{P}_n = \mathbb{P}_{n-1} \otimes \Lambda_n$ for $n \geq 2$. The requirement that each Λ_n be a probability kernel ensures that $\{\mathbb{P}_n : n \in \mathbb{N}\}$ is a consistent family of finite dimensional distributions.

<49> **Theorem.** *Suppose $\{\mathbb{P}_n : n \in \mathbb{N}\}$ be the consistent family of finite dimensional distributions defined via probability kernels, $\mathbb{P}_n = \mathbb{P}_{n-1} \otimes \Lambda_n$ for each n, as in <48>. Then the $\mathbb{P}$ defined by <47> has an extension to a countably additive probability measure on the product sigma-field $\mathcal{F}$.*

Proof. We have only to establish the σ-smoothness at 0. Equivalently, suppose we have a sequence $\{h_i : i \in \mathbb{N}\}$ in $\mathcal{H}^+$ with $1 \geq h_i \downarrow h$, but for which $\inf_i \mathbb{P}h_i > \epsilon$ for some $\epsilon > 0$. We need to show that h is not the zero function. That is, we need to find a point $\bar{\omega} = (\bar{x}_1, \bar{x}_2, \ldots,)$ in Ω for which $h(\bar{\omega}) > 0$.

Construct $\bar{\omega}$ one coordinate at a time. With no loss of generality we may assume h_n depends on only the first n coordinates, that is, $h_n(\omega) = g_n(\omega_n)$, so that

<50> $$\mathbb{P}h_n = \mathbb{P}_n g_n = (\mathbb{P}_1 \otimes \Lambda_2 \otimes \ldots \otimes \Lambda_n)\, g_n(x_1, x_2, \ldots, x_n) > \epsilon$$

The product $\Lambda_2 \otimes \ldots \otimes \Lambda_n$ defines a probability kernel from $\mathcal{X}_1$ to $\mathsf{X}_{2 \leq i \leq n} \mathcal{X}_i$. Define functions f_n on $\mathcal{X}_1$ by

$$f_n(x_1) := \left(\lambda_{x_1} \otimes \Lambda_3 \otimes \ldots \otimes \Lambda_n\right) g_n(x_1, x_2, \ldots, x_n) \qquad \text{for } n \geq 2,$$

with $f_1(x_1) = g_1(x_1)$. Then $\mathbb{P}_1 f_n = \mathbb{P}_n g_n > \epsilon$ for each n. The assumed monotonicity,

<51> $$g_n(\omega|n) = h_n(\omega) \geq h_{n+1}(\omega) = g_{n+1}(\omega|n+1) \qquad \text{for all } \omega,$$

implies that $\{f_n : n \in \mathbb{N}\}$ is a decreasing sequence of measurable functions. By Dominated Convergence, $\epsilon < \mathbb{P}_1 f_n \downarrow \mathbb{P}_1 \inf_{i \in \mathbb{N}} f_i$, from which we may deduce existence of at least one value $\bar{x}_1$ for which $f_n(\bar{x}_1) > \epsilon$ for all n. In particular, $g_1(\bar{x}_1) > \epsilon$.

Hold $\bar{x}_1$ fixed for the rest of the argument. The defining property of $\bar{x}_1$ becomes

$$\left(\lambda_{\bar{x}_1} \otimes \Lambda_3 \otimes \ldots \otimes \Lambda_n\right) g_n(\bar{x}_1, x_2, \ldots, x_n) > \epsilon \qquad \text{for } n \geq 2,$$

Notice the similarity to the final inequality in <50>, with the role of $\mathbb{P}_1$ taken over by $\lambda_{\bar{x}_1}$. Repeating the argument, we find an $\bar{x}_2$ for which

$$\left(\lambda_{(\bar{x}_1, \bar{x}_2)} \otimes \Lambda_4 \otimes \ldots \otimes \Lambda_n\right) g_n(\bar{x}_1, \bar{x}_2, x_3 \ldots, x_n) > \epsilon \qquad \text{for } n \geq 3,$$

and with $g_2(\bar{x}_1, \bar{x}_2) > \epsilon$.

And so on. In this way construct an $\bar{\omega} = (\bar{x}_1, \bar{x}_2, \ldots)$ for which $h_n(\bar{\omega}) = g_n(\bar{\omega}|n) > \epsilon$ for all n. In the limit we have $h(\bar{\omega}) \geq \epsilon$, which ensures that h is not the zero function. Sigma-smoothness at zero, and the asserted countable additivity for $\mathbb{P}$, follow. □

<52> **Corollary.** *For probability measures P_i on arbitrary measure spaces $(\mathcal{X}_i, \mathcal{A}_i)$, there exists a probability measure $\mathbb{P}$ (also denoted by $\otimes_{i \in \mathbb{N}} P_i$) for which $\mathbb{P}(A_1 \times A_2 \times \ldots \times A_k) = \prod_{i \leq k} P_i A_i$, for all measurable rectangles.*

□ *Proof.* Specialize to the case where $\lambda_i(w_{i-1}, \cdot) \equiv P_i$ for all $w_{i-1} \in \Omega_{i-1}$.

The proof for existence of a countably additive $\mathbb{P}$ under topological assumptions is similar, with only a slight change in the definition of $\mathcal{H}^+$.

<53> **Theorem.** *(Daniell/Kolmogorov extension theorem) Suppose $\{\mathbb{P}_n : n \in \mathbb{N}\}$ is a consistent family of finite dimensional distributions. If each $\mathcal{X}_i$ is a separable metric space, equipped with its Borel sigma-field $\mathcal{A}_i$, and if each $\mathbb{P}_n$ is tight, then the $\mathbb{P}$ defined by <47> has an extension to a tight probability measure on the product sigma-field $\mathcal{F}$.*

REMARK. As you will see in Chapter 5, the tightness assumption actually ensures that the $\mathbb{P}_n$ satsify the conditions of Theorem <49>. However, the construction of the probability kernels requires even more topological work than the direct proof of Theorem <53> given below.

Proof. As shown in Section A.6 of Appendix A, we may simplify definition <47> by restricting it to bounded, continuous, nonnegative g_n. Countable additivity of $\mathbb{P}$ is implied by its σ-smoothness at zero for the smaller $\mathcal{H}^+$ class. That is, it suffices to prove that $\mathbb{P}h_n \downarrow 0$ for every sequence $\{h_n : n \in \mathbb{N}\}$ from $\mathcal{H}^+$ that decreases pointwise to the zero function.

As in the proof of Theorem <49>, we may consider a sequence $\{h_i : i \in \mathbb{N}\}$ in $\mathcal{H}^+$ with $1 \geq h_i \downarrow h$, but for which $\inf_i \mathbb{P}h_i > \epsilon$ for some $\epsilon > 0$. Once again, with no loss of generality we may assume h_n depends on only the first n coordinates, that is, $h_n(\omega) = g_n(\omega|n)$; but now the g_n functions are also continuous. Again we need to find a point $\bar{\omega} = (\bar{x}_1, \bar{x}_2, \ldots,)$ in Ω for which $h(\bar{\omega}) > 0$.

The tightness assumption lets us construct compact subsets $K_n \subseteq \Omega_n$ for which $\mathbb{P}_n K_n > 1 - \epsilon/2$, with the added property that $K_n \times \mathcal{X}_{n+1} \supseteq K_{n+1}$, for every n: use the fact that $\sup_K \mathbb{P}_{n+1}\left((K_n \times \mathcal{X}_{n+1}) \cap K\right) = \mathbb{P}_{n+1}\left(K_n \times \mathcal{X}_{n+1}\right) = \mathbb{P}_n K_n$, with the supremum running over all compact subsets of Ω_{n+1}. The construction of K_n ensures that $\mathbb{P}_n g_n(\omega_n)\{\omega_n \in K_n\} > \epsilon/2$, so the compact set $L_n := \{\omega_n \in K_n : g_n(\omega_n) \geq \epsilon/2\}$ is nonempty. Moreover, $L_n \times \mathcal{X}_{n+1} \supseteq L_{n+1}$.

By a Cantor diagonalization argument (Problem [30]), $\cap_{n \in \mathbb{N}} \left(L_n \times \mathsf{X}_{i>n} \mathcal{X}_i\right) \neq \emptyset$. That is, there exists an $\bar{\omega}$ in Ω for which $\bar{\omega}|n \in K_n$ and $h_n(\bar{\omega}) \geq \epsilon/2$ for every n. It follows that $h(\bar{\omega}) \geq \epsilon/2$. The probability measure $\mathbb{P}$ gives mass $\geq 1 - \epsilon/2$ to the set $\cap_{n \in \mathbb{N}} \left(K_n \times \mathsf{X}_{i>n} \mathcal{X}_i\right)$, which Problem [30] shows is compact. □

9. Problems

[1] Let $B_1, B_2, \ldots$ be independent events for which $\sum_{i=1}^{\infty} \mathbb{P}B_i = \infty$. Show that $\mathbb{P}\{B_i \text{ infinitely often }\} = 1$ by following these steps.

(i) Show that $\mathbb{P}\left(B_1^c B_2^c \ldots B_n^c\right) \leq \exp\left(-\sum_{i=1}^{n} \mathbb{P}B_i\right) \to 0$.

(ii) Deduce that $\prod_{i=1}^{\infty} B_i^c = 0$ almost surely.

(iii) Deduce that $\sum_{i=1}^{\infty} B_i \geq 1$ almost surely.

(iv) Deduce that $\sum_{i=m}^{\infty} B_i \geq 1$ almost surely, for each finite m. Hint: The events $B_m, B_{m+1}, \ldots$ are independent.

(v) Complete the proof.

Remark: This result is a converse to the Borel-Cantelli Lemma discussed in Section 2.6. A stronger converse was established in the Problems to Chapter 2.

[2] In Example <26> we needed the function $g(x,y) = py^{p-1}\{f(x) > y > 0\}$ to be product measurable if f is $\mathcal{A}$-measurable. Prove this fact by following these steps.

(i) The map $\phi : \mathbb{R}^2 \to \mathbb{R}$ defined by $\phi(s,t) = \{s > t\}$ is $\mathcal{B}(\mathbb{R}^2)\backslash\mathcal{B}(\mathbb{R})$-measurable. (What do you know about inverse images for ϕ?)

(ii) Prove that the map $(x,y) \mapsto (f(x), y)$ is $\mathcal{A} \otimes \mathcal{B}(\mathbb{R})\backslash\mathcal{B}(\mathbb{R}) \otimes \mathcal{B}(\mathbb{R})$-measurable.

(iii) Show that the composition of measurable functions is measurable, if the various sigma-fields fit together in the right way.

(iv) Complete the argument.

[3] Show that every continuous real function on a topological space $\mathcal{X}$ is $\mathcal{B}(\mathcal{X})\backslash\mathcal{B}(\mathbb{R})$-measurable. Hint: What do you know about inverse images of open sets?

[4] Let $\mathcal{X}$ be a topological space. Say that its topology is ***countably generated*** if there exists a countable class of open sets $\mathcal{G}_0$ such that $G = \bigcup\{G_0 : G \supseteq G_0 \in \mathcal{G}_0\}$ for each open set G. Show that such a $\mathcal{G}_0$ generates the Borel sigma-field on $\mathcal{X}$.

[5] Let $(\mathcal{X}, d)$ be a separable metric space with a countable dense subset $\mathcal{X}_0$. Show that its topology is countably generated. [Lindelöf's theorem.] Hint: Let $\mathcal{G}_0$ be the countable class of all open balls of rational radius centered at a point of $\mathcal{X}_0$. If $x \in G$, find a rational r such that G contains the ball of radius $2r$ centered at x, then find a point of $\mathcal{X}_0$ lying within a distance r of x.

[6] Let $\mathcal{X}$ and $\mathcal{Y}$ be topological spaces equipped with their Borel sigma-fields $\mathcal{B}(\mathcal{X})$ and $\mathcal{B}(\mathcal{Y})$. Equip $\mathcal{X} \times \mathcal{Y}$ with the product topology and its Borel sigma-field $\mathcal{B}(\mathcal{X} \times \mathcal{Y})$. (The open sets in the product space are, by definition, all possible unions of sets $G \times H$, with G open in $\mathcal{X}$ and H open in $\mathcal{Y}$.)

(i) Show that $\mathcal{B}(\mathcal{X}) \otimes \mathcal{B}(\mathcal{Y}) \subseteq \mathcal{B}(\mathcal{X} \times \mathcal{Y})$.

(ii) If both $\mathcal{X}$ and $\mathcal{Y}$ have countably generated topologies, prove equality of the two sigma-fields on the product space.

(iii) Show that $\mathcal{B}(\mathbb{R}^n) = \mathcal{B}(\mathbb{R}^k) \otimes \mathcal{B}(\mathbb{R}^{n-k})$.

[7] Let $\mathcal{X}$ be a set with cardinality greater than $2^{\aleph_0}$, the cardinality of the set of all sequences of 0's and 1's. Equip $\mathcal{X}$ with the trivial topology for which all subsets are open.

(i) Show that the Borel sigma-field $\mathcal{B}(\mathcal{X} \times \mathcal{X})$ consists of all subsets of $\mathcal{X} \times \mathcal{X}$.

(ii) Show that $\mathcal{B}(\mathcal{X}) \otimes \mathcal{B}(\mathcal{X})$ equals

$$\bigcup\{\sigma(\mathcal{E}) : \mathcal{E} \text{ a countable class of measurable rectangles}\}.$$

(iii) For a given countable class of subsets $\mathcal{C}$ of $\mathcal{X}$ define an equivalence relation $x \sim y$ if and only if $\{x \in C\} = \{y \in C\}$ for all C in $\mathcal{C}$. Show that there are at most $2^{\aleph_0}$ equivalence classes. Deduce that there exists at least one pair of distinct points x_0 and y_0 such that $x_0 \sim y_0$. Deduce that $\sigma(\mathcal{C})$ cannot separate the pair (x_0, y_0).

(iv) Show that the diagonal $\Delta = \{(x,y) \in \mathcal{X} \times \mathcal{X} : x = y\}$ cannot belong to the product sigma-field $\mathcal{B}(\mathcal{X}) \otimes \mathcal{B}(\mathcal{X})$, which is therefore a proper sub-sigma-field of $\mathcal{B}(\mathcal{X}\times\mathcal{X})$. Hint: Suppose $\Delta \in \sigma(\mathcal{E})$ for some countable class $\mathcal{E}$ of measurable rectangles. Find a pair of distinct points x_0, y_0 such that no member of $\mathcal{E}$, or of $\sigma(\mathcal{E})$, can extract a proper subset of $F = \{(x_0, x_0), (y_0, x_0), (x_0, y_0), (y_0, y_0)\}$, but $F\Delta = \{(x_0, x_0), (y_0, x_0)\}$.

[8] Prove Theorem <9> by following these steps.

(i) For fixed $E_2, \ldots, E_n$ in $\mathcal{E}_2, \ldots, \mathcal{E}_n$, define $\mathcal{D}_1$ as the class of all events D for which $\mathbb{P}(DE_2 \ldots E_n) = (\mathbb{P}D)(\mathbb{P}E_2)\ldots(\mathbb{P}E_n)$. Show that $\mathcal{D}_1$ is a λ-class.

(ii) Deduce that $\mathcal{D}_1 \supseteq \sigma(\mathcal{E}_1)$. That is, that the analog of the hypothesized factorization with $\mathcal{E}_1$ replaced by $\sigma(\mathcal{E}_1)$ also holds.

(iii) Argue similarly that each subsequent $\mathcal{E}_i$ can also be replaced by its $\sigma(\mathcal{E}_i)$.

[9] Let $\bar{Z}_n = (Z_1 + \ldots + Z_n)/n$, for a sequence $\{Z_i\}$ of independent random variables.

(i) Use the Kolmogorov zero-one law (Example <12>) to show that the set $\{\limsup \bar{Z}_n > r\}$ is a tail event for each constant r, and hence it has probability either zero or one. Deduce that $\limsup \bar{Z}_n = c_0$ almost surely, for some constant c_0 (possibly $\pm\infty$).

(ii) If $\bar{Z}_n$ converges to a finite limit (possibly random) at each ω in a set A with $\mathbb{P}A > 0$, show that in fact there must exists a finite constant c_0 for which $\bar{Z}_n \to c_0$ almost surely.

[10] Let X and Y be independent, real-valued random variables for which $\mathbb{P}(XY)$ is well defined, that is, either $\mathbb{P}(XY)^+ < \infty$ or $\mathbb{P}(XY)^- < \infty$. Suppose neither X nor Y is degenerate (equal to zero almost surely). Show that both $\mathbb{P}X$ and $\mathbb{P}Y$ are well defined and $\mathbb{P}(XY) = (\mathbb{P}X)(\mathbb{P}Y)$. Hint: What would you learn from $\infty > \mathbb{P}(XY)^+ = \mathbb{P}X^+Y^+ + \mathbb{P}X^-\mathbb{P}Y^-$?

[11] For sigma-finite measures μ and λ, on sigma-fields $\mathcal{A}$ and $\mathcal{B}$, show that the only measure Γ on $\mathcal{A}\otimes\mathcal{B}$ for which $\Gamma(A \times B) = (\mu A)(\lambda B)$ for all $A \in \mathcal{A}$ and $B \in \mathcal{B}$ is the product measure $\mu \otimes \lambda$. Hint: Use the π–λ theorem when both measures are finite. Extend to the sigma-finite case by breaking the underlying product space into a countable union of measurable rectangles, $A_i \times B_j$, with $\mu A_i < \infty$ and $\lambda B_j < \infty$, for all i, j.

[12] Let $\mathfrak{m}$ denote Lebesgue measure and λ denote counting measure (that is, λA equals the number of points in A, possibly infinite), both on $\mathcal{B}(\mathbb{R})$. Let $f(x, y) = \{x = y\}$. Show that $\mathfrak{m}^x\lambda^y f(x, y) = \infty$ but $\lambda^y\mathfrak{m}^x f(x, y) = 0$. Why does the Tonelli Theorem not apply?

[13] Let λ and μ both denote counting measure on the sigma-field of all subsets of $\mathbb{N}$. Let $f(x, y) := \{y = x\} - \{y = x + 1\}$. Show that $\mu^x\lambda^y f(x, y) = 0$ but $\lambda^y\mu^x f(x, y) = 1$. Why does the Fubini Theorem not apply?

[14] For nonnegative random variables $Z_1, \cdots, Z_m$, show that $\mathbb{P}(Z_1Z_2\cdots Z_m)$ is no greater than $\int_0^1 Q_{Z_1}(u)Q_{Z_2}(u)\cdots Q_{Z_m}(u)du$, with Q_{Z_i} the quantile function for Z_i.

[15] Let $\mathfrak{m}$ denote Lebesgue measure on $\mathbb{R}^k$. For each fixed α in $\mathbb{R}^k$, define $\mathfrak{m}_\alpha f = \mathfrak{m}^x f(x + \alpha)$. Check that $\mathfrak{m}_\alpha$ is a measure on $\mathcal{B}(\mathbb{R}^k)$ for which $\mathfrak{m}_\alpha[a, b] = \mathfrak{m}[a, b]$,

for each rectangle $[a,b] = \times_i [a_i, b_i]$. Deduce that $\mathfrak{m}_\alpha = \mathfrak{m}$. That is, $\mathfrak{m}$ is invariant under translation: $\mathfrak{m}_\alpha f = \int f(x+\alpha)\,dx = \int f(x)\,dx = \mathfrak{m} f$.

[16] Let μ and ν be finite measures on $\mathcal{B}(\mathbb{R})$. Define distribution functions $F(t) := \mu(-\infty, t]$ and $G(t) := \nu(-\infty, t]$.

(i) Show that there are at most countably many points x, atoms, for which $\mu\{x\} > 0$.

(ii) Show that
$$\mu^t G(t) + \nu^t F(t) = \mu(\mathbb{R})\nu(\mathbb{R}) + \sum\nolimits_i \mu\{x_i\}\nu\{x_i\},$$
where $\{x_i : i \in \mathbb{N}\}$ contains all the atoms for both measures.

(iii) Explain how (ii) is related to the integration-by-parts formula:
$$\int F(t)\frac{dG(t)}{dt}\,dt = F(\infty)G(\infty) - \int G(t)\frac{dF(t)}{dt}\,dt$$
Hint: Read Section 3.4.

[17] In the notation of Example <32>, show that $\mathbb{P}\Psi(|X-Y|) \ge \widetilde{\mathbb{P}}\Psi(|\widetilde{X}-\widetilde{Y}|)$, for every convex, increasing function Ψ on $\mathbb{R}^+$. Hint: Use the fact that $\Psi(x) = \Psi(0) + \int_0^x H(t)\,dt$, where H is an increasing right-continuous function on $\mathbb{R}^+$. Represent $H(t)$ as $\mu(0,t]$ for some measure μ. Consider $\mathbb{P}\mathfrak{m}^s\mu^t\{0 < t \le s \le \Delta\}$, remembering inequality <35>.

[18] Let $\mathbb{P} = \otimes_{i\le n} P_i$ and $\mathbb{Q} = \otimes_{i\le n} Q_i$, where P_i and Q_i are defined on the same sigma-field $\mathcal{A}_i$. Show that
$$H^2(\mathbb{P},\mathbb{Q}) = 2 - 2\prod\nolimits_{i\le n}\left(1 - \tfrac{1}{2}H^2(P_i,Q_i)\right) \le \sum\nolimits_{i\le n} H^2(P_i,Q_i),$$
where H^2 denotes squared Hellinger distance, as defined in Section 3.3. Hint: For the equality, factorize Hellinger affinities calculated using dominating measures $\lambda_i = (P_i+Q_i)/2$. For the inequality, establish the identity $\sum_{i\le n} y_i + \prod_{i\le n}(1-y_i) \ge 1$ for all $0 \le y_i \le 1$.

[19] (One-sided version of <38>) Let $Z_1, \ldots, Z_N$ be independent random variables, and ϵ_1, ϵ_2, and β be nonnegative constants for which $\mathbb{P}\{Z_i + Z_{i+1} + \ldots + Z_N \ge -\epsilon_2\} \ge 1/\beta$ for each i. Show that $\mathbb{P}\left\{\max_{i\le N}(Z_1 + \ldots + Z_i) > \epsilon_1 + \epsilon_2\right\} \le \beta\mathbb{P}\{Z_1 + \ldots + Z_N > \epsilon_1\}$.

[20] Let $X_1, X_2, \ldots$ be independent, identically distributed, random variables with $\mathbb{P}X_i^+ = \infty > \mathbb{P}X_i^-$. Show that $\sum_{i\le n} X_i/n \to \infty$ almost surely. Hint: Apply Theorem <3> to the random variables $\{X_i^-\}$ and $\{m \wedge X_i^+\}$, for positive constants m. Note that $\mathbb{P}(m \wedge X_1^+) \to \infty$ as $m \to \infty$.

[21] Let $X_1, X_2, \ldots$ be independent, identically distributed, random variables with $\mathbb{P}|X_i| = \infty$. Let $S_n = X_1 + \ldots + X_n$. Show that S_n/n cannot converge almost surely to a finite value. Hint: If $S_n/n \to c$, show that $(S_{n+1} - S_n)/n \to 0$ almost surely. Deduce from Problem [1] that $\sum_n \mathbb{P}\{|X_n| \ge n\} < \infty$. Argue for a contradiction by showing that $\mathbb{P}|X_1| \le 1 + \sum_{n=1}^\infty \mathbb{P}\{|X_n| \ge n\}$.

[22] (Kronecker's lemma) Let $\{b_i\}$ and $\{x_i\}$ be sequences of real numbers for which $0 < b_1 \le b_2 \le \ldots \to \infty$ and $\sum_{i=1}^\infty x_i$ is convergent (with a finite limit). Show that $\sum_{i=1}^n b_i x_i / b_n \to 0$ as $n \to \infty$, by following these steps.

(i) Express b_i as $\alpha_1 + \ldots + \alpha_i$, for a sequence $\{\alpha_i\}$ of nonnegative numbers. By a change in the order of summation, show, for $m < n$, that $\sum_{i=1}^n b_i x_i$ equals

$$C_m + \sum_{i,j}\{1 \le j \le n\}\alpha_j\{\max(m,j) \le i \le n\}x_i \qquad \text{where } C_m = \sum_{i=1}^{m-1} b_i x_i.$$

(ii) Given $\epsilon > 0$, find an m such that $\left|\sum_{i=p}^n x_i\right| < \epsilon$ whenever $m \le p \le n$.

(iii) With m as in (ii), and $n \ge m$, show that $\left|\sum_{i=1}^n b_i x_i\right| \le |C_m| + \epsilon \sum_{j=1}^n \alpha_j$.

(iv) Deduce the asserted convergence.

[23] Let $Y_1, Y_2, \ldots$ be independent, identically distributed random variables, with $\mathbb{P}|Y_1|^\alpha < \infty$ for some fixed α with $0 < \alpha < 1$. Define $\beta = 1/\alpha$ and $Z_i = Y_i\{|Y_i|^\alpha \le i\}$.

(i) Show that $\sum_{i=1}^\infty \mathbb{P}Z_i^2/i^{2\beta} < \infty$ and $\sum_{i=1}^\infty \mathbb{P}|Z_i|/i^\beta < \infty$.

(ii) Deduce that $\sum_{i=1}^\infty Z_i/i^\beta$ is convergent almost surely.

(iii) Show that $\sum_{i=1}^\infty \mathbb{P}\{|Y_i|^\alpha > i\} < \infty$.

(iv) Deduce that $\sum_{i=1}^\infty Y_i/i^\beta$ is convergent almost surely.

(v) Deduce via Kronecker's Lemma (Problem [22]) that $n^{-1/\alpha}(Y_1 + \ldots + Y_n)$ converges to 0 almost surely.

[24] Let $S_n = X_1 + \ldots + X_n$, a sum of independent, identically distributed random variables with $\mathbb{P}X_1^{2k} < \infty$ for some positive integer k and $\mathbb{P}X_i = 0$. Show that $\mathbb{P}S_n^{2k} = O(n^k)$ as $n \to \infty$.

[25] Establish inequality <44>. Hint: First show that $\sum_{i=k}^\infty i^{-2}$ decreases like k^{-1}, for $k = 2, 3, \ldots$, by comparing with $\int_{k-1}^\infty y^{-2}\,dy$. Then convert to a continuous range. It helps to consider the case $|x| < 2$ separately, to avoid inadvertent division by zero.

[26] (Etemadi 1981) Let $X_1, X_2, \ldots$ be independent, integrable random variables, with common expected value μ. Give another proof of the SLLN by following these steps. As in Section 7, define $Y_i = X_i\{|X_i| \le i\}$ and $\mu_i = \mathbb{P}Y_i$. Let $T_n = \sum_{i \le n} Y_i$.

(i) Argue that there is no loss of generality in assuming $X_i \ge 0$. Hint: Consider positive and negative parts.

(ii) Show that $\mathbb{P}T_n/n \to \mu$ as $n \to \infty$ and $(S_n - T_n)/n \to 0$ almost surely.

(iii) Show that $\text{var}(T_n) \le \sum_i \mathbb{P}X_1^2\{X_1 \le i \le n\}$.

(iv) For a fixed $\rho > 1$, let $\{k_n\}$ be an increasing sequence of positive integers such that $k_n/\rho^n \to 1$. Show that $\sum_n \{i \le k_n\}/k_n^2 \le C/i^2$ for each positive integer i, for some constant C.

(v) Use parts (iii) and (iv), and Problem [25], to show that

$$\sum_n \mathbb{P}\{|T_{k_n} - \mathbb{P}T_{k(n)}| > \epsilon k_n\} \le C\epsilon^{-2}\sum_i \mathbb{P}X_1^2\{X_1 \le i\}/i^2 < \infty.$$

Deduce via the Borel-Cantelli lemma that $(T_{k_n} - \mathbb{P}T_{k_n})/k_n \to 0$ almost surely.

(vi) Deduce that $S_{k_n}/k_n \to \mu$ almost surely, as $n \to \infty$.

(vii) For each $\rho' > \rho$, show that

$$\frac{S_{k_n}}{\rho' k_n} \le \frac{S_m}{m} \le \rho' \frac{S_{k_{n+1}}}{k_{n+1}} \qquad \text{for } k_n \le m \le k_{n+1},$$

when n is large enough.

(viii) Deduce that $\limsup S_m/m$ and $\liminf S_m/m$ both lie between μ/ρ and $\mu\rho$, with probability one.

(ix) Cast out a sequence of negligible sets as ρ decreases to 1 to deduce that $S_m/m \to \mu$ almost surely.

[27] Let P be a probability measure on $(\Omega, \mathcal{F})$. Suppose $\mathcal{X}$ is a (possibly nonmeasurable) subset of Ω with outer probability 1, that is, $PF = 1$ for every set with $\mathcal{X} \subseteq F \in \mathcal{F}$.

(i) If F_1 and F_2 are sets in $\mathcal{F}$ for which $F_1\mathcal{X} = F_2\mathcal{X}$, show that $PF_1 = PF_2$.

(ii) Write $\mathcal{A}$ for the collection of all sets of the form $F\mathcal{X}$, with $F \in \mathcal{F}$. Show that $\mathcal{A}$ is a sigma-field (the so-called ***trace*** sigma-field).

(iii) Show that $\mathcal{M}^+(\mathcal{X}, \mathcal{A}) = \{\bar{f}(x)\big|_{\mathcal{X}} : \bar{f} \in \mathcal{M}^+(\Omega, \mathcal{F})\}$.

(iv) Show that $Q(F\mathcal{X}) := PF$ is a well defined (by (i)) probability measure on $\mathcal{A}$.

(v) Show that $Qf = P\bar{f}$, for each $\bar{f} \in \mathcal{M}^+(\Omega, \mathcal{F})$ whose restriction to $\mathcal{X}$ equals f.

[28] Let $\mathfrak{m}$ denote Lebesgue measure on the Borel sigma-field of $[0, 1)$. Define an equivalence relation on $[0, 1)$ by $x \sim y$ if $x - y$ is rational. Let A_0 be any subset containing exactly one point from each equivalence class. Let $\{r_i : i \in \mathbb{N}\}$ be an enumeration for the set of all rational numbers in $(0, 1)$. Write A_i for set of all numbers of the form $x + r_i$, with $x \in A_0$ and addition carried out modulo 1.

(i) Show that the sets A_i, for $i \in \overline{\mathbb{N}}$, are disjoint, and $\cup_{i\geq 0} A_i = [0, 1)$.

(ii) Let D_0 be a Borel measurable subset of A_0. Define D_i as the set of points $x + r_i$, with $x \in D_0$ and addition carried out modulo 1. Show $\mathfrak{m}D_0 = \mathfrak{m}D_i$ for each i, and $1 \geq \sum_{i\geq 0} \mathfrak{m}D_i$. Deduce that $\mathfrak{m}D_0 = 0$.

(iii) Deduce that A_0 cannot be measurable (not even for the completion of the Borel sigma-field), for otherwise $[0, 1)$ would be a countable union of Lebesgue negligible sets.

(iv) Suppose D is a Borel measurable subset of $\cup_{i\leq n} A_i$ for some finite n. Show that $[0, 1)$ contains countably many disjoint translations of D. Deduce that $\mathfrak{m}D = 0$.

(v) Define $\mathcal{X}_n = \cup_{i\geq n} A_i$. Deduce from (iv) that each $\mathcal{X}_n$ has Lebesgue outer measure 1 (that is, $\mathfrak{m}B = 1$ for each Borel set $B \supseteq \mathcal{X}_n$), but $\cap_n \mathcal{X}_n = \emptyset$.

[29] Let $\mathfrak{m}$ denote Lebesgue measure on $[0, 1)$, and let $\{\mathcal{X}_n : n \in \mathbb{N}\}$ be a decreasing sequence of (nonmeasurable) subsets with $\cap_n \mathcal{X}_n = \emptyset$ but with each $\mathcal{X}_n$ having outer measure 1, as in Problem [28]. Write $\mathcal{A}_n$ for the trace of $\mathcal{B} := \mathcal{B}[0, 1)$ on $\mathcal{X}_n$, as in Problem [27]. Write Ω_n for $\mathsf{X}_{i\leq n} \mathcal{X}_i$ and $\mathcal{F}_n$ for $\otimes_{i\leq n} \mathcal{A}_i$, as in Section 8.

(i) Show that each function f in $\mathcal{M}^+(\Omega_n, \mathcal{F}_n)$ is the restriction of a function $\bar{f}$ in $\mathcal{M}^+([0, 1)^n, \mathcal{B}^n)$ to Ω_n.

(ii) Show that $\mathbb{P}_n f := \mathfrak{m}^t \bar{f}(t, t, \ldots, t)$, for $f \in \mathcal{M}^+(\Omega_n, \mathcal{F}_n)$, defines a consistent family of finite dimensional distributions.

(iii) Let g_n denote the indicator function of $\{\omega_n \in \Omega_n : x_1 = x_2 = \ldots = x_n\}$, and let $f_n(\omega) = g_n(\omega|n)$ be the corresponding functions in $\mathcal{H}^+$. Show that $\mathbb{P}_n g_n = 1$ for all n, even though $f_n \downarrow 0$ pointwise.

(iv) Deduce that the functional $\mathbb{P}$, as defined by <47>, is not sigma-smooth at zero.

[30] Let $\Omega = \mathsf{X}_{i\in\mathbb{N}}\mathcal{X}_i$ be a product of metric spaces. Define $\Omega_n = \mathsf{X}_{i\le n}\mathcal{X}_i$ and $S_n = \mathsf{X}_{i\ge n}\mathcal{X}_i$. For each n let K_n be a compact subset of Ω_n, with the property that $H_n := \cap_{i\le n}(K_i \times S_{i+1}) \neq \emptyset$ for each finite n. Write π_i for the projection map from Ω onto Ω_i. Show that $H := \cap_{i\in\mathbb{N}} H_i$ is a nonempty compact subset of Ω, by following these steps. (Remember, for metric spaces, compactness is equivalent to the property that each sequence has a convergent subsequence.)

(i) Let $\{z_n : n \in \mathbb{N}\}$ be a sequence with $z_n \in H_n$. Use compactness of each K_i to find subsequences $\mathbb{N}_1 \supseteq \mathbb{N}_2 \supseteq \mathbb{N}_3 \supseteq \ldots$ of $\mathbb{N}$ for which $y(i) := \lim_{n\in\mathbb{N}_i} \pi_i z_n$ exists, as a point of K_i. Define $\mathbb{N}_\infty$ to be the subsequence whose ith member equals the ith member of $\mathbb{N}_i$. Show that $\lim_{n\in\mathbb{N}_\infty} \pi_i z_n = y(i)$ for every i.

(ii) Show that the first i components of $y(i+1)$ coincide with $y(i)$ for each i. Deduce that there exists a y in Ω for which $\pi_i y = y(i) \in K_i$ for every i. Deduce that $y \in H$, and hence $H \neq \emptyset$.

(iii) Specialize to the case where $z_n \in H$ for every n. Show that there is a subsequence that converges to a point of H.

[31] Let T be an uncountable index set, and let $\Omega = \mathsf{X}_{t\in T}\mathcal{X}_t$ denote the set of all functions $\omega : T \to \cup_{t\in T}\mathcal{X}_t$ with $\omega(t) \in \mathcal{X}_t$ for each t. Let $\mathcal{A}_t$ be a sigma-field on $\mathcal{X}_t$. For each $S \subseteq T$, define $\mathcal{F}_S$ be the smallest sigma-field for which each of the maps $\omega \mapsto \omega(s)$, for $s \in S$, is $\mathcal{F}\backslash\mathcal{A}_S$-measurable. The product sigma-field $\otimes_{t\in T}\mathcal{A}_t$ is defined to equal $\mathcal{F}_T$.

(i) Show that $\mathcal{F} = \cup_S \mathcal{F}_S$, the union running over all countable subsets S of T.

(ii) For each countable $S \subset T$, let $\mathbb{P}_S$ be a probability measure on $\mathcal{F}_S$, with the property that $\mathbb{P}_S$ equals the restriction of $\mathbb{P}_{S'}$ to $\mathcal{F}_S$ if $S \subseteq S'$. Show that $\mathbb{P}F := \mathbb{P}_S F$ for $F \in \mathcal{F}_S$ defines a countably additive probability measure on $\mathcal{F}_T$.

10. Notes

According to the account by Hawkins (1979, pages 154–162), the Fubini result (in a 1907 paper, discussed by Hawkins) was originally stated only for products of Lebesgue measure on bounded intervals. I do not personally know whether the appelation *Tonelli's Theorem* is historically justified, but Dudley (1989, page 113) cited a 1909 paper of Tonelli as correcting an error in the Fubini paper. As noted by Hawkins, Lebesgue also has a claim to being the inventor of the measure theoretic version of the theorem for iterated integrals: in his thesis (pages 44–51 of Lebesgue 1902) he established a form of the theorem for bounded measurable functions defined on a rectangle. He expressed the two-dimensional Lebesgue integral as an iterated inner or outer one-dimensional integral. With modern hindsight, his result essentially contains the Fubini Theorem for Lebesgue measure on intervals, but the reformulation involves some later refinements. Royden (1968, Section 12.4) gave an excellent discussion of the distinctions between the two theorems, Tonelli and Fubini, and the need for something like sigma-finiteness in the Tonelli theorem.

I do not know, in general, whether my definition of sigma-finiteness of a kernel, in the sense of Definition <21>, is equivalent, to the apparently weaker property of sigma-finiteness for each measure λ_x, for $x \in \mathcal{X}$.

The inequality between $\mathcal{L}^2$ norms in Example <32> was noted by Fréchet (1957). He cited earlier works of Salvemini, Bass, and Dall'Aglio (none of which I have seen) containing more specialized forms of the result, based on Fréchet (1951). The $\mathcal{L}^1$ version of the result is also in the literature (see the comments by Dudley 1989, page 342). I do not know whether the general version of the inequality, as in Problem [17], has been stated before.

The Maximal Inequality <38> is usually attributed to a 1939 paper of Ottaviani, which I have not seen. Theorem <39> is due to Kolmogorov (1928). Theorem <3> was stated without proof by Kolmogorov (1933, p 57; English edition p 69), with the remark that the proof had not been published. However, the necessary techniques for the proof were already contained in his earlier papers (Kolmogorov 1928, 1930). Problem [26] presents a slight repackaging of an alternative method of proof due to Etemadi (1981). By splitting summands into positive and negative parts, he was able to greatly simplify the method for handling blocks of partial sums.

Daniell (1919) constructed measures on countable products of bounded subintervals of the real line. Kolmogorov (1933, Section III.4), apparently unaware of Daniell's work, proved the extension theorem for arbitrary products of real lines. As shown by Problem [31], the extension from countable to uncountable products is almost automatic. Theorem <49> is due to Ionescu Tulcea (1949). See Doob (1953, 613–615) or Neveu (1965, Section 5.1) for different arrangements of the proof. Apparently (Doob 1953, p 639) there was quite a history of incorrect assertions before the question of existence of measures on infinite product spaces was settled. Andersen & Jessen (1948), as well as providing a counterexample (Problem [29]) to the general analog of the Kolmogorov extension theorem, also suggested that the Ionescu Tulcea form was more widely known:

> In the terminology of the theory of probability this means, that the case of dependent variables cannot be treated for abstract variables in the same manner as for unrestricted real variables. Professor Doob has kindly pointed out, what was also known to us, that this case may be dealt with along similar lines as the case of independent variables (product measures) when conditional probability measures are supposed to exist. This question will be treated in a forthcoming paper by Doob and Jessen.

References

Andersen, E. S. & Jessen, B. (1948), 'On the introduction of measures in infinite product sets', *Danske Vid. Selsk. Mat.-Fys. Medd.*

Daniell, P. J. (1919), 'Functions of limited variation in an infinite number of dimensions', *Annals of Mathematics (series 2)* **21**, 30–38.

Doob, J. L. (1953), *Stochastic Processes*, Wiley, New York.

Dudley, R. M. (1989), *Real Analysis and Probability*, Wadsworth, Belmont, Calif.

Etemadi, N. (1981), 'An elementary proof of the strong law of large numbers', *Zeitschrift für Wahrscheinlichkeitstheorie und Verwandte Gebiete* **55**, 119–122.

Fréchet, M. (1951), 'Sur les tableaux de corrélation dont les marges sont données', *Annales de l'Université de Lyon* **14**, 53–77.

Fréchet, M. (1957), 'Sur la distance de deux lois de probabilité', *Comptes Rendus de l'Academie des Sciences, Paris, Ser. I Math* **244**, 689–692.

Hawkins, T. (1979), *Lebesgue's Theory of Integration: Its Origins and Development*, second edn, Chelsea, New York.

Ionescu Tulcea, C. T. (1949), 'Mesures dans les espaces produits', *Lincei–Rend. Sc. fis. mat. e nat.* **7**, 208–211.

Kolmogorov, A. (1928), 'Über die Summen durch den Zufall bestimmter unabhängiger Größen', *Mathematische Annalen* **99**, 309–319. Corrections: same journal, volume 102, 1929, pages 484–488.

Kolmogorov, A. (1930), 'Sur la loi forte des grands nombres', *Comptes Rendus de l'Academie des Sciences, Paris* **191**, 910–912.

Kolmogorov, A. N. (1933), *Grundbegriffe der Wahrscheinlichkeitsrechnung*, Springer-Verlag, Berlin. Second English Edition, *Foundations of Probability* 1950, published by Chelsea, New York.

Lebesgue, H. (1902), *Intégrale, longueur, aire.* Doctoral dissertation, submitted to Faculté des Sciences de Paris. Published separately in Ann. Mat. Pura Appl. 7. Included in the first volume of his *Œuvres Scientifiques*, published in 1972 by L'Enseignement Mathématique.

Neveu, J. (1965), *Mathematical Foundations of the Calculus of Probability*, Holden-Day, San Francisco.

Royden, H. L. (1968), *Real Analysis*, second edn, Macmillan, New York.

Wald, A. (1949), 'Note on the consistency of the maximum likelihood estimate', *Annals of Mathematical Statistics* **20**, 595–601.

Chapter 5

Conditioning

SECTION 1 considers the elementary case of conditioning on a map that takes only finitely many different values, as motivation for the general definition.

SECTION 2 defines conditional probability distributions for conditioning on the value of a general measurable map.

SECTION 3 discusses existence of conditional distributions by means of a slightly more general concept, disintegration, which is essential for the understanding of general conditional densities.

SECTION 4 defines conditional densities. It develops the general analog of the elementary formula for a conditional density: (joint density)/(marginal density).

*SECTION *5 illustrates how conditional distributions can be identified by symmetry considerations. The classical Borel paradox is presented as a warning against the misuse of symmetry.*

SECTION 6 discusses the abstract Kolmogorov conditional expectation, explaining why it is natural to take the conditioning information to be a sub-sigma-field.

*SECTION *7 discusses the statistical concept of sufficiency.*

1. Conditional distributions: the elementary case

In introductory probability courses, conditional probabilities of events are defined as ratios, $\mathbb{P}(A \mid B) = \mathbb{P}AB/\mathbb{P}B$, provided $\mathbb{P}B \neq 0$. The division by $\mathbb{P}B$ ensures that $\mathbb{P}(\cdot \mid B)$ is also a probability measure, which puts zero mass outside the set B, that is, $\mathbb{P}(B^c \mid B) = 0$. The conditional expectation of a random variable X is defined as its expectation with respect to $\mathbb{P}(\cdot \mid B)$, or, more succinctly, $\mathbb{P}(X \mid B) = \mathbb{P}(XB)/\mathbb{P}B$. If $\mathbb{P}B = 0$, the conditional probabilities and conditional expectations are either left undefined or are extracted by some heuristic limiting argument. For example, if Y is a random variable with $\mathbb{P}\{Y = y\} = 0$ for each possible value y, one hopes that something like $\mathbb{P}(A \mid Y = y) = \lim_{\delta \to 0} \mathbb{P}(A \mid y \leq Y \leq y + \delta)$ exists and is a probability measure for each fixed y. Rigorous proofs lie well beyond the scope of the typical introductory course.

In applications of conditioning, the definitions get turned around, to derive probabilities and expectations from conditional distributions constructed by appeals to symmetry or modelling assumptions. The typical calculation starts from a partition of the sample space Ω into finitely many disjoint events, such as the sets $\{T = t\}$ where some random variable T takes each of its possible values $1, 2, \ldots, n$. From the probabilities $\mathbb{P}\{T = t\}$ and the conditional distributions $\mathbb{P}(\cdot \mid T = t)$, for each t, one calculates expected values as weighted averages.

<1>
$$\mathbb{P}X = \sum\nolimits_t \mathbb{P}(X\{T = t\}) = \sum\nolimits_t \mathbb{P}(X \mid T = t)\mathbb{P}\{T = t\}.$$

Notice that the weights $\mathbb{Q}\{t\} := \mathbb{P}\{T = t\}$ define a probability measure $\mathbb{Q}$ on the range space $\mathcal{T} = \{1, 2, \ldots, n\}$, the distribution of T under $\mathbb{P}$. (That is $\mathbb{Q} = T\mathbb{P}$.) Also, if there is no ambiguity about the choice of T, it helps notationally to abbreviate the conditional distribution to $\mathbb{P}_t(\cdot)$, writing $\mathbb{P}_t(X)$ instead of $\mathbb{P}(X \mid T = t)$. The probability measure $\mathbb{P}_t$ lives on Ω, with $\mathbb{P}_t\{T \neq t\} = 0$ for each t in $\mathcal{T}$. With these simplifications in notation, formula <1> can be rewritten more concisely as

<2>
$$\mathbb{P}X = \sum\nolimits_t \mathbb{Q}\{t\}\mathbb{P}_t X = \mathbb{Q}^t \mathbb{P}_t^\omega X(\omega),$$

with the interpretation that the probability measure $\mathbb{P}$ is a weighted average of the family of probability measures $\mathcal{P} = \{\mathbb{P}_t : t \in \mathcal{T}\}$. The new formula also has a suggestive interpretation as a two-step method for generating an observation ω from $\mathbb{P}$:

(i) First generate a t from the distribution $\mathbb{Q}$ on $\mathcal{T}$.

(ii) Given the value t from step (i), generate ω from the distribution $\mathbb{P}_t$.

Notice that $\mathbb{P}_t$ concentrates on the set of ω for which $T(\omega) = t$. The value of $T(\omega)$ from step (ii) must therefore equal the t from step (i).

<3> **Example.** Suppose a deck of 26 red and 26 black cards is well shuffled, that is, all 52! permutations of the cards are equally likely. Let A denote the event {top and bottom cards red}, and let T be the map into $\mathcal{T} = \{\text{red, black}\}$ that gives the color of the top card. Then T has distribution $\mathbb{Q}$ given by

$$\mathbb{Q}\{\text{red}\} = \mathbb{P}\{T = \text{red}\} = 1/2 \qquad \text{and} \qquad \mathbb{Q}\{\text{black}\} = \mathbb{P}\{T = \text{black}\} = 1/2.$$

By symmetry, the conditional distribution $\mathbb{P}_t$ gives equal probability to all permutations of the remaining 51 cards. In particular

$$\mathbb{P}_{\text{red}}A = \mathbb{P}\{\text{top and bottom cards red} \mid T = \text{red}\} = 25/51,$$
$$\mathbb{P}_{\text{black}}A = \mathbb{P}\{\text{top and bottom cards red} \mid T = \text{black}\} = 0/51,$$

from which we deduce that

$$\mathbb{P}A = \mathbb{Q}\{\text{red}\}\,(25/51) + \mathbb{Q}\{\text{black}\}\,(0/51) = (25/102)\,.$$

Notice how we were able to assign $\mathbb{P}_t$ probabilities by appeals to symmetry, rather than by a direct calculation of a ratio. □

Section 2 will describe the extension of formula <2> to more general families of conditional distributions $\{\mathbb{P}_t : t \in \mathcal{T}\}$. Unfortunately, for subtle technical reasons (discussed in Appendix F), conditional distributions do not always exist. In such situations we must settle for a weaker concept of conditional expectation, following an approach introduced by Kolmogorov (1933, Chapter 5), as explained in Section 6.

REMARK. I claim that it is usually easier to think in terms of conditional distributions, despite the technical caveats regarding existence. Kolmogorov's abstract conditional expectations can be thought of as rescuing just one of the desirable properties of conditional distributions in situations where the full conditional distribution does not exist. The rescue comes at the cost of some loss in intuitive appeal, and with some undesirable side effects. Not all intuitively desirable properties of conditioning survive the nonexistence of a conditional probability distribution. Section 7 will provide an example, where the abstract approach to conditioning allows some counterintuitive cases to slip through the definition of sufficiency.

In some situations—such as the study of martingales—we need conditional expectations for only a small collection of random variables. In those situations we do not need the full conditional distribution and so the abstract Kolmogorov approach suffices.

2. Conditional distributions: the general case

With some small precautions about negligible sets, the representation of $\mathbb{P}$ as a weighted average of distributions living on the level sets $\{T = t\}$, as in <2>, makes sense in a more general setting.

<4> **Definition.** *Let T be an $\mathcal{F}\backslash\mathcal{B}$-measurable map from a probability space $(\Omega, \mathcal{F}, \mathbb{P})$ into a measurable space $(\mathcal{T}, \mathcal{B})$. Let $\mathbb{Q}$ equal $T\mathbb{P}$, the distribution of T under $\mathbb{P}$. Call a family $\mathcal{P} = \{\mathbb{P}_t : t \in \mathcal{T}\}$ of probability measures on $\mathcal{F}$ the* ***conditional probability distribution*** *of $\mathbb{P}$ given T if*

(i) $\mathbb{P}_t\{T \neq t\} = 0$ for $\mathbb{Q}$ almost all t in $\mathcal{T}$,

(ii) the map $t \mapsto \mathbb{P}_t^\omega f(\omega)$ is $\mathcal{B}$-measurable and $\mathbb{P}^\omega f(\omega) = \mathbb{Q}^t \mathbb{P}_t^\omega f(\omega)$, for each f in $\mathcal{M}^+(\Omega, \mathcal{F})$.

In the language of Chapter 4, the family $\mathcal{P}$ is a probability kernel from $(\mathcal{T}, \mathcal{B})$ to $(\Omega, \mathcal{F})$. The fine print about an exceptional $\mathbb{Q}$-negligible set in (i) protects us against those t not in the range of T; if $\{T = t\}$ were empty then $\mathbb{P}_t$ would have nowhere to live. We could equally well escape embarrassment by allowing $\mathbb{P}_t$ to have total mass not equal to 1 for a $\mathbb{Q}$-negligible set of t.

The Definition errs slightly in referring to *the* conditional probability distribution, as if it were unique. Clearly we could change $\mathbb{P}_t$ on a $\mathbb{Q}$-negligible set of t and still have the two defining properties satisfied. Under mild conditions, that is the extent of the possible nonuniqueness: see Theorem <9>.

REMARK. Many authors work with the slightly weaker concept of a ***regular conditional distribution***, substituting a requirement such as

$$\mathbb{P}^\omega h(T\omega)X(\omega) = \mathbb{Q}^t h(t)\mathbb{P}_t X \qquad \text{for all } h \text{ in } \mathcal{M}^+(\mathcal{T}, \mathcal{B})$$

for the concentration property (i). As you will see in Section 3, the difference between the two concepts comes down to little more than a question of measurability of a particular subset of $\Omega \times \mathcal{T}$.

In some problems, where intuitively obvious candidates for the conditional distributions exist, it is easy to check directly properties (i) and (ii) of the Definition.

<5> **Exercise.** Let $\mathbb{P}$ denote Lebesgue measure on $\mathcal{B}\left([0,1]^2\right)$. Let $T(x,y) = \max(x,y)$. Show that the conditional probability distributions $\{\mathbb{P}_t\}$ for $\mathbb{P}$ given T are uniform on the sets $\{T = t\}$.

SOLUTION: Write m for Lebesgue measure on $\mathcal{B}[0,1]$. For $0 < t \le 1$, formalize the idea of a uniform distribution on the set

$$\{T = t\} = \{(x,t) : 0 < x \le t\} \cup \{(t,y) : 0 < y \le t\}$$

by defining

$$\mathbb{P}_t f = \frac{1}{2t}\mathrm{m}^x\left(f(x,t)\{0 < x \le t\}\right) + \frac{1}{2t}\mathrm{m}^y\left(f(t,y)\{0 < y \le t\}\right).$$

You should check that $\mathbb{P}_t\{T \ne t\} = 0$ by direct substitution. The definition of $\mathbb{P}_t$ for $t = 0$ will not matter, because $\mathbb{P}\{T = 0\} = 0$. Tonelli gives measurability of $t \mapsto \mathbb{P}_t f$ for each f in $\mathcal{M}^+([0,1]^2)$.

The image measure $\mathbb{Q}$ is determined by the values it gives to the generating class of all intervals $[0,t]$,

$$\mathbb{Q}[0,t] = \mathbb{P}\{(x,y) : \max(x,y) \le t\} = \mathbb{P}[0,t]^2 = t^2 \qquad \text{for } 0 \le t \le 1.$$

That is, $\mathbb{Q}$ is the measure that has density $2t$ with respect to m.

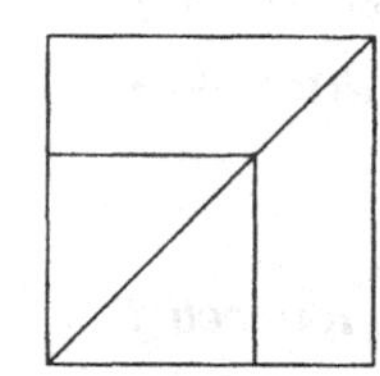

It remains to show that the $\{\mathbb{P}_t\}$ satisfy property (ii) required of a conditional distribution.

$$\begin{aligned}\mathbb{Q}^t\mathbb{P}_t f &= \mathrm{m}^t 2t\left(\frac{1}{2t}\mathrm{m}^x\left(f(x,t)\{0 < x \le t\}\right) + \frac{1}{2t}\mathrm{m}^y\left(f(t,y)\{0 < y \le t\}\right)\right) \\ &= \mathrm{m}^t\mathrm{m}^x\left(f(x,t)\{0 < x \le t\}\right) + \mathrm{m}^t\mathrm{m}^y\left(f(t,y)\{0 < y \le t\}\right).\end{aligned}$$

Replace the dummy variables t, y in the last iterated integral by new dummy variables x, t to see that the sum is just a decomposition of $\mathbb{P}f = \mathrm{m} \otimes \mathrm{m} f$ into contributions from the two triangular regions $\{x \le t\}$ and $\{t \le x\}$. The overlap between the two regions, and the missing edges, all have zero $\mathrm{m} \otimes \mathrm{m}$ measure. □

The decomposition of the uniform distribution on the square, into an average of uniform conditional distributions on the boundaries of subsquares, has an extension to more general sets, but the conditonal distributions need no longer be uniform.

<6> **Example.** Let K be a compact subset of $\mathbb{R}^d$, ***star-shaped*** around the origin. That is, if $y \in K$ and $0 \le t \le 1$ then $ty \in K$.

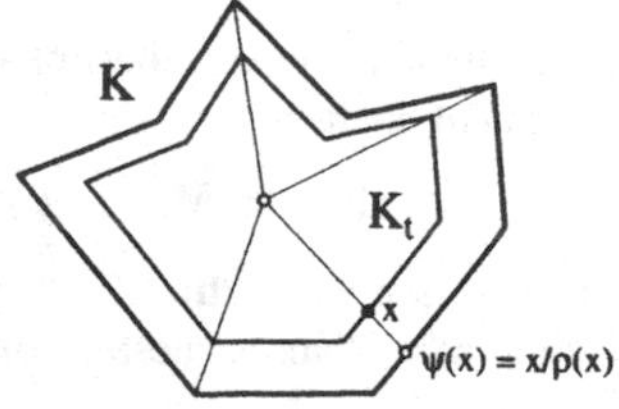

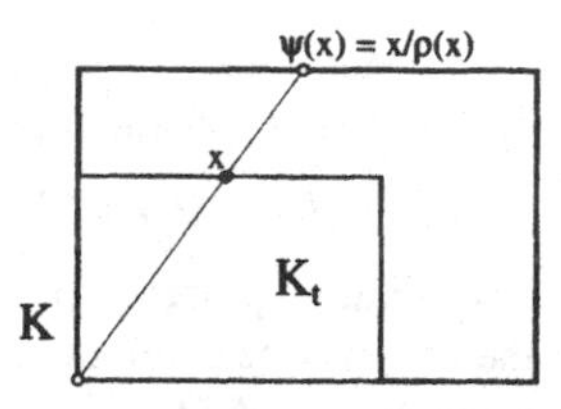

For each $t > 0$, let K_t denote the compact subset $\{ty : y \in K\}$. The sets $\{K_t : t \ge 0\}$ are nested, shrinking to $\{0\}$ as t decreases to zero. The sets define a function $\rho : \mathbb{R}^d \to \mathbb{R}^+$, by $\rho(x) := \inf\{t > 0 : x \in K_t\} = \inf\{t : x/t \in K\}$. In fact, the infimum is achieved for each x in $K\backslash\{0\}$, with $1 \ge \rho(x) > 0$. Only when $x = 0$ do we have $\rho(x) = 0$. The set $\{x : \rho(x) = 1\}$ is a subset of the boundary, ∂K, a proper subset unless 0 lies in the interior of K. For each x in $\mathbb{R}^d\setminus\{0\}$, the point $\psi(x) := x/\rho(x)$ lies in ∂K. Notice that $\psi(x/t) = \psi(x)$ for each $t > 0$.

Let P denote the uniform distribution on K, defined by

$$P^x f(x) = \frac{\mathrm{m}^x\left(f(x)\{x \in K\}\right)}{\mathrm{m}K} \qquad \text{where m is Lebesgue measure on } \mathbb{R}^d.$$

The scaling property of Lebesgue measure under the transformation $x \mapsto x/t$,

$$\mathrm{m}^x f(x/t) = t^d \mathrm{m}^y f(y) \qquad \text{for } f \in \mathcal{M}^+(\mathbb{R}^d) \text{ and } t > 0,$$

will imply independence of $\rho(x)$ and $\psi(x)$ under P. For $0 < t \le 1$ and $g \in \mathcal{M}^+(\mathbb{R}^d)$,

$$\begin{aligned} P^x g(\psi(x))\{\rho(x) \le t\} &= \mathrm{m}^x g(\psi(x/t))\{x/t \in K\}/\mathrm{m}K \\ &= t^d \mathrm{m}^y g(\psi(y))\{y \in K\}/\mathrm{m}K \\ &= t^d P^y g(\psi(y)). \end{aligned}$$

In particular, $P\{\rho(x) \le t\} = t^d$ for $0 < t \le 1$ and

$$P^x g(\psi(x))\{\rho(x) \le t\} = \left(P^x g(\psi(x))\right)\left(P\{\rho(x) \le t\}\right).$$

A generating class argument then leads to the factorization needed for independence.

Write μ for the distribution of $\psi(x)$, a probability measure concentrated on ∂K, and Q for the distribution with density dt^{d-1} with respect to Lebesgue measure on $[0, 1]$. The image of μ under the map $x \mapsto xr$, for a fixed r in $[0, 1]$, defines a probability measure P_r concentrated on the set $\{x : \rho(x) = r\} \subseteq \partial K_r$. The defining equality $x = \rho(x)\psi(x)$ then has the interpretation: if R has distribution Q, independently of Y, which has distribution μ, then RY has distribution P, that is,

$$P^x f(x) = Q^r \mu^y f(ry) = Q^r P_r^x f(x) \qquad \text{for } f \in \mathcal{M}^+(\mathbb{R}^d).$$

The probability kernel $\{P_r : 0 \le r \le 1\}$ is the conditional distribution for P given $\rho(x)$. □

<7> **Example.** Consider $\mathbb{P} := P^n$, the joint distribution for n independent observations $x_1, \dots, x_n$ from P, with P equal to the uniform distribution on a compact, star-shaped subset K of $\mathbb{R}^d$, as in Example <6>. Define $T(x) := \max_{i \le n} \rho(x_i)$. We may generate each x_i from a pair of independent variables, $x_i = r_i y_i$, with $y_i \in \partial K$ having distribution μ, and $r_i \in [0, 1]$ having distribution Q. More formally,

$$\mathbb{P} f(x_1, \dots, x_n) = \mathbb{Q}^r \mathbb{M}^y f(r_1 y_1, \dots, y_n y_n),$$

where $r := (r_1, \dots, r_n)$ and $y := (y_1, \dots, y_n)$, with $\mathbb{M} := \mu^n$, the product measure on $(\partial K)^n$, and $\mathbb{Q} := Q^n$, the product measure on $[0, 1]^n$. Then we have $T = \max_{i \le n} r_i$, with distribution ν for which $\nu[0, t] = (Q[0, t])^n = t^{nd}$ for $0 \le t \le 1$. Thus ν has density ndt^{nd-1} with respect to Lebesgue measure on $[0, 1]$.

Problem [2] shows how to construct the conditional distribution $\mathbb{Q}_t$, for $\mathbb{Q}$ given $T = t$, namely, generate $n - 1$ independent observations $\{s_2, \dots, s_n\}$ from the conditional distribution $Q_t := Q(\cdot \mid y < t)$, then take $(r_1, \dots, r_n)$ as a random permutation of $(t, s_2, \dots, s_n)$. The representation for $\mathbb{P}$ becomes

$$\mathbb{P} f(x_1, \dots, x_n) = \nu^t \mathbb{Q}_t^r \mathbb{M}^y f(r_1 y_1, \dots, r_n y_n).$$

Thus the conditional distribution $\mathbb{P}_t$ equals $\mathbb{Q}_t \otimes \mathbb{M}$, for each t in $[0, 1]$. Less formally, to generate a random sample $x := (x_1, \ldots, x_n)$ from the conditional probability distribution $\mathbb{P}_t$

(i) independently generate w from μ and $y_2, \ldots, y_n$ from P^{n-1}

(ii) take x as a random permutation of $(tw, ty_2, \ldots, ty_n)$.

□

REMARK. The two step construction lets us reduce calculations involving n independent observations from P to the special case where one of the points lies on the boundary ∂K. The conditioning machinery provides a rigorous setting for ***Crofton's theorem*** from geometric probability.

The theorem dates from an 1885 article by M. F. Crofton in the *Encyclopedia Britannica*. As noted by Eisenberg & Sullivan (2000), existing statements and proofs (such as those in Chapter 2 of Kendall & Moran 1963, or Chapter 5 of Solomon 1978) of the theorem are rather on the heuristic side. Eisenberg and Sullivan derived a version of the theorem by means of Kolmogorov conditional expectations, with supplementary smoothness assumptions. If the boundary is smooth, the measure μ is absolutely continuous with respect to surface measure, with a density expressible in terms of quantities familiar from differential geometry (as discussed by Baddeley 1977).

Of course there are many situations where one cannot immediately guess the form of the conditional distributions, and then one must rely on systematic methods such as those to be discussed in Sections 4 and 5.

3. Integration and disintegration

Suppose $\mathbb{P}$ has conditional distribution $\mathcal{P} = \{\mathbb{P}_t : t \in \mathcal{T}\}$ given T, as in Definiton <4>. As shown in Section 4.3, from the probability kernel $\mathcal{P}$ and the distribution $\mathbb{Q}$, it is possible to construct a probability measure $\mathbb{Q} \otimes \mathcal{P}$ on the product sigma-field by means of an iterated integral, $(\mathbb{Q} \otimes \mathcal{P})(g) := \mathbb{Q}^t \left(\mathbb{P}_t^x g(\omega, t)\right)$ for $g \in \mathcal{M}^+(\Omega \times \mathcal{T}, \mathcal{F} \otimes \mathcal{B})$. This measure has marginal distributions $\mathbb{P}$ and $\mathbb{Q}$, as may be seen by restricting g to functions of ω alone or t alone. The concentration property (i) of the Definition ensures that $\mathbb{P}_t^\omega g(\omega, t) = \mathbb{P}_t^\omega g(\omega, T\omega)$ for $\mathbb{Q}$ almost all t, which, by property (ii) of the Definition, leads to

<8>
$$(\mathbb{Q} \otimes \mathcal{P})(g) = \mathbb{Q}^t \left(\mathbb{P}_t^x g(\omega, t)\right) = \mathbb{Q}^t \mathbb{P}_t^\omega g(\omega, T\omega) = \mathbb{P}^\omega g(\omega, T\omega).$$

That is, $\mathbb{Q} \otimes \mathcal{P}$ is the joint distribution of ω and $T\omega$, the image of $\mathbb{P}$ under the map $\gamma : \omega \mapsto (\omega, T\omega)$, which "lifts $\mathbb{P}$ up to live on the graph."

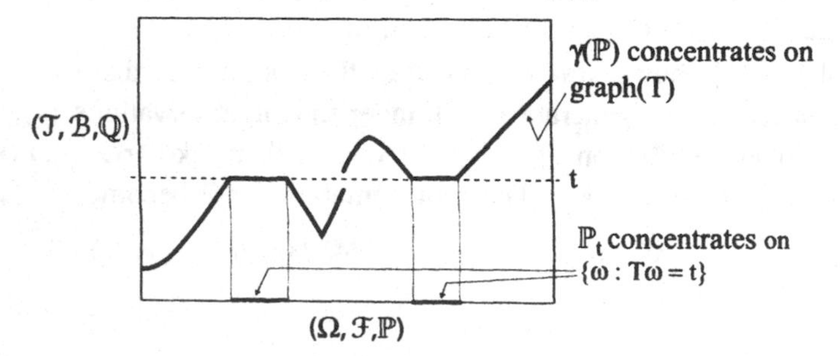

That is, the joint distribution concentrates on the set

$$\text{graph}(T) := \{(\omega, t) \in \Omega \times \mathcal{T} : t = T\omega\},$$

the ***graph*** of the map T. Indeed, provided the graph is a product measurable set, existence of the conditional distribution is equivalent to the representation of the image measure $\gamma(\mathbb{P})$ as $\mathbb{Q} \otimes \mathcal{P}$.

Existence of conditional distributions follows as a special case of a more general decomposition. The following Theorem is proved in Appendix F, where it is also shown that condition (ii) is satisfied by every sigma-finite Borel measure on a complete separable metric space.

<9> **Theorem.** *Let $\mathcal{X}$ be a metric space equipped with its Borel sigma-field $\mathcal{A}$, and let T be a measurable map from $\mathcal{X}$ into a space $\mathcal{T}$, equipped with a sigma-field $\mathcal{B}$. Suppose λ is a sigma-finite measure on $\mathcal{A}$ and μ is a sigma-finite measure on $\mathcal{B}$ dominating the image measure $T(\lambda)$. Suppose:*

(i) the graph of T is $\mathcal{A} \otimes \mathcal{B}$-measurable;

(ii) the measure λ is expressible as a countable sum of finite measures, each with compact support.

Then there exists a kernel $\Lambda = \{\lambda_t : t \in \mathcal{T}\}$ from $(\mathcal{T}, \mathcal{B})$ to $(\mathcal{X}, \mathcal{A})$ for which the image of λ under the map $x \mapsto (x, Tx)$ equals $\mu \otimes \Lambda$, that is,

<10> $$\lambda^x g(x, Tx) = (\mu \otimes \Lambda)(g) := \mu^t \lambda_t^x g(x, t) \qquad \text{for } g \in \mathcal{M}^+(\mathcal{T} \times \mathcal{X}, \mathcal{B} \otimes \mathcal{A}).$$

Moreover, property <10> is equivalent to the two requirements

(iii) $\lambda_t\{T \neq t\} = 0$ for μ almost all t,

(iv) $\lambda^x f(x) = \mu^t \lambda_t^x f(x)$, for each f in $\mathcal{M}^+(\mathcal{X}, \mathcal{A})$,

or to the assertion that $Tx = t$ for $\mu \otimes \Lambda$ almost all (x, t). The kernel Λ is unique up to a μ almost sure equivalence.

REMARK. The uniqueness assertion is quite strong. If $\Lambda = \{\lambda_t : t \in \mathcal{T}\}$ and $\tilde{\Lambda} = \{\tilde{\lambda}_t : t \in \mathcal{T}\}$ are two kernels with the stated properties, it requires $\lambda_t = \tilde{\lambda}_t$, as measures on $\mathcal{A}$, for μ almost all t, and not just that $\lambda_t A = \tilde{\lambda}_t A$ a.e. $[\mu]$, for each A.

A kernel Λ with the properties described by the Theorem is called a (T, μ)-***disintegration*** of λ. The construction of a disintegration is a sort of reverse operation to the "integration" methods used in Section 4.3 to construct measures on product spaces. As you will see in the next Section, disintegrations are essential for the understanding of general conditional densities.

You should not worry too much about the details of Theorem <9>. I have stated it in detail so that you can see how existence of disintegrations involves topological assumptions (or topologically inspired assumptions, as in Pachl 1978). Dellacherie & Meyer (1978, page 78) lamented that "The theorem on disintegration of measures has a bad reputation, and probabilists often try to avoid the use of conditional distributions ... But it really is simple and easy to prove." Perhaps the unpopularity is due, in part, to the role of topology in the proof. Many probabilists seem to regard topology as completely extraneous to any discussion of conditioning, or even to any discussion of abstract probability theory. Nevertheless, it is a sad reality of measure theoretic life that the axioms for countable additivity are not well suited to dealing with uncountable families of negligible sets, and that occasionally they need a little topological help.

REMARK. Topological ideas also come to the rescue of countable additivity in the modern theory of stochastic processes in continuous time. I think it is no accident that the abstract theory of (stochastic) processes has flourished more readily amongst the probabilists who are influenced by the French approach to measure theory, where measures are linear functionals and topological requirements are accepted as natural.

<11> **Example.** Let λ denote Lebesgue measure on $\mathcal{B}(\mathbb{R}^2)$, let X denote the map that projects $\mathbb{R}^2$ onto the first coordinate axis, and let μ denote the one-dimensional Lebesgue measure on that axis. Write λ_x for one-dimension Lebesgue measure transplanted to live on $\{x\} \otimes \mathbb{R}$. That is, λ_x is defined on $\mathcal{B}(\mathcal{R}^2)$ but it concentrates all its mass along a line orthogonal to the first coordinate axis. By the Tonelli Theorem, $\{\lambda_x : x \in \mathbb{R}\}$ is an (X, μ)-disintegration of λ. □

REMARK. Many authors (including Dellacherie & Meyer 1978, Section III-70) require Λ to be a probability kernel, a restriction that excludes interesting and useful cases, such as the decomposition of Lebesgue measure from Example <11>. As shown by Problem [3], Λ can be chosen as a probability kernel if and only if μ is equal to the image measure $T\lambda$. In particular, because the image of two-dimensional Lebesgue measure under a coordinate projection is not a sigma-finite measure, there is no way we could have chosen Λ as a probability kernel in Example <11>.

4. Conditional densities

In introductory courses one learns to calculate conditional densities by dividing marginal densities into joint densities. Such a calculation has a natural generalization for conditional distributions: it is merely a matter of reinterpreting the meanings of the joint and marginal densities.

Recall the elementary construction. Suppose P has density $p(x, y)$ with respect to Lebesgue measure λ on $\mathcal{B}(\mathbb{R}^2)$. Then the X-marginal has density $q(x) := \int p(x, y)\, dy$ with respect to Lebesgue measure μ on the X-axis, and the conditional distribution $P_x(\cdot)$ is given by the conditional density $p(y \mid x) := p(x, y)/q(x)$. Typically one does not worry too hard about how to define conditional distributions when $q(x) = 0$ or $q(x) = \infty$, but maybe one should.

The dy in the definition of $q(x)$ corresponds to the disintegrating Lebesgue measure λ_x from Example <11>. The marginal density $q(x)$ equals $\lambda_x p$. It is the density, with respect to Lebesgue measure μ on $\mathbb{R}$, of the image of P under the coordinate map X. The probability distribution $P_x(\cdot)$ is absolutely continuous with respect to λ_x, with density $p(\cdot \mid x)$ standardized to integrate to 1.

To generalize the elementary formula, suppose a probability measure P is dominated by a sigma-finite measure λ, which has a (T, μ)-disintegration $\Lambda = \{\lambda_t : t \in \mathcal{T}\}$. The density $p(x) := dP/d\lambda$ corresponds to the *joint density*, and $q(t) := \lambda_t^x p(x)$ corresponds to the *marginal density*. If the analogy is appropriate, the ratio $p_t(x) := p(x)/q(t)$ should correspond to the *conditional density*, with λ_t

as the dominating measure. In fact, if we took a little care to avoid $0/0$ or ∞/∞ by inserting an indicator function $\{0 < q < \infty\}$ into the definition, the elementary formula would indeed carry over to the general setting. For some purposes (such as the discussion of sufficiency in Section 7) it is better not to force $p_t(x)$ to be zero when $q(t) = 0$ or $q(t) = \infty$. The wording in part (iii) of the next Theorem is designed to accommodate a slightly more flexible choice for the conditional density.

<12> **Theorem.** *Suppose P is a probability measure with density p with respect to a sigma-finite measure λ that has a (T, μ)-disintegration $\Lambda = \{\lambda_t : t \in \mathcal{T}\}$. Then*

(i) the image measure $Q = TP$ has density $q(t) := \lambda_t p$ with respect to μ;

(ii) the set $\{(x, t) : q(t) = \infty \text{ or } q(t) = 0 < p(x)\}$ has zero $\mu \otimes \Lambda$ measure.

Let $p_t(x)$ be an $\mathcal{A}\otimes\mathcal{B}$-measurable function for which $q(t)p_t(x) = p(x)$ a. e. $[\mu\otimes\Lambda]$. Then

(iii) the P_t defined by $dP_t/d\lambda_t := p_t(\cdot)$ is a probability measure, for Q almost all t, and P has conditional probability distributions $\{P_t : t \in \mathcal{T}\}$ given T.

Proof. For each h in $\mathcal{M}^+(\mathcal{T})$,

$$\begin{aligned} Qh &= Ph(Tx) && \text{image measure} \\ &= \lambda^x(p(x)h(Tx)) && \text{density } p = dP/d\lambda \\ &= \mu^t\left(h(t)\lambda_t^x p(x)\right) && \text{by <10> with } g(x,t) = p(x)h(t). \end{aligned}$$

Thus, Q has density $q(t) := \lambda_t^x p(x)$ with respect to μ.

From the fact that $\mu q = 1$ we have $\mu\{q = \infty\} = 0$. Also, if $q(t) = 0$ for some t then $p(x) = 0$ for λ_t almost all x. Assertion (ii) follows.

For Assertion (iii), first note that P_t concentrates on $\{T = t\}$, for μ almost all t (and hence Q almost all t), because it is absolutely continuous with respect to λ_t. The set $\{t : q(t) = 0 \text{ or } \infty\}$ has zero Q measure. When $0 < q(t) < \infty$ we have $\lambda_t^x p_t(x) = \lambda_t^x p(x)/q(t) = 1$. Also, for $f \in \mathcal{M}^+(\mathcal{X})$,

$$\begin{aligned} Pf &= \lambda^x(p(x)f(x)) && \text{density } p = dP/d\lambda \\ &= \mu^t\lambda_t^x(p(x)f(x)) && \text{disintegration of } \lambda \\ &= \mu^t\lambda_t^x\left(q(t)p_t(x)f(x)\right) && \text{assumption on } p_t \\ &= \mu^t\left(q(t)\lambda_t^x(p_t(x)f(x))\right) = Q^t\left(P_t^x f(x)\right), \end{aligned}$$

□ as required for the conditional distribution.

REMARK. Notice that I did not prove that every P_t is a probability measure. In fact, the best we can assert, as in Problem [3], is that Q-almost all P_t are probability measures. Of course we have no control over $p_t(x)$ when $q(t)$ is zero or infinite. For maximum precision you might like to make appropriate almost sure modifications to Definition <4>.

<13> **Example.** Let $\{\mathbb{P}_\theta : \theta \in \Theta\}$ be a family of probability measures on $\mathcal{X}$. If the map $\theta \mapsto \mathbb{P}_\theta f$ is measurable for $f \in \mathcal{M}^+(\mathcal{X})$, and if π is a probability (a Bayesian ***prior distribution***) on Θ, then $\mathbb{Q} = \pi \otimes \mathcal{P}$ is a probability measure on $\mathcal{X} \otimes \Theta$. The coordinate maps X (onto $\mathcal{X}$) and T (onto Θ) have $\mathbb{Q}$ as their joint distribution. The conditional distribution of X given $T = \theta$ is $\mathbb{P}_\theta$. The conditional distribution $\mathbb{Q}_x(\cdot) = \mathbb{Q}(\cdot \mid X = x)$ is called the Bayesian ***posterior distribution***.

If $\mathbb{P}_\theta$ has a density $p(x,\theta)$ with respect to a sigma-finite μ on $\mathcal{X}$, then

$$\mathbb{Q}f = \pi^\theta \mu^x \left(p(x,\theta) f(x,\theta)\right) \qquad \text{for } f \in \mathcal{M}^+(\mathcal{X} \otimes \Theta).$$

That is, $\mathbb{Q}$ has density $p(x,\theta)$ with respect to the product measure $\mu \otimes \pi$. The product measure has the trivial disintegration $(\mu \otimes \pi)_x = \pi$, if we regard π to be a measure living on $\{x\} \otimes \Theta$. It follows from Theorem <12> that the posterior distribution $\mathbb{Q}_x$ has density $p(x,\theta)/\pi^\theta p(x,\theta)$ with respect to π. Why would a Bayesian probably not worry about the negligible sets where such a ratio is not well defined? □

<14> **Example.** Let P and Q be probability measures on $(\mathcal{X}, \mathcal{A})$, with densities p and q with respect to a sigma-finite measure λ. The Hellinger distance $H(P,Q)$ was defined in Section 3.3 as the $\mathcal{L}^2(\lambda)$ distance between $\sqrt{p}$ and $\sqrt{q}$. Suppose λ has a (T,μ)-disintegration $\Lambda = \{\lambda_t : t \in \mathcal{T}\}$, with T a measurable map from $\mathcal{X}$ into $\mathcal{T}$.

The image measures TP and TQ have densities $\tilde{p}(t) = \lambda_t(p)$ and $\tilde{q}(t) = \lambda_t(q)$ with respect to μ. By the Cauchy-Schwarz inequality, $\left(\lambda_t\sqrt{pq}\right)^2 \le \tilde{p}\tilde{q}$, and hence

$$\lambda\sqrt{pq} = \mu^t \lambda_t^x \sqrt{p(x)q(x)} \le \mu^t \sqrt{\tilde{p}(t)\tilde{q}(t)}.$$

That is, the Hellinger affinity between P and Q is smaller than the Hellinger affinity between TP and TQ. Equivalently, $H^2(TP,TQ) \le H^2(P,Q)$. □

<15> **Example.** Let $\{P_\theta : \theta \in \Theta\}$ be a family of probability measures with densities $p(x,\theta)$ with respect to a sigma-finite measure λ on $(\mathcal{X}, \mathcal{A})$. The method of maximum likelihood promises good properties for the estimator defined by maximizing $p(x,\theta)$ as a function of θ, for each observed x.

If x itself is not known, but instead the value of a measurable function $T(x)$, taking values in some space $(\mathcal{T}, \mathcal{B})$, is observed, the method still applies. If Q_θ, the distribution of T under P_θ, has density $q(t,\theta)$ with respect to some sigma-finite measure μ, the estimate of θ is obtained by maximizing $q(t,\theta)$ as a function of θ, for the observed value $t = T(x)$.

REMARK. Often we can conceive of x as a one-to-one function of some pair of statistics $(S(x), T(x))$. By observing only $T(x)$ we find ourselves working with *incomplete data*.

If direct maximization of $\theta \mapsto q(t,\theta)$ is awkward, there is an alternative iterative method, due to Dempster, Laird & Rubin (1977), which at least offers a way of finding a sequence of θ values with $q(t,\theta)$ increasing at each iteration. Their method is called the ***EM-algorithm***, the ***E*** standing for estimation, the ***M*** for maximization.

Each iteration consists of a two-step procedure to replace an initial guess θ_0 by a new guess θ_1.

(E-step) Calculate $G(\theta) = P_{\theta_0}(\log p(x,\theta) \mid T = t)$.

(M-step) Maximize G, or at least find a θ_1 for which $G(\theta_1) > G(\theta_0)$.

It is easiest to analyze the algorithm when λ has a (T,μ)-disintegration, so that $dQ_\theta/d\mu = q(t,\theta) = \lambda_t^x p(x,\theta)$. Throughout the calculation t is held fixed, so there is no ambiguity in writing ν for the measure λ_t. Similarly, it is cleaner to write π_i

for the density of $P_{\theta_i}(\cdot \mid T = t)$ with respect to ν and q_i instead of $q(t, \theta_i)$, for $i = 1, 2$. Then $p(x, \theta_i) = \pi_i(x) q_i$, and

$$0 < G(\theta_1) - G(\theta_0) = \nu^x \pi_0(x) \log\left(\frac{\pi_1(x) q_1}{\pi_0(x) q_0}\right)$$

which implies $\log(q_1/q_0) > \nu^x \left(\pi_0(x) \log\left(\pi_1(x)/\pi_0(x)\right)\right)$. The last integral defines the Kullback-Leibler distance, which is always nonnegative, by Jensen's inequality. Perhaps more informative is the lower bound $(\nu|\pi_1 - \pi_0|)^2/2$, proved in Section 3.3, which quantifies the improvement in $q(t, \theta)$ from the EM iteration. □

REMARK. I paid no attention to possible division by zero in the calculations leading to the lower bound for $\log(q_1/q_0)$. You might like to provide appropriate integrability assumptions, and insert indicator functions to guard against 0/0. It would also be helpful to ignore the story about missing data and conditioning—as the simplifications in notation should make clear, the EM algorithm works for any $q(\theta)$ expressible as $\nu^x p(x, \theta)$, for any measure ν and any ν-integrable, nonnegative p. Some regularity assumptions are needed to ensure finiteness of the various integrals appearing in the Example.

*5. Invariance

Formal verification of the disintegration property can sometimes be reduced to a symmetry, or invariance, argument, as in the following Exercise.

<16> **Exercise.** Let P be a rotationally symmetric probability measure on $\mathcal{B}(\mathbb{R}^2)$, such as the standard bivariate normal distribution. That is, if S_θ denotes the map for rotation through an angle θ about the origin, then $S_\theta P = P$ for all θ. Define $R(x)$ to equal $(x_1^2 + x_2^2)^{1/2}$, the distance from the point $x = (x_1, x_2)$ to the origin. Show that P has conditional probability distributions P_r uniform on the sets $\{R = r\}$. That is, show that the polar angle $\theta(x)$ is uniformly distributed on $[0, 2\pi)$, independently of R.

SOLUTION: Let Q denote the distribution of R under P. Let λ_r denote the uniform probability measure on $\{R = r\}$, which can be characterized by invariance, $S_\theta \lambda_r = \lambda_r$, for θ in a countable, dense set Θ_0 of rotational angles (Problem [14]). We need to show that $P_r = \lambda_r$, for Q-almost all r.

The problem lies in the transfer of rotational invariance from P to each of the conditional probabilities. Consider a fixed θ in the countable Θ_0 and a fixed f in $\mathcal{M}^+(\mathbb{R}^2)$. Then

$$\begin{aligned} Pf &= (S_\theta P)^y f(y) && \text{invariance of } P \\ &= P^x f(S_\theta x) && \text{image measure} \\ &= Q^r P_r^x f(S_\theta x) && \text{disintegration} \\ &= Q^r (S_\theta P_r)^y f(y) && \text{image measure.} \end{aligned}$$

When P_r concentrates on the set $\{R = r\}$, so does $S_\theta P_r$. It follows that the family $\{S_\theta P_r : r \in \mathbb{R}^+\}$ is another disintegration for P. By uniqueness of disintegrations, there exists a Q-negligible set $\mathcal{N}_\theta$ such that $S_\theta P_r = P_r$ for all r in $\mathcal{N}_\theta^c$. Cast out a sequence of negligible sets to deduce that, for Q almost all r, the probability measure concentrates on $\{R = r\}$ and it is invariant under all Θ_0 rotations, which implies that $P_r = \lambda_r$, as asserted.

Invariance of P_r implies that $\theta(x)$ has the uniform conditional distribution, $\mathfrak{m} :=$ Uniform$[0, 2\pi)$, under P_r. Thus, for $g \in \mathcal{M}^+(\mathbb{R}^+)$ and $h \in \mathcal{M}^+[0, 2\pi)$,

$$P^x g(R(x))h(\theta(x)) = Q^r P_r^x g(r)h\big(\theta(x)\big) = \big(Q^r g(r)\big)\left(\mathfrak{m}^\theta h(\theta)\right).$$

By a generating class argument it follows that the joint distribution of $R(x)$ and $\theta(x)$ equals the product $Q \otimes \mathfrak{m}$ on $\mathcal{B}(\mathbb{R}^+) \otimes \mathcal{B}[0, 2\pi)$, which implies independence. □

REMARK. A similar argument would work for any sigma-finite measure λ that is invariant under a group $\mathbb{G}$ of measurable transformations on $\mathcal{X}$, if the level sets $\{T = t\}$ are also invariant. Let $\mathbb{G}_0$ be a countable subset of $\mathbb{G}$. Then the measures $\{\lambda_t\}$ must be invariant under each transformation in $\mathbb{G}_0$, except possibly for those t in a negligible set that depends on $\mathbb{G}_0$. Often invariance under a suitable $\mathbb{G}_0$ will characterize the $\{\lambda_t\}$ up to multiplicative constants.

Appeals to symmetry can be misleading if not formalized as invariance arguments. The classical Borel paradox, in the next Example, is a case in point.

<17> **Example.** Let x be a point chosen at random (from the uniform distribution P) on the surface of the Earth. Intuitively, if the point lies somewhere along the equator then the longitude should be uniformly distributed over the range $[-180°, 180°]$. But what is so special about the equator? Given that the point lies on any particular great circle, its position should be uniformly distributed around that circle. In particular, for a great circle through the poles (that is, conditional on the longitude) there should be conditional probability 1/4 that the point lies north of latitude 45°N. Average out over the longitude to deduce that the point has probability 1/4 of lying in the spherical cap extending from the north pole down to the 45° parallel of latitude. Unfortunately, that cap does not cover 1/4 of the earth's surface area, as one would require for a point uniformly distributed over the whole surface. That is Borel's paradox.

The apparent paradox does not mean that one cannot argue by symmetry in assigning conditional distributions. As Kolmogorov (1933, page 51) put it, "the concept of a conditional probability with regard to an isolated hypothesis whose probability equals 0 is inadmissible. For we can obtain a probability distribution for [the latitude] on the meridean circle only if we regard this circle as an element of the decomposition of the entire spherical surface into meridian circles with the given poles." In other words, the conditional distributions on circles are not defined unless the disintegration in which they are level sets is specified. Even then, the conditional distributions are only determined up to an almost sure equivalence.

We can, however, argue by invariance for a uniform conditional distribution on almost all circles of constant latitude in their roles as level sets for a latitude disintegration. Suppose T picks off the latitude of the point on the surface. The sets of constant T value are the parallels of constant latitude. Each such parallel is mapped onto itself by rotations about the polar axis; the sets $\{T = t\}$ are invariant under each member of that group $\mathcal{G}$ of rotations. The uniform measure P is also invariant under such rotations. As in Example <16>, it follows that almost all P_t measures must be invariant under $\mathcal{G}$. That is, almost all the conditional probability distributions are uniform around circles of constant latitude. It follows that almost all *great* circles of constant latitude must carry a uniform conditional distribution. Unfortunately, the equator is the only such great circle; it corresponds to a negligible set of T values. Thus there is no harm in assuming that P_0, the conditional distribution around the equator, is uniform, just as there is no harm in assuming it to carry any other conditional distribution. A uniform conditional distribution around the equator is not forced by the T disintegration.

What happens if we change T to pick off the longitude of the random point on the surface? It would be nonsense to argue that the conditional distributions stay the same—the level sets supporting those distributions are not even the same as before. Couldn't we just argue again by invariance? Certainly not. The great circles of constant T are no longer invariant under a group of rotations; the disintegrating measures need no longer be invariant; the conditional distributions around the great circles of constant longitude are not uniform. □

There is a general lesson to be learned from the last Example. Even if almost all level sets $\{T = t\}$ in one disintegration must carry particular distributions, it does not follow that similar sets in a different disintegration must carry the same distributions. It is an even worse error to assume that a conditional distribution that *could* be assigned to a region as a typical level set of one disintegration *must* be the same as the conditional distribution similar regions *must* carry if they appear as level sets of another disintegration.

6. Kolmogorov's abstract conditional expectation

Let $(\Omega, \mathcal{F}, \mathbb{P})$ be a probability space, and T be an $\mathcal{F}\backslash\mathcal{B}$-measurable map into a set $\mathcal{T}$ equipped with a sigma-field $\mathcal{B}$. Write $\mathbb{Q}$ for the image measure $T\mathbb{P}$.

Suppose the conditional distribution $\{\mathbb{P}_t : t \in \mathcal{T}\}$ of $\mathbb{P}$ given T exists. For a fixed X in $\mathcal{M}^+(\Omega, \mathcal{F})$, the conditional expectation $g_X(t) := \mathbb{P}_t^\omega X(\omega)$ is a function in $\mathcal{M}^+(\mathcal{T}, \mathcal{B})$, with the property that $\mathbb{P}_t^\omega (\alpha(T\omega)X(\omega)) = \alpha(t)g_X(t)$ a.e. $[\mathbb{Q}]$, and hence

<18> $$\mathbb{P}^\omega (\alpha(T\omega)X(\omega)) = \mathbb{Q}^t (\alpha(t)g_X(t)) \qquad \text{for each } \alpha \in \mathcal{M}^+(\mathcal{T}, \mathcal{B}).$$

Even when the conditional distribution does not exist, it turns out that it is still possible to find a function $t \mapsto g(t, X)$ satisfying an analog of <18>, for each fixed X. Kolmogorov (1933) suggested that such a function should be interpreted as the conditional expectation of X given $T = t$. Most authors write $\mathbb{E}(X \mid T = t)$ for $g(t, X)$. This abstract conditional expectation has many—but not all—of the properties associated with an expectation with respect to a conditional distribution. Kolmogorov's suggestion has the great merit that it imposes no extra regularity assumptions on the underlying probability spaces and measures but at the cost of greater abstraction.

The properties of the Kolmogorov conditional expectation $X \mapsto g(t, X)$ are analogous to the properties of increasing linear functionals with the Monotone Convergence property, except for omnipresent a.e. $[\mathbb{Q}]$ qualifiers.

<19> **Theorem.** *There is a map $X \mapsto g(t, X)$ from $\mathcal{M}^+(\Omega, \mathcal{F})$ to $\mathcal{M}^+(\mathcal{T}, \mathcal{B})$ with the property that $\mathbb{P}^\omega(\alpha(T\omega)X(\omega)) = \mathbb{Q}^t(\alpha(t)g(t, X))$ for each $\alpha \in \mathcal{M}^+(\mathcal{T}, \mathcal{B})$. For each X, the function $t \mapsto g(t, X)$ is unique up to a $\mathbb{Q}$-equivalence. The map has the following properties.*

(i) *If $X = h(T)$ for some h in $\mathcal{M}^+(\mathcal{T}, \mathcal{B})$ then $g(t, X) = h(t)$ a.e. $[\mathbb{Q}]$.*

(ii) *For X_1, X_2 in $\mathcal{M}^+(\Omega, \mathcal{F})$ and h_1, h_2 in $\mathcal{M}^+(\mathcal{T}, \mathcal{B})$,*

$$g\left(t, h_1(T)X_1 + h_2(T)X_2\right) = h_1(t)g(t, X_1) + h_2(t)g(t, X_2) \qquad \text{a.e. } [\mathbb{Q}].$$

(iii) *If $0 \le X_1 \le X_2$ a.e. $[\mathbb{P}]$ then $g(t, X_1) \le g(t, X_2)$ a.e. $[\mathbb{Q}]$.*

(iv) *If $0 \le X_1 \le X_2 \le \ldots \uparrow X$ a.e. $[\mathbb{P}]$ then $g(t, X_n) \uparrow g(t, X)$ a.e. $[\mathbb{Q}]$.*

To establish the existence and the four properties of the function $g(t, X)$, we must systematically translate to corresponding assertions about averages, by means of the following simple result.

<20> **Lemma.** *Let h_1 and h_2 be functions in $\mathcal{M}^+(\mathcal{T}, \mathcal{B})$ for which*

$$\mathbb{Q}^t(\alpha(t)h_1(t)) \le \mathbb{Q}^t(\alpha(t)h_2(t)) \qquad \text{for all } \alpha \in \mathcal{M}^+(\mathcal{T}, \mathcal{B}).$$

Then $h_1 \le h_2$ a.e. $[\mathbb{Q}]$. If the inequality is replaced by an equality, that is, if $\mathbb{Q}^t(\alpha(t)h_1(t)) = \mathbb{Q}^t(\alpha(t)h_2(t))$ for all α, then $h_1 = h_2$ a.e. $[\mathbb{Q}]$.

Proof. For a positive rational number r, let $A_r := \{t : h_2(t) \le r < h_1(t)\}$. There are no $\infty - \infty$ problems in subtracting to get $0 \le \mathbb{Q}^t\left((h_2(t) - h_1(t))\{t \in A_r\}\right)$, which forces $\mathbb{Q}A_r = 0$ because $h_2 - h_1$ is strictly negative on A_r. Take the union over all positive rational r to deduce that $\mathbb{Q}\{h_2 < h_1\} = 0$. Reverse the roles of h_1 and h_2 to get the companion assertion, $\mathbb{Q}\{h_1 < h_2\} = 0$, for the case where the inequality is replaced by an equality. □

Proof of Theorem <19>. As before, to simplify notation, abbreviate $\mathcal{M}^+(\Omega, \mathcal{F})$ to $\mathcal{M}^+(\Omega)$ and $\mathcal{M}^+(\mathcal{T}, \mathcal{B})$ to $\mathcal{M}^+(\mathcal{T})$.

Fix an X in $\mathcal{M}^+(\Omega)$. For each $n \in \mathbb{N}$, define an increasing linear functional ν_n on $\mathcal{M}^+(\mathcal{T})$ by $\nu_n(\alpha) := \mathbb{P}^\omega((X(\omega) \wedge n)\alpha(T\omega))$ for $\alpha \in \mathcal{M}^+(\mathcal{T})$. It is easy to check that $\nu_n(\alpha) \le n\mathbb{Q}(\alpha)$ and that ν_n has the Monotone Convergence property. Thus each ν_n corresponds to a finite measure on $\mathcal{B}$, absolutely continuous with respect to $\mathbb{Q}$. By the Radon-Nikodym Theorem (Section 3.1), there exist bounded densities γ_n in $\mathcal{M}^+(\mathcal{T})$ for which $\nu_n\alpha = \mathbb{Q}(\gamma_n\alpha)$ for each α in $\mathcal{M}^+(\mathcal{T})$. By Lemma <20>, the inequality $\mathbb{Q}(\gamma_n\alpha) = \nu_n\alpha \le \nu_{n+1}\alpha = \mathbb{Q}(\gamma_{n+1}\alpha)$ for all $\alpha \in \mathcal{M}^+(\mathcal{T}, \mathcal{B})$ implies that $g_n \le g_{n+1}$ a.e. $[\mathbb{Q}]$. That is, $\{\gamma_n\}$ is increasing a.e. $[\mathbb{Q}]$ to $\gamma := \limsup_n \gamma_n \in \mathcal{M}^+(\mathcal{T})$. Two appeals to Monotone Convergence then give the desired equality,

$$\mathbb{P}^\omega(\alpha(T\omega)X(\omega)) = \lim_{n\to\infty} \nu_n^t\alpha(t) = \lim_{n\to\infty} \mathbb{Q}^t(\gamma_n(t)\alpha(t)) = \mathbb{Q}^t(\gamma(t)\alpha(t)),$$

for $\alpha \in \mathcal{M}^+(\mathcal{T})$. The uniqueness of γ up to $\mathbb{Q}$-equivalence follows directly from Lemma <20>. Arbitrarily choose from the $\mathbb{Q}$-equivalence class of all possible γ's one member, and call it $g(t, X)$.

Again by Lemma <20>, the first three assertions are equivalent to the relationships, for all $\alpha \in \mathcal{M}^+(\mathcal{T}, \mathcal{B})$,

(i) $\mathbb{Q}\big(\alpha(t)g(t, X)\big) = \mathbb{Q}\big(\alpha(t)h(t)\big)$,

(ii) $\mathbb{Q}\big(\alpha(t)g(t, h_1(T)X_1 + h_2(T)X_2)\big) = \mathbb{Q}\big(\alpha(t)h_1(t)g(t, X_1) + h_2(t)g(t, X_2)\big)$,

(iii) $\mathbb{Q}\big(\alpha(t)g(t, X_1)\big) \le \mathbb{Q}\big(\alpha(t)g(t, X_2)\big)$.

Systematically replace all expressions like $\mathbb{Q}\big(F(t)g(t, Z)\big)$ by the corresponding $\mathbb{P}\big(F(T)Z\big)$, and expressions like $\mathbb{Q}G(t)$ by the corresponding $\mathbb{P}G(T)$, to get further equivalent relationships,

(i) $\mathbb{P}\big(\alpha(T)X\big) = \mathbb{P}\big(\alpha(T)h(T)\big)$,

(ii) $\mathbb{P}\big(\alpha(T)(h_1(T)X_1 + h_2(T)X_2)\big) = \mathbb{P}\big(\alpha(T)h_1(T)X_1 + h_2(T)X_2\big)$,

(iii) $\mathbb{P}\big(\alpha(T)X_1\big) \le \mathbb{P}\big(\alpha(T)X_2\big)$.

The first three assertions follow.

For the fourth assertion, note that (iii) implies $g(t, X_n) \uparrow \gamma(t) := \limsup_n g(t, X_n)$ a.e. $[\mathbb{Q}]$. Then apply Monotone Convergence twice, with α in $\mathcal{M}^+(\mathcal{T})$, to get

$$\mathbb{Q}\big(\alpha(t)\gamma(t)\big) = \lim_n \mathbb{Q}\big(\alpha(t)g(t, X_n)\big) = \lim_n \mathbb{P}\big(\alpha(T)X_n\big) = \mathbb{P}\big(\alpha(T)X\big) = \mathbb{Q}\big(\alpha(t)g(t, X)\big),$$

□ from which (iv) follows via Lemma <20>.

At the risk of misinterpreting the symbol as an expectation of X with respect to a conditional probabilty measure $\mathbb{P}(\cdot \mid T = t)$, I will write $\mathbb{P}(X \mid T = t)$ instead of $g(t, X)$ for the Kolmogorov conditional expectation, by analogy with the traditional $\mathbb{E}(X \mid T = t)$.

If it were possible to combine the negligible sets corresponding to the a.e. $[\mathbb{Q}]$ qualifiers on each of the assertions (ii), (iii), and (iv) of the Theorem into a single $\mathbb{Q}$-negligible set $\mathcal{N}$, the maps $\{g(t, \cdot) : t \notin \mathcal{N}\}$ would define a family of increasing linear functional on $\mathcal{M}^+(\Omega, \mathcal{F})$, each with the Monotone Convergence property. Moreover, (ii) would even allow us to treat functions of T as constants when conditioning on T. Such functionals would then define the conditional distribution as a family of probability measures on $\mathcal{F}$. Unfortunately, we accumulate uncountably many negligible sets as we cycle (ii)—(iv) through all the required combinations of functions in $\mathcal{M}^+(\mathcal{X})$ and positive constants. If we could somehow reduce (ii)—(iv) to a countable collection of requirements, each involving the exclusion of a single μ-negligible set, then the task of constructing the $\mathcal{N}$ would become easy. The topological assumptions in Theorem <9> are used precisely to achieve this effect.

REMARK. The accumulation of negligible sets causes no difficulty when we deal with only countably many random variables X. In such circumstances, the Kolmogorov conditional expectation effectively has the properties of an expectation with respect to a conditional distribution.

For general $\mathbb{P}$-integrable X, define

$$\mathbb{P}(X \mid T = t) := \mathbb{P}(X^+ \mid T = t) - \mathbb{P}(X^- \mid T = t),$$

with the conditional expectations for positive and negative parts of X defined as in Theorem <19>. Notice that both those conditional expectations can be taken (almost) everywhere finite, because they have finite expectations: choose $\alpha \equiv 1$ in defining equality to get $\mathbb{P}\big(\mathbb{P}(X^\pm \mid T = t)\big) = \mathbb{P}X^\pm < \infty$. To avoid any problem with infinite expectations, we could also restrict the α to be a *bounded* $\mathcal{G}$-measurable random variable, and define $\mathbb{P}(X \mid T = t)$ for integrable X as the $\mathbb{Q}$-integrable random variable $g(t, X)$ for which

<21> $$\mathbb{P}(\alpha(T)X) = \mathbb{Q}(\alpha(t)g(t, X)) \qquad \text{for all bounded, } \mathcal{G}\text{-measurable } \alpha$$

Brave readers might also want to dabble with conditional expectation for other cases where only one of $\mathbb{P}X^+$ or $\mathbb{P}X^-$ is finite. Beware of conditional $\infty - \infty$, whatever that might mean!

There are conditional analogs of the Fatou Lemma (Problem [7]), Dominated Convergence (Problem [8]), Jensen's inequality (Problem [13]), and so on. The derivations are essentially the same as for ordinary expectations, because the accumulation of countably many $\mathbb{Q}$-negligible sets causes no difficulty when we deal with only countably many random variables X.

Conditioning on a sub-sigma-field

The choice of $\mathbb{Q}$ as the image measure lets us rewrite $\mathbb{Q}^t(\alpha(t)g(t, X))$ as $\mathbb{P}(\alpha(T)g(T, X))$. Both $g(T, X)$ and $\alpha(T)$ are $\sigma(T)$-measurable functions on Ω. Indeed (recall the Problems to Chapter 2), every $\sigma(T)\backslash\mathcal{B}[0, \infty]$-measurable map into $[0, \infty]$ must be of the form $\alpha(T)$ for some α in $\mathcal{M}^+(\mathcal{T}, \mathcal{B})$. The defining property may be recast as: to each X in $\mathcal{M}^+(\Omega, \mathcal{F})$ there exists an $X_T := g(T, X)$ in $\mathcal{M}^+(\Omega, \sigma(T))$ for which

<22> $$\mathbb{P}(WX) = \mathbb{P}(WX_T) \qquad \text{for each } W \text{ in } \mathcal{M}^+(\Omega, \sigma(T)).$$

The random variable X_T, which is also denoted by $\mathbb{P}(X \mid T)$ (or $\mathbb{E}(X \mid T)$, in traditional notation), is unique up to a $\mathbb{P}$ almost sure equivalence.

REMARK. Be sure that you understand the distinction between the functions $g(t) := g(t, X)$ and $g(T) = g \circ T$; they live on different spaces, $\mathcal{T}$ and Ω. If we write $\mathbb{P}(X \mid T = t)$ for $g(t)$ it is tempting to write the nonsensical $\mathbb{P}(X \mid T = T)$ for $g(T)$, instead of $\mathbb{P}(X \mid T)$, or even $\mathbb{P}\big(X \mid \sigma(T)\big)$ (see below).

The conditional expectation $\mathbb{P}(X \mid T)$ depends on T only through the sigma-field $\sigma(T)$. If S were another random element for which $\sigma(T) = \sigma(S)$, the conditional expectation $\mathbb{P}(X \mid S)$ would be defined by the same requirement <22>. That is, except for the unavoidable nonuniqueness within a $\mathbb{P}$-equivalence class of random variables, we have $\mathbb{P}(X \mid T) = \mathbb{P}(X \mid S)$. We could regard the conditioning information as coming from a sub-sigma-field of $\mathcal{F}$ rather than from a T that generates the sub-sigma-field.

<23> **Definition.** *Let X belong to $\mathcal{M}^+(\Omega, \mathcal{F})$ and $\mathbb{P}$ be a probability measure on $\mathcal{F}$. For each sub-sigma-field $\mathcal{G}$ of $\mathcal{F}$, the conditional expectation $\mathbb{P}(X \mid \mathcal{G})$ is the random variable $X_{\mathcal{G}}$ in $\mathcal{M}^+(\mathcal{X}, \mathcal{G})$ for which*

<24> $$\mathbb{P}(gX) = \mathbb{P}(gX_{\mathcal{G}}) \qquad \text{for each } g \text{ in } \mathcal{M}^+(\Omega, \mathcal{G}).$$

*The variable $X_{\mathcal{G}}$ is called the **conditional expectation of X given the sub-sigma-field** $\mathcal{G}$. It is unique up to $\mathbb{P}$-equivalence.*

Be careful that you remember to check that $X_{\mathcal{G}}$ is $\mathcal{G}$-measurable. Otherwise you might be tempted to leap to the conclusion that $X_{\mathcal{G}}$ equals X, as a trivial solution to the equality <24>. Such an identification is valid only when X itself is $\mathcal{G}$-measurable. That is, $\mathbb{P}(X \mid \mathcal{G}) = X$ when X is $\mathcal{G}$-measurable (cf. part (i) of Theorem <19>).

REMARK. The traditional notation for $\mathbb{P}(X \mid \mathcal{G})$ is $\mathbb{E}(X \mid \mathcal{G})$. It would perhaps be more precise to speak of *a conditional expectation*, to stress the nonuniqueness. But, as with most situations involving an almost sure equivalence, the tradition is to ignore such niceties, except for the occasional almost sure reminder.

Some authors write the conditional expectation $\mathbb{P}(X \mid \mathcal{G})$ as $\mathcal{G}X$, a prefix notation that stresses its role as a linear map from random variables to random variables. If I were not such a traditionalist, I might adopt this more concise notation, which has much to recommend it.

The existence and almost sure uniqueness of $\mathbb{P}(X \mid \mathcal{G})$ present no new challenges: it is just $\mathbb{P}(X \mid T)$ for the identity map T from $(\Omega, \mathcal{F})$ onto $(\Omega, \mathcal{G})$. Theorem <19> specializes easily to give us "measure-like" properties of $\mathbb{P}(\cdot \mid \mathcal{G})$.

<25> **Theorem.** *For a fixed sub-sigma-field $\mathcal{G}$ of $\mathcal{F}$, conditional expectations have the following properties.*

(i) $\mathbb{P}(X \mid \mathcal{G}) = X$ *a.e.* $[\mathbb{P}]$, *for each $\mathcal{G}$-measurable X.*

(ii) *For X_1, X_2 in $\mathcal{M}^+(\mathcal{F})$ and g_1, g_2 in $\mathcal{M}^+(\mathcal{G})$,*

$$\mathbb{P}(g_1 X_1 + g_2 X_2 \mid \mathcal{G}) = g_1 \mathbb{P}(X_1 \mid \mathcal{G}) + g_2 \mathbb{P}(X_2 \mid \mathcal{G}) \qquad \text{a.e. } [\mathbb{P}].$$

(iii) *If $0 \le X_1 \le X_2$ then $\mathbb{P}(X_1 \mid \mathcal{G}) \le \mathbb{P}(X_2 \mid \mathcal{G})$ a.e. $[\mathbb{P}]$.*

(iv) *If $0 \le X_1 \le X_2 \le \ldots \uparrow X$ then $\mathbb{P}(X_n \mid \mathcal{G}) \uparrow \mathbb{P}(X \mid \mathcal{G})$ a.e. $[\mathbb{P}]$.*

REMARK. You might find it helpful to think of $\mathcal{G}$ as partial information—the information one obtains about ω by learning the value $g(\omega)$ for each $\mathcal{G}$-measurable random variable g. The value of the conditional expectation $\mathbb{P}(X \mid \mathcal{G})$ is determined by the "$\mathcal{G}$-information;" it is a prediction of $X(\omega)$ based on partial information. (The conditional expectation can also be interpreted as the fair price to pay for the random return $X(\omega)$ after one learns the $\mathcal{G}$-information about ω, as with the fair-price interpretation described in Section 1.5.) Whatever you prefer as a guide to intuition, be warned: you should not take the interpretation too literally, because there are examples (Problem [5]) where one can determine ω precisely from the values of all $\mathcal{G}$-measurable g, even with $\mathcal{G}$ a proper sub-sigma-field of $\mathcal{F}$.

<26> **Example.** Suppose an X in $\mathcal{M}^+(\Omega, \mathcal{F})$ is independent of all $\mathcal{G}$-measurable random variables. (That is, $\sigma(X)$ is independent of $\mathcal{G}$.) What does the partial information heuristic for conditioning on sigma-fields suggest for the value $\mathbb{P}(X \mid \mathcal{G})$?

Information from $\mathcal{G}$ leaves us just as ignorant about X as when we started, when the prediction of X was the constant $\mathbb{P}X$. It would seem, that we should have $\mathbb{P}(X \mid \mathcal{G}) = \mathbb{P}X$ when X is independent of $\mathcal{G}$. Indeed, the constant random variable $C = \mathbb{P}X$ is $\mathcal{G}$-measurable and, for all g in $\mathcal{M}^+(\mathcal{G})$, independence gives $\mathbb{P}(gX) = (\mathbb{P}g)(\mathbb{P}X) = \mathbb{P}(gC)$, as required to show that $C = \mathbb{P}(X \mid \mathcal{G})$. □

<27> **Example.** Suppose $\mathcal{G}_0$ and $\mathcal{G}_1$ are sub-sigma-fields of $\mathcal{F}$ with $\mathcal{G}_0 \subseteq \mathcal{G}_1$, and let $X \in \mathcal{M}^+(\Omega, \mathcal{F})$. Let $X_i = \mathbb{P}(X \mid \mathcal{G}_i)$, for $i = 1, 2$. That is, $X_i \in \mathcal{M}^+(\Omega, \mathcal{G}_i)$ and $\mathbb{P}(g_i X) = \mathbb{P}(g_i X_i)$ for each g_i in $\mathcal{M}^+(\Omega, \mathcal{G}_i)$. In particular, $\mathbb{P}(g_0 X_1) = \mathbb{P}(g_0 X) = \mathbb{P}(g_0 X_0)$ for each g_0 in $\mathcal{M}^+(\Omega, \mathcal{G}_0)$, because $\mathcal{M}^+(\Omega, \mathcal{G}_0) \subseteq \mathcal{M}^+(\Omega, \mathcal{G}_1)$. That is, the random variable X_0 has the two properties characterizing $\mathbb{P}(X_1 \mid \mathcal{G}_0)$. Put another way, $\mathbb{P}(\mathbb{P}(X \mid \mathcal{G}_1) \mid \mathcal{G}_0) = \mathbb{P}(X \mid \mathcal{G}_0)$ almost surely. □

<28> **Example.** When $\mathbb{P}X^2 < \infty$, the defining property of the conditional expectation $X_{\mathcal{G}} = \mathbb{P}(X \mid \mathcal{G})$ may be written as $\mathbb{P}\left(X - X_{\mathcal{G}}\right) g = 0$ for all bounded, $\mathcal{G}$-measurable g. A generating class argument extends the equality to all square-integrable, $\mathcal{G}$-measurable Z. That is, $X - X_{\mathcal{G}}$ is orthogonal to $\mathcal{L}^2(\Omega, \mathcal{G}, \mathbb{P})$. The (equivalence class of the) conditional expectation is just the projection of X onto $L^2(\Omega, \mathcal{G}, \mathbb{P})$, as a closed subspace of $L^2(\Omega, \mathcal{F}, \mathbb{P})$.

Some abstract conditioning results are readily explained in this setting. For example, if $\mathcal{G}_0 \subseteq \mathcal{G}_1$, for two sub-sigma-fields of $\mathcal{F}$, then the assertion

$$\mathbb{P}\left(\mathbb{P}(X \mid \mathcal{G}_1) \mid \mathcal{G}_0\right) = \mathbb{P}(X \mid \mathcal{G}_0) \qquad \text{almost surely}$$

from Example <27> corresponds to the fact that $\pi_0 \circ \pi_1 = \pi_0$ where π_i denotes the projection map from $L^2(\Omega, \mathcal{F}, \mathbb{P})$ onto $\mathbb{H}_i := L^2(\Omega, \mathcal{G}_i, \mathbb{P})$. The π_0 on the left-hand side kills any component in $\mathbb{H}_1$ that is orthogonal to $\mathbb{H}_0$, leaving only the component in $\mathbb{H}_0$. □

*7. Sufficiency

The intuitive definition of sufficiency says that a statistic T (a measurable map into some space $\mathcal{T}$) is sufficient for a family of probability measures $\mathcal{P} = \{\mathbb{P}_\theta : \theta \in \Theta\}$ if the conditional distributions given T do not depend on θ. The value of $T(\omega)$ carries all the "information about θ" there is to learn from an observation ω on $\mathbb{P}_\theta$.

There are (at least) two ways of making the definition precise, using either the Kolmogorov notion of conditional expectation or the properties of conditional distributions in the sense of Section 2. To distinguish between the two approaches, I will use nonstandard terminology.

<29> **Definition.** *Say that T is **strongly sufficient** for a family of probability measures $\mathcal{P}_\Theta = \{\mathbb{P}_\theta : \theta \in \Theta\}$ on $(\Omega, \mathcal{F})$ if there is a probability kernel $\{P_t : t \in \mathcal{T}\}$ that serves as a conditional distribution for each $\mathbb{P}_\theta$ given T.*

*Say that T is **weakly sufficient** for $\mathcal{P}_\Theta$ if for each X in $\mathcal{M}^+(\Omega, \mathcal{F})$ there exists a version of the Kolmogorov conditional expectation $\mathbb{P}_\theta(X \mid T = t)$ that does not depend on θ. Say that a sub-sigma-field $\mathcal{G}$ of $\mathcal{F}$ is weakly sufficient for $\mathcal{P}_\Theta$ if, for each X in $\mathcal{M}^+(\Omega, \mathcal{F})$ there exists a version of $\mathbb{P}_\theta(X \mid \mathcal{G})$ that does not depend on θ.*

REMARK. I add the subscript to $\mathcal{P}_\Theta$ to ensure that there is no confusion with $\mathcal{P}$ as a probability kernel from $\mathcal{T}$ to Ω.

The concept of strong sufficiency places more restrictions on the underlying probability spaces, but those restrictions have the added benefit of making some arguments more intuitive. The weaker concept has the advantage that it requires only the Kolmogorov definition of conditioning. However, it also has the drawback of allowing some counterintuitive consequences. For example, if a sub-sigma-field $\mathcal{G}_0$ is weakly sufficient for a family $\mathcal{P}_\Theta$ then, in the intuitive sense explained in Section 6, it provides all the information available about θ from an observation on an unknown $\mathbb{P}_\theta$ in $\mathcal{P}_\Theta$. If $\mathcal{G}_1$ is another sub-sigma-field, with $\mathcal{G}_0 \subseteq \mathcal{G}_1$, then intuitively $\mathcal{G}_1$ should provide even more information about θ, which should make $\mathcal{G}_1$ weakly sufficient for $\mathcal{P}_\Theta$. Unfortunately, as the next Example shows, the intuition is not correct.

<30> **Example.** Let Ω be the real line, equipped with its Borel sigma-field $\mathcal{F}$. For each $\theta \geq 0$ define $\mathbb{P}_\theta$ as the probability measure on $\mathcal{F}$ that puts mass 1/2 at each of the two points $\pm\theta$. Say that a set B is symmetric if $-\omega \in B$ for each ω in B. Let S be a fixed symmetric set not in $\mathcal{F}$, and not containing the origin. Write $\mathcal{G}_0$ for the sub-sigma-field of all symmetric Borel sets, and $\mathcal{G}_1$ for the sub-sigma-field of all Borel sets for which BS^c is symmetric.

A simple generating class argument shows that a Borel measurable function g is $\mathcal{G}_0$-measurable if and only if $g(\omega) = g(-\omega)$ for all ω, and it is $\mathcal{G}_1$-measurable if and only if $g(\omega) = g(-\omega)$ for all ω in S^c.

Consider an X in $\mathcal{M}^+(\mathcal{F})$. The symmetric function $X_0(\omega) := \frac{1}{2}(X(\omega) + X(-\omega))$ is $\mathcal{G}_0$-measurable. For all $\theta \geq 0$ and all g_0 in $\mathcal{M}^+(\mathcal{G}_0)$,

$$\mathbb{P}_\theta (Xg_0) = \tfrac{1}{2}\left(X(\theta)g_0(\theta) + X(-\theta)g_0(-\theta)\right) = X_0(\theta)g_0(\theta) = \mathbb{P}_\theta (X_0 g_0).$$

That is, X_0 is a version of $\mathbb{P}_\theta(X \mid \mathcal{G}_0)$ that does not depend on θ. The sub-sigma-field $\mathcal{G}_0$ is weakly sufficient for $\mathcal{P}_\Theta = \{\mathbb{P}_\theta : \theta \geq 0\}$. In fact, a stronger assertion is possible. The sigma-field $\mathcal{G}_0$ is generated by the map $T(\omega) = |\omega|$. For each θ we can take the conditional distribution P_t as the probability measure that places mass 1/2 at $\pm t$. The statistic T is strongly sufficient.

Now consider X equal to the indicator of the half-line $H = [0, \infty)$. Could there be a $\mathcal{G}_1$-measurable function X_1 for which $\mathbb{P}_\theta (Xg_1) = \mathbb{P}_\theta (X_1 g_1)$ for all g_1 in $\mathcal{M}^+(\mathcal{G}_1)$? If such an X_1 existed we would have

$$\tfrac{1}{2}g_1(\theta) = \mathbb{P}_\theta^\omega (X(\omega)g_1(\omega)) = X_1(\theta)g_1(\theta) + X_1(-\theta)g_1(-\theta) \qquad \text{for all } \theta \geq 0$$

For θ in $H \cap S$, take g_1 as the indicator function of the singleton set $\{\theta\}$, and then as the indicator function of the singleton set $\{-\theta\}$, to deduce that $X_1(\theta) = X_1(-\theta) = 1$. For θ in $H \backslash S$ take $g_1 \equiv 1$ to deduce (via the $\mathcal{G}_1$-measurability property $X_1(\theta) = X_1(-\theta)$) that $X_1(\theta) = X_1(-\theta) = 1/2$. That is, $\{X_1 = 1\} = S$, which would contract the Borel measurability of X_1 and the nonmeasurability of S. There can be no $\mathcal{G}_1$-measurable function X_1 that is a version of $\mathbb{P}_\theta(X \mid \mathcal{G}_1)$ for all θ. The sub-sigma-field $\mathcal{G}_1$ is not weakly sufficient for $\mathcal{P}_\Theta$. □

REMARK. The failure of weak sufficiency for $\mathcal{G}_1$ is due to the fact that the $\mathcal{P}_\Theta$ is not dominated: that is, there exists no sigma-finite measure domininating each $\mathbb{P}_\theta$. Problem [17] shows that the failure cannot occur for dominated families.

The concepts of strong and weak sufficiency coincide when Ω is a finite set and $\mathbb{P}_\theta\{\omega\} > 0$ for each ω in Ω and each θ in Θ. If $\{\omega : T\omega = t\}$ is nonempty then the sum $g(t, \theta) := \sum_{\omega'}\{T(\omega') = t\}\mathbb{P}_\theta\{\omega'\}$ is nonzero and

$$\mathbb{P}_\theta\{\omega \mid T = t\} = \begin{cases} \mathbb{P}_\theta\{\omega\}/g(t,\theta) & \text{if } T(\omega) = t, \\ 0 & \text{otherwise,} \end{cases}$$

If T is sufficient, the left-hand side is a function of ω and t alone. If we write it as $H(\omega, t)$, then sufficiency implies

$$\mathbb{P}_\theta\{\omega\} = \begin{cases} g(t,\theta)H(\omega,t) & \text{if } T(\omega) = t, \\ 0 & \text{otherwise.} \end{cases}$$

or, more succinctly, $\mathbb{P}_\theta\{\omega\} = g(T(\omega), \theta)h(\omega)$ where $h(\omega) = H(\omega, T(\omega))$. Conversely, if $\mathbb{P}_\theta\{\omega\}$ has such a factorization then, for ω with $T(\omega) = t$,

$$\mathbb{P}_\theta\{\omega \mid T = t\} = \frac{g(t,\theta)h(\omega)}{\sum_{\omega'}\{T(\omega') = t\}g(t,\theta)h(\omega')} = \frac{h(\omega)}{\sum_{\omega'}\{T(\omega') = t\}h(\omega')}.$$

Thus T is sufficient if $\mathbb{P}_\theta\{\omega\}$ factorizes in the way shown.

The factorization criterion also extends to more general settings. The proof of the analog for weak sufficiency, due to Halmos & Savage (1949), appears in the textbook of Lehmann (1959, Section 2.6). The corresponding proof for strong sufficiency is easier to recognize as a direct generalization of the argument for finite Ω.

<31> **Example.** Suppose each $\mathbb{P}_\theta$ has a density $p(\omega,\theta)$ with respect to a sigma-finite measure λ, and that λ has a (T,μ)-disintegration $\{\lambda_t : t \in \mathcal{T}\}$. If the density factorizes, $p(\omega,\theta) = g(T\omega,\theta)h(\omega)$, then Theorem <12> will show that T is strongly sufficient for $\mathcal{P}$.

Ignoring problems related to division by zero, we would expect the conditional distribution $\mathbb{P}_t$ to be dominated by λ_t, with density

$$\frac{p(\omega,\theta)}{\lambda_t^\omega p(\omega,\theta)} = \frac{g(T\omega,\theta)h(\omega)}{\lambda_t^\omega g(T\omega,\theta)h(\omega)}.$$

For λ_t almost all ω, the $T\omega$ in the last numerator and denominator are both equal to t, allowing us to cancel out a $g(t,\theta)$ factor, leaving a conditional density that doesn't depend on θ.

Let us try to be more careful about division by zero. Define

$$p_t(\omega) = \frac{h(\omega)}{H(t)}\{0 < H(t) < \infty\} \qquad \text{where } H(t) := \lambda_t h.$$

A proof that p_t is a conditional density for each $\mathbb{P}_\theta$ requires careful handling of contributions from the sets where H is zero or infinite. According to Theorem <12>, it suffices to show, for each fixed θ, that

$$q(t,\theta)p_t(\omega) = p(\omega,\theta) \qquad \text{a.e. } [\mu \otimes \Lambda],$$

where

$$q(t,\theta) = \lambda_t^\omega\left(g(T\omega,\theta)h(\omega)\right) = g(t,\theta)H(t) \qquad \text{a.e. } [\mu].$$

The aberrant negligible sets are allowed to depend on θ. Because $T\omega = t$ a.e. $[\mu\otimes\Lambda]$, the task reduces to showing that

$$g(t,\theta)h(\omega)\{0 < H(t) < \infty\} = g(t,\theta)h(\omega) \qquad \text{a.e. } [\mu \otimes \Lambda].$$

We need to show that the contributions to the right-hand side from the sets where H is zero or infinite vanish a.e. $[\mu \otimes \Lambda]$. Equivalently, we need to show

$$\mu^t\lambda_t^x\left(g(t,\theta)h(\omega)\{H(t) = 0 \text{ or } \infty\}\right) = 0.$$

The $g(t,\theta)\{H(t) = 0 \text{ or } \infty\}$ factor slips outside the innermost integral with respect to λ_t, leaving $\mu^t\left(g(t,\theta)H(t)\{H(t) = 0 \text{ or } \infty\}\right)$. Clearly the product $g(t,\theta)H(t)$ is zero on the set $\{H = 0\}$. That $g(t,\theta)$ must be zero almost everywhere on the set $\{H = \infty\}$ follows from the fact that $q(t,\theta)$ is a probability density:

$$1 = \mu^t q(t,\theta) = \mu^t\left(g(t,\theta)H(t)\right).$$

The strong sufficiency follows.

See Problem [18] for a converse, where strong sufficiency implies existence of a factorizable density. □

<32> **Example.** Let $\mathbb{P}_\theta$ denote the uniform distribution on $[0,\theta]^2$, for $\theta > 0$. Let $T(x,y) = \max(x,y)$. As in Example <5>, the conditional probability distributions $\mathbb{P}_\theta$ given T are uniform on the sets $\{T = t\}$. That is T is a strongly sufficient statistic for the family $\{\mathbb{P}_\theta : \theta > 0\}$. The same conclusion follows directly from the factorization criterion, because $\mathbb{P}_\theta$ has density $p(x,y,\theta) = \theta^{-2}\{T(x,y) \le \theta\}$ with respect to Lebesgue measure on $(\mathbb{R}^+)^2$. □

8. Problems

[1] Suppose g_1 and g_2 are maps from $(\mathcal{X}, \mathcal{A})$ to $(\mathcal{Y}, \mathcal{B})$, with product-measurable graphs. Define $\psi_i(x) = (x, g_i(x))$. Let $P_i = \psi_i(\mu_i)$, for probability measures μ_i on $\mathcal{A}$. Let $P = \alpha_1 P_1 + \alpha_2 P_2$, for constants $\alpha_i > 0$ with $\alpha_1 + \alpha_2 = 1$. Let X denote the coordinate map onto the $\mathcal{X}$ space. Show that the conditional probability distribution P_x concentrates on the points $(x, g_1(x))$ and $(x, g_2(x))$. Find the conditional probabilities assigned to each point. Hint: Consider the density of μ_i with respect to $\alpha_1\mu_1 + \alpha_2\mu_2$.

[2] Let Q be a nonatomic probability measure on $\mathcal{B}(\mathbb{R})$ with corresponding distribution function F. Let $\mathbb{Q} = Q^n$, the joint distribution of n independent observations $x_1, \dots, x_n$ from Q. Define $T(x_1, \dots, x_n) := \max_{i \le n} x_i$.

(i) Show that T has distribution ν, defined by $\nu(-\infty, t] = F(t)^n$.

(ii) For each fixed $t \in \mathbb{R}$, show that

$$\begin{aligned}\nu(-\infty,t] &= \sum\nolimits_{i\le n} \mathbb{Q}\{x_1 \le t, \dots, x_n \le t, T = x_i\}\\ &= n\mathbb{Q}\{x_1 \le t, x_2 < x_1, x_3 < x_1, \dots, x_n < x_1\}\\ &= Q^{x_1}\left(nF(x_1)^{n-1}\{x_1 \le t\}\right).\end{aligned}$$

Deduce that ν has density nF^{n-1} with respect to Q.

(iii) Write Q_t for the distribution of x_1 conditional on $x_1 < t$. That is, $Q_t^x g(x) = Q^x\left(g(x)\{x < t\}\right)/F(t)$. Write ϵ_t for the probability measure degenerate at the ponit t. Show that the conditional distribution $\mathbb{Q}_t$, for $\mathbb{Q}$ given $T = t$, equals

$$\mathbb{Q}_t = n^{-1}\sum\nolimits_{i\le n}\left(Q_t^{i-1} \otimes \epsilon_t \otimes Q_t^{n-i}\right)$$

That is, to generate a sample $x_1, \dots, x_n$ from $\mathbb{Q}_t$, select an x_i to equal t (with probability n^{-1} for each i), then generate the remaining $n-1$ observations independently from the conditional distribution Q_t.

[3] Show that a $(T, \mu))$ disintegration $\Lambda = \{\lambda_t : t \in \mathcal{T}\}$ of a sigma-finite measure λ can be chosen as a probability kernel if and only if the sigma-finite measure μ is equal to the image measure $T\lambda$. Hint: Write $\ell(t)$ for $\lambda_t\mathcal{X}$. Show that $(T\lambda)g = \mu^t\left(g(t)\ell(t)\right)$ for all g in $\mathcal{M}^+(\mathcal{B})$. Prove that the right-hand side reduces to μg for all g if and only if $\ell(t) = 1$ for μ almost all t.

[4] For families of probability measures $\mathcal{P}$ and $\mathcal{Q}$ defined on the same sigma-field, define $\alpha_2(\mathcal{P}, \mathcal{Q})$ to equal $\sup\{\alpha_2(P, Q) : P \in \mathcal{P},\ Q \in \mathcal{Q}\}$, where α_2 denotes the Hellinger affinity, as in Section 3.3. Write $\mathcal{P} \otimes \mathcal{P}$ for $\{P_1 \otimes P_2 : P_i \in \mathcal{P}\}$, and $\text{co}(\mathcal{P} \otimes \mathcal{P})$ for its convex hull. Show that $\alpha_2(\text{co}(\mathcal{P} \otimes \mathcal{P}), \text{co}(\mathcal{Q} \otimes \mathcal{Q})) \le \alpha_2(\text{co}(\mathcal{P}), \text{co}(\mathcal{Q}))^2$ by the following steps. Write A for $\alpha_2(\text{co}(\mathcal{P}), \text{co}(\mathcal{Q}))$.

(i) For given $\mathbb{P} = \sum_i \beta_i P_{1i} \otimes P_{2i}$ in $\text{co}(\mathcal{P})$ and $\mathbb{Q} = \sum_j \gamma_j Q_{1j} \otimes Q_{2j}$ in $\text{co}(\mathcal{Q})$, let λ be any probability measure dominating all the component measures, with corresponding densities $p_{1i}(x)$, and so on. Show that $\mathbb{P}$ has density $p(x, y) := \sum_i \beta_i p_{1i}(x) p_{2i}(y)$ with respect to $\lambda \otimes \lambda$, with a similar expression for the density $q(x, y)$ of $\mathbb{Q}$.

(ii) Write X for the projection map of $\mathcal{X} \times \mathcal{X}$ onto its first coordinate. Show that $X\mathbb{P}$ has density $p(x) := \sum_i \beta_i p_{1i}(x)$ with respect to λ, with a similar expression for the density $q(x)$ of $X\mathbb{Q}$. Deduce that $X\mathbb{P} \in \text{co}(\mathcal{P})$ and $X\mathbb{Q} \in \text{co}(\mathcal{Q})$.

(iii) Define $p_x(y) := \sum_i \beta_i p_{1i}(x)/p(x)$ on the set $\{x : p(x) > 0\}$. Define $q_x(y)$ analogously. Show that $\lambda^y \sqrt{p_x(y) q_x(y)} \le A$ for all x.

(iv) Show that $\alpha_2(\mathbb{P}, \mathbb{Q}) = \lambda^x \left(\sqrt{p(x)q(x)}\lambda^y \sqrt{p_x(y)q_x(y)}\right) \le A^2$.

[5] Let $\mathbb{P}$ be Lebesgue measure on $\mathcal{F}$, the Borel sigma-field of $[0, 1]$. Let $\mathcal{G}$ denote the sigma-field generated by all the singletons in $[0, 1]$.

(i) Show that each member of $\mathcal{G}$ has probability either zero or one.

(ii) Deduce that $\mathbb{P}(X \mid \mathcal{G}) = \mathbb{P}X$ for each X in $\mathcal{M}^+(\mathcal{F})$.

(iii) Show for each Borel measurable X that $X(\omega)$ is uniquely determined once we know the values of all $\mathcal{G}$-measurable random variables.

[6] Let X be an integrable random variable, and Z be a $\mathcal{G}$-measurable random variable for which XZ is integrable. Show that $\mathbb{P}(XZ) = \mathbb{P}(YZ)$ where $Y = \mathbb{P}(X \mid \mathcal{G})$.

[7] Suppose $X_n \in \mathcal{M}^+(\mathcal{F})$. Show that $\mathbb{P}(\liminf_n X_n \mid \mathcal{G}) \le \liminf_n \mathbb{P}(X_n \mid \mathcal{G})$ almost surely, for each sub-sigma-field $\mathcal{G}$. Hint: Derive from the conditional form of Monotone Convergence. Imitate the proof of Fatou's Lemma.

[8] Suppose $X_n \to X$ almost surely, with $|X_n| \le H$ for an integrable H. Show that $\mathbb{P}(X_n \mid \mathcal{G}) \to \mathbb{P}(X \mid \mathcal{G})$ almost surely. Hint: Derive from the conditional form of Fatou's Lemma (Problem [7]). Imitate the proof of Dominated Convergence.

[9] Suppose $\mathbb{P}X^2 < \infty$. Show that $\text{var}(X) = \text{var}(\mathbb{P}(X \mid \mathcal{G})) + \mathbb{P}(\text{var}(X \mid \mathcal{G}))$.

[10] Let X be an integrable random variable on a probability space $(\Omega, \mathcal{F}, \mathbb{P})$. Let $\mathcal{G}$ be a sub-sigma-field of $\mathcal{F}$ containing all $\mathbb{P}$-negligible sets. Show that X is $\mathcal{G}$-measurable if and only if $\mathbb{P}(XW) = 0$ for every bounded random variable W with $\mathbb{P}(W \mid \mathcal{G}) = 0$ almost surely. *(Compare with the corresponding statement for random variables that are square integrable: $Z \in \mathcal{L}^2(\mathcal{G})$ if and only if it is orthogonal to every square integrable W that is orthogonal to $\mathcal{L}^2(\mathcal{G})$.)* Hint: For a fixed real t with $\mathbb{P}\{X = t\} = 0$ define $Z_t = \mathbb{P}(X > t \mid \mathcal{G})$ and $W_t = \{X > t\} - Z_t$. Show that $(X - t)W_t \ge 0$ almost surely, but $\mathbb{P}((X - t)W_t) = 0$. Deduce that $W_t = 0$ almost surely, and hence W_t is $\mathcal{G}$-measurable.

[11] Let $S_0, S_1, \ldots, S_N$ be random vectors, and $\mathcal{F}_0 \subseteq \mathcal{F}_1 \subseteq \ldots \subseteq \mathcal{F}_N \subseteq \mathcal{F}$ be sigma-fields such that S_i is $\mathcal{F}_i$-measurable for each i. Let ϵ, α, and β be positive numbers for which $\beta\mathbb{P}\{|S_N - S_i| \le (1-\alpha)|S_i| \mid \mathcal{F}_i\} \ge \{|S_i| > \epsilon\}$ almost surely. Show that $\mathbb{P}\{\max_i |S_i| > \epsilon\} \le \beta\mathbb{P}\{|S_N| > \alpha\epsilon\}$. Hint: Let τ denote the first i for which $|S_i| > \epsilon$. Show that $\{\tau = i\} \in \mathcal{F}_i$. Use the definition of conditional expectation to bound $\beta\mathbb{P}\{|S_i| > \epsilon, \tau = i\}$ by $\mathbb{P}\{|S_N| > \alpha\epsilon, \tau = i\}$.

[12] Suppose P is a probability measure on $(\mathcal{X}, \mathcal{A})$ with density $p(x)$ with respect to a probability measure λ. Let Q equal the image TP, and μ equal $T\lambda$, for a measurable map T into $(\mathcal{T}, \mathcal{B})$. Without assuming existence of disintegrations or conditional distributions, show

(i) Q has density $q(t) := \lambda(p \mid T = t)$ with respect to μ.

(ii) For $X \in \mathcal{M}^+(\mathcal{X})$, show that $P(X \mid T = t)$, the Kolmogorov conditional expectation, is given by $\{0 < q(t) < \infty\}\lambda(Xp \mid T = t)/q(t)$, up to a Q-equivalence.

[13] For each convex function ψ on the real line there exists a countable family of linear functions for which $\psi(x) = \sup_{i\in\mathbb{N}}(a_i + b_i x)$ for all x (see Appendix C). Use this representation to prove the conditional form of Jensen's inequality: if X and $\psi(X)$ are both integrable then $\mathbb{P}(\Psi(X) \mid \mathcal{G}) \ge \Psi(\mathbb{P}(X \mid \mathcal{G}))$ almost surely. Hint: For each i argue that $\mathbb{P}(\Psi(X) \mid \mathcal{G}) \ge a_i + b_i\mathbb{P}(X \mid \mathcal{G})$ almost surely. Question: Is integrability of $\psi(X)$ really needed?

[14] Let P be a probability measure on $\mathcal{B}(\mathbb{R}^2)$ that is invariant under rotations S_θ for a dense set Θ_0 of angles. Show that it is invariant under all rotations. Hint: For bounded, continuous f, if $\theta_i \to \theta$ then $P^x f(S_{\theta_i}x) \to P^x f(S_\theta x)$.

[15] Find an invariance argument that produces the conditional distributions from Exercise <5>. Hint: Consider transformations g_θ that map (x, y) into (x, y_θ), where $y_\theta = y + \theta$ if $y + \theta \le x$ and $y + \theta - x$ otherwise.

[16] Let $\mathcal{Q}$ be a family of probability measures on a sigma-field $\mathcal{A}$ dominated by a fixed sigma-finite measure λ. That is, each Q in $\mathcal{Q}$ has a density δ_Q with respect to λ. Show that $\mathcal{Q}$ is also dominated by a probability measure of the form $\nu = \sum_{j=1}^{\infty} 2^{-j} Q_j$, with $Q_j \in \mathcal{Q}$ for each j, by following these steps.

(i) Show that there is no loss of generality in assuming that λ is a finite measure. Hint: Express λ as a sum of finite measures $\sum_i \lambda_i$ then choose positive constants α_i so that $\sum_i \alpha_i\lambda_i$ has finite total mass.

(ii) For each countable subfamily S of $\mathcal{Q}$ define $L(S) := \lambda\left(\cup_{Q\in S}\{\delta_Q > 0\}\right)$. Define $L := \sup\{L(S) : \mathcal{Q} \supseteq S \text{ countable}\}$. Find countable subsets S_n for which $L(S_n) \uparrow L$. Show that $L = L(S^*)$, where $S^* = \cup_n S_n$.

(iii) Write $\mathcal{X}_0$ for $\cup_{Q\in S^*}\{\delta_Q > 0\}$. For each Q_0 in $\mathcal{Q}$, show that $\lambda\left(\{\delta_{Q_0} > 0\}\backslash\mathcal{X}_0\right) = 0$. Hint: $L(S^* \cup \{Q_0\}) \le L(S^*)$.

(iv) Enumerate S^* as $\{Q_j : j \in \mathbb{N}\}$. Define $\nu = \sum_{j=1}^{\infty} 2^{-j} Q_j$. If $f \in \mathcal{M}^+$ and $\nu f = 0$, show that $\lambda(f\mathcal{X}_0) = 0$. Deduce that $Q_0 f = 0$ for all $Q_0 \in \mathcal{Q}$. That is, ν dominates $\mathcal{Q}$.

[17] Let $\mathcal{P}_\Theta = \{\mathbb{P}_\theta : \theta \in \Theta\}$ be a family of probability measures on $\mathcal{F}$, dominated by a sigma-finite measure λ. Suppose $\mathcal{G}_0$ is a weakly sufficient sub-sigma-field, and $\mathcal{G}_0 \subseteq \mathcal{G}_1$ for another sigma-field $\mathcal{G}_1 \subseteq \mathcal{F}$. Show that $\mathcal{G}_1$ is also weakly sufficient, by following these steps.

(i) With no loss of generality (Problem [16]), suppose $\lambda = \sum_{j\in\mathbb{N}} 2^{-j} P_{\theta_j}$. Write f_θ for the ($\mathcal{G}_0$-measurable) density of $\mathbb{P}_\theta|_{\mathcal{G}_0}$ with respect to $\lambda|_{\mathcal{G}_0}$. For an X in $\mathcal{M}^+(\mathcal{F})$, write X_0 for the version of $\mathbb{P}_\theta(X \mid \mathcal{G}_0)$ that doesn't depend on θ. Show that X_0 is also a version of $\lambda(X \mid \mathcal{G}_0)$. Deduce that $\mathbb{P}_\theta X = \mathbb{P}_\theta X_0 = \lambda(f_\theta X_0) = \lambda(f_\theta X)$, that is, f_θ is also the density of $\mathbb{P}_\theta$ with respect to λ, as measures on $\mathcal{F}$.

(ii) For an X in $\mathcal{M}^+(\mathcal{F})$, write X_1 for $\lambda(X \mid \mathcal{G}_1)$. For $g_1 \in \mathcal{M}^+(\mathcal{G}_1)$, show that $\mathbb{P}_\theta(g_1 X) = \lambda(f_\theta g_1 X) = \lambda(f_\theta g_1 X_1) = \mathbb{P}_\theta(g_1 X_1)$. Deduce that X_1 is a version of $\mathbb{P}_\theta(X \mid \mathcal{G}_1)$ that doesn't depend on θ.

[18] Let $\mathcal{P}_\Theta = \{\mathbb{P}_\theta : \theta \in \Theta\}$ be a family of probability measures dominated by a sigma-finite measure λ. Suppose T is a strongly sufficient statistic for $\mathcal{P}$, meaning that there exists a probability kernel $\{P_t : t \in \mathcal{T}\}$ that is a conditional distribution for each $\mathbb{P}_\theta$ given T. Show that there exist versions of the densities $d\mathbb{P}_\theta/d\lambda$ of the form $g(T\omega, \theta)h(\omega)$, by following these steps.

(i) From Problem [16], there exists a dominating probability measure for $\mathcal{P}_\Theta$ of the form $\mathbb{P} = \sum_i 2^{-i}\mathbb{P}_{\theta_i}$, for some countable subfamily $\{\mathbb{P}_{\theta_i}\}$ of $\mathcal{P}$. Show that $\{P_t : t \in \mathcal{T}\}$ is also a conditional distribution for $\mathbb{P}$ given T.

(ii) By part (i) of Theorem <12>, $\mathbb{Q}_\theta := T\mathbb{P}_\theta$ is dominated by $\mathbb{Q} := T\mathbb{P}$. Write $g(t, \theta)$ for some choice of the density $d\mathbb{Q}_\theta/d\mathbb{Q}$. Show that $\mathbb{P}_\theta^\omega f(\omega) = \mathbb{P}^\omega\left(g(T\omega, \theta)f(\omega)\right)$ for $f \in \mathcal{M}^+(\Omega)$.

(iii) Deduce that

$$\frac{d\mathbb{P}_\theta}{d\lambda} = g(T\omega, \theta)\frac{d\mathbb{P}}{d\lambda} \qquad \text{a.e. } [\lambda],$$

in the sense that the right-hand side can be taken as a density for $\mathbb{P}_\theta$.

[19] Let $\mathcal{P}_\Theta = \{\mathbb{P}_\theta : \theta \in \Theta\}$ be a family of probability measures dominated by a sigma-finite measure λ. Suppose T is a strongly sufficient statistic for $\mathcal{P}$. Suppose T^* is another statistic, taking values in a set $\mathcal{T}^*$ equipped with a countably generated sigma-field $\mathcal{B}^*$ containing the singletons, such that $\sigma(T) \subseteq \sigma(T^*)$. Show that T can be written as a measurable function of T^*. Deduce via the factorization criterion that T^* is also strongly sufficient.

[20] Let $(\mathcal{X}, \mathcal{A}, P)$ be a probability space, and let $\mathbb{P}$ equal P^n, the n-fold product measure on $\mathcal{A}^\mathbb{N}$. For each x in $\mathcal{X}$, let $\delta(x)$ denote the point mass at x. For $\mathbf{x}$ in $\mathcal{X}^n$, let $T(\mathbf{x})$ denote the so-called ***empirical measure***, $n^{-1}\sum_{i\le n}\delta(x_i)$. Intuitively, if we are given the measure $T(\mathbf{x})$, the conditional distribution for $\mathbb{P}$ should give mass $1/n!$ to each of the $n!$ permutations $(x_1, \ldots, x_n)$. Formalize this intuition by constructing a $(T, T\mathbb{P})$-disintegration for $\mathbb{P}$. Warning: Not as easy as it seems. What is the $(\mathcal{T}, \mathcal{B})$ for this problem? What happens if the empirical measure is not supported by n

distinct points? How should you define $\mathbb{P}_t$ when t does not correspond to a possible realization of an empirical measure?

9. Notes

The idea of defining abstract conditional probabilities and expectations as Radon-Nikodym derivatives is due to Kolmogorov (1933, Chapter 5).

(Regular) conditional distributions have a more complicated history. Loève (1978, Section 30.2) mentioned that the problem of existence was "investigated principally by Doob," but he cited no specific reference. Doob (1953, page 624) cited a counterexample to the unrestricted existence of a regular conditional probability, which also appears in the exercises to Section 48 of the 1969 printing of Halmos (1950). Doob's remarks suggest that the original edition of the Halmos book contained a slightly weaker form of the counterexample. Doob also noted that the counterexample destroyed a claim made in (Doob 1938), an error pointed out by Dieudonné (no citation) and Andersen & Jessen (1948). Blackwell (1956) cited Dieudonné (1948) as the source of a counterexample for unrestricted existence of a regular conditional probabilities. Blackwell also proved existence of regular conditional distributions for (what are now known as) Blackwell spaces.

In point process theory, disintegrations appear as Palm distributions—conditional distributions given a point of the process at a particular position (Kallenberg 1969).

Pfanzagl (1979) gave conditions under which a regular conditional distribution can be obtained by means of the elementary limit of ratio of probabilities. The existence of limits of carefully chosen ratios can also be established by martingale methods (see Chapter 6).

The Barndorff-Nielsen, Blaesild & Eriksen (1989) book contains much material on the invariance properties of conditional distributions.

Halmos & Savage (1949) cited a 1935 paper by Neyman, which I have not seen, as the source of the factorization criterion for (weak) sufficiency. See *Bahadur (1954) for a detailed discussion of sufficiency. Example <30> is based on* a meatier counterexample of Burkholder 1961. The traditional notion of sufficiency (what I have called weak sufficiency) has strange behavior for undominated families.

I learned much about the subtleties of conditioning while working on the paper Chang & Pollard (1997), where we explored quite a range of statistical applications.

The result in Problem [4] is taken from Le Cam (1973) (see also Le Cam 1986, Section 16.4), who used it to establish asymptotic results in the theory of estimation and testing. Donoho & Liu (1991) adapted Le Cam's ideas to establish results about achievability of lower bounds for minimax rates of convergence of estimators.

References

Andersen, E. S. & Jessen, B. (1948), 'On the introduction of measures in infinite product sets', *Danske Vid. Selsk. Mat.-Fys. Medd.*

Baddeley, A. (1977), 'Integrals on a moving manifold and geometrical probability', *Advances in Applied Probability* **9**, 588–603.

Bahadur, R. R. (1954), 'Sufficiency and statistical decision functions', *Annals of Mathematical Statistics* **25**, 423–462.

Barndorff-Nielsen, O. E., Blaesild, P. & Eriksen, P. S. (1989), *Decomposition and Invariance of Measures, and Statistical Transformation Models*, Vol. 58 of *Springer Lecture Notes in Statistics*, Springer-Verlag, New York.

Blackwell, D. (1956), On a class of probability spaces, *in* J. Neyman, ed., 'Proceedings of the Third Berkeley Symposium on Mathematical Statistics and Probability', Vol. I, University of California Press, Berkeley, pp. 1–6.

Burkholder, D. L. (1961), 'Sufficiency in the undominated case', *Annals of Mathematical Statistics* **32**, 1191–1200.

Chang, J. & Pollard, D. (1997), 'Conditioning as disintegration', *Statistica Neerlandica* **51**, 287–317.

Dellacherie, C. & Meyer, P. A. (1978), *Probabilities and Potential*, North-Holland, Amsterdam.

Dempster, A. P., Laird, N. M. & Rubin, D. B. (1977), 'Maximum likelihood estimation from incomplete data via the EM algorithm (with discussion)', *Journal of the Royal Statistical Society, Series B* **39**, 1–38.

Dieudonné, J. (1948), 'Sur le théorème de Lebesgue-Nikodym, III', *Ann. Univ. Grenoble* **23**, 25–53.

Donoho, D. L. & Liu, R. C. (1991), 'Geometrizing rates of convergence, II', *Annals of Statistics* **19**, 633–667.

Doob, J. L. (1938), 'Stochastic processes with integral-valued parameter', *Transactions of the American Mathematical Society* **44**, 87–150.

Doob, J. L. (1953), *Stochastic Processes*, Wiley, New York.

Eisenberg, B. & Sullivan, R. (2000), 'Crofton's differential equation', *American Mathematical Monthly* pp. 129–139.

Halmos, P. R. (1950), *Measure Theory*, Van Nostrand, New York, NY. July 1969 reprinting.

Halmos, P. R. & Savage, L. J. (1949), 'Application of the Radon-Nikodym theorem to the theory of sufficient statistics', *Annals of Mathematical Statistics* **20**, 225–241.

Kallenberg, O. (1969), *Random Measures*, Akademie-Verlag, Berlin. US publisher: Academic Press.

Kendall, M. G. & Moran, P. A. P. (1963), *Geometric Probability*, Griffin.

Kolmogorov, A. N. (1933), *Grundbegriffe der Wahrscheinlichkeitsrechnung*, Springer-Verlag, Berlin. Second English Edition, *Foundations of Probability* 1950, published by Chelsea, New York.

Le Cam, L. (1973), 'Convergence of estimates under dimensionality restrictions', *Annals of Statistics* **1**, 38–53.

Le Cam, L. (1986), *Asymptotic Methods in Statistical Decision Theory*, Springer-Verlag, New York.

Lehmann, E. L. (1959), *Testing Statistical Hypotheses*, Wiley, New York. Later edition published by Chapman and Hall.

Loève (1978), *Probability Theory*, Springer, New York. Fourth Edition, Part II.

Pachl, J. (1978), 'Disintegration and compact measures', *Mathematica Scandinavica* **43**, 157–168.

Pfanzagl, J. (1979), 'Conditional distributions as derivatives', *Annals of Probability* **7**, 1046–1050.

Solomon, H. (1978), *Geometric Probability*, NSF-CBMS Regional Conference Series in Applied Mathematics, Society for Industrial and Applied Mathematics.

Chapter 6

Martingale et al.

SECTION 1 gives some examples of martingales, submartingales, and supermartingales.
SECTION 2 introduces stopping times and the sigma-fields corresponding to "information available at a random time." A most important Stopping Time Lemma is proved, extending the martingale properties to processes evaluted at stopping times.
SECTION 3 shows that positive supermartingales converge almost surely.
SECTION 4 presents a condition under which a submartingale can be written as a difference between a positive martingale and a positive supermartingale (the Krickeberg decomposition). A limit theorem for submartingales then follows.
*SECTION *5 proves the Krickeberg decomposition.*
*SECTION *6 defines uniform integrability and shows how uniformly integrable martingales are particularly well behaved.*
*SECTION *7 show that martingale theory works just as well when time is reversed.*
*SECTION *8 uses reverse martingale theory to study exchangeable probability measures on infinite product spaces. The de Finetti representation and the Hewitt-Savage zero-one law are proved.*

1. What are they?

The theory of martingales (and submartingales and supermartingales and other related concepts) has had a profound effect on modern probability theory. Whole branches of probability, such as stochastic calculus, rest on martingale foundations. The theory is elegant and powerful: amazing consequences flow from an innocuous assumption regarding conditional expectations. Every serious user of probability needs to know at least the rudiments of martingale theory.

A little notation goes a long way in martingale theory. A fixed probability space $(\Omega, \mathcal{F}, \mathbb{P})$ sits in the background. The key new ingredients are:

(i) a subset T of the extended real line $\overline{\mathbb{R}}$;

(ii) a ***filtration*** $\{\mathcal{F}_t : t \in T\}$, that is, a collection of sub-sigma-fields of $\mathcal{F}$ for which $\mathcal{F}_s \subseteq \mathcal{F}_t$ if $s < t$;

(iii) a family of integrable random variables $\{X_t : t \in T\}$ ***adapted*** to the filtration, that is, X_t is $\mathcal{F}_t$-measurable for each t in T.

The set T has the interpretation of time, the sigma-field $\mathcal{F}_t$ has the interpretation of *information available at time* t, and X_t denotes some random quantity whose value $X_t(\omega)$ is revealed at time t.

<1> **Definition.** *A family of integrable random variables $\{X_t : t \in T\}$ adapted to a filtration $\{\mathcal{F}_t : t \in T\}$ is said to be a* ***martingale*** *(for that filtration) if*

(MG) $\qquad X_s \underset{a.s.}{=} \mathbb{P}(X_t \mid \mathcal{F}_s) \qquad$ for all $s < t$.

Equivalently, the random variables should satisfy

(MG)′ $\qquad \mathbb{P}X_s F = \mathbb{P}X_t F \qquad$ for all $F \in \mathcal{F}_s$, all $s < t$.

REMARK. Often the filtration is fixed throughout an argument, or the particular choice of filtration is not important for some assertion about the random variables. In such cases it is easier to talk about a martingale $\{X_t : t \in T\}$ without explicit mention of that filtration. If in doubt, we could always work with the ***filtration!natural***, $\mathcal{F}_t := \sigma\{X_s : s \le t\}$, which takes care of adaptedness, by definition.

Analogously, if there is a need to identify the filtration explicitly, it is convenient to speak of a martingale $\{(X_t, \mathcal{F}_t) : t \in T\}$, and so on.

Property (MG) has the interpretation that X_s is the best predictor for X_t based on the information available at time s. The equivalent formulation (MG)′ is a minor repackaging of the definition of the conditional expectation $\mathbb{P}(X_t \mid \mathcal{F}_s)$. The $\mathcal{F}_s$-measurability of X_s comes as part of the adaptation assumption. Approximation by simple functions, and a passage to the limit, gives another equivalence,

(MG)″ $\qquad \mathbb{P}X_s Z = \mathbb{P}X_t Z \qquad$ for all $Z \in \mathcal{M}_{bdd}(\mathcal{F}_s)$, all $s < t$,

where $\mathcal{M}_{bdd}(\mathcal{F}_s)$ denotes the set of all bounded, $\mathcal{F}_s$-measurable random variables. The formulations (MG)′ and (MG)″ have the advantage of removing the slippery concept of conditioning on sigma-fields from the definition of a martingale. One could develop much of the basic theory without explicit mention of conditioning, which would have some pedagogic advantages, even though it would obscure one of the important ideas behind the martingale concept.

Several of the desirable properties of martingales are shared by families of random variables for which the defining equalities (MG) and (MG)′ are relaxed to inequalities. I find that one of the hardest things to remember about these martingale relatives is which name goes with which direction of the inequality.

<2> **Definition.** *A family of integrable random variables $\{X_t : t \in T\}$ adapted to a filtration $\{\mathcal{F}_t : t \in T\}$ is said to be a* ***submartingale*** *(for that filtration) if it satisfies any (and hence all) of the following equivalent conditions:*

(subMG) $\qquad X_s \le \mathbb{P}(X_t \mid \mathcal{F}_s) \qquad$ for all $s < t$, $\qquad$ almost surely

(subMG)′ $\qquad \mathbb{P}X_s F \le \mathbb{P}X_t F \qquad$ for all $F \in \mathcal{F}_s$, all $s < t$.

(subMG)″ $\qquad \mathbb{P}X_s Z \le \mathbb{P}X_t Z \qquad$ for all $Z \in \mathcal{M}^+_{bdd}(\mathcal{F}_s)$, all $s < t$,

The family is said to be a ***supermartingale*** *(for that filtration) if $\{-X_t : t \in T\}$ is a submartingale. That is, the analogous requirements (superMG), (superMG)′, and (superMG)″ reverse the direction of the inequalities.*

REMARK. It is largely a matter of taste, or convenience of notation for particular applications, whether one works primarily with submartingales or supermartingales.

For most of this Chapter, the index set T will be ***discrete***, either finite or equal to $\mathbb{N}$, the set of positive integers, or equal to one of

$$\mathbb{N}_0 := \{0\} \cup \mathbb{N} \qquad \text{or} \qquad \overline{\mathbb{N}} := \mathbb{N} \cup \{\infty\} \qquad \text{or} \qquad \overline{\mathbb{N}}_0 := \{0\} \cup \mathbb{N} \cup \{\infty\}.$$

For some purposes it will be useful to have a distinctively labelled first or last element in the index set. For example, if a limit $X_\infty := \lim_{n \in \mathbb{N}} X_n$ can be shown to exist, it is natural to ask whether $\{X_n : n \in \overline{\mathbb{N}}\}$ also has sub- or supermartingale properties. Of course such a question only makes sense if a corresponding sigma-field $\mathcal{F}_\infty$ exists. If it is not otherwise defined, I will take $\mathcal{F}_\infty$ to be the sigma-field $\sigma\left(\cup_{i<\infty}\mathcal{F}_i\right)$.

Continuous time theory, where T is a subinterval of $\overline{\mathbb{R}}$, tends to be more complicated than discrete time. The difficulties arise, in part, from problems related to management of uncountable families of negligible sets associated with uncountable collections of almost sure equality or inequality assertions. A nontrivial part of the continuous time theory deals with sample path properties, that is, with the behavior of a process $X_t(\omega)$ as a function of t for fixed ω or with properties of X as a function of two variables. Such properties are typically derived from probabilistic assertions about finite or countable subfamilies of the $\{X_t\}$ random variables. An understanding of the discrete-time theory is an essential prerequisite for more ambitious undertakings in continuous time—see Appendix E.

For discrete time, the (MG)′ property becomes

$$\mathbb{P}X_n F = \mathbb{P}X_m F \qquad \text{for all } F \in \mathcal{F}_n, \text{ all } n < m.$$

It suffices to check the equality for $m = n + 1$, with $n \in \mathbb{N}_0$, for then repeated appeals to the special case extend the equality to $m = n + 2$, then $m = n + 3$, and so on. A similar simplification applies to submartingales and supermartingales.

<3> **Example.** Martingales generalize the theory for sums of independent random variables. Let $\xi_1, \xi_2, \ldots$ be independent, integrable random variables with $\mathbb{P}\xi_n = 0$ for $n \geq 1$. Define $X_0 := 0$ and $X_n := \xi_1 + \ldots + \xi_n$. The sequence $\{X_n : n \in \mathbb{N}_0\}$ is a martingale with respect to the natural filtration, because for $F \in \mathcal{F}_{n-1}$,

$$\mathbb{P}(X_n - X_{n-1})F = (\mathbb{P}\xi_n)(\mathbb{P}F) = 0 \qquad \text{by independence.}$$

You could write F as a measurable function of $X_1, \ldots, X_{n-1}$, or of $\xi_1, \ldots, \xi_{n-1}$, if you prefer to work with random variables. □

<4> **Example.** Let $\{X_n : n \in \mathbb{N}_0\}$ be a martingale and let Ψ be a convex function for which each $\Psi(X_n)$ is integrable. Then $\{\Psi(X_n) : n \in \mathbb{N}_0\}$ is a submartingale: the required almost sure inequality, $\mathbb{P}(\Psi(X_n) \mid \mathcal{F}_{n-1}) \geq \Psi(X_{n-1})$, is a direct application of the conditional expectation form of Jensen's inequality.

The companion result for submartingales is: if the convex Ψ function is increasing, if $\{X_n\}$ is a submartingale, and if each $\Psi(X_n)$ is integrable, then $\{\Psi(X_n) : n \in \mathbb{N}_0\}$ is a submartingale, because

$$\mathbb{P}(\Psi(X_n) \mid \mathcal{F}_{n-1}) \underset{a.s.}{\geq} \Psi(\mathbb{P}(X_n \mid \mathcal{F}_{n-1}) \underset{a.s.}{\geq} \Psi(X_{n-1}).$$

Two good examples to remember: if $\{X_n\}$ is a martingale and each X_n is square integrable then $\{X_n^2\}$ is a submartingale; and if $\{X_n\}$ is a submartingale then $\{X_n^+\}$ is also a submartingale. □

<5> **Example.** Let $\{X_n : n \in \mathbb{N}_0\}$ be a martingale written as a sum of increments, $X_n := X_0 + \xi_1 + \ldots + \xi_n$. Not surprisingly, the $\{\xi_i\}$ are called ***martingale differences***. Each ξ_n is integrable and $\mathbb{P}(\xi_n \mid \mathcal{F}_{n-1}) \underset{a.s.}{=} 0$ for $n \in \mathbb{N}_0^+$.

A new martingale can be built by weighting the increments using ***predictable*** functions $\{H_n : n \in \mathbb{N}\}$, meaning that each H_n should be an $\mathcal{F}_{n-1}$-measurable random variable, a more stringent requirement than adaptedness. The value of the weight becomes known before time n; it is known before it gets applied to the next increment.

If we assume that each $H_n\xi_n$ is integrable then the sequence

$$Y_n := X_0 + H_1\xi_1 + \ldots + H_n\xi_n$$

is both integrable and adapted. It is a martingale, because

$$\mathbb{P}H_i\xi_i F = \mathbb{P}(X_i - X_{i-1})(H_i F),$$

which equals zero by a simple generalization of (MG)″. (Use Dominated Convergence to accommodate integrable Z.) If $\{X_n : n \in \mathbb{N}_0\}$ is just a submartingale, a similar argument shows that the new sequence is also a submartingale, provided the predictable weights are also nonnegative. □

<6> **Example.** Suppose X is an integrable random variable and $\{\mathcal{F}_t : t \in T\}$ is a filtration. Define $X_t := \mathbb{P}(X \mid \mathcal{F}_t)$. Then the family $\{X_t : t \in T\}$ is a martingale with respect to the filtration, because for $s < t$,

$$\begin{aligned} \mathbb{P}(X_t F) &= \mathbb{P}(XF) && \text{if } F \in \mathcal{F}_t \\ &= \mathbb{P}(X_s F) && \text{if } F \in \mathcal{F}_s \end{aligned}$$

(We have just reproved the formula for conditioning on nested sigma-fields.) □

<7> **Example.** Every sequence $\{X_n : n \in \mathbb{N}_0\}$ of integrable random variables adapted to a filtration $\{\mathcal{F}_n : n \in \mathbb{N}_0\}$ can be broken into a sum of a martingale plus a sequence of accumulated conditional expectations. To establish this fact, consider the increments $\xi_n := X_n - X_{n-1}$. Each ξ_n is integrable, but it need not have zero conditional expectation given $\mathcal{F}_{n-1}$, the property that characterizes martingale differences. Extraction of the martingale component is merely a matter of recentering the increments to zero conditional expectations. Define $\eta_n := \mathbb{P}(\xi_n \mid \mathcal{F}_{n-1})$ and

$$M_n := X_0 + (\xi_1 - \eta_1) + \ldots + (\xi_n - \eta_n)$$
$$A_n := \eta_1 + \ldots + \eta_n.$$

Then $X_n = M_n + A_n$, with $\{M_n\}$ a martingale and $\{A_n\}$ a predictable sequence.

Often $\{A_n\}$ will have some nice behavior, perhaps due to the smoothing involved in the taking of a conditional expectation, or perhaps due to some other special property of the $\{X_n\}$. For example, if $\{X_n\}$ were a submartingale the η_i would all be nonnegative (almost surely) and $\{A_n\}$ would be an increasing sequence of random variables. Such properties are useful for establishing limit theory and inequalities—see Example <18> for an illustration of the general method. □

REMARK. The representation of a submartingale as a martingale plus an increasing, predictable process is sometimes called the ***Doob decomposition***. The corresponding representation for continuous time, which is exceedingly difficult to establish, is called the ***Doob-Meyer decomposition***.

2. Stopping times

The martingale property requires equalities $\mathbb{P}X_sF = \mathbb{P}X_tF$, for $s < t$ and $F \in \mathcal{F}_s$. Much of the power of the theory comes from the fact that analogous inequalities hold when s and t are replaced by certain types of random times. To make sense of the broader assertion, we need to define objects such as $\mathcal{F}_\tau$ and X_τ for random times τ.

<8> **Definition.** *A random variable τ taking values in $\overline{T} := T \cup \{\infty\}$ is called a* ***stopping time*** *for a filtration $\{\mathcal{F}_t : t \in T\}$ if $\{\tau \le t\} \in \mathcal{F}_t$ for each t in T.*

In discrete time, with $T = \mathbb{N}_0$, the defining property is equivalent to

$$\{\tau = n\} \in \mathcal{F}_n \qquad \text{for each } n \text{ in } \overline{\mathbb{N}}_0,$$

because $\{\tau \le n\} = \bigcup_{i \le n}\{\tau = i\}$ and $\{\tau = n\} = \{\tau \le n\}\{\tau \le n-1\}^c$.

<9> **Example.** Let $\{X_n : n \in \mathbb{N}_0\}$ be adapted to a filtration $\{\mathcal{F}_n : n \in \mathbb{N}_0\}$, and let B be a Borel subset of $\mathbb{R}$. Define $\tau(\omega) := \inf\{n : X_n(\omega) \in B\}$, with the interpretation that the infimum of the empty set equals $+\infty$. That is, $\tau(\omega) = +\infty$ if $X_n(\omega) \notin B$ for all n. The extended-real-valued random variable τ is a stopping time because

$$\{\tau \le n\} = \bigcup_{i \le n}\{X_i \in B\} \in \mathcal{F}_n \qquad \text{for } n \in \mathbb{N}_0.$$

It is called the ***first hitting time*** of the set B. Do you see why it is convenient to allow stopping times to take the value $+\infty$? □

If $\mathcal{F}_i$ corresponds to the information available up to time i, how should we define a sigma-field $\mathcal{F}_\tau$ to correspond to information available up to a random time τ? Intuitively, on the part of Ω where $\tau = i$ the sets in the sigma-field $\mathcal{F}_\tau$ should be the same as the sets in the sigma-field $\mathcal{F}_i$. That is, we could hope that

$$\{F\{\tau = i\} : F \in \mathcal{F}_\tau\} = \{F\{\tau = i\} : F \in \mathcal{F}_i\} \qquad \text{for each } i.$$

These equalities would be suitable as a definition of $\mathcal{F}_\tau$ in discrete time; we could define $\mathcal{F}_\tau$ to consist of all those F in $\mathcal{F}$ for which

$$F\{\tau = i\} \in \mathcal{F}_i \qquad \text{for all } i \in \overline{\mathbb{N}}_0. \tag{10}$$

For continuous time such a definition could become vacuous if all the sets $\{\tau = t\}$ were negligible, as sometimes happens. Instead, it is better to work with a definition that makes sense in both discrete and continuous time, and which is equivalent to <10> in discrete time.

<11> **Definition.** *Let τ be a stopping time for a filtration $\{\mathcal{F}_t : t \in T\}$, taking values in $\overline{T} := T \cup \{\infty\}$. If the sigma-field $\mathcal{F}_\infty$ is not already defined, take it to be $\sigma\left(\cup_{t \in T}\mathcal{F}_t\right)$. The pre-$\tau$ sigma-field $\mathcal{F}_\tau$ is defined to consist of all F for which $F\{\tau \le t\} \in \mathcal{F}_t$ for all $t \in \overline{T}$.*

The class $\mathcal{F}_\tau$ would not be a sigma-field if τ were not a stopping time: the property $\Omega \in \mathcal{F}_\tau$ requires $\{\tau \le t\} \in \mathcal{F}_t$ for all t.

REMARK. Notice that $\mathcal{F}_\tau \subseteq \mathcal{F}_\infty$ (because $F\{\tau \le \infty\} \in \mathcal{F}_\infty$ if $F \in \mathcal{F}_\tau$), with equality when $\tau \equiv \infty$. More generally, if τ takes a constant value, t, then $\mathcal{F}_\tau = \mathcal{F}_t$. It would be very awkward if we had to distinguish between random variables taking constant values and the constants themselves.

<12> **Example.** The stopping time τ is measurable with respect to $\mathcal{F}_\tau$, because, for each $\alpha \in \mathbb{R}^+$ and $t \in T$,

$$\{\tau \le \alpha\}\{\tau \le t\} = \{\tau \le \alpha \wedge t\} \in \mathcal{F}_{\alpha\wedge t} \subseteq \mathcal{F}_t.$$

That is, $\{\tau \le \alpha\} \in \mathcal{F}_\tau$ for all $\alpha \in \mathbb{R}^+$, from which the $\mathcal{F}_\tau$-measurability follows by the usual generating class argument. It would be counterintuitive if the information corresponding to the sigma-field $\mathcal{F}_\tau$ did not include the value taken by τ itself. □

<13> **Example.** Suppose σ and τ are both stopping times, for which $\sigma \le \tau$ always. Then $\mathcal{F}_\sigma \subseteq \mathcal{F}_\tau$ because

$$F\{\tau \le t\} = \big(F\{\sigma \le t\}\big)\{\tau \le t\} \qquad \text{for all } t \in \overline{T},$$

and both sets on the right-hand side are $\mathcal{F}_t$-measurable if $F \in \mathcal{F}_\sigma$. □

<14> **Exercise.** Show that a random variable Z is $\mathcal{F}_\tau$-measurable if and only if $Z\{\tau \le t\}$ is $\mathcal{F}_t$-measurable for all t in $\overline{T}$.

SOLUTION: For necessity, write Z as a pointwise limit of $\mathcal{F}_\tau$-measurable simple functions Z_n, then note that each $Z_n\{\tau \le t\}$ is a linear combination of indicator functions of $\mathcal{F}_t$-measurable sets.

For sufficiency, it is enough to show that $\{Z > \alpha\} \in \mathcal{F}_\tau$ and $\{Z < -\alpha\} \in \mathcal{F}_\tau$, for each $\alpha \in \mathbb{R}^+$. For the first requirement, note that $\{Z > \alpha\}\{\tau \le t\} = \{Z\{\tau \le t\} > \alpha\}$, which belongs to $\mathcal{F}_t$ for each t, because $Z\{\tau \le t\}$ is assumed to be $\mathcal{F}_t$-measurable. Thus $\{Z > \alpha\} \in \mathcal{F}_\tau$. Argue similarly for the other requirement. □

The definition of X_τ is almost straightforward. Given random variables $\{X_t : t \in T\}$ and a stopping time τ, we should define X_τ as the function taking the value $X_t(\omega)$ when $\tau(\omega) = t$. If τ takes only values in T there is no problem. However, a slight embarrassment would occur when $\tau(\omega) = \infty$ if ∞ were not a point of T, for then $X_\infty(\omega)$ need not be defined. In the happy situation when there is a natural candidate for X_∞, the embarrassment disappears with little fuss; otherwise it is wiser to avoid the difficulty altogether by working only with the random variable $X_\tau\{\tau < \infty\}$, which takes the value zero when τ is infinite.

Measurability of $X_\tau\{\tau < \infty\}$, even with respect to the sigma-field $\mathcal{F}$, requires further assumptions about the $\{X_t\}$ for continuous time. For discrete time the task is much easier. For example, if $\{X_n : n \in \mathbb{N}_0\}$ is adapated to a filtration $\{\mathcal{F}_n : n \in \mathbb{N}_0\}$, and τ is a stopping time for that filtration, then

$$X_\tau\{\tau < \infty\}\{\tau \le t\} = \sum_{i\in\mathbb{N}_0} X_i\{i = \tau \le t\}.$$

For $i > t$ the ith summand is zero; for $i \le t$ it equals $X_i\{\tau = i\}$, which is $\mathcal{F}_i$-measurable. The $\mathcal{F}_\tau$-measurability of $X_\tau\{\tau < \infty\}$ then follows by Exercise <14>

The next Exercise illustrates the use of stopping times and the σ-fields they define. The discussion does not directly involve martingales, but they are lurking in the background.

<15> **Exercise.** A deck of 52 cards (26 reds, 26 blacks) is dealt out one card at a time, face up. Once, and only once, you will be allowed to predict that the next card will be red. What strategy will maximize your probability of predicting correctly?

SOLUTION: Write R_i for the event $\{i\text{th card is red}\}$. Assume all permutations of the deck are equally likely initially. Write $\mathcal{F}_n$ for the σ-field generated by $R_1, \dots, R_n$. A strategy corresponds to a stopping time τ that takes values in $\{0, 1, \dots, 51\}$: you should try to maximize $\mathbb{P}R_{\tau+1}$.

Surprisingly, $\mathbb{P}R_{\tau+1} = 1/2$ for all such stopping rules. The intuitive explanation is that you should always be indifferent, given that you have observed cards $1, 2, \dots, \tau$, between choosing card $\tau + 1$ or choosing card 52. That is, it should be true that $\mathbb{P}(R_{\tau+1} \mid \mathcal{F}_\tau) = \mathbb{P}(R_{52} \mid \mathcal{F}_\tau)$ almost surely; or, equivalently, that $\mathbb{P}R_{\tau+1}F = \mathbb{P}R_{52}F$ for all $F \in \mathcal{F}_\tau$; or, equivalently, that

$$\mathbb{P}R_{k+1}F\{\tau = k\} = \mathbb{P}R_{52}F\{\tau = k\} \qquad \text{for all } F \in \mathcal{F}_\tau \text{ and } k = 0, 1, \dots, 51.$$

We could then deduce that $\mathbb{P}R_{\tau+1} = \mathbb{P}R_{52} = 1/2$. Of course, we only need the case $F = \Omega$, but I'll carry along the general F as an illustration of technique while proving the assertion in the last display. By definition of $\mathcal{F}_\tau$,

$$F\{\tau = k\} = F\{\tau \le k - 1\}^c\{\tau \le k\} \in \mathcal{F}_k.$$

That is, $F\{\tau = k\}$ must be of the form $\{(R_1, \dots, R_k) \in B\}$ for some Borel subset B of $\mathbb{R}^k$. Symmetry of the joint distribution of $R_1, \dots, R_{52}$ implies that the random vector $(R_1, \dots, R_k, R_{k+1})$ has the same distribution as the random vector $(R_1, \dots, R_k, R_{52})$, whence

$$\mathbb{P}R_{k+1}\{(R_1, \dots, R_k) \in B\} = \mathbb{P}R_{52}\{(R_1, \dots, R_k) \in B\}.$$

□

See Section 8 for more about symmetry and martingale properties.

The hidden martingale in the previous Exercise is X_n, the proportion of red cards remaining in the deck after n cards have been dealt. You could check the martingale property by first verifying that $\mathbb{P}(R_{n+1} \mid \mathcal{F}_n) = X_n$ (an equality that is obvious if one thinks in terms of conditional distributions), then calculating

$$(52-n-1)\mathbb{P}(X_{n+1} \mid \mathcal{F}_n) = \mathbb{P}\left((52-n)X_n - R_{n+1} \mid \mathcal{F}_n\right) = (52-n)X_n - \mathbb{P}(R_{n+1} \mid \mathcal{F}_n).$$

The problem then asks for the stopping time to maximaize

$$\begin{aligned} \mathbb{P}R_{\tau+1} &= \textstyle\sum_{i=0}^{51} \mathbb{P}(R_{i+1}\{\tau = i\}) \\ &= \textstyle\sum_{i=0}^{51} \mathbb{P}(X_i\{\tau = i\}) \qquad \text{because } \{\tau = i\} \in \mathcal{F}_i \\ &= \mathbb{P}X_\tau. \end{aligned}$$

The martingale property tells us that $\mathbb{P}X_0 = \mathbb{P}X_i$ for $i = 1, \dots, 51$. If we could extend the equality to random i, by showing that $\mathbb{P}X_\tau = \mathbb{P}X_0$, then the surprising conclusion from the Exercise would follow.

Clearly it would be useful if we could always assert that $\mathbb{P}X_\sigma = \mathbb{P}X_\tau$ for every martingale, and every pair of stopping times. Unfortunately (Or should I say fortunately?) the result is not true without some extra assumptions. The simplest and most useful case concerns finite time sets. If σ takes values in a finite set T, and if each X_t is integrable, then $|X_\sigma| \le \sum_{t\in T} |X_t|$, which eliminates any integrability difficulties. For an infinite index set, the integrability of X_σ is not automatic.

<16> **Stopping Time Lemma.** *Suppose σ and τ are stopping times for a filtration $\{\mathcal{F}_t : t \in T\}$, with T finite. Suppose both stopping times take only values in T. Let F be a set in $\mathcal{F}_\sigma$ for which $\sigma(\omega) \le \tau(\omega)$ when $\omega \in F$. If $\{X_t : t \in T\}$ is a submartingale, then $\mathbb{P}X_\sigma F \le \mathbb{P}X_\tau F$. For supermartingales, the inequality is reversed. For martingales, the inequality becomes an equality.*

Proof. Consider only the submartingale case. For simplicity of notation, suppose $T = \{0, 1, \dots, N\}$. Write each X_n as a sum of increments, $X_n = X_0 + \xi_1 + \dots + \xi_n$. The inequality $\sigma \le \tau$, on F, lets us write

$$X_\tau F - X_\sigma F = \left(X_0 F + \sum_{1\le i\le N} \{i \le \tau\} F\xi_i\right) - \left(X_0 F + \sum_{1\le i\le N} \{i \le \sigma\} F\xi_i\right)$$
$$= \sum_{1\le i\le N} \{\sigma < i \le \tau\} F\xi_i.$$

Note that $\{\sigma < i \le \tau\}F = \left(\{\sigma \le i-1\}F\right)\{\tau \le i-1\}^c \in \mathcal{F}_{i-1}$. The expected value of each summand is nonnegative, by (subMG)′. □

REMARK. If $\sigma \le \tau$ everywhere, the inequality for all F in $\mathcal{F}_\sigma$ implies that $X_\sigma \le \mathbb{P}\left(X_\tau \mid \mathcal{F}_\sigma\right)$ almost surely. That is, the submartingale (or martingale, or supermartingale) property is preserved at bounded stopping times.

The Stopping Time Lemma, and its extensions to various cases with infinite index sets, is basic to many of the most elegant martingale properties. Results for a general stopping time τ, taking values in $\overline{\mathbb{N}}$ or $\overline{\mathbb{N}}_0$, can often be deduced from results for $\tau \wedge N$, followed by a passage to the limit as N tends to infinity. (The random variable $\tau \wedge N$ is a stopping time, because $\{\tau \wedge N \le n\}$ equals the whole of Ω when $N \le n$, and equals $\{\tau \le n\}$ when $N > n$.) As Problem [1] shows, the finiteness assumption on the index set T is not just a notational convenience; the Lemma <16> can fail for infinite T.

It is amazing how many of the classical inequalities of probability theory can be derived by invoking the Lemma for a suitable martingale (or submartingale or supermartingale).

<17> **Exercise.** Let $\xi_1, \dots, \xi_N$ be independent random variables (or even just martingale increments) for which $\mathbb{P}\xi_i = 0$ and $\mathbb{P}\xi_i^2 < \infty$ for each i. Define $S_i := \xi_1 + \dots + \xi_i$. Prove the maximal inequality ***Kolmogorov inequality***: for each $\epsilon > 0$,

$$\mathbb{P}\left\{\max_{1\le i\le N} |S_i| \ge \epsilon\right\} \le \mathbb{P}S_N^2/\epsilon^2.$$

SOLUTION: The random variables $X_i := S_i^2$ form a submartingale, for the natural filtration. Define stopping times $\tau \equiv N$ and $\sigma :=$ first i such that $|S_i| \ge \epsilon$, with the convention that $\sigma = N$ if $|S_i| < \epsilon$ for every i. Why is σ a stopping time? Check the pointwise bound,

$$\epsilon^2\{\max_i |S_i| \ge \epsilon\} = \epsilon^2\{X_\sigma \ge \epsilon^2\} \le X_\sigma.$$

What happens in the case when σ equals N because $|S_i| < \epsilon$ for every i? Take expectations, then invoke the Stopping Time Lemma (with $F = \Omega$) for the submartingale $\{X_i\}$, to deduce

$$\epsilon^2 \mathbb{P}\{\max_i |S_i| \geq \epsilon\} \leq \mathbb{P}X_\sigma \leq \mathbb{P}X_\tau = \mathbb{P}S_N^2,$$

□ as asserted.

Notice how the Kolmogorov inequality improves upon the elementary bound $\mathbb{P}\{|S_N| \geq \epsilon\} \leq \mathbb{P}S_N^2/\epsilon^2$. Actually it is the same inequality, applied to S_σ instead of S_N, supplemented by a useful bound for $\mathbb{P}S_\sigma^2$ made possible by the submartingale property. Kolmogorov (1928) established his inequality as the first step towards a proof of various convergence results for sums of independent random variables. More versatile maximal inequalities follow from more involved appeals to the Stopping Time Lemma. For example, a strong law of large numbers can be proved quite efficiently (Bauer 1981, Section 6.3) by an appeal to the next inequality.

<18> **Exercise.** Let $0 = S_0, \ldots, S_N$ be a martingale with $v_i := \mathbb{P}(S_i - S_{i-1})^2 < \infty$ for each i. Let $\gamma_1 \geq \gamma_2 \geq \ldots \geq \gamma_N$ be nonnegative constants. Prove the ***Hájek-Rényi inequality***:

<19>
$$\mathbb{P}\left\{\max_{1\leq i\leq N} \gamma_i |S_i| \geq 1\right\} \leq \sum_{1\leq i\leq N} \gamma_i^2 v_i.$$

SOLUTION: Define $\mathcal{F}_i := \sigma(S_1, \ldots, S_i)$. Write $\mathbb{P}_i$ for $\mathbb{P}(\cdot \mid \mathcal{F}_i)$. Define $\eta_i := \gamma_i^2 S_i^2 - \gamma_{i-1}^2 S_{i-1}^2$, and $\Delta_i := \mathbb{P}_{i-1}\eta_i$. By the Doob decomposition from Example <7>, the sequence $M_k := \sum_{i=1}^k (\eta_i - \Delta_i)$ is a martingale with respect to the filtration $\{\mathcal{F}_i\}$; and $\gamma_k^2 S_k^2 = (\Delta_1 + \ldots + \Delta_k) + M_k$. Define stopping times $\sigma \equiv 0$ and

$$\tau = \begin{cases} \text{first } i \text{ such that } \gamma_i |S_i| \geq 1, \\ N \text{ if } \gamma_i |S_i| < 1 \text{ for all } i. \end{cases}$$

The main idea is to bound each $\Delta_1 + \ldots + \Delta_k$ by a single random variable Δ, whose expectation will become the right-hand side of the asserted inequality.

Construct Δ from the martingale differences $\xi_i := S_i - S_{i-1}$ for $i = 1, \ldots, N$. For each i, use the fact that S_{i-1} is $\mathcal{F}_{i-1}$ measurable to bound the contribution of Δ_i:

$$\begin{aligned} \Delta_i &= \mathbb{P}_{i-1}\left(\gamma_i^2 S_i^2 - \gamma_{i-1}^2 S_{i-1}^2\right) \\ &= \gamma_i^2 \mathbb{P}_{i-1}\left(\xi_i^2 + 2\xi_i S_{i-1} + S_{i-1}^2\right) - \gamma_{i-1}^2 S_{i-1}^2 \\ &= \gamma_i^2 \mathbb{P}_{i-1}\xi_i^2 + 2\gamma_i^2 S_{i-1} \mathbb{P}_{i-1}\xi_i + (\gamma_i^2 - \gamma_{i-1}^2) S_{i-1}^2. \end{aligned}$$

The middle term on the last line vanishes, by the martingale difference property, and the last term is negative, because $\gamma_i^2 \leq \gamma_{i-1}^2$. The sum of the three terms is less than the nonnegative quantity $\gamma_i^2 \mathbb{P}(\xi_i^2 \mid \mathcal{F}_{i-1})$, and

$$\Delta := \sum_{i\leq N} \gamma_i^2 \mathbb{P}_{i-1}\xi_i^2 \geq \sum_{i\leq k} \Delta_i,$$

for each k, as required.

The asserted inequality now follows via the Stopping Time Lemma:

$$
\begin{aligned}
\mathbb{P}\{\max_i \gamma_i |S_i| \geq 1\} &= \mathbb{P}\{\gamma_\tau |S_\tau| \geq 1\} \\
&\leq \mathbb{P}\gamma_\tau^2 S_\tau^2 \\
&= \mathbb{P}M_\tau + \mathbb{P}(\Delta_1 + \ldots + \Delta_\tau) \\
&\leq \mathbb{P}\Delta,
\end{aligned}
$$

because $\Delta_1 + \ldots + \Delta_\tau \leq \Delta$ and $\mathbb{P}M_\tau = \mathbb{P}M_\sigma = 0$. □

The method of proof in Example <18> is worth remembering; it can be used to derive several other bounds.

3. Convergence of positive supermartingales

In several respects the theory for positive (meaning nonnegative) supermartingales $\{X_n : n \in \mathbb{N}_0\}$ is particularly elegant. For example (Problem [5]), the Stopping Time Lemma extends naturally to pairs of unbounded stopping times for positive supermartingales. Even more pleasantly surprising, positive supermartingales converge almost surely, to an integrable limit—as will be shown in this Section.

The key result for the proof of convergence is an elegant lemma (Dubins's Inequality) that shows why a positive supermartingale $\{X_n\}$ cannot oscillate between two levels infinitely often.

For fixed constants α and β with $0 \leq \alpha < \beta < \infty$ define increasing sequences of random times at which the process might drop below α or rise above β:

$$
\begin{aligned}
\sigma_1 &:= \inf\{i \geq 0 : X_i \leq \alpha\}, & \tau_1 &:= \inf\{i \geq \sigma_1 : X_i \geq \beta\}, \\
\sigma_2 &:= \inf\{i \geq \tau_1 : X_i \leq \alpha\}, & \tau_2 &:= \inf\{i \geq \sigma_2 : X_i \geq \beta\},
\end{aligned}
$$

and so on, with the convention that the infimum of an empty set is taken as $+\infty$.

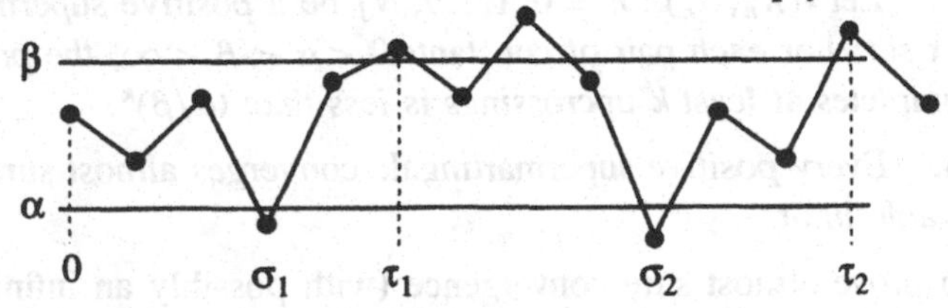

Because the $\{X_i\}$ are adapted to $\{\mathcal{F}_i\}$, each σ_i and τ_i is a stopping time for the filtration. For example,

$$
\{\tau_1 \leq k\} = \{X_i \leq \alpha,\ X_j \geq \beta \text{ for some } i \leq j \leq k\},
$$

which could be written out explicitly as a finite union of events involving only $X_0, \ldots, X_k$.

When τ_k is finite, the segment $\{X_i : \sigma_k \leq i \leq \tau_k\}$ is called the kth ***upcrossing*** of the interval $[\alpha, \beta]$ by the process $\{X_n : n \in \mathbb{N}_0\}$. The event $\{\tau_k \leq N\}$ may be described, slightly informally, by saying that the process completes at least k upcrossings of $[\alpha, \beta]$ up to time N.

<20> **Dubins's inequality.** *For a positive supermartingale $\{(X_n, \mathcal{F}_n) : n \in \mathbb{N}_0\}$ and constants $0 \le \alpha < \beta < \infty$, and stopping times as defined above,*

$$\mathbb{P}\{\tau_k < \infty\} \le (\alpha/\beta)^k \qquad \text{for } k \in \mathbb{N}.$$

Proof. Choose, and temporarily hold fixed, a finite positive integer N. Define τ_0 to be identically zero. For $k \ge 1$, using the fact that $X_{\tau_k} \ge \beta$ when $\tau_k < \infty$ and $X_{\sigma_k} \le \alpha$ when $\sigma_k < \infty$, we have

$$\begin{aligned}\mathbb{P}\left(\beta\{\tau_k \le N\} + X_N\{\tau_k > N\}\right) &\le \mathbb{P}X_{\tau_k \wedge N} \\ &\le \mathbb{P}X_{\sigma_k \wedge N} \qquad \text{Stopping Time Lemma} \\ &\le \mathbb{P}\left(\alpha\{\sigma_k \le N\} + X_N\{\sigma_k > N\}\right),\end{aligned}$$

which rearranges to give

$$\begin{aligned}\beta\mathbb{P}\{\tau_k \le N\} &\le \alpha\mathbb{P}\{\sigma_k \le N\} + \mathbb{P}X_N\left(\{\sigma_k > N\} - \{\tau_k > N\}\right) \\ &\le \alpha\mathbb{P}\{\tau_{k-1} \le N\} \qquad \text{because } \tau_{k-1} \le \sigma_k \le \tau_k \text{ and } X_N \ge 0.\end{aligned}$$

That is,

$$\mathbb{P}\{\tau_k \le N\} \le \frac{\alpha}{\beta}\mathbb{P}\{\tau_{k-1} \le N\} \qquad \text{for } k \ge 1.$$

Repeated appeals to this inequality, followed by a passage to the limit as $N \to \infty$, leads to Dubins's Inequality. □

REMARK. When $0 = \alpha < \beta$ we have $\mathbb{P}\{\tau_1 < \infty\} = 0$. By considering a sequence of β values decreasing to zero, we deduce that on the set $\{\sigma_1 < \infty\}$ we must have $X_n = 0$ for all $n \ge \sigma_1$. That is, if a positive supermartingale hits zero then it must stay there forever.

Notice that the main part of the argument, before N was sent off to infinity, involved only the variables $X_0, \ldots, X_N$. The result may fruitfully be reexpressed as an assertion about positive supermartingales with a finite index set.

<21> **Corollary.** *Let $\{(X_n, \mathcal{F}_n) : n = 0, 1, \ldots, N\}$ be a positive supermartingale with a finite index set. For each pair of constants $0 < \alpha < \beta < \infty$, the probability that the process completes at least k upcrossings is less than $(\alpha/\beta)^k$.*

<22> **Theorem.** *Every positive supermartingale converges almost surely to a nonnegative, integrable limit.*

Proof. To prove almost sure convergence (with possibly an infinite limit) of the sequence $\{X_n\}$, it is enough to show that the event

$$D = \{\omega : \limsup X_n(\omega) > \liminf X_n(\omega)\}$$

is negligible. Decompose D into a countable union of events

$$D_{\alpha,\beta} = \{\limsup X_n > \beta > \alpha > \liminf X_n\},$$

with α, β ranging over all pairs of rational numbers. On $D_{\alpha,\beta}$ we must have $\tau_k < \infty$ for every k. Thus $\mathbb{P}D_{\alpha,\beta} \le (\alpha/\beta)^k$ for every k, which forces $\mathbb{P}D_{\alpha,\beta} = 0$, and $\mathbb{P}D = 0$.

The sequence X_n converges to $X_\infty := \liminf X_n$ on the set D^c. Fatou's lemma, and the fact that $\mathbb{P}X_n$ is nonincreasing, ensure that X_∞ is integrable. □

<23> **Exercise.** Suppose $\{\xi_i\}$ are independent, identically distributed random variables ξ_i with $\mathbb{P}\{\xi_i = +1\} = p$ and $\mathbb{P}\{\xi_i = -1\} = 1 - p$. Define the partial

sums $S_0 = 0$ and $S_i = \xi_1 + \ldots + \xi_i$ for $i \geq 1$. For $1/2 \leq p < 1$, show that $\mathbb{P}\{S_i = -1 \text{ for at least one } i\} = (1-p)/p$.

SOLUTION: Consider a fixed p with $1/2 < p < 1$. Define $\theta = (1-p)/p$. Define $\tau = \inf\{i \in \mathbb{N} : S_i = -1\}$. We are trying to show that $\mathbb{P}\{\tau < \infty\} = \theta$. Observe that $X_n = \theta^{S_n}$ is a positive martingale with respect to the filtration $\mathcal{F}_n = \sigma(\xi_1, \ldots, \xi_n)$: by independence and the equality $\mathbb{P}\theta^{\xi_i} = 1$,

$$\mathbb{P}X_n F = \mathbb{P}\theta^{\xi_n}\theta^{S_{n-1}} F = \mathbb{P}\theta^{\xi_n}\mathbb{P}X_{n-1}F = \mathbb{P}X_{n-1}F \qquad \text{for } F \text{ in } \mathcal{F}_{n-1}.$$

The sequence $\{X_{\tau\wedge n}\}$ is a positive martingale (Problem [3]). It follows that there exists an integrable X_∞ such that $X_{\tau\wedge n} \to X_\infty$ almost surely. The sequence $\{S_n\}$ cannot converge to a finite limit because $|S_n - S_{n-1}| = 1$ for all n. On the set where $\tau = \infty$, convergence of θ^{S_n} to a finite limit is possible only if $S_n \to \infty$ and $\theta^{S_n} \to 0$. Thus,

$$X_{\tau\wedge n} \to \theta^{-1}\{\tau < \infty\} + 0\{\tau = \infty\} \qquad \text{almost surely.}$$

The bounds $0 \leq X_{\tau\wedge n} \leq \theta^{-1}$ allow us to invoke Dominated Convergence to deduce that $1 = \mathbb{P}X_{\tau\wedge n} \to \theta^{-1}\mathbb{P}\{\tau < \infty\}$.

□ Monotonicity of $\mathbb{P}\{\tau < \infty\}$ as a function of p extends the solution to $p = 1/2$.

The almost sure limit X_∞ of a positive supermartingale $\{X_n\}$ satisfies the inequality $\liminf \mathbb{P}X_n \geq \mathbb{P}X_\infty$, by Fatou. The sequence $\{\mathbb{P}X_n\}$ is decreasing. Under what circumstances do we have it converging to $\mathbb{P}X_\infty$? Equality certainly holds if $\{X_n\}$ converges to X_∞ in L^1 norm. In fact, convergence of expectations is equivalent to L^1 convergence, because

$$\begin{aligned}\mathbb{P}|X_n - X_\infty| &= \mathbb{P}(X_\infty - X_n)^+ + \mathbb{P}(X_\infty - X_n)^- \\ &= 2\mathbb{P}(X_\infty - X_n)^+ - (\mathbb{P}X_\infty - \mathbb{P}X_n).\end{aligned}$$

On the right-hand side the first contribution tends to zero, by Dominated Convergence, because $X_\infty \geq (X_\infty - X_n)^+ \to 0$ almost surely. (I just reproved Scheffé's lemma.)

<24> **Corollary.** *A positive supermartingale $\{X_n\}$ converges in L^1 to its limit X_∞ if and only if $\mathbb{P}X_n \to \mathbb{P}X_\infty$.*

<25> **Example.** Female bunyips reproduce once every ten years, according to a fixed offspring distribution P on $\mathbb{N}$. Different bunyips reproduce independently of each other. What is the behavior of the number Z_n of nth generation offspring from Lucy bunyip, the first of the line, as n gets large? (The process $\{Z_n : n \in \mathbb{N}_0\}$ is usually called a ***branching process***.)

Write μ for the expected number of offspring for a single bunyip. If reproduction went strictly according to averages, the nth generation size would equal μ^n. Intuitively, if $\mu > 1$ there could be an explosion of the bunyip population; if $\mu < 1$ bunyips would be driven to extinction; if $\mu = 1$, something else might happen. A martingale argument will lead rigorously to a similar conclusion.

Given $Z_{n-1} = k$, the size of the nth generation is a sum of k independent random variables, each with distribution P. Perhaps we could write $Z_n = \sum_{i=1}^{Z_{n-1}} \xi_{ni}$, with the $\{\xi_{ni} : i = 1, \ldots, Z_{n-1}\}$ (conditionally) independently distributed like P. I have a few difficulties with that representation. For example, where is ξ_{n3} defined?

Just on $\{Z_{n-1} \geq 3\}$? On all of Ω? Moreover, the notation invites the blunder of ignoring the randomness of the range of summation, leading to an absurd assertion that $\mathbb{P}\sum_{i=1}^{Z_{n-1}} \xi_{ni}$ equals $\sum_{i=1}^{Z_{n-1}} \mathbb{P}\xi_{ni} = Z_{n-1}\mu$. The corresponding assertion for an expectation conditional on Z_{n-1} is correct, but some of the doubts still linger.

It is much better to start with an entire family $\{\xi_{ni} : n \in \mathbb{N}, i \in \mathbb{N}\}$ of independent random variables, each with distribution P, then define $Z_0 = 1$ and $Z_n := \sum_{i\in\mathbb{N}} \xi_{ni}\{i \leq Z_{n-1}\}$ for $n \geq 1$. The random variable Z_n is measurable with respect to the sigma-field $\mathcal{F}_n = \sigma\{\xi_{ki} : k \leq n, i \in \mathbb{N}\}$, and, almost surely,

$$
\begin{aligned}
\mathbb{P}(Z_n \mid \mathcal{F}_{n-1}) &= \sum_{i\in\mathbb{N}} \mathbb{P}(\xi_{ni}\{i \leq Z_{n-1}\} \mid \mathcal{F}_{n-1}) \\
&= \sum_{i\in\mathbb{N}} \{i \leq Z_{n-1}\}\mathbb{P}(\xi_{ni} \mid \mathcal{F}_{n-1}) && \text{because } Z_{n-1} \text{ is } \mathcal{F}_{n-1}\text{-measurable} \\
&= \sum_{i\in\mathbb{N}} \{i \leq Z_{n-1}\}\mathbb{P}(\xi_{ni}) && \text{because } \xi_{ni} \text{ is independent of } \mathcal{F}_{n-1} \\
&= Z_{n-1}\mu.
\end{aligned}
$$

If $\mu \leq 1$, the $\{Z_n\}$ sequence is a positive supermartingale with respect to the $\{\mathcal{F}_n\}$ filtration. By Theorem <22>, there exists an integrable random variable Z_∞ with $Z_n \to Z_\infty$ almost surely.

A sequence of integers $Z_n(\omega)$ can converge to a finite limit k only if $Z_n(\omega) = k$ for all n large enough. If $k > 0$, the convergence would imply that, with nonzero probability, only finitely many of the independent events $\{\sum_{i\leq k} \xi_{ni} \neq k\}$ can occur. By the converse to the Borel-Cantelli lemma, it would follow that $\sum_{i\leq k} \xi_{ni} = k$ almost surely, which can happen only if $P\{1\} = 1$. In that case, $Z_n \equiv 1$ for all n. If $P\{1\} < 1$, then Z_n must converge to zero, with probability one, if $\mu \leq 1$. The bunyips die out if the average number of offspring is less than or equal to 1.

If $\mu > 1$ the situation is more complex. If $P\{0\} = 0$ the population cannot decrease and Z_n must then diverge to infinity with probability one. If $P\{0\} > 0$ the convex function $g(t) := P^x t^x$ must have a unique value θ with $0 < \theta < 1$ for which $g(\theta) = \theta$: the strictly convex function $h(t) := g(t) - t$ has $h(0) = P\{0\} > 0$ and $h(1) = 0$, and its left-hand derivative $P^x(xt^{x-1} - 1)$ converges to $\mu - 1 > 0$ as t increases to 1. The sequence $\{\theta^{Z_n}\}$ is a positive martingale:

$$
\begin{aligned}
\mathbb{P}(\theta^{Z_n} \mid \mathcal{F}_{n-1}) &= \{Z_{n-1} = 0\} + \sum_{k\geq 1} \mathbb{P}\left(\theta^{\xi_{n,1}+\ldots+\xi_{n,k}}\{Z_{n-1} = k\} \mid \mathcal{F}_{n-1}\right) \\
&= \{Z_{n-1} = 0\} + \sum_{k\geq 1} \{Z_{n-1} = k\}\mathbb{P}(\theta^{\xi_{n1}+\ldots+\xi_{n,k-1}} \mid \mathcal{F}_{n-1}) \\
&= \sum_{k\in\mathbb{N}_0} \{Z_{n-1} = k\} g(\theta)^k \\
&= g(\theta)^{Z_{n-1}} = \theta^{Z_{n-1}} \qquad \text{because } g(\theta) = \theta.
\end{aligned}
$$

The positive martingale $\{\theta^{Z_n}\}$ has an almost sure limit, W. The sequence $\{Z_n\}$ must converge almost surely, with an infinite limit when $W = 0$. As with the situation when $\mu \leq 1$, the only other possible limit for Z_n is 0, corresponding to $W = 1$. Because $0 \leq \theta^{Z_n} \leq 1$ for every n, Dominated Convergence and the martingale property give $\mathbb{P}\{W = 1\} = \lim_{n\to\infty} \mathbb{P}\theta^{Z_n} = \mathbb{P}\theta^{Z_0} = \theta$.

In summary: On the set $D := \{W = 1\}$, which has probability θ, the bunyip population eventually dies out; on D^c, the population explodes, that is, $Z_n \to \infty$.

It is possible to say a lot more about the almost sure behavior of the process $\{Z_n\}$ when $\mu > 1$. For example, the sequence $X_n := Z_n/\mu^n$ is a positive martingale, which must converge to an integrable random variable X. On the set $\{X > 0\}$, the process $\{Z_n\}$ grows geometrically fast, like μ^n. On the set D we must have $X = 0$, but it is not obvious whether or not we might have $X = 0$ for some realizations where the process does not die out.

There is a simple argument to show that, in fact, either $X = 0$ almost surely or $X > 0$ almost surely on D. With a little extra bookkeeping we could keep track of the first generation ancestor of each bunyip in the later generations. If we write $Z_n^{(j)}$ for the members of the nth generation descended from the jth (possibly hypothetical) member of the first generation, then $Z_n = \sum_{j\in\mathbb{N}} Z_n^{(j)}\{j \le Z_1\}$. The $Z_n^{(j)}$, for $j = 1, 2, \ldots$, are independent random variables, each with the same distribution as Z_{n-1}, and each independent of Z_1. In particular, for each j, we have $Z_n^{(j)}/\mu^{n-1} \to X^{(j)}$ almost surely, where the $X^{(j)}$, for $j = 1, 2, \ldots$, are independent random variables, each distributed like X, and

$$\mu X = \textstyle\sum_{j\in\mathbb{N}} X^{(j)}\{j \le Z_1\} \qquad \text{almost surely.}$$

Write ϕ for $\mathbb{P}\{X = 0\}$. Then, by independence,

$$\mathbb{P}\{X = 0 \mid Z_1 = k\} = \prod_{j=1}^{k} \mathbb{P}\{X^{(j)} = 0\} = \phi^k,$$

whence $\phi = \sum_{k\in\mathbb{N}_0} \phi^k \mathbb{P}\{Z_1 = k\} = g(\phi)$. We must have either $\phi = 1$, meaning that $X = 0$ almost surely, or else $\phi = \theta$, in which case $X > 0$ almost surely on D^c. The latter must be the case if X is nondegenerate, that is, if $\mathbb{P}\{X > 0\} > 0$, which happens if and only if $P^x(x\log(1 + x)) < \infty$—see Problem [14]. □

4. Convergence of submartingales

Theorem <22> can be extended to a large class of submartingales by means of the following decomposition Theorem, whose proof appears in the next Section.

<26> **Krickeberg decomposition.** *Let $\{S_n : n \in \mathbb{N}_0\}$ be a submartingale for which $\sup_n \mathbb{P}S_n^+ < \infty$. Then there exists a positive martingale $\{M_n\}$ and a positive supermartingale $\{X_n\}$ such that $S_n = M_n - X_n$ almost surely, for each n.*

<27> **Corollary.** *A submartingale with $\sup_n \mathbb{P}S_n^+ < \infty$ converges almost surely to an integrable limit.*

For a direct proof of this convergence result, via an upcrossing inequality for supermartingales that are not necessarily nonnegative, see Problem [11].

REMARK. Finiteness of $\sup_n \mathbb{P}S_n^+$ is equivalent to the finiteness of $\sup_n \mathbb{P}|S_n|$, because $|S_n| = 2S_n^+ - (S_n^+ - S_n^-)$ and by the submartingale property, $\mathbb{P}(S_n^+ - S_n^-) = \mathbb{P}S_n$ increases with n.

<28> **Example.** (Section 68 of Lévy 1937.) Let $\{M_n : n \in \mathbb{N}_0\}$ be a martingale such that $|M_n - M_{n-1}| \le 1$, for all n, and $M_0 = 0$. In order that $M_n(\omega)$ converges to a

finite limit, it is necessary that $\sup_n M_n(\omega)$ be finite. In fact, it is also a sufficicient condition. More precisely

$$\{\omega : \lim_n M_n(\omega) \text{ exists as a finite limit}\} = \{\omega : \sup M_n(\omega) < \infty\} \qquad \text{almost surely.}$$

To establish that the right-hand side is (almost surely) a subset of the left-hand side, for a fixed positive C define τ as the first n for which that $M_n > C$, with $\tau = \infty$ when $\sup_n M_n \leq C$. The martingale $X_n := M_{\tau \wedge n}$ is bounded above by the constant $C + 1$, because the increment (if any) that pushes M_n above C cannot be larger than 1. In particular, $\sup_n \mathbb{P}X_n^+ < \infty$, which ensures that $\{X_n\}$ converges almost surely to a finite limit. On the set $\{\sup_n M_n \leq C\}$ we have $M_n = X_n$ for all n, and hence M_n also converges almost surely to a finite limit on that set. Take a union over a sequence of C values increasing to ∞ to complete the argument.

REMARK. Convergence of $M_n(\omega)$ to a finite limit also implies that $\sup_n |M_n(\omega)| < \infty$. The result therefore contains the surprising asssertion that, almost surely, finiteness of $\sup_n M_n(\omega)$ implies finiteness of $\sup_n |M_n(\omega)|$.

As a special case, consider a sequence $\{A_n\}$ of events adapted to a filtration $\{\mathcal{F}_n\}$. The martingale $M_n := \sum_{i=1}^n (A_i - \mathbb{P}(A_i \mid \mathcal{F}_{i-1}))$ has increments bounded in absolute value by 1. For almost all ω, finiteness of $\sum_{n \in \mathbb{N}} \{\omega \in A_n\}$ implies $\sup_n M_n(\omega) < \infty$, and hence convergence of the sum of conditional probabilities. Argue similarly for the martingale $\{-M_n\}$ to conclude that

$$\{\omega : \textstyle\sum_{n=1}^\infty A_n < \infty\} = \{\omega : \sum_{n=1}^\infty \mathbb{P}(A_n \mid \mathcal{F}_{n-1}) < \infty\} \qquad \text{almost surely,}$$

a remarkable generalization of the Borel-Cantelli lemma for sequences of independent events. □

*5. Proof of the Krickeberg decomposition

It is easiest to understand the proof by reinterpreting the result as assertion about measures. To each integrable random variable X on $(\Omega, \mathcal{F}, \mathbb{P})$ there corresponds a signed measure μ defined on $\mathcal{F}$ by $\mu F := \mathbb{P}(XF)$ for $F \in \mathcal{F}$. The measure can also be written as a difference of two nonnegative measures μ^+ and μ^-, defined by $\mu^+ F := \mathbb{P}(X^+ F)$ and $\mu^- F := \mathbb{P}(X^- F)$, for $f \in \mathcal{F}$.

By equivalence (MG)′, a sequence of integrable random variables $\{X_n : n \in \mathbb{N}_0\}$ adapted to a filtration $\{\mathcal{F}_n : n \in \mathbb{N}_0\}$ is a martingale if and only if the corresponding sequence of measures $\{\mu_n\}$ on $\mathcal{F}$ has the property

<29> $$\mu_{n+1}\big|_{\mathcal{F}_n} = \mu_n\big|_{\mathcal{F}_n} \qquad \text{for each } n,$$

where, in general, $\nu\big|_{\mathcal{G}}$ denotes the restriction of a measure ν to a sub-sigma-field $\mathcal{G}$. Similarly, the defining inequality (subMG)′ for a submartingale, $\mu_{n+1} F := \mathbb{P}(X_{n+1} F) \geq \mathbb{P}X_n F =: \mu_n F$ for all $F \in \mathcal{F}_n$, is equivalent to

<30> $$\mu_{n+1}\big|_{\mathcal{F}_n} \geq \mu_n\big|_{\mathcal{F}_n} \qquad \text{for each } n.$$

Now consider the submartingale $\{S_n : n \in \mathbb{N}_0\}$ from the statement of the Krickeberg decomposition. Define an increasing functional $\lambda : \mathcal{M}^+(\mathcal{F}) \to [0, \infty]$ by

$$\lambda f := \limsup_n \mathbb{P}(S_n^+ f) \qquad \text{for } f \in \mathcal{M}^+(\mathcal{F}).$$

Notice that $\lambda 1 = \limsup_n \mathbb{P}S_n^+$, which is finite, by assumption. The functional also has a property analogous to absolute continuity: if $\mathbb{P}f = 0$ then $\lambda f = 0$.

Write λ_k for the restriction of λ to $\mathcal{M}^+(\mathcal{F}_k)$. For f in $\mathcal{M}^+(\mathcal{F}_k)$, the submartingale property for $\{S_n^+\}$ ensures that $\mathbb{P}S_n^+ f$ increases with n for $n \geq k$. Thus

<31> $$\lambda_k f := \lambda f = \lim_n \mathbb{P}(S_n^+ f) = \sup_{n \geq k} \mathbb{P}(S_n^+ f) \qquad \text{if } f \in \mathcal{M}^+(\mathcal{F}_k).$$

The increasing functional λ_k is linear (because linearity is preserved by limits), and it inherits the Monotone Convergence property from $\mathbb{P}$: for functions in $\mathcal{M}^+(\mathcal{F}_k)$ with $0 \leq f_i \uparrow f$,

$$\sup_i \lambda_k f_i = \sup_i \sup_{n \geq k} \mathbb{P}\left(S_n^+ f_i\right) = \sup_{n \geq k} \sup_i \mathbb{P}\left(S_n^+ f_i\right) = \sup_{n \geq k} \mathbb{P}\left(S_n^+ f\right) = \lambda_k f.$$

It defines a finite measure on $\mathcal{F}_k$ that is absolutely continuous with respect to $\mathbb{P}|_{\mathcal{F}_k}$. Write M_k for the corresponding density in $\mathcal{M}^+(\Omega, \mathcal{F}_k)$.

The analog of <29> identifies $\{M_k\}$ as a nonnegative martingale, because $\lambda_{k+1}|_{\mathcal{F}_k} = \lambda|_{\mathcal{F}_k} = \lambda_k$. Moreover, $M_k \geq S_k^+$ almost surely because

$$\mathbb{P}M_k\{M_k < S_k^+\} = \lambda_k\{M_k < S_k^+\} \geq \mathbb{P}S_k^+\{M_k < S_k^+\},$$

the last inequality following from <31> with $f := \{M_k < S_k^+\}$. The random variables $X_k := M_k - S_k$ are almost surely nonnegative. Also, for $F \in \mathcal{F}_k$,

$$\mathbb{P}X_k F = \mathbb{P}M_k F - \mathbb{P}S_k F \geq \mathbb{P}M_{k+1} F - \mathbb{P}S_{k+1} F = \mathbb{P}X_{k+1} F,$$

because $\{M_k\}$ is a martingale and $\{S_k\}$ is a submartingale. It follows that $\{X_k\}$ is a supermartingale, as required for the Krickeberg decomposition.

*6. Uniform integrability

Corollary <27> gave a sufficient condition for a submartingale $\{X_n\}$ to converge almost surely to an integrable limit X_∞. If $\{X_n\}$ happens to be a martingale, we know that $X_n = \mathbb{P}(X_{n+m} \mid \mathcal{F}_n)$ for arbitrarily large m. It is tempting to leap to the conclusion that

<32> $$X_n \stackrel{?}{=} \mathbb{P}(X_\infty \mid \mathcal{F}_n),$$

as suggested by a purely formal passage to the limit as m tends to infinity. One should perhaps look before one leaps.

<33> **Example.** Reconsider the limit behavior of the partial sums $\{S_n\}$ from Example <23> but with $p = 1/3$ and $\theta = 2$. The sequence $X_n = 2^{S_n}$ is a positive martingale. By the strong law of large numbers, $S_n/n \to -1/3$ almost surely, which gives $S_n \to -\infty$ almost surely and $X_\infty = 0$ as the limit of the martingale. Clearly X_n is not equal to $\mathbb{P}(X_\infty \mid \mathcal{F}_n)$. □

REMARK. The branching process of Example <25> with $\mu = 1$ provides another case of a nontrivial martingale converging almost surely to zero.

As you will learn in this Section, the condition for the validity of <32> (without the cautionary question mark) is ***uniform integrability***. Remember that a family of random variables $\{Z_t : t \in T\}$ is said to be uniformly integrable if $\sup_{t \in T} \mathbb{P}|Z_t|\{|Z_t| > M\} \to 0$ as $M \to \infty$. Remember also the following characterization of $\mathcal{L}^1$ convergence, which was proved in Section 2.8.

<34> **Theorem.** *Let $\{Z_n : n \in \mathbb{N}\}$ be a sequence of integrable random variables. The following two conditions are equivalent.*

(i) *The sequence is uniformly integrable and it converges in probability to a random variable Z_∞, which is necessarily integrable.*

(ii) *The sequence converges in $\mathcal{L}^1$ norm, $\mathbb{P}|Z_n - Z_\infty| \to 0$, to an integrable random variable Z_∞.*

The necessity of uniform integrability for <32> follows immediately from a general property of conditional expectations.

<35> **Lemma.** *For a fixed integrable random variable Z, the family of all conditional expectations $\{\mathbb{P}(Z \mid \mathcal{G}) : \mathcal{G}$ a sub-sigma-field of $\mathcal{F}\}$ is uniformly integrable.*

Proof. Write $Z_\mathcal{G}$ for $\mathbb{P}(Z \mid \mathcal{G})$. With no loss of generality, we may suppose $Z \geq 0$, because $|Z_\mathcal{G}| \leq \mathbb{P}\left(|Z| \mid \mathcal{G}\right)$. Invoke the defining property of the conditional expectation, and the fact that $\{Z_\mathcal{G} > M^2\} \in \mathcal{G}$, to rewrite $\mathbb{P}Z_\mathcal{G}\{Z_\mathcal{G} > M^2\}$ as

$$\mathbb{P}Z\{Z_\mathcal{G} > M^2\} \leq M\mathbb{P}\{Z_\mathcal{G} > M^2\} + \mathbb{P}Z\{Z > M\}.$$

The first term on the right-hand side is less than $M\mathbb{P}Z_\mathcal{G}/M^2 = \mathbb{P}Z/M$, which tends to zero as $M \to \infty$. The other term also tends to zero, because Z is integrable. □

More generally, if X is an integrable random variable and $\{\mathcal{F}_n : n \in \mathbb{N}_0\}$ is a filtration then $X_n := \mathbb{P}(X \mid \mathcal{F}_n)$ defines a uniformly integrable martingale. In fact, every uniformly integrable martingale must be of this form.

<36> **Theorem.** *Every uniformly integrable martingale $\{X_n : n \in \mathbb{N}_0\}$ converges almost surely and in $\mathcal{L}^1$ to an integrable random variable X_∞, for which $X_n = \mathbb{P}(X_\infty \mid \mathcal{F}_n)$. Moreover, if $X_n := \mathbb{P}(X \mid \mathcal{F}_n)$ for some integrable X then $X_\infty = \mathbb{P}(X \mid \mathcal{F}_\infty)$, where $\mathcal{F}_\infty := \sigma\left(\cup_{n \in \mathbb{N}} \mathcal{F}_n\right)$.*

Proof. Uniform integrability implies finiteness of $\sup_n \mathbb{P}|X_n|$, which lets us deduce via Corollary <27> the almost sure convergence to the integrable limit X_∞. Almost sure convergence implies convergence in probability, which uniform integrability and Theorem <34> strengthen to $\mathcal{L}^1$ convergence. To show that $X_n = \mathbb{P}(X_\infty \mid \mathcal{F}_n)$, fix an F in $\mathcal{F}_n$. Then, for all positive m,

$$|\mathbb{P}X_\infty F - \mathbb{P}X_n F| \leq \mathbb{P}|X_\infty - X_{n+m}| + |\mathbb{P}X_{n+m} F - \mathbb{P}X_n F|.$$

The $\mathcal{L}^1$ convergence makes the first term on the right-hand side converge to zero as m tends to infinity. The second term is zero for all positive m, by the martingale property. Thus $\mathbb{P}X_\infty F = \mathbb{P}X_n F$ for every F in $\mathcal{F}_n$.

If $\mathbb{P}(X \mid \mathcal{F}_n) = X_n = \mathbb{P}(X_\infty \mid \mathcal{F}_n)$ then $\mathbb{P}XF = \mathbb{P}X_\infty F$ for each F in $\mathcal{F}_n$. A generating class argument then gives the equality for all F in $\mathcal{F}_\infty$, which characterizes the $\mathcal{F}_\infty$-measurable random variable X_∞ as the conditional expectation $\mathbb{P}(X \mid \mathcal{F}_\infty)$. □

REMARK. More concisely: the uniformly integrable martingales $\{X_n : n \in \mathbb{N}\}$ are precisely those that can be extended to martingales $\{X_n : n \in \overline{\mathbb{N}}\}$. Such a martingale is sometimes said to be ***closed on the right***.

<37> **Example.** Classical statistical models often consist of a parametric family $\mathcal{P} = \{\mathbb{P}_\theta : \theta \in \Theta\}$ of probability measures that define joint distributions of infinite sequences $\omega := (\omega_1, \omega_2, \ldots)$ of possible observations. More formally, each $\mathbb{P}_\theta$ could be thought of as a probability measure on $\mathbb{R}^{\mathbb{N}}$, with random variables X_i as the coordinate maps.

For simplicity, suppose Θ is a Borel subset of a Euclidean space. The parameter θ is said to be ***consistently estimated*** by a sequence of measurable functions $\widehat{\theta}_n = \widehat{\theta}_n(\omega_0, \ldots, \omega_n)$ if

$$\mathbb{P}_\theta\{\,|\widehat{\theta}_n - \theta| > \epsilon\} \to 0 \qquad \text{for each } \epsilon > 0 \text{ and each } \theta \text{ in } \Theta. \tag{38}$$

A Bayesian would define a joint distribution $\mathbb{Q} := \pi \otimes \mathcal{P}$ for θ and ω by equipping Θ with a prior probability distribution π. The conditional distributions $\mathbb{Q}_{n,t}$ given the random vectors $T_n := (X_0, \ldots, X_n)$ are called posterior distributions. We could also regard $\pi_{n\omega}(\cdot) := \mathbb{Q}_{n,T_n(\omega)}(\cdot)$ as random probability measures on the product space. An expectation with respect to $\pi_{n\omega}$ is a version of a conditional expectation given the the sigma-field $\mathcal{F}_n := \sigma(X_1, \ldots, X_n)$.

A mysterious sounding result of Doob (1949) asserts that mere existence of some consistent estimator for θ ensures that the $\pi_{n\omega}$ distributions will concentrate around the right value, in the delicate sense that for π-almost all θ, the $\pi_{n\omega}$ measure of each neighborhood of θ tends to one for $\mathbb{P}_\theta$-almost all ω.

The mystery dissipates when one understands the role of the consistent estimator. When averaged out over the prior, property <38> implies (via Dominated Convergence) that $\mathbb{Q}\{(\theta, \omega) : |\widehat{\theta}_n(\omega) - \theta| > \epsilon\} \to 0$. A $\mathbb{Q}$-almost surely convergent subsequence identifies θ as an $\mathcal{F}_\infty$-measurable random variable, $\tau(\omega)$, on the product space, up to a $\mathbb{Q}$ equivalence. That is, $\theta = \tau(\omega)$ a.e. $[\mathbb{Q}]$.

Let $\mathcal{U}$ be a countable collection of open sets generating the topology of Θ. That is, each open set should equal the union of the $\mathcal{U}$-sets that it contains. For each U in $\mathcal{U}$, the sequence of posterior probabilities $\pi_{n\omega}\{\theta \in U\} = \mathbb{Q}\{\theta \in U \mid \mathcal{F}_n\}$ defines a uniformly integrable martingale, which converges $\mathbb{Q}$-almost surely to

$$\mathbb{Q}\{\theta \in U \mid \mathcal{F}_\infty\} = \{\theta \in U\} \qquad \text{because } \{\theta \in U\} \underset{a.s.}{=} \{\tau(\omega) \in U\} \in \mathcal{F}_\infty.$$

Cast out a sequence of $\mathbb{Q}$-negligible sets, leaving a set E with $\mathbb{Q}E = 1$ and $\pi_{n\omega}\{\theta \in U\} \to \{\theta \in U\}$ for all U in $\mathcal{U}$, all $(\theta, \omega) \in E$, which implies Doob's result. □

*7. Reversed martingales

Martingale theory gets easier when the index set T has a largest element, as in the case $T = -\mathbb{N}_0 := \{-n : n \in \mathbb{N}_0\}$. Equivalently, one can reverse the "direction of time," by considering families of integrable random variables $\{X_t : t \in T\}$ adapted to ***decreasing filtrations***, families of sub-sigma-fields $\{\mathcal{G}_t : t \in T\}$ for which $\mathcal{G}_s \supseteq \mathcal{G}_t$ when $s < t$. For such a family, it is natural to define $\mathcal{G}_\infty := \cap_{t \in T} \mathcal{G}_t$ if it is not already defined.

<39> **Definition.** *Let $\{X_n : n \in \mathbb{N}_0\}$ be a sequence of integrable random variables, adapted to a decreasing filtration $\{\mathcal{G}_n : n \in \mathbb{N}_0\}$. Call $\{(X_n, \mathcal{G}_n) : n \in \mathbb{N}_0\}$ a **reversed supermartingale** if $\mathbb{P}(X_n \mid \mathcal{G}_{n+1}) \le X_{n+1}$ almost surely, for each n. Define reversed submartingales and reversed martingales analogously.*

That is, $\{(X_n, \mathcal{G}_n) : n \in \mathbb{N}_0\}$ is a reversed supermartingale if and only if $\{(X_{-n}, \mathcal{G}_{-n}) : n \in -\mathbb{N}_0\}$ is a supermartingale. In particular, for each fixed N, the finite sequence $X_N, X_{N-1}, \ldots, X_0$ is a supermartingale with respect to the filtration $\mathcal{G}_N \subseteq \mathcal{G}_{N-1} \subseteq \ldots \subseteq \mathcal{G}_0$.

<40> **Example.** If $\{\mathcal{G}_n : n \in \mathbb{N}_0\}$ is a decreasing filtration and X is an integrable random variable, the sequence $X_n := \mathbb{P}(X \mid \mathcal{G}_n)$ defines a uniformly integrable, (Lemma <35>) reversed martingale. □

The theory for reversed positive supermartingales is analogous to the theory from Section 3, except for the slight complication that the sequence $\{\mathbb{P}X_n : n \in \mathbb{N}_0\}$ might be increasing, and therefore it is not automatically bounded.

<41> **Theorem.** *For every reversed, positive supermartingale $\{(X_n, \mathcal{G}_n) : n \in \mathbb{N}_0\}$:*

(i) there exists an X_∞ in $\mathcal{M}^+(\Omega, \mathcal{G}_\infty)$ for which $X_n \to X_\infty$ almost surely;

(ii) $\mathbb{P}(X_n \mid \mathcal{G}_\infty) \uparrow X_\infty$ almost surely;

(iii) $\mathbb{P}|X_n - X_\infty| \to 0$ if and only if $\sup_n \mathbb{P}X_n < \infty$.

Proof. The Corollary <21> to Dubins's Inequality bounds by $(\alpha/\beta)^k$ the probability that $X_N, X_{N-1}, \ldots, X_0$ completes at least k upcrossings of the interval $[\alpha, \beta]$, no matter how large we take N. As in the proof of Theorem <22>, it then follows that $\mathbb{P}\{\limsup X_n > \beta > \alpha > \liminf X_n\} = 0$, for each pair $0 \le \alpha < \beta < \infty$, and hence X_n converges almost surely to a nonnegative limit X_∞, which is necessarily $\mathcal{G}_\infty$-measurable.

I will omit most of the "almost sure" qualifiers for the remainder of the proof. Temporarily abbreviate $\mathbb{P}(\cdot \mid \mathcal{G}_n)$ to $\mathbb{P}_n(\cdot)$, for $n \in \overline{\mathbb{N}}_0$, and write Z_n for $\mathbb{P}_\infty X_n$. From the reversed supermartingale property, $\mathbb{P}_{n+1} X_n \le X_{n+1}$, and the rule for iterated conditional expectations we get

$$Z_n = \mathbb{P}_\infty X_n = \mathbb{P}_\infty (\mathbb{P}_{n+1} X_n) \le \mathbb{P}_\infty X_{n+1} = Z_{n+1}.$$

Thus $Z_n \uparrow Z_\infty := \limsup_n Z_n$, which is $\mathcal{G}_\infty$-measurable.

For (ii) we need to show $Z_\infty = X_\infty$, almost surely. Equivalently, as both variables are $\mathcal{G}_\infty$-measurable, we need to show $\mathbb{P}(Z_\infty G) = \mathbb{P}(X_\infty G)$ for each G in $\mathcal{G}_\infty$. For such a G,

$$
\begin{aligned}
\mathbb{P}(Z_\infty G) &= \sup_{n \in \mathbb{N}_0} \mathbb{P}(Z_n G) && \text{Monotone Convergence} \\
&= \sup_n \mathbb{P}(X_n G) && \text{definition of } Z_n := \mathbb{P}_\infty X_n \\
&= \sup_n \sup_{m \in \mathbb{N}} \mathbb{P}\left((X_n \wedge m) G\right) && \text{Monotone Convergence, for fixed } n \\
&= \sup_m \sup_n \mathbb{P}\left((X_n \wedge m) G\right).
\end{aligned}
$$

The sequence $\{X_n \wedge m : n \in \mathbb{N}_0\}$ is a uniformly bounded, reversed positive supermartingale, for each fixed m. Thus $\mathbb{P}((X_n \wedge m)G)$ increases with n, and, by Dominated Convergence, its limit equals $\mathbb{P}((X_\infty \wedge m)G)$. Thus

$$\mathbb{P}(Z_\infty G) = \sup_m \mathbb{P}((X_\infty \wedge m)G) = \mathbb{P}(X_\infty G),$$

the final equality by Monotone Convergence. Assertion (ii) follows.

Monotone Convergence and (ii) imply that $\mathbb{P}X_\infty = \sup_n \mathbb{P}X_n$. Finiteness of the supremum is equivalent to integrability of X_∞. The sufficiency in assertion (iii) follows by the usual Dominated Convergence trick (also known as Scheffé's lemma):

$$\mathbb{P}|X_\infty - X_n| = 2\mathbb{P}(X_\infty - X_n)^+ - (\mathbb{P}X_\infty - \mathbb{P}X_n) \to 0.$$

For the necessity, just note that an $\mathcal{L}^1$ limit of a sequence of integrable random variables is integrable. □

The analog of the Krickeberg decomposition extends the result to reversed submartingales $\{(X_n, \mathcal{G}_n) : n \in \mathbb{N}_0\}$. The sequence $M_n := \mathbb{P}(X_0^+ \mid \mathcal{G}_n)$ is a reversed positive martingale for which $M_n \geq \mathbb{P}(X_0 \mid \mathcal{G}_n) \geq X_n$ almost surely. Thus $X_n = M_n - (M_n - X_n)$ decomposes X_n into a difference of a reversed positive martingale and a reversed positive supermartingale.

<42> **Corollary.** *Every reversed submartingale $\{(X_n, \mathcal{G}_n) : n \in \mathbb{N}_0\}$ converges almost surely. The limit is integrable, and the sequence also converges in the $\mathcal{L}^1$ sense, if $\inf_n \mathbb{P}X_n > -\infty$.*

Proof. Apply Theorem <41> to both reversed positive supermartingales $\{M_n - X_n\}$ and $\{M_n\}$, noting that $\sup_n \mathbb{P}M_n = \mathbb{P}X_0^+$ and $\sup_n \mathbb{P}(M_n - X_n) = \mathbb{P}X_0^+ - \inf_n \mathbb{P}X_n$. □

<43> **Corollary.** *Every reversed martingale $\{(X_n, \mathcal{G}_n) : n \in \mathbb{N}_0\}$ converges almost surely, and in $\mathcal{L}^1$, to the limit $X_\infty := \mathbb{P}(X_0 \mid \mathcal{G}_\infty)$, where $\mathcal{G}_\infty := \cap_{n \in \mathbb{N}_0} \mathcal{G}_n$.*

Proof. The identification of the limit as the conditional expectation follows from the facts that $\mathbb{P}(X_0 G) = \mathbb{P}(X_n G)$, for each n, and $|\mathbb{P}(X_n G) - \mathbb{P}(X_\infty G)| \leq \mathbb{P}|X_n - X_\infty| \to 0$, for each G in $\mathcal{G}_\infty$. □

Reversed martingales arise naturally from symmetry arguments.

<44> **Example.** Let $\{\xi_i : i \in \mathbb{N}\}$ be a sequence of independent random elements taking values in a set $\mathcal{X}$ equipped with a sigma-field $\mathcal{A}$. Suppose each ξ_i induces the same distribution P on $\mathcal{A}$, that is, $\mathbb{P}f(\xi_i) = Pf$ for each f in $\mathcal{M}^+(\mathcal{X}, \mathcal{A})$. For each n define the ***empirical measure*** $P_{n,\omega}$ (or just P_n, if there is no need to emphasize the dependence on ω) on $\mathcal{X}$ as the probability measure that places mass n^{-1} at each of the points $\xi_1(\omega), \ldots, \xi_n(\omega)$. That is, $P_{n,\omega} f := n^{-1} \sum_{1 \leq i \leq n} f(\xi_i(\omega))$.

Intuitively speaking, knowledge of P_n tells us everything about the values $\xi_1(\omega), \ldots, \xi_n(\omega)$ except for the order in which they were produced. Conditionally on P_n, we know that ξ_1 should be one of the points supporting P_n, but we don't know which one. The conditional distribution of ξ_1 given P_n should put probability n^{-1} at each support point, and it seems we should then have

<45>
$$\mathbb{P}(f(\xi_1) \mid P_n) = P_n f.$$

REMARK. Here I am arguing, heuristically, assuming P_n concentrates on n distinct points. A similar heuristic could be developed when there are ties, but there is no point in trying to be too precise at the moment. The problem of ties would disappear from the formal argument.

Similarly, if we knew all P_i for $i \geq n$ then we should be able to locate $\xi_i(\omega)$ exactly for $i \geq n+1$, but the values of $\xi_1(\omega), \ldots, \xi_n(\omega)$ would still be known only up to random relabelling of the support points of P_n. The new information would tell us no more about ξ_1 than we already knew from P_n. In other words, we should have

<46> $$\mathbb{P}(f(\xi_1) \mid \mathcal{G}_n) = \ldots = \mathbb{P}(f(\xi_n) \mid \mathcal{G}_n) = P_n f \qquad \text{where } \mathcal{G}_n \overset{?}{:=} \sigma(P_n, P_{n+1}, \ldots),$$

which would then give

$$\mathbb{P}(P_{n-1} f \mid \mathcal{G}_n) = \frac{1}{n-1} \sum_{i=1}^{n-1} \mathbb{P}(f(\xi_i) \mid \mathcal{G}_n) = \frac{1}{n-1} \sum_{i=1}^{n-1} P_n f = P_n f.$$

That is, $\{(P_n f, \mathcal{G}_n) : n \in \mathbb{N}\}$ would be a reversed martingale, for each fixed f.

It is possible to define $\mathcal{G}_n$ rigorously, then formalize the preceeding heuristic argument to establish the reversed martingale property. I will omit the details, because it is simpler to replace $\mathcal{G}_n$ by the closely related ***n-symmetric sigma-field*** $\mathcal{S}_n$, to be defined in the next Section, then invoke the more general symmetry arguments (Example <50>) from that Section to show that $\{(P_n f, \mathcal{S}_n) : n \in \mathbb{N}\}$ is a reversed martingale for each P-integrable f.

Corollary <43> ensures that $P_n f \to \mathbb{P}(f(\xi_1) \mid \mathcal{S}_\infty)$ almost surely. As you will see in the next Section (Theorem <51>, to be precise), the sigma-field $\mathcal{S}_\infty$ is trivial—it contains only events with probability zero or one—and $\mathbb{P}(X \mid \mathcal{S}_\infty) = \mathbb{P}X$, almost surely, for each integrable random variable X. In particular, for each P-integrable function f we have

$$\frac{f(\xi_1) + \ldots + f(\xi_n)}{n} = P_n f \to Pf \qquad \text{almost surely.}$$

The special case $\mathcal{X} := \mathbb{R}$ and $\mathbb{P}|\xi_1| < \infty$ and $f(x) \equiv x$ recovers the Strong Law of Large Numbers (SLLN) for independent, identically distributed summands.

In statistical problems it is sometimes necessary to prove a uniform analog of the SLLN (a USLLN):

$$\Delta_n := \sup_\theta |P_n f_\theta - P f_\theta| \to 0 \qquad \text{almost surely,}$$

where $\{f_\theta : \theta \in \Theta\}$ is a class of P-integrable functions on $\mathcal{X}$. Corollary <41> can greatly simplify the task of establishing such a USLLN.

To avoid measurability difficulties, let me consider only the case where Θ is countable. Write $X_{n,\theta}$ for $P_n f_\theta - P f_\theta$. Also, assume that the ***envelope*** $F := \sup_\theta |f_\theta|$ is P-integrable, so that $\mathbb{P}\Delta_n \leq \mathbb{P}(P_n F + PF) = 2PF < \infty$.

For each fixed θ, we know that $\{(X_{n,\theta}, \mathcal{S}_n) : n \in \mathbb{N}\}$ is a reversed martingale, and hence

$$\mathbb{P}(\Delta_n \mid \mathcal{S}_{n+1}) = \mathbb{P}\left(\sup_\theta |X_{n,\theta}| \mid \mathcal{S}_{n+1}\right) \geq \sup_\theta \left|\mathbb{P}(X_{n,\theta} \mid \mathcal{S}_{n+1})\right| = \Delta_{n+1}.$$

That is, $\{(\Delta_n, \mathcal{S}_n) : n \in \mathbb{N}\}$ is a reversed submartingale. From Corollary <41>, Δ_n converges almost surely to a $\mathcal{S}_\infty$-measurable random variable Δ_∞, which by the triviality of $\mathcal{S}_\infty$ (Theorem <51> again) is (almost surely) constant. To prove the USLLN, we have only to show that the constant is zero. For example, it would suffice to show $\mathbb{P}\Delta_n \to 0$, a great simplification of the original task. See Pollard (1984, Section II.5) for details. □

*8. Symmetry and exchangeability

The results in this Section involve probability measures on infinite product spaces. You might want to consult Section 4.8 for notation and the construction of product measures.

The symmetry arguments from Example <44> did not really require an assumption of independence. The reverse martingale methods can be applied to more general situations where probability distributions have symmetry properties.

Rather than work with random elements of a space $(\mathcal{X}, \mathcal{A})$, it is simpler to deal with their joint distribution as a probability measure on the product sigma-field $\mathcal{A}^{\mathbb{N}}$ of the product space $\mathcal{X}^{\mathbb{N}}$, the space of all sequences, $\mathbf{x} := (x_1, x_2, \ldots)$, on $\mathcal{X}$. We can think of the coordinate maps $\xi_i(\mathbf{x}) := x_i$ as a sequence of random elements of $(\mathcal{X}, \mathcal{A})$, when it helps.

<47> **Definition.** *Call a one-to-one map π from $\mathbb{N}$ onto itself an **n-permutation** if $\pi(i) = i$ for $i > n$. Write $\mathcal{R}(n)$ for the set of all $n!$ distinct n-permutations. Call $\cup_{n\in\mathbb{N}}\mathcal{R}(n)$ the set of all **finite permutations** of $\mathbb{N}$. Write S_π for the map, from $\mathcal{X}^{\mathbb{N}}$ back onto itself, defined by*

$$\begin{aligned} S_\pi(x_1, x_2, \ldots, x_n, \ldots) &:= (x_{\pi(1)}, x_{\pi(2)}, \ldots, x_{\pi(n)}, \ldots) \\ &:= (x_{\pi(1)}, x_{\pi(2)}, \ldots, x_{\pi(n)}, x_{n+1}, \ldots) \qquad \text{if } \pi \in \mathcal{R}(n). \end{aligned}$$

*Say that a function h on $\mathcal{X}^{\mathbb{N}}$ is **n-symmetric** if it is unaffected all n-permutations, that is, if $h_\pi(\mathbf{x}) := h(S_\pi \mathbf{x}) = h(\mathbf{x})$ for every n-permutation π.*

<48> **Example.** Let f be a real valued function on $\mathcal{X}$. Then the function $\sum_{i=1}^m f(x_i)$ is n-symmetric for every $m \geq n$, and the function $\limsup_{m\to\infty} \sum_{i=1}^m f(x_i)/m$ is n-symmetric for every n.

Let g be a real valued function on $\mathcal{X} \otimes \mathcal{X}$. Then $g(x_1, x_2) + g(x_2, x_1)$ is 2-symmetric. More generally, $\sum_{1\leq i\neq j\leq m} g(x_i, x_j)$ is n-symmetric for every $m \geq n$.

For every real valued function f on $\mathcal{X}^{\mathbb{N}}$, the function

$$F(\mathbf{x}) := \frac{1}{n!}\sum_{\pi\in\mathcal{R}(n)} f_\pi(\mathbf{x}) = \frac{1}{n!}\sum_{\pi\in\mathcal{R}(n)} f(x_{\pi(1)}, x_{\pi(2)}, \ldots, x_{\pi(n)}, x_{n+1}, \ldots)$$

is n-symmetric. □

<49> **Definition.** *A probability measure $\mathbb{P}$ on $\mathcal{A}^{\mathbb{N}}$ is said to be **exchangeable** if it is invariant under S_π for every finite permutation π, that is, if $\mathbb{P}h = \mathbb{P}h_\pi$ for every h in $\mathcal{M}^+(\mathcal{X}^{\mathbb{N}}, \mathcal{A}^{\mathbb{N}})$ and every finite permutation π. Equivalently, under $\mathbb{P}$ the random vector $(\xi_{\pi(1)}, \xi_{\pi(2)}, \ldots, \xi_{\pi(n)})$ has the same distribution as $(\xi_1, \xi_2, \ldots, \xi_n)$, for every n-permutation π, and every n.*

The collection of all sets in $\mathcal{A}^{\mathbb{N}}$ whose indicator functions are n-symmetric forms a sub-sigma-field $\mathcal{S}_n$ of $\mathcal{A}^{\mathbb{N}}$, the ***n*-symmetric sigma-field**. The $\mathcal{S}_n$-measurable functions are those $\mathcal{A}^{\mathbb{N}}$-measurable functions that are n-symmetric. The sigma-fields $\{\mathcal{S}_n : n \in \mathbb{N}\}$ form a decreasing filtration on $\mathcal{X}^{\mathbb{N}}$, with $\mathcal{S}_1 = \mathcal{A}^{\mathbb{N}}$.

<50> **Example.** Suppose $\mathbb{P}$ is exchangeable. Let f be a fixed $\mathbb{P}$-integrable function on $\mathcal{X}^{\mathbb{N}}$. Then a symmetry argument will show that

$$\mathbb{P}(f \mid \mathcal{S}_n) = \frac{1}{n!} \sum_{\pi \in \mathcal{R}(n)} f_\pi(\mathbf{x}).$$

The function—call it $F(\mathbf{x})$—on the right-hand side is n-symmetric, and hence $\mathcal{S}_n$-measurable. Also, for each bounded, $\mathcal{S}_n$-measurable function H,

$$\begin{aligned} \mathbb{P}(f(\mathbf{x})H(\mathbf{x})) &= \mathbb{P}(f_\pi H_\pi) \qquad \text{for all } \pi, \text{ by exchangeability} \\ &= \mathbb{P}(f_\pi H) \qquad \text{for all } \pi \text{ in } \mathcal{R}(n) \\ &= \frac{1}{n!} \sum_{\pi \in \mathcal{R}(n)} \mathbb{P}(f_\pi(\mathbf{x})H(\mathbf{x})) = \mathbb{P}(F(\mathbf{x})H(\mathbf{x})). \end{aligned}$$

As a special case, if f depends only on the first coordinate then we have

$$\mathbb{P}(f(x_1) \mid \mathcal{S}_n)) = \frac{1}{n!} \sum_{\pi \in \mathcal{R}(n)} f\left(x_{\pi(1)}\right) = P_n f,$$

□ where P_n denotes the empirical measure, as in Example <44>.

When the coordinate maps are independent under an exchangeable $\mathbb{P}$, the symmetric sigma-field $\mathcal{S}_\infty$ becomes trivial, and conditional expectations (such as $\mathbb{P}(f(x_1) \mid \mathcal{S}_\infty)$) reduce to constants.

<51> **Theorem.** *(Hewitt-Savage zero-one law) If $\mathbb{P} = P^{\mathbb{N}}$, the symmmetric sigma-field $\mathcal{S}_\infty$ is trivial: for each F in $\mathcal{S}_\infty$, either $\mathbb{P}F = 0$ or $\mathbb{P}F = 1$.*

Proof. Write $h(\mathbf{x})$ for the indicator function of F, a set in $\mathcal{S}_\infty$. By definition, $h_\pi = h$ for every finite permutation. Equip $\mathcal{X}^{\mathbb{N}}$ with the filtration $\mathcal{F}_n = \sigma\{x_i : i \le n\}$. Notice that $\mathcal{F}_\infty := \sigma\left(\cup_{n \in \mathbb{N}} \mathcal{F}_n\right) = \mathcal{A}^{\mathbb{N}} = \mathcal{S}_1$.

The martingale $Y_n := \mathbb{P}(F \mid \mathcal{F}_n)$ converges almost surely to $\mathbb{P}(F \mid \mathcal{F}_\infty) = F$, and also, by Dominated Convergence, $\mathbb{P}|h - Y_n|^2 \to 0$.

The $\mathcal{F}_n$-measurable random variable Y_n may be written as $h_n(x_1, \ldots, x_n)$, for some $\mathcal{A}^n$-measurable h_n on $\mathcal{X}^n$. The random variable $Z_n := h_n(x_{n+1}, \ldots, x_{2n})$ is independent of Y_n, and it too converges in $\mathcal{L}^2$ to h: if π denotes the $2n$-permutation that interchanges i and $i+n$, for $1 \le i \le n$, then, by exchangeability,

$$\mathbb{P}|h(\mathbf{x}) - Z_n|^2 = \mathbb{P}|h_\pi(\mathbf{x}) - h_n\left(x_{\pi(n+1)}, \ldots, x_{\pi(2n)}\right)|^2 = \mathbb{P}|h(\mathbf{x}) - Y_n|^2 \to 0.$$

The random variables Z_n and Y_n are independent, and they both converge in $\mathcal{L}^2(\mathbb{P})$-norm to F. Thus

$$0 = \lim_{n\to\infty} \mathbb{P}|Y_n - Z_n|^2 = \lim_{n\to\infty} \left(\mathbb{P}Y_n^2 - 2(\mathbb{P}Y_n)(\mathbb{P}Z_n) + \mathbb{P}Z_n^2\right) = \mathbb{P}F - 2(\mathbb{P}F)^2 + \mathbb{P}F.$$

□ It follows that either $\mathbb{P}F = 0$ or $\mathbb{P}F = 1$.

In a sense made precise by Problem [17], the product measures $P^{\mathbb{N}}$ are the extreme examples of exchangeable probability measures—they are the extreme points in the convex set of all exchangeable probability measures on $\mathcal{A}^{\mathbb{N}}$. A celebrated result of de Finetti (1937) asserts that all the exchangeable probabilities can be built up from mixtures of product measures, in various senses. The simplest general version of the de Finetti result is expressed as an assertion of conditional independence.

<52> **Theorem.** *Under an exchangeable probability distribution $\mathbb{P}$ on $(\mathcal{X}^{\mathbb{N}}, \mathcal{A}^{\mathbb{N}})$, the coordinate maps are conditionally independent given the symmetric sigma-field $\mathcal{S}_\infty$. That is, for all sets A_i in $\mathcal{A}$,*

$$\mathbb{P}(x_1 \in A_1, \ldots, x_m \in A_m \mid \mathcal{S}_\infty) = \mathbb{P}(x_1 \in A_1 \mid \mathcal{S}_\infty) \times \ldots \times \mathbb{P}(x_m \in A_m \mid \mathcal{S}_\infty)$$

almost surely, for every m.

Proof. Consider only the typical case where $m = 3$. The proof of the general case is similar. Write f_i for the indicator function of A_i. Abbreviate $\mathbb{P}(\cdot \mid \mathcal{S}_n)$ to $\mathbb{P}_n$, for $n \in \overline{\mathbb{N}}$. From Example <50>, for $n \geq 3$,

$$n^3 (\mathbb{P}_n f_1(x_1)) (\mathbb{P}_n f_2(x_2)) (\mathbb{P}_n f_3(x_3)) = \sum \{1 \leq i, j, k \leq n\} f_1(x_i) f_2(x_j) f_3(x_k).$$

On the right-hand side, there are $n(n-1)(n-2)$ triples of distinct subscripts (i, j, k), leaving $O(n^2)$ of them with at least one duplicated subscript. The latter contribute a sum bounded in absolute value by a multiple of n^2; the former appear in the sum that Example <50> identifies as $\mathbb{P}_n\left(f_1(x_1) f_2(x_2) f_3(x_3)\right)$. Thus

$$(\mathbb{P}_n f_1(x_1)) (\mathbb{P}_n f_2(x_2)) (\mathbb{P}_n f_3(x_3)) = \frac{n(n-1)(n-2)}{n^3} \mathbb{P}_n\left(f_1(x_1) f_2(x_2) f_3(x_3)\right) + O(n^{-1}).$$

By the convergence of reverse martingales, in the limit we get

$$(\mathbb{P}_\infty f_1(x_1)) (\mathbb{P}_\infty f_2(x_2)) (\mathbb{P}_\infty f_3(x_3)) = \mathbb{P}_\infty\left(f_1(x_1) f_2(x_2) f_3(x_3)\right),$$

the desired factorization. □

When conditional distributions exist, it is easy to extract from Theorem <52> the representation of $\mathbb{P}$ as a mixture of product measures.

<53> **Theorem.** *Let $\mathcal{A}$ be the Borel sigma-field of a separable metric space $\mathcal{X}$. Let $\mathbb{P}$ be an exchangeable probability measure on $\mathcal{A}^{\mathbb{N}}$, under which the distribution P of x_1 is tight. Then there exists an $\mathcal{S}_\infty$-measurable map T into $[0, 1]^{\mathbb{N}}$, with distribution $\mathbb{Q}$, for which conditional distributions $\{\mathbb{P}_t : t \in \mathcal{T}\}$ exist, and $\mathbb{P}_t = P_t^{\mathbb{N}}$, a product measure, for $\mathbb{Q}$ almost all t.*

Proof. Let $\mathcal{E} := \{E_i : i \in \mathbb{N}\}$ be a countable generating class for the sigma-field $\mathcal{A}$, stable under finite intersections and containing $\mathcal{X}$. For each i let $T_i(\mathbf{x})$ be a version of $\mathbb{P}(x_1 \in E_i \mid \mathcal{S}_\infty)$. By symmetry, $T_i(\mathbf{x})$ is also a version of $\mathbb{P}\left(x_j \in E_i \mid \mathcal{S}_\infty\right)$, for every j. Define T as the map from $\mathcal{X}^{\mathbb{N}}$ into $\mathcal{T} := [0, 1]^{\mathbb{N}}$ for which $T(\mathbf{x})$ has ith coordinate $T_i(\mathbf{x})$.

The joint distribution of x_1 and T is a probability measure Γ on the product sigma-field of $\mathcal{X} \times \mathcal{T}$, with marginals P and $\mathbb{Q}$. As shown in Section 1 of Appendix F, the assumptions on P ensure existence of a probability kernel $\mathcal{P} := \{P_t : t \in \mathcal{T}\}$ for which

$$\mathbb{P} g(x_1, T) = \Gamma^{x,t} g(x, t) = \mathbb{Q}^t P_t^x g(x, t) \qquad \text{for all } g \text{ in } \mathcal{M}^+(\mathcal{X} \times \mathcal{T}).$$

In particular, by definition of T_i and the $\mathcal{S}_\infty$-measurability of T,

$$\mathbb{Q}^t \left(t_i h(t)\right) = \mathbb{P}\left(T_i h(T)\right) = \mathbb{P}\left(\{x_1 \in E_i\} h(T)\right) = \mathbb{Q}^t \left(h(t) P_t E_i\right)$$

for all h in $\mathcal{M}^+(\mathcal{T})$, which implies that $P_t E_i = t_i$ a.e. $[\mathbb{Q}]$, for each i.

For every finite subcollection $\{E_{i_1}, \ldots, E_{i_n}\}$ of $\mathcal{E}$, Theorem <52> asserts

$$\mathbb{P}\{x_1 \in E_{i_1}, \ldots, x_n \in E_{i_n} \mid \mathcal{S}_\infty\} = \prod\nolimits_{j=1}^n \mathbb{P}\{x_j \in E_{i_j} \mid \mathcal{S}_\infty\} = \prod\nolimits_{j=1}^n T_{i_j}(\mathbf{x}) \qquad \text{a.e. } [\mathbb{P}],$$

which integrates to

$$\begin{aligned}\mathbb{P}\{x_1 \in E_{i_1}, \ldots, x_n \in E_{i_n}\} &= \mathbb{P}\left(\prod\nolimits_{j=1}^n T_{i_j}(\mathbf{x})\right) \\ &= \mathbb{Q}^t\left(\prod\nolimits_{j=1}^n t_{i_j}\right) \\ &= \mathbb{Q}^t\left(\prod\nolimits_{j=1}^n P_t E_{i_j}\right) = \mathbb{Q}^t \mathbb{P}_t\{x_1 \in E_{i_1}, \ldots, x_n \in E_{i_n}\}.\end{aligned}$$

□ A routine generating class argument completes the proof.

9. Problems

[1] Follow these steps to construct an example of a martingale $\{Z_i\}$ and a stopping time τ for which $\mathbb{P}Z_0 \neq \mathbb{P}Z_\tau\{\tau < \infty\}$.

(i) Let $\xi_1, \xi_2, \ldots$ be independent, identically distributed random variables with $\mathbb{P}\{\xi_i = +1\} = 1/3$ and $\mathbb{P}\{\xi_i = -1\} = 2/3$. Define $X_0 = 0$ and $X_i := \xi_1 + \ldots + \xi_i$ and $Z_i := 2^{X_i}$, for $i \geq 1$. Show that $\{Z_i\}$ is a martingale with respect to an appropriate filtration.

(ii) Define $\tau := \inf\{i : X_i = -1\}$. Show that τ is a stopping time, finite almost everywhere. Hint: Use SLLN.

(iii) Show that $\mathbb{P}Z_0 > \mathbb{P}Z_\tau$. (Should you worry about what happens on the set $\{\tau = \infty\}$?)

[2] Let τ be a stopping time for the natural filtration generated by a sequence of random variables $\{Z_n : n \in \mathbb{N}\}$. Show that $\mathcal{F}_\tau = \sigma\{Z_{\tau\wedge n} : n \in \mathbb{N}\}$.

[3] Let $\{(Z_n, \mathcal{F}_n) : n \in \mathbb{N}_0\}$ be a (sub)martingale and τ be a stopping time. Show that $\{(Z_{\tau\wedge n}, \mathcal{F}_n) : n \in \mathbb{N}_0\}$ is also a (sub)martingale. Hint: For F in $\mathcal{F}_{n-1}$, consider separately the contributions to $\mathbb{P}Z_{n\wedge\tau}F$ and $\mathbb{P}Z_{(n-1)\wedge\tau}F$ from the regions $\{\tau \leq n-1\}$ and $\{\tau \geq n\}$.

[4] Let τ be a stopping time for a filtration $\{\mathcal{F}_i : i \in \overline{\mathbb{N}}_0\}$. For an integrable random variable X, define $X_i := \mathbb{P}(X \mid \mathcal{F}_i)$. Show that

$$\mathbb{P}(X \mid \mathcal{F}_\tau) = \sum_{i \in \overline{\mathbb{N}}_0} \{\tau = i\} X_i = X_\tau \qquad \text{almost surely.}$$

Hint: Start with $X \geq 0$, so that there are no convergence problems.

[5] Let $\{(X_n, \mathcal{F}_n) : n \in \mathbb{N}_0\}$ be a positive supermartingale, and let σ and τ be stopping times (not necessarily bounded) for which $\sigma \leq \tau$ on a set F in $\mathcal{F}_\sigma$. Show that $\mathbb{P}X_\sigma\{\sigma < \infty\}F \geq \mathbb{P}X_\tau\{\tau < \infty\}F$. Hint: For each positive integer N, show that $F_N := F\{\sigma \leq N\} \in \mathcal{F}_{\sigma\wedge N}$. Use the Stopping Time Lemma to prove that $\mathbb{P}X_{\sigma\wedge N}F_N \geq \mathbb{P}X_{\tau\wedge N}F_N \geq \mathbb{P}X_\tau\{\tau \leq N\}F$, then invoke Monotone Convergence.

[6] For each positive supermartingale $\{(X_n, \mathcal{F}_n) : n \in \mathbb{N}_0\}$, and stopping times $\sigma \le \tau$, show that $\mathbb{P}(X_\tau\{\tau < \infty\} \mid \mathcal{F}_\sigma) \le X_\sigma\{\sigma < \infty\}$ almost surely.

[7] (Kolmogorov 1928) Let $\xi_1, \ldots, \xi_n$ be independent random variables with $\mathbb{P}\xi_i = 0$ and $|\xi_i| \le 1$ for each i. Define $X_i := \xi_1 + \ldots + \xi_i$ and $V_i := \mathbb{P}X_i^2$. For each $\epsilon > 0$ show that $\mathbb{P}\{\max_{i\le n} |X_i| \le \epsilon\} \le (1+\epsilon)^2/V_n$. Note the direction of the inequalities. Hint: Define a stopping time τ for which $V_n\{\max_{i\le n} |X_i| \le \epsilon\} \le V_\tau\{\tau = n\}$. Show that $\mathbb{P}V_\tau = \mathbb{P}X_\tau^2 \le (1+\epsilon)^2$.

[8] (Birnbaum & Marshall 1961) Let $0 = X_0, X_1, \ldots$ be nonnegative integrable random variables that are adapted to a filtration $\{\mathcal{F}_i\}$. Suppose there exist constants θ_i, with $0 \le \theta_i \le 1$, for which

$$(*) \qquad \mathbb{P}(X_i \mid \mathcal{F}_{i-1}) \ge \theta_i X_{i-1} \qquad \text{for } i \ge 1.$$

Let $C_1 \ge C_2 \ge \ldots \ge C_{N+1} = 0$ be constants. Prove the inequality

$$(**) \qquad \mathbb{P}\{\max_{i\le N} C_i X_i \ge 1\} \le \sum_{i=1}^{N} (C_i - \theta_{i+1}C_{i+1})\mathbb{P}X_i,$$

by following these steps.

(i) Interpret $(*)$ to mean that there exist nonnegative, $\mathcal{F}_{i-1}$-measurable random variables Y_{i-1} for which $\mathbb{P}(X_i \mid \mathcal{F}_{i-1}) = Y_{i-1} + \theta_i X_{i-1}$ almost surely. Put $Z_i := X_i - Y_{i-1} - \theta_i X_{i-1}$. Show that $C_i X_i \le C_{i-1}X_{i-1} + C_i Z_i + C_i Y_{i-1}$ almost surely.

(ii) Deduce that $C_i X_i \le M_i + \Delta$, where M_i is a martingale with $M_0 = 0$ and $\Delta := \sum_{i=1}^N C_i Y_{i-1}$.

(iii) Show that the left-hand side of inequality $(**)$ is less than $\mathbb{P}C_\tau X_\tau$ for an appropriate stopping time τ, then rearrange the sum for $\mathbb{P}\Delta$ to get the asserted upper bound.

[9] (Doob 1953, page 317) Suppose $S_1, \ldots, S_n$ is a nonnegative submartingale, with $\mathbb{P}S_i^p < \infty$ for some fixed $p > 1$. Let $q > 1$ be defined by $p^{-1} + q^{-1} = 1$. Show that $\mathbb{P}(\max_{i\le n} S_i^p) \le q^p \mathbb{P}S_n^p$, by following these steps.

(i) Write M_n for $\max_{i\le n} S_i$. For fixed $x > 0$, and an appropriate stopping time τ, apply the Stopping Time Lemma to show that

$$x\mathbb{P}\{M_n \ge x\} \le \mathbb{P}S_\tau\{S_\tau \ge x\} \le \mathbb{P}S_n\{M_n \ge x\}.$$

(ii) Show that $\mathbb{P}X^p = \int_0^\infty px^{p-1}\mathbb{P}\{X \ge x\}\,dx$ for each nonnegative random variable X.

(iii) Show that $\mathbb{P}M_n^p \le q\mathbb{P}S_n M_n^{p-1}$.

(iv) Bound the last product using Hölder's inequality, then rearrange to get the stated inequality. (Any problems with infinite values?)

[10] Let $(\Omega, \mathcal{F}, \mathbb{P})$ be a probability space such that $\mathcal{F}$ is *countably generated*: that is, $\mathcal{F} = \sigma\{B_1, B_2, \ldots\}$ for some sequence of sets $\{B_i\}$. Let μ be a finite measure on $\mathcal{F}$, dominated by $\mathbb{P}$. Let $\mathcal{F}_n := \sigma\{B_1, \ldots, B_n\}$.

(i) Show that there is a partition π_n of Ω into at most 2^n disjoint sets from $\mathcal{F}_n$ such that each F in $\mathcal{F}_n$ is a union of sets from π_n.

(ii) Define $\mathcal{F}_n$-measurable random variables X_n by: for $\omega \in A \in \pi_n$,

$$X_n(\omega) = \begin{cases} \mu A/\mathbb{P}A & \text{if } \mathbb{P}A > 0, \\ 0 & \text{otherwise.} \end{cases}$$

Show that $\mathbb{P}X_n F = \mu F$ for all F in $\mathcal{F}_n$.

(iii) Show that $(X_n, \mathcal{F}_n)$ is a positive martingale.

(iv) Show that $\{X_n\}$ is uniformly integrable. Hint: What do you know about $\mu\{X_n \geq M\}$?

(v) Let X_∞ denote the almost sure limit of the $\{X_n\}$. Show that $\mathbb{P}X_\infty F = \mu F$ for all F in $\mathcal{F}$. That is, show that X_∞ is a density for μ with respect to $\mathbb{P}$.

[11] Let $\{(X_n, \mathcal{F}_n) : n \in \mathbb{N}_0\}$ be a submartingale. For fixed constants $\alpha < \beta$ (not necessarily nonnegative), define stopping times $\sigma_1 \leq \tau_1 \leq \sigma_2 \leq \ldots$, as in Section 3. Establish the upcrossing inequality,

$$\mathbb{P}\{\tau_k \leq N\} \leq \frac{\mathbb{P}(X_N - \alpha)^+}{k(\beta - \alpha)}$$

for each positive integer N, by following these steps.

(i) Show that $Z_n = (X_n - \alpha)^+$ is a positive submartingale, with $Z_{\sigma_i} = 0$ if $\sigma_i < \infty$ and $Z_{\tau_i} \geq \beta - \alpha$ if $\tau_i < \infty$.

(ii) For each i, show that $Z_{\tau_i \wedge N} - Z_{\sigma_i \wedge N} \geq (\beta - \alpha)\{\tau_i \leq N\}$. Hint: Consider separately the three cases $\sigma_i > N$, $\sigma_i \leq N < \tau_i$, and $\tau_i \leq N$.

(iii) Show that $-\mathbb{P}Z_{\sigma_1 \wedge N} + \mathbb{P}Z_{\tau_k \wedge N} \geq k(\beta - \alpha)\mathbb{P}\{\tau_k \leq N\}$. Hint: Take expectations then sum over i in the inequality from part (ii). Use the Stopping Time Lemma for submartingales to prove $\mathbb{P}Z_{\tau_i \wedge N} - \mathbb{P}Z_{\sigma_{i+1} \wedge N} \leq 0$.

(iv) Show that $\mathbb{P}Z_{\tau_k \wedge N} \leq \mathbb{P}Z_N = \mathbb{P}(X_N - \alpha)^+$.

[12] Reprove Corollary <27> (a submartingale $\{X_n : n \in \mathbb{N}_0\}$ converges almost surely to an integrable limit if $\sup_n \mathbb{P}X_n^+ < \infty$) by following these steps.

(i) For fixed $\alpha < \beta$, use the upcrossing inequality from Problem [11] to prove that

$$\mathbb{P}\{\liminf_n X_n < \alpha < \beta < \limsup_n X_n\} = 0$$

(ii) Deduce that $\{X_n\}$ converges almost surely to a limit random variable X that might take the values $\pm\infty$.

(iii) Prove that $\mathbb{P}|X_n| \leq 2\mathbb{P}X_n^+ - \mathbb{P}X_1$ for every n. Deduce via Fatou's lemma that $\mathbb{P}|X| < \infty$.

[13] Suppose the offspring distribution in Example <25> has finite mean $\mu > 1$ and variance σ^2.

(i) Show that $\text{var}(Z_n) = \sigma^2\mu^{n-1} + \mu^2\text{var}(Z_{n-1})$.

(ii) Write X_n for the martingale Z_n/μ^n. Show that $\sup_n \text{var}(X_n) < \infty$.

(iii) Deduce that X_n converges both almost surely and in $\mathcal{L}^1$ to the limit X, and hence $\mathbb{P}X = 1$. In particular, the limit X cannot be degenerate at 0.

[14] Suppose the offspring distribution P from Example <25> has has finite mean $\mu > 1$. Write X_n for the martingale Z_n/μ^n, which converges almost surely to an integrable limit random variable X. Show that the limit X is nondegenerate if and only if the condition

$$(\text{XLOGX}) \qquad P^x\,(x\log(1+x)) < \infty,$$

holds. Follow these steps. Write μ_n for $P^x\left(x\{x \le \mu^n\}\right)$ and $\mathbb{P}_n(\cdot)$ for expectations conditional on $\mathfrak{F}_n$.

(i) Show that $\sum_n(\mu - \mu_n) = P^x\left(x\sum_n\{x > \mu^n\}\right)$, which converges to a finite limit if and only if (XLOGX) holds.

(ii) Define $\widetilde{X}_n := \mu^{-n}\sum_i \xi_{n,i}\{\xi_{n,i} \le \mu^n\}\{i \le Z_{n-1}\}$. Show that $\mathbb{P}_{n-1}\widetilde{X}_n = \mu_n X_{n-1}/\mu$ almost surely. Show also that

$$X - X_N = \sum_{n=N+1}^{\infty}(X_n - X_{n-1}) \ge \sum_{n=N+1}^{\infty}\left(\widetilde{X}_n - X_{n-1}\right) \qquad \text{almost surely.}$$

(iii) Show that, for some constant C_1,

$$\sum_n \mathbb{P}\{\widetilde{X}_n \ne X_n\} \le \sum_n \mu^{n-1}P\{x > \mu^n\} \le C_1\mu < \infty.$$

Deduce that $\sum_n\left(\widetilde{X}_n - X_n\right)$ converges almost surely to a finite limit.

(iv) Write var_{n-1} for the conditional variance corresponding to $\mathbb{P}_{n-1}$. Show that

$$\text{var}_{n-1}(\widetilde{X}_n) = \mu^{-2n}\sum_i\{i \le Z_{n-1}\}\text{var}_{n-1}\left(\xi_{n,i}\{\xi_{n,i} \le \mu^n\}\right).$$

Deduce, via (ii), that

$$\sum_n \mathbb{P}\left(\widetilde{X}_n - \mu_n X_{n-1}/\mu\right)^2 \le \sum_n \mu^{-n-1}Px^2\{x \le \mu^n\} \le C_2\mu < \infty,$$

for some constant C_2. Conclude that $\sum_n\left(\widetilde{X}_n - \mu_n X_{n-1}/\mu\right)$ is a martingale, which converges both almost surely and in $\mathcal{L}^1$.

(v) Deduce from (iii), (iv), and the fact that $\sum_n(X_n - X_{n-1})$ converges almost surely, that $\sum_n X_{n-1}(1 - \mu_n/\mu)$ converges almost surely to a finite limit.

(vi) Suppose $\mathbb{P}\{X > 0\} > 0$. Show that there exists an ω for which both $\sum_n X_{n-1}(\omega)(1 - \mu_n/\mu) < \infty$ and $\lim X_{n-1}(\omega) > 0$. Deduce via (i) that (XLOGX) holds.

(vii) Suppose (XLOGX) holds. From (i) deduce that $\mathbb{P}\left(\sum_n X_{n-1}(1 - \mu_n/\mu)\right) < \infty$. Deduce via (iv) that $\sum_n(\widetilde{X}_n - X_{n-1})$ converges in $\mathcal{L}^1$. Deduce via (ii) that $\mathbb{P}X \ge \mathbb{P}\left(X_N + \sum_{n=N+1}^{\infty}(\widetilde{X}_n - X_{n-1})\right) = 1 - o(1)$ as $N \to \infty$, from which it follows that X is nondegenerate. (In fact, $\mathbb{P}|X_n - X| \to 0$. Why?)

[15] Let $\{\xi_i : i \in \mathbb{N}\}$ be a martingale difference array for which $\sum_{i\in\mathbb{N}}\mathbb{P}\left(\xi_i^2/i^2\right) < \infty$.

(i) Define $X_n := \sum_{i=1}^n \xi_i/i$. Show that $\sup_n \mathbb{P}X_n^2 < \infty$. Deduce that $X_n(\omega)$ converges to a finite limit for almost all ω.

(ii) Invoke Kronecker's lemma to deduce that $n^{-1}\sum_{i=1}^n \xi_i \to 0$ almost surely.

[16] Suppose $\{X_n : n \in \mathbb{N}\}$ is an exchangeable sequence of square-integrable random variables. Show that $\text{cov}(X_1, X_2) \ge 0$. Hint: Each X_i must have the same variance, V; each pair X_i, X_j, for $i \ne j$, must have the same covariance, C. Consider $\text{var}\left(\sum_{i\le n} X_i\right)$ for arbitrarily large n.

[17] (Hewitt & Savage 1955, Section 5) Let $\mathbb{P}$ be exchangeable, in the sense of Definition <49>.

(i) Let f be a bounded, $\mathcal{A}^n$-measurable function on $\mathcal{X}^n$. Define $X := f(x_1, \ldots, x_n)$ and $Y := f(x_{n+1}, \ldots, x_{2n})$. Use Problem [16] to show that $\mathbb{P}(XY) \geq (\mathbb{P}X)(\mathbb{P}Y)$, with equality if $\mathbb{P}$ is a product measure.

(ii) Suppose $\mathbb{P} = \alpha_1\mathbb{Q}_1 + \alpha_2\mathbb{Q}_2$, with $\alpha_i > 0$ and $\alpha_1 + \alpha_2 = 1$, where $\mathbb{Q}_1$ and $\mathbb{Q}_2$ are distinct exchangeable probability measures. Let f be a bounded, measurable function on some $\mathcal{X}^n$ for which $\mu_1 := \mathbb{Q}_1 f(x_1, \ldots, x_n) \neq \mathbb{Q}_2 f(x_1, \ldots, x_n) =: \mu_2$. Define X and Y as in part (i). Show that $\mathbb{P}(XY) > (\mathbb{P}X)(\mathbb{P}Y)$. Hint: Use strict convexity of the square function to show that $\alpha_1\mu_1^2 + \alpha_2\mu_2^2 > (\alpha_1\mu_1 + \alpha_2\mu_2)^2$. Deduce that $\mathbb{P}$ is not a product measure.

(iii) Suppose $\mathbb{P}$ is not a product measure. Explain why there exists an $E \in \mathcal{A}^n$ and a bounded measurable function g for which

$$\mathbb{P}\left(\{\mathbf{z} \in E\}g(x_{n+1}, x_{n+2}, \ldots)\right) \neq (\mathbb{P}\{\mathbf{z} \in E\})\left(\mathbb{P}g(x_{n+1}, x_{n+2}, \ldots)\right),$$

where $\mathbf{z} := (x_1, \ldots, x_n)$. Define $\alpha = \mathbb{P}\{\mathbf{z} \in E\}$. Show that $0 < \alpha < 1$. For each $h \in \mathcal{M}^+(\mathcal{X}^{\mathbb{N}}, \mathcal{A}^{\mathbb{N}})$, define

$$\mathbb{Q}_1 h := \mathbb{P}\left(\{\mathbf{z} \in E\}h(x_{n+1}, x_{n+2}, \ldots)\right)/\alpha,$$
$$\mathbb{Q}_2 h := \mathbb{P}\left(\{\mathbf{z} \in E^c\}h(x_{n+1}, x_{n+2}, \ldots)\right)/(1-\alpha).$$

Show that $\mathbb{Q}_1$ and $\mathbb{Q}_2$ correspond to distinct exchangeable probability measures for which $\mathbb{P} = \alpha\mathbb{Q}_1 + (1-\alpha)\mathbb{Q}_2$. That is, $\mathbb{P}$ is not an extreme point of the set of all exchangeable probability measures on $\mathcal{A}^{\mathbb{N}}$.

10. Notes

De Moivre used what would now be seen as a martingale method in his solution of the gambler's ruin problem. (Apparently first published in 1711, according to Thatcher (1957). See pages 51–53 of the 1967 reprint of the third edition of de Moivre (1718).)

The name *martingale* is due to Ville (1939). Lévy (1937, chapter VIII), expanding on earlier papers (Lévy 1934, 1935*a*, 1935*b*), had treated martingale differences, identifying them as sequences satisfying his condition ($\mathbb{C}$). He extended several results for sums of independent variables to martingales, including Kolmogorov's maximal inequality and strong law of large numbers (the version proved in Section 4.6), and even a central limit theorem, extending Lindeberg's method (to be discussed, for independent summands, in Section 7.2). He worked with martingales stopped at random times, in order to have sums of conditional variances close to specified constant values.

Doob (1940) established convergence theorems (without using stopping times) for martingales and reversed martingales, calling them sequences with "property $\mathcal{E}$." He acknowledged (footnote to page 458) that the basic maximal inequalities were "implicit in the work of Ville" and that the method of proof he used "was used by Lévy (1937), in a related discussion." It was Doob, especially with his stochastic processes book (Doob 1953—see, in particular the historical notes to Chapter VII, starting page 629), who was the major driving force behind the recognition of martingales as one of the most important tools of probability theory. See Lévy's comments in Note II of the 1954 edition of Lévy (1937) and in Lévy (1970, page 118) for the relationship between his work and Doob's.

I first understood some martingale theory by reading the superb text of Ash (1972, Chapter 7), and from conversations with Jim Pitman. The material in Section 3 on positive supermartingales was inspired by an old set of notes for lectures given by Pitman at Cambridge. I believe the lectures were based in part on the original French edition of the book Neveu (1975). I have also borrowed heavily from that book, particularly so for Theorems <26> and <41>. The book of Hall & Heyde (1980), although aimed at central limit theory and its application, contains much about martingales in discrete time. Dellacherie & Meyer (1982, Chapter V) covered discrete-time martingales as a preliminary to the detailed study of martingales in continuous time.

Exercise <15> comes from Aldous (1983, p. 47).

Inequality <20> is due to Dubins (1966). The upcrossing inequality of Problem [11] comes from the same paper, slightly weakening an analogous inequality of Doob (1953, page 316). Krickeberg (1963, Section IV.3) established the decomposition (Theorem <26>) of submartingales as differences of positive supermartingales.

I adapted the branching process result of Problem [14], which is due to Kesten & Stigum (1966), from Asmussen & Hering (1983, Chapter II).

The reversed submartingale part of Example <44> comes from Pollard (1981). The zero-one law of Theorem <51> for symmetric events is due to Hewitt & Savage (1955). The study of exchangeability has progressed well beyond the original representation theorem. Consult Aldous (1983) if you want to know more.

References

Aldous, D. (1983), 'Exchangeability and related topics', *Springer Lecture Notes in Mathematics* **1117**, 1–198.

Ash, R. B. (1972), *Real Analysis and Probability*, Academic Press, New York.

Asmussen, S. & Hering, H. (1983), *Branching processes*, Birkhäuser.

Bauer, H. (1981), *Probability Theory and Elements of Measure Theory*, second english edn, Academic Press.

Birnbaum, Z. W. & Marshall, A. W. (1961), 'Some multivariate Chebyshev inequalities with extensions to continuous parameter processes', *Annals of Mathematical Statistics* pp. 687–703.

de Finetti, B. (1937), 'La prévision: ses lois logiques, ses sources subjectives', *Annales de l'Institut Henri Poincaré* **7**, 1–68. English translation by H. Kyburg in Kyberg & Smokler 1980.

de Moivre, A. (1718), *The Doctrine of Chances*, first edn. Second edition 1738. Third edition, from 1756, reprinted in 1967 by Chelsea, New York.

Dellacherie, C. & Meyer, P. A. (1982), *Probabilities and Potential B: Theory of Martingales*, North-Holland, Amsterdam.

Doob, J. L. (1940), 'Regularity properties of certain families of chance variables', *Transactions of the American Mathematical Society* **47**, 455–486.

Doob, J. L. (1949), 'Application of the theory of martingales', *Colloques Internationaux du Centre National de la Recherche Scientifique* pp. 23–27.

Doob, J. L. (1953), *Stochastic Processes*, Wiley, New York.

Dubins, L. E. (1966), 'A note on upcrossings of semimartingales', *Annals of Mathematical Statistics* **37**, 728.

Hall, P. & Heyde, C. C. (1980), *Martingale Limit Theory and Its Application*, Academic Press, New York, NY.

Hewitt, E. & Savage, L. J. (1955), 'Symmetric measures on cartesian products', *Transactions of the American Mathematical Society* **80**, 470–501.

Kesten, H. & Stigum, B. P. (1966), 'Additional limit theorems for indecomposable multidimensional Galton-Watson process', *Annals of Mathematical Statistics* **37**, 1463–1481.

Kolmogorov, A. (1928), 'Über die Summen durch den Zufall bestimmter unabhängiger Größen', *Mathematische Annalen* **99**, 309–319. Corrections: same journal, volume 102, 1929, pages 484–488.

Krickeberg, K. (1963), *Wahrscheinlichkeitstheorie*, Teubner. English translation, 1965, Addison-Wesley.

Kyberg, H. E. & Smokler, H. E. (1980), *Studies in Subjective Probability*, second edn, Krieger, Huntington, New York. Reprint of the 1964 Wiley edition.

Lévy, P. (1934), 'L'addition de variables aléatoires enchaînées et la loi de Gauss', *Bull. Soc. Math. France* **62**, 42–43.

Lévy, P. (1935*a*), 'Propriétés asymptotiques des sommes de variables aléatoires enchaînées', *Comptes Rendus de l'Academie des Sciences, Paris* **199**, 627–629.

Lévy, P. (1935*b*), 'Propriétés asymptotiques des sommes de variables aléatoires enchaînées', *Bull. Soc. math* **59**, 1–32.

Lévy, P. (1937), *Théorie de l'addition des variables aléatoires*, Gauthier-Villars, Paris. Second edition, 1954.

Lévy, P. (1970), *Quelques Aspects de la Pensée d'un Mathématicien*, Blanchard, Paris.

Neveu, J. (1975), *Discrete-Parameter Martingales*, North-Holland, Amsterdam.

Pollard, D. (1981), 'Limit theorems for empirical processes', *Zeitschrift für Wahrscheinlichkeitstheorie und Verwandte Gebiete* **57**, 181–195.

Pollard, D. (1984), *Convergence of Stochastic Processes*, Springer, New York.

Thatcher, A. R. (1957), 'A note on the early solutions of the problem of the duration of play', *Biometrika* **44**, 515–518.

Ville, J. (1939), *Etude Critique de la Notion de Collectif*, Gauthier-Villars, Paris.

Chapter 7

Convergence in distribution

SECTION 1 defines the concepts of weak convergence for sequences of probability measures on a metric space, and of convergence in distribution for sequences of random elements of a metric space and derives some of their consequences. Several equivalent definitions for weak convergence are noted.
SECTION 2 establishes several more equivalences for weak convergence of probability measures on the real line, then derives some central limit theorems for sums of independent random variables by means of Lindeberg's substitution method.
SECTION 3 explains why the multivariate analogs of the methods from Section 2 are not often explicitly applied.
SECTION 4 develops the calculus of stochastic order symbols.
*SECTION *5 derives conditions under which sequences of probability measures have weakly convergent subsequences.*

1. Definition and consequences

Roughly speaking, central limit theorems give conditions under which sums of random variable have approximate normal distributions. For example:

> If $\xi_1, \ldots, \xi_n$ are independent random variables with $\mathbb{P}\xi_i = 0$ for each i and $\sum_i \text{var}(\xi_i) = 1$, and if none of the ξ_i makes too large a contribution to their sum, then $\sum_i \xi_i$ is approximately $N(0, 1)$ distributed.

The traditional way to formalize approximate normality requires, for each real x, that $\mathbb{P}\{\sum_i \xi_i \le x\} \approx \mathbb{P}\{Z \le x\}$ where Z has a $N(0, 1)$ distribution. Of course the variable Z is used just as a convenient way to describe a calculation with the $N(0, 1)$ probability measure; Z could be replaced by any other random variable with the same distribution. The assertion does not mean that $\sum_i \xi_i \approx Z$, as functions defined on a common probability space. Indeed, the Z need not even live on the same space as the $\{\xi_i\}$. We could remove the temptation to misinterpret the approximation by instead writing $\mathbb{P}\{\sum_i \xi_i \le x\} \approx P(-\infty, x]$ where P denotes the $N(0, 1)$ probability measure.

Assertions about approximate distributions of random variables are usually expresssed as limit theorems. For example, the sum could be treated as one of a sequence of such sums, with the approximation interpreted as an assertion of convergence to a limit. We thereby avoid all sorts of messy details about the size of error terms, replacing them by a less specific assurance that the errors all disappear in the limit. Explicit approximations would be better, but limit theorems are often easier to work with.

In this Chapter you will learn about the notion of convergence traditionally used for central limit theorems. To accommodate possible extensions (such as the theories for convergence in distribution of stochastic processes, as in Pollard 1984), I will start from a more general concept of convergence in distribution for random elements of a general metric space, then specialize to the case of real random variables. I must admit I am motivated not just by a desire for added generality. I also wish to discourage my readers from clinging to inconvenient, old-fashioned definitions involving pointwise convergence of distribution functions (at points of continuity of the limit function).

Let $\mathcal{X}$ be a metric space, with metric $d(\cdot,\cdot)$, equipped with its Borel σ-field $\mathcal{B} := \mathcal{B}(\mathcal{X})$. A ***random element of*** $\mathcal{X}$ is just an $\mathcal{F}\backslash\mathcal{B}(\mathcal{X})$-measurable map from some probability space $(\Omega, \mathcal{F}, \mathbb{P})$ into $\mathcal{X}$. Remember that the image measure $X\mathbb{P}$ is called the distribution of X under $\mathbb{P}$.

The concept of convergence in distribution of a sequence of random elements $\{X_n\}$ depends on X_n only through its distribution. It is really a concept of convergence for probability measures, the image measures $X\mathbb{P}_n$. There are many equivalent ways to define convergence of a sequence of probability measures $\{P_n\}$, all defined on $\mathcal{B}(\mathcal{X})$. I feel that it is best to start from a definition that is easy to work with, and from which useful conclusions can be drawn quickly.

<1> **Definition.** *A real-valued function ℓ on a metric space $\mathcal{X}$ is said to satisfy a* ***Lipschitz condition*** *if there exists a finite constant K for which*

$$|\ell(x) - \ell(y)| \le K d(x,y) \qquad \text{for all } x \text{ and } y \text{ in } \mathcal{X}.$$

Write $BL(\mathcal{X})$ for the vector space of all bounded Lipschitz functions on $\mathcal{X}$.

The space $BL(\mathcal{X})$ has a simple characterization via the quantity $\|f\|_{BL}$, defined for all real valued functions f on $\mathcal{X}$ by

$$\|f\|_{BL} := \max(K_1, 2K_2)$$

<2>
$$\text{where } K_1 := \sup_{x \neq y} \frac{|f(x) - f(y)|}{d(x,y)} \text{ and } K_2 := \sup_x |f(x)|.$$

I have departed slightly from the usual definition of $\|\cdot\|_{BL}$, in order to get the neat bound

<3>
$$|f(x) - f(y)| \le \|f\|_{BL}\left(1 \wedge d(x,y)\right) \qquad \text{for all } x, y \in \mathcal{X}.$$

The space $BL(\mathcal{X})$ consists precisely of those functions ℓ for which $\|\ell\|_{BL} < \infty$. It is easy to show that $\|\cdot\|_{BL}$ is a norm when restricted to $BL(\mathcal{X})$. Moreover, slightly tedious checking of various pointwise cases leads to inequalities such as

$$\|f_1 \vee f_2\|_{BL} \le \max\left(\|f_1\|_{BL}, \|f_2\|_{BL}\right),$$

which implies that $BL(\mathcal{X})$ is stable under the formation of pointwise maxima of pairs of functions. (Replace f_i by $-f_i$ to deduce the same bound for $\|f_1 \wedge f_2\|_{BL}$, and hence stability under pairwise minima.)

REMARK. It is also easy to show that $BL(\mathcal{X})$ is complete under $\|\cdot\|_{BL}$: that is, if $\{\ell_n : n \in \mathbb{N}\} \subseteq BL(\mathcal{X})$ and $\|\ell_n - \ell_m\|_{BL} \to 0$ as $\min(m, n) \to \infty$ then $\|\ell_n - \ell\|_{BL} \to 0$ for a uniquely determined ℓ in $BL(\mathcal{X})$. In fact, $BL(\mathcal{X})$ is a ***Banach lattice***, but that fact will play no explicit role in this book.

It is easy to manufacture useful members of $BL(\mathcal{X})$ by means of the distance function

$$d(x, B) := \inf\{d(x, y) : y \in B\} \qquad \text{for } B \subseteq \mathcal{X}.$$

Problem [1] shows that $|d(x, B) - d(y, B)| \leq d(x, y)$. Thus functions such as $\ell_{\alpha,\beta,,B}(x) := \alpha \wedge \beta d(x, B)$, for positive constants α and β, and $B \subseteq \mathcal{X}$, all belong to $BL(\mathcal{X})$.

<4> **Definition.** *Say that a sequence of probability measures $\{P_n\}$, defined on $\mathcal{B}(\mathcal{X})$, **converges weakly** to a probability measure P, on $\mathcal{B}(\mathcal{X})$, if $P_n\ell \to P\ell$ for all ℓ in $BL(\mathcal{X})$. Write $P_n \rightsquigarrow P$ to denote weak convergence.*

REMARK. Functional analytically minded readers might prefer the term weak-$*$ convergence. Many authors use the symbol $\Rightarrow$ to denote weak convergence. I beg my students to avoid this notation, because it too readily leads to indecipherable homework assertions, such as $P_n \Rightarrow P \Rightarrow TP_n \Rightarrow TP$. (Which $\Rightarrow$ is an implication sign?)

<5> **Definition.** *Say that a sequence $X_1, X_2, \ldots$ of random elements of $\mathcal{X}$ of $\mathcal{X}$ **converges in distribution** to a probability measure P on $\mathcal{B}(\mathcal{X})$ if their distributions converge weakly to P. Denote this convergence by $X_n \rightsquigarrow P$. If X is a random element with distribution P, write $X_n \rightsquigarrow X$ to mean $X_n \rightsquigarrow P$.*

REMARK. Convergence in distribution is also called convergence in law (the word *law* is a synonym for distribution) or weak convergence. For the study of abstract empirical process it was necessary (Hoffmann-Jørgensen 1984, Dudley 1985) to extend the definition to nonmeasurable maps X_n into $\mathcal{X}$. For that case, the concept of *distribution* for X_n is not defined. It turns out that the most natural and successful definition requires convergence of outer expectations, $\mathbb{P}^*h(X_n) \to Ph$, for bounded, continuous, functions h. That is, the convergence $X_n \rightsquigarrow P$ becomes the primary concept, with no corresponding generalization needed for weak convergence of probability measures.

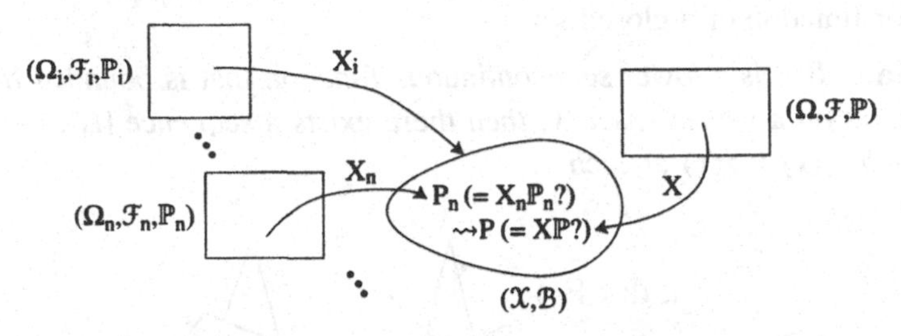

It is important to remember that convergence in distribution, in general, says nothing about pointwise convergence of the X_n as functions. Indeed, each X_n might be defined on a different probability space, $(\Omega_n, \mathcal{F}_n, \mathbb{P}_n)$ so that the very concept of pointwise is void. In that case, $X_n \rightsquigarrow P$ means that $X_n(\mathbb{P}_n) \rightsquigarrow P$, that is,

$$\mathbb{P}_n\ell(X_n) := (X_n\mathbb{P}_n)(\ell) \to P\ell \qquad \text{for all } \ell \text{ in } BL(\mathcal{X});$$

and $X_n \rightsquigarrow X$, with X defined on $(\Omega, \mathcal{F}, \mathbb{P})$, means $\mathbb{P}\ell(X_n) \to \mathbb{P}\ell(X)$ for all ℓ in $BL(\mathcal{X})$.

Similarly, convergence in probability need not be well defined if the X_n live on different probability spaces. There is, however, one important exceptional case. If $X_n \rightsquigarrow P$ with P a probability measure putting mass 1 at a single point x_0 in $\mathcal{X}$, then, for each $\epsilon > 0$,

$$\mathbb{P}_n\{d(X_n, x_0) \geq \epsilon\} \leq \mathbb{P}_n\ell_\epsilon(X_n) \to P\ell_\epsilon = 0 \qquad \text{where } \ell_\epsilon(x) := 1 \wedge (d(x, x_0)/\epsilon).$$

That is, X_n converges in probability to x_0.

<6> **Example.** Suppose $X_n \rightsquigarrow P$ and $\{X'_n\}$ is another sequence for which $d(X_n, X'_n)$ converges to zero in probability. Then $X'_n \rightsquigarrow P$. For if $\|\ell\|_{BL} := K < \infty$,

$$\begin{aligned}|\mathbb{P}_n\ell(X_n) - \mathbb{P}_n\ell(X'_n)| &\leq \mathbb{P}_n|\ell(X_n) - \ell(X'_n)| \\ &\leq K\mathbb{P}_n\left(1 \wedge d(X_n, X'_n)\right) \qquad \text{by <3>} \\ &\leq K\mathbb{P}\{d(X_n, X'_n) > \epsilon\} + K\epsilon \qquad \text{for each } \epsilon > 0.\end{aligned}$$

The first term in the final bound tends to zero as $n \to \infty$, for each $\epsilon > 0$, by the assumed convergence in probability. It follows that $\mathbb{P}_n\ell(X'_n) \to P\ell$. □

REMARK. A careful probabilist would worry about measurability of $d(X_n, X'_n)$. If $\mathcal{X}$ were a separable metric space, Problem [6] would ensure measurability. In general, one could reinterpret the assertion of convergence in probability to mean: there exists a sequence of measurable functions $\{\Delta_n\}$ with $\Delta_n \geq d(X_n, X'_n)$ and $\Delta_n \to 0$ in probabilty. The argument for the Example would be scarcely affected.

Convergence for expectations of the functions in $BL(\mathcal{X})$ will lead to convergence for expectations of other types of function, by means of various approximation schemes. The argument for semicontinuous functions is typical. Recall that a function $g : \mathcal{X} \to \mathbb{R}$ is said to be ***lower semicontinuous*** (LSC) if $\{x : g(x) > t\}$ is an open set for each fixed t. Similary, a function $f : \mathcal{X} \to \mathbb{R}$ is said to be ***upper semicontinuous*** (USC) if $\{x : f(x) < t\}$ is an open set for each fixed t. (That is, f is USC if and only if $-f$ is LSC.) If a function is both LSC and UCS then it is continuous. The prototypical example for lower semicontinuity is the indicator function of an open set; the prototypical example for upper semicontinuity is the indicator function of a closed set.

<7> **Lemma.** *If g is a lower semicontinuous function that is bounded from below by a constant, on a metric space $\mathcal{X}$, then there exists a sequence $\{\ell_i : i \in \mathbb{N}\}$ in $BL(\mathcal{X})$ for which $\ell_i(x) \uparrow g(x)$ at each x.*

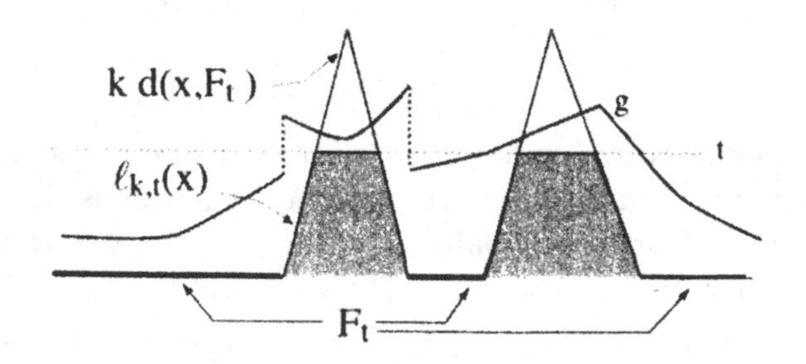

Proof. With no loss of generality, we may assume $g \geq 0$. For each $t > 0$, the set $F_t := \{g \leq t\}$ is closed. The sequence of nonnegative BL functions $\ell_{k,t}(x) := t \wedge (kd(x, F_t))$ for $k \in \mathbb{N}$ increases pointwise to $t\{g > t\}$, because $d(x, F_t) > 0$ if and only if $g(x) > t$. (Compare with Problem [3].)

The countable collection $\mathcal{G}$ of all $\ell_{k,t}$ functions, for $k \in \mathbb{N}$ and positive rational t has pointwise supremum equal to g. Enumerate $\mathcal{G}$ as $\{h_1, h_2, \ldots\}$, then define $\ell_i := \max_{j \leq i} h_j$. □

<8> **Theorem.** *Suppose $P_n \rightsquigarrow P$. Then*

(i) $\liminf_{n\to\infty} P_n g \geq Pg$ *for each lower semicontinuous function g that is bounded from below by a constant,*

(ii) $\limsup_{n\to\infty} P_n f \leq Pf$ *for each upper semicontinuous function f that is bounded from above by a constant.*

Proof. For a given LSC g, bounded from below, invoke Lemma <7> to find an increasing sequence $\{\ell_i\}$ in $BL(\mathcal{X})$ with $\ell_i \uparrow g$ pointwise. Then, for fixed i,

$$\liminf_n P_n g \geq \liminf_n P_n \ell_i = P\ell_i \qquad \text{because } P_n \ell_i \to P\ell_i.$$

Take the supremum over i, using Monotone Convergence to show $\sup_i P\ell_i = Pg$, to obtain (ii). Put $g = -f$ to deduce (ii). □

When specialized to the case of indicator functions, we have

$$\text{<9>} \qquad \left. \begin{array}{l} \liminf_n P_n G \geq PG \text{ for all open } G \\ \limsup_n P_n F \leq PF \text{ for all closed } F \end{array} \right\} \quad \text{if } P_n \rightsquigarrow P.$$

<10> **Example.** Let f be a bounded, measurable function on $\mathcal{X}$. The collection of all lower semicontinuous functions g with $g \leq f$ has a largest member, because the supremum of any family $\mathcal{G}$ of LSC functions is also LSC: $\{x : \sup_{g\in\mathcal{G}} g(x) > t\} = \cup_{g\in\mathcal{G}}\{x : g(x) > t\}$, a union of open sets. By analogy with the notation for the interior of a set, write $\mathring{f}$ for this largest LSC $\leq f$. The analogy is helpful, because $\mathring{f}$ equals the indicator function of $\mathring{B}$ when f equals the indicator of a set B. Similarly, there is a smallest USC function $\bar{f}$ that is everywhere $\geq f$, and $\bar{f}$ is the indicator of $\bar{B}$ when f is the indicator of B.

For simplicity suppose $0 \leq f \leq 1$. We have $\bar{f}(x) \geq f(x) \geq \mathring{f}(x)$ for all x. At a point x where $\mathring{f}(x) = f(x)$ the set $\{y : \mathring{f}(y) > f(x) - \epsilon\}$ is an open neighborhood of x within which $f(y) > f(x) - \epsilon$. Similarly, if $\bar{f}(x) = f(x)$, there is a neighborhood of x on which $f(y) < f(x) + \epsilon$. In short, f is continuous at each point of the set $\{x : \bar{f}(x) = \mathring{f}(x)\}$. Conversely, if f is continuous at a point x then, for each $\epsilon > 0$, there is some open neighborhood G of x for which $|f(x) - f(y)| < \epsilon$ for $y \in G$. We may assume $\epsilon < 1$. Then the function f is sandwiched between a LSC function and an USC function,

$$(f(x) - \epsilon)\{y \in G\} - 2\{y \notin G\} \leq f(y) \leq (f(x) + \epsilon)\{y \in G\} + 2\{y \notin G\},$$

which differ by only 2ϵ at x, thereby implying that $\bar{f}(x) = \mathring{f}(x)$. The Borel measurable set $C_f := \{x : \bar{f}(x) = \mathring{f}(x)\}$ is precisely the set of all points at which f is continuous.

Now suppose $P_n \rightsquigarrow P$. From Theorem <8>, and the inequality $\bar{f} \geq f \geq \mathring{f}$, we have

$$P\bar{f} \geq \limsup_n P_n\bar{f} \geq \limsup_n P_n f \geq \liminf_n P_n f \geq \liminf_n P_n \mathring{f} \geq P\mathring{f}.$$

If $P\bar{f} = P\mathring{f}$, which happens if $PC_f = 1$, then we have convergence, $P_n f \to Pf$. That is, for a bounded, Borel measurable, real function f on $\mathcal{X}$,

<11> $$P_n f \to Pf \qquad \text{if } P_n \rightsquigarrow P \text{ and } f \text{ is continuous } P \text{ almost everywhere.}$$

When specialized to the case of an indicator function of a set $B \in \mathcal{B}(\mathcal{X})$, we have

<12> $$P_n B \to PB \qquad \text{if } P_n \rightsquigarrow P \text{ and } P(\partial B) = 0,$$

because the discontinuities of the indicator function of a set occur only at its boundary. A set with zero P measure on its boundary is called a ***P*-continuity set**. For example, an interval $(-\infty, x]$ on the real line is a continuity set for every probability measure that puts zero mass at the point x. When specialized to real random variables, assertion <12> gives the traditional convergence of distribution functions at continuity points of the limit distribution function. □

REMARK. Intuitively speaking, the closeness of probability measures in a weak convergence sense is not sensitive to changes that have only small effects on functions in $BL(\mathcal{X})$: arbitrary relocations of small amounts of mass (because the functions in $BL(\mathcal{X})$ are bounded), or small relocations of arbitrarily large amounts of mass (because the functions in $BL(\mathcal{X})$ satisfy a Lipschitz condition). The P-continuity condition, $P(\partial B) = 0$, ensures that small rearrangements of P masses near the boundary of B cannot have much effect on PB. See Problem [4] for a way of making this idea precise, by constructing sequences $P_n' \rightsquigarrow P$ and $P_n'' \rightsquigarrow P$ for which $P_n' B \to P\bar{B}$ and $P_n'' B \to P\mathring{B}$.

Problem <10> shows that convergence for all P-continuity sets implies the convergence $P_n f \to Pf$ for all bounded, measurable functions f that are continuous a.e. $[P]$, and, in particular, for all f in $BL(\mathcal{X})$. Thus, any one of the assertions in the following summary diagram of equivalences could be taken as the definition of weak convergence, and then the other equivalences would become theorems. It is largely a matter of taste, or convenience, which equivalent form one chooses as the definition. It is is worth noting that, because of the equivalences, the concept of weak convergence does not depend on the particular choice for the metric: all metrics generating the same topology lead to the same concept.

REMARK. Billingsley (1968, Section 2) applied the name Portmanteau theorem to a subset of the equivalences shown in the following diagram. The circle of ideas behind these equivalences goes back to Alexandroff (1940–43), who worked in an abstract, nontopological setting. Prohorov (1956) developed a very useful theory for weak convergence in complete separable metric spaces. Independently, Le Cam (1957) developed an analogous theory for more general topological spaces. (See also Varadarajan 1961.) For arbitrary topological spaces, Topsøe (1970, page 41) chose the semicontinuity assertions (or more precisely, their analogs for generalized sequences) to define weak convergence. Such a definition is needed to build a nonvacuous theory, because there exist nontrivial spaces for which the only continuous functions are the constants.

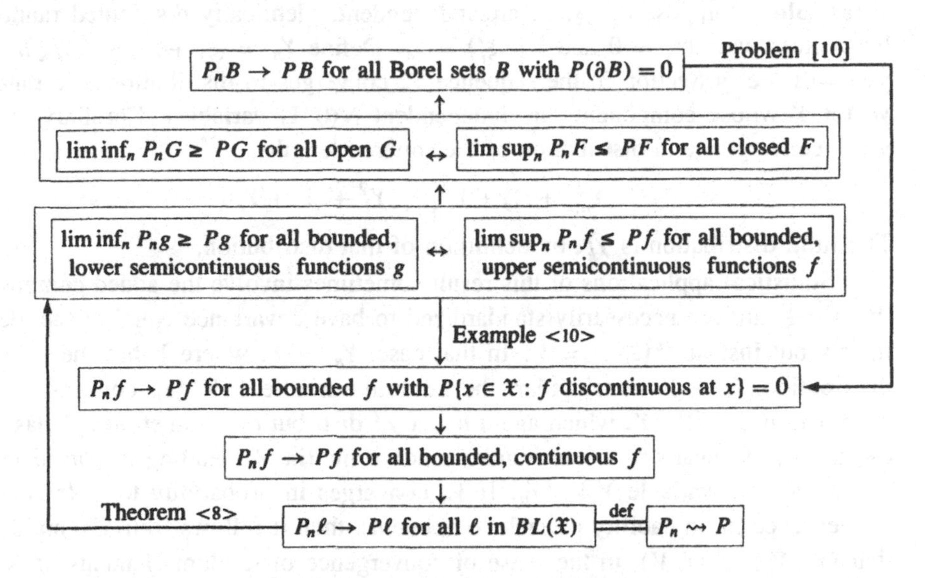

Equivalences for weak convergence of Borel probability measures on a general metric space. Further equivalences, for the special case of probability measures on the real line, are stated in the next Section.

The following consequence of <11> is often referred to as the ***Continuous Mapping Theorem***, even though the mapping in question need not be continuous. It would be more accurate, if clumsier, to call it the Almost-Surely-Continuous Mapping Theorem.

<13> **Corollary.** *Let T be a $\mathcal{B}(\mathcal{X})\backslash\mathcal{B}(\mathcal{Y})$-measurable map from $\mathcal{X}$ into another metric space $\mathcal{Y}$, which is continuous at each point of a measurable subset C_T. If $P_n \rightsquigarrow P$ and $PC_T = 1$ then $T(P_n) \rightsquigarrow T(P)$.*

Proof. For $\ell \in BL(\mathcal{Y})$, the composition $f := \ell \circ T$ is continuous at each point of C_T. From <11>, we have $(TP_n)(\ell) := P_n\ell(T) \to P\ell(T) =: (TP)(\ell)$. □

REMARK. The equivalent assertion for random elements is: if $X_n \rightsquigarrow X$ and T is continuous at almost all realizations $X(\omega)$ then $T(X_n) \rightsquigarrow T(X)$.

<14> **Example.** Suppose $Y_n \rightsquigarrow Y$, as random elements of a metric space $(\mathcal{Y}, d_{\mathcal{Y}})$, and $Z_n \to z_0$ in probability, as random elements of a metric space $(\mathcal{Z}, d_{\mathcal{Z}})$. Equip $\mathcal{X} := \mathcal{Y} \times \mathcal{Z}$ with the metric

$$d(x_1, x_2) := \max\left(d_{\mathcal{Y}}(y_1, y_2), d_{\mathcal{Z}}(z_1, z_2)\right) \qquad \text{where } x_i := (y_i, z_i).$$

If $\ell \in BL(\mathcal{Y} \times \mathcal{Z})$ then the function $y \mapsto \ell(y, z_0)$ belongs to $BL(\mathcal{Y})$, and hence $\mathbb{P}_n\ell(Y_n, z_0) \to \mathbb{P}\ell(Y, z_0)$. That is, the random elements $X'_n := (Y_n, z_0)$ converge in distribution to (Y, z_0). The random element $X_n := (Y_n, Z_n)$ is close to X'_n: in fact, $d(X_n, X'_n) = d_{\mathcal{Z}}(Z_n, z_0) \to 0$ in probability. By Example <6>, it follows that $X_n \rightsquigarrow (Y, z_0)$. If T is a measurable function of (y, z) that is continuous at almost all realizations $(Y(\omega), z_0)$ of the limit process, Corollary <13> then gives $T(Y_n, Z_n) \rightsquigarrow T(Y, z_0)$. The special cases where $\mathcal{Y} = \mathcal{Z} = \mathbb{R}$ and $T(y, z) = y + z$ or $T(y, z) = yz$ are sometimes referred to as ***Slutsky's theorem***. □

<15> **Example.** Suppose $\xi_1, \xi_2, \ldots$ are independent, identically distributed random $\mathbb{R}^k$-vectors with $\mathbb{P}\xi_i = 0$ and $\mathbb{P}(\xi_i\xi_i') = I_k$. Define $Y_n = (\xi_1 + \ldots + \xi_n)/\sqrt{n}$. As you will see in Section 3, the sequence Y_n converges in distribution to a random vector Y whose components are independent $N(0,1)$ variables. Corollary <13> gives convergence in distribution of the squared lengths,

$$Y_{n,1}^2 + \ldots + Y_{n,k}^2 \rightsquigarrow Y_1^2 + \ldots + Y_k^2.$$

The limit distribution is χ_k^2, by definition of that distribution.

Statistical applications of this result sometimes involve the added complication that the ξ_i are not necessarily standardized to have covariance equal to the identity matrix but instead $\mathbb{P}(\xi_i\xi_i') = V$. In that case, $Y_n \rightsquigarrow Y$, where Y has the $N(0, V)$ distribution. If V is nonsingular, the random variable $Y_n'V^{-1}Y_n$ converges in distribution to $Y'V^{-1}Y$, which again has a χ_k^2 distribution. Sometimes V has to be estimated, by means of a nonsingular random matrix V_n, leading to consideration of the random variable $Y_n'V_n^{-1}Y_n$. If V_n converges in probability to V (meaning convergence in probability of each component), then it follows from Example <14> that $(Y_n, V_n) \rightsquigarrow (Y, V)$, in the sense of convergence of random elements of $\mathbb{R}^{k+k^2}$. Corollary <13> then gives $Y_n'V_n^{-1}Y_n \rightsquigarrow Y'V^{-1}Y$, because the map $(y, A) \mapsto y'A^{-1}y$ is continous at each point of $\mathbb{R}^{k+k^2}$ where the matrix A is nonsingular. □

2. Lindeberg's method for the central limit theorem

The first Section showed what rewards we reap once we have established convergence in distribution of a sequence of random elements of a metric space. I will now explain a method to establish such convergence, for sums of real random variables, using the central limit theorem (CLT) as the motivating example. It will be notationally convenient to work directly with convergence in distribution of random variables, rather than weak convergence of probability measures. To simplify even further, let us assume all random variables are defined on a single probability space $(\Omega, \mathcal{F}, \mathbb{P})$.

The CLT, in its various forms, gives conditions under which a sum of independent variables $\xi_1 + \ldots + \xi_k$ has an approximate normal distribution, in the sense that $\mathbb{P}\ell(\xi_1 + \ldots + \xi_k)$ for ℓ in $BL(\mathbb{R})$ is close to the corresponding expectation for the normal distribution. Lindeberg's (1922) method transforms the sum into a sum of normal increments by successive replacement of each ξ_i by a normally distributed η_i with the same expected value and variance. The errors accumulated during the replacements are bounded using a Taylor expansion of the function f. Of course the method can work only if f is smooth enough to allow the Taylor expansion. More precisely, it requires functions with bounded derivatives up to third order. We therefore need first to check that convergence of expectations for such smooth functions suffices to establish convergence in distribution. In fact, convergence for an even smaller class of functions will suffice.

<16> **Lemma.** *If $\mathbb{P}f(X_n) \to \mathbb{P}f(X)$ for each f in the class $\mathcal{C}^\infty(\mathbb{R})$ of all bounded functions with bounded derivatives of all orders, then $X_n \rightsquigarrow X$.*

Proof. Let Z have a $N(0,1)$ distribution. For a fixed $\ell \in BL(\mathbb{R})$ and $\sigma > 0$, define a smoothed function by convolution,

$$\ell_\sigma(x) := \mathbb{P}\ell(x+\sigma Z) := (2\pi\sigma^2)^{-1/2}\int_{-\infty}^{\infty} \exp\left(-\tfrac{1}{2}(y-x)^2/\sigma^2\right)\ell(y)\,dy.$$

The function ℓ_σ belongs to $\mathcal{C}^\infty(\mathbb{R})$: a Dominated Convergence argument justifies repeated differentiation under the integral sign (Problem [15]). As σ tends to zero, ℓ_σ converges uniformly to ℓ, because

$$|\ell_\sigma(x) - \ell(x)| \le \mathbb{P}|\ell(x+\sigma Z) - \ell(x)| \le \|\ell\|_{BL}\mathbb{P}(1 \wedge \sigma|Z|) \to 0,$$

again by Dominated Convergence.

Given $\epsilon > 0$, fix a σ for which $\sup_x |\ell_\sigma(x) - \ell(x)| \le \epsilon$. Then observe that $|\mathbb{P}\ell(X_n) - \mathbb{P}\ell(X)|$ lies within 2ϵ of $|\mathbb{P}\ell_\sigma(X_n) - \mathbb{P}\ell_\sigma(X)|$, which converges to zero as $n \to \infty$, because $\ell_\sigma \in BL(\mathbb{R})$. □

REMARK. There is nothing special about the choice of the $N(0,1)$ distribution for Z in the construction of ℓ_σ. It matters only that the distribution have a density with bounded derivatives of all orders, and that differentiation under the integral sign in the convolution integral can be justified.

For weak convergence of probability measures on the real line, we can augment the equivalences from Section 1 by a further collection involving specific properties of the real line.

$P_n(-\infty, x] \to P(-\infty, x]$ for all $x \in \mathbb{R}$ with $P\{x\} = 0$

$\Updownarrow$

$P_n \rightsquigarrow P$

$\Updownarrow$

$P_n f \to Pf$ for all f in $\mathcal{C}^3(\mathbb{R})$

$\Updownarrow$

$P_n f \to Pf$ for all f in $\mathcal{C}^\infty(\mathbb{R})$

$\Updownarrow$

$P_n^x e^{ixt} \to P^x e^{ixt}$ for all real t

Further equivalences for weak convergence of probability measures on the real line. Lemma <16> handles the implications leading from $\mathcal{C}^\infty(\mathbb{R})$ to weak convergence. Example <10> then gives the convergence of distribution functions. Problem [5] gives the approximation arguments leading from convergence of distribution functions (in fact, it even treats the messier case for $\mathbb{R}^2$) to weak convergence. The final equivalence, involving the complex exponentials, will be explained in Chapter 8.

Lindeberg's method needs only bounded derivatives up to third order. Define $\mathcal{C}^3(\mathbb{R})$ to to be the class of all such bounded functions. Of course convergence of expectations for functions in $\mathcal{C}^3(\mathbb{R})$ is more than enough to establish convergence in distribution, because $\mathcal{C}^3(\mathbb{R}) \supseteq \mathcal{C}^\infty(\mathbb{R})$.

For a fixed f in $\mathcal{C}^3(\mathbb{R})$, define

$$C = \sup_x \tfrac{1}{6}|f'''(x)|,$$

which is finite by assumption. By Taylor's Theorem,

$$f(x+y) = f(x) + yf'(x) + \tfrac{1}{2}y^2 f''(x) + R(x,y),$$

where $R(x,y) = \frac{1}{6}y^3 f'''(x^*)$ for some x^* lying between x and $x+y$. Consequently,

$$|R(x,y)| \le C|y|^3 \qquad \text{for all } x \text{ and } y.$$

If X and Y are independent random variables, with $\mathbb{P}|Y|^3 < \infty$, then

$$\mathbb{P}f(X+Y) = \mathbb{P}f(X) + \mathbb{P}\left(Yf'(X)\right) + \tfrac{1}{2}\mathbb{P}\left(Y^2 f''(X)\right) + \mathbb{P}R(X,Y).$$

Using independence to factorize two of the terms and bounding $|R(X,Y)|$ by $C|Y|^3$, we get

$$|\mathbb{P}f(X+Y) - \mathbb{P}f(X) - (\mathbb{P}Y)\left(\mathbb{P}f'(X)\right) - \tfrac{1}{2}(\mathbb{P}Y^2)\left(\mathbb{P}f''(X)\right)| \le C\mathbb{P}|Y|^3.$$

Suppose Z is another random variable independent of X, with $\mathbb{P}|Z|^3 < \infty$ and $\mathbb{P}Z = \mathbb{P}Y$ and $\mathbb{P}Z^2 = \mathbb{P}Y^2$. Subtract, cancelling out the first and second moment contributions, to leave

<17> $$|\mathbb{P}f(X+Y) - \mathbb{P}f(X+Z)| \le C\left(\mathbb{P}|Y|^3 + \mathbb{P}|Z|^3\right).$$

For the particular case where Z has a $N(\mu, \sigma^2)$ distribution, with $\mu := \mathbb{P}Y$ and $\sigma^2 := \text{var}(Y)$, the third moment bound simplifies slightly. For convenience, write W for $(Z-\mu)/\sigma$, which has a $N(0,1)$ distribution. Then

$$\begin{aligned}
\mathbb{P}|Z|^3 &\le 8|\mu|^3 + 8\sigma^3\mathbb{P}|W|^3 \\
&\le 8(\mathbb{P}|Y|)^3 + 8(\mathbb{P}|Y|^2)^{3/2}\mathbb{P}|W|^3 \\
&\le \left(8 + 8\mathbb{P}|W|^3\right)\mathbb{P}|Y|^3 \qquad \text{by Jensen's inequality.}
\end{aligned}$$

The right-hand side of <17> is therefore less than $C_1\mathbb{P}|Y|^3$ where C_1 denotes the constant $(9 + 8\mathbb{P}|W|^3)C$.

Now consider independent random variables $\xi_1, \ldots, \xi_k$ with

$$\mu_i := \mathbb{P}\xi_i, \qquad \sigma_i^2 := \text{var}(\xi_i), \qquad \mathbb{P}|\xi_i|^3 < \infty.$$

Independently of all the $\{\xi_i\}$, generate η_i distributed $N(\mu_i, \sigma_i^2)$, for $i = 1, \ldots, k$. Choose the $\{\eta_i\}$ so that all $2k$ variables are independent. Define

$$S := \xi_1 + \ldots + \xi_k \qquad \text{and} \qquad T := \eta_1 + \ldots + \eta_k.$$

The sum T has a normal distribution with

$$\mathbb{P}T = \mu_1 + \ldots + \mu_k \qquad \text{and} \qquad \text{var}(T) = \sigma_1^2 + \ldots + \sigma_k^2.$$

Repeated application of inequality <17> will lead to the third moment bound for the difference $\mathbb{P}f(S) - \mathbb{P}f(T)$.

For each i define

$$\begin{aligned}
X_i &:= \xi_1 + \ldots + \xi_{i-1} + \qquad + \eta_{i+1} + \ldots + \eta_k \\
Y_i &:= \qquad\qquad\qquad \xi_i \\
Z_i &:= \qquad\qquad\qquad \eta_i
\end{aligned}$$

The variables X_i, Y_i, and Z_i are independent. From <17> with the upper bound simplified for the normally distributed Z_i,

$$|\mathbb{P}f(X_i + Y_i) - \mathbb{P}f(X_i + Z_i)| \le C_1 \mathbb{P}|\xi_1|^3,$$

for each i. At $i = k$ and $i = 1$ we recover the two sums of interest, $X_k + Y_k = \xi_1 + \ldots + \xi_k = S$ and $X_1 + Z_1 = \eta_1 + \ldots + \eta_k = T$. Each substitution of a Z_i for a Y_i replaces one more ξ_i by the corresponding η_i; the k substitutions replace all the ξ_i by the η_i. The accumulated change in the expectations is bounded by a sum of third moment terms,

<18> $$|\mathbb{P}f(S) - \mathbb{P}f(T)| \le C_1 \left(\mathbb{P}|\xi_1|^3 + \ldots + \mathbb{P}|\xi_k|^3\right).$$

We have only to add an extra subscript to get the basic central limit theorem.

It is cleanest to state the theorem in terms of a ***triangular array*** of random variables,

$$\begin{array}{l} \xi_{1,1}, \xi_{1,2}, \ldots, \xi_{1,k(1)} \\ \xi_{2,1}, \xi_{2,2}, \quad \ldots, \xi_{2,k(2)} \\ \xi_{3,1}, \xi_{3,2}, \qquad \ldots, \xi_{3,k(3)} \\ \qquad \vdots \end{array}$$

The variables within each row are assumed independent. Nothing need be assumed about the relationship between variables in different rows; all calculations are carried out for a fixed row. By working with triangular arrays we eliminate various centering and scaling constants that might otherwise be needed.

<19> **Theorem.** *Let $\xi_{n,1}, \ldots, \xi_{n,k(n)}$, for $n = 1, 2, \ldots$, be a triangular array of random variables, independent within rows, such that*

(i) $\sum_i \mathbb{P}\xi_{n,i} \to \mu$, with μ finite,

(ii) $\sum_i \text{var}(\xi_{n,i}) \to \sigma^2 < \infty$,

(iii) $\sum_i \mathbb{P}|\xi_{n,i}|^3 \to 0$.

Then $\sum_{i \le k(n)} \xi_{n,i} \rightsquigarrow N(\mu, \sigma^2)$ as $n \to \infty$.

Proof. Choose f in $\mathcal{C}^3(\mathbb{R})$. Apply inequality <18> and (iii) to show that $\mathbb{P}f(\sum_i \xi_{n,i})$ equals $\mathbb{P}f(T_n) + o(1)$, where T_n is $N(\mu_n, \sigma_n^2)$ distributed with $\mu_n \to \mu$ by (i) and $\sigma_n^2 \to \sigma^2$ by (ii). Deduce (see Problem [11]) that $T_n \rightsquigarrow N(\mu, \sigma^2)$, whence $\mathbb{P}f(\sum_i \xi_{n,i}) \to \mathbb{P}f\left(N(\mu, \sigma^2)\right)$. □

<20> **Exercise.** If X_n has a Bin (n, p_n) distribution and if $np_n(1 - p_n) \to \infty$, show that

$$\frac{X_n - np_n}{\sqrt{np_n(1 - p_n)}} \rightsquigarrow N(0, 1).$$

SOLUTION: Manufacture a random variable with the same distribution as the standardized X_n as follows. Let $A_{n,1}, \ldots, A_{n,n}$ be independent events with $\mathbb{P}A_{n,i} = p_n$ for $i = 1, \ldots, n$. Define

$$\xi_{n,i} := \frac{A_{n,i} - p_n}{\sigma_n} \qquad \text{where } \sigma_n := \sqrt{np_n(1 - p_n)}.$$

Then $\sum_{i \le n} \xi_{n,i}$ has the same distribution as $(X_n - np_n)/\sigma_n$.

Check the conditions of Theorem <19> with $\mu := 0$ and $\sigma^2 := 1$ for these $\xi_{n,i}$. The centering was chosen to given $\mathbb{P}\xi_{n,i} = 0$, so (i) holds. By direct calculation,

$$\text{var}(\xi_{n,i}) = \mathbb{P}\xi_{n,i}^2 = p_n \frac{(1-p_n)^2}{\sigma_n^2} + (1-p_n)\frac{(-p_n)^2}{\sigma_n^2} = \frac{1}{n},$$

so $\sum_i \text{var}(\xi_{n,i}) = 1$. Requirement (ii) holds. Finally, because $|A_{n,i} - p_n| \le 1$,

$$\textstyle\sum_i \mathbb{P}|\xi_{n,i}|^3 \le \sigma_n^{-1} \sum_i \mathbb{P}|\xi_{n,i}|^2 = \sigma_n^{-1} \to 0.$$

□ It follows that $\sum_{i\le n} \xi_{n,i} \rightsquigarrow N(0,1)$, as required.

Theorem <19> can be extended to random variables that don't satisfy the moment conditions—indeed, to variables that don't even have finite moments—by means of truncation arguments. The theorem for independent, identically distributed random variables with finite variances illustrates the truncation technique well. The method of proof will delight all fans of Dominated Convergence, which once again emerges as the right tool to handle truncation arguments for identically distributed variables.

<21> **Theorem.** *Let $X_1, X_2, \ldots$ be independent, identically distributed random variables with $\mathbb{P}X_i = 0$ and $\mathbb{P}X_i^2 = 1$. Then $(X_1 + \ldots + X_n)/\sqrt{n} \rightsquigarrow N(0,1)$.*

Proof. The argument will depend on three applications of Dominated Convergence, with dominating variable X_1^2:

$$\mathbb{P}X_1^2\{|X_1| \le \sqrt{n}\} \to 1,$$
$$\mathbb{P}X_1^2\{|X_1| > \sqrt{n}\} \to 0,$$
$$\mathbb{P}X_1^2\left(1 \wedge (|X_1|/\sqrt{n})\right) \to 0.$$

Apply Theorem <19> to the variables $\xi_{n,i} := X_i\{|X_i| \le \sqrt{n}\}/\sqrt{n}$, for $i = 1, \ldots, n$. Notice that

$$\mu_n := \textstyle\sum_i \mathbb{P}\xi_{n,i} = n\mathbb{P}\xi_{n,1} = -n\mathbb{P}X_1\{|X_1| > \sqrt{n}\}/\sqrt{n} \qquad \text{because } \mathbb{P}X_i = 0,$$

which gives the bound

$$|\mu_n| \le \sqrt{n}\,\mathbb{P}|X_1|\{|X_1| > \sqrt{n}\} \le \mathbb{P}|X_1|^2\{|X_1| > \sqrt{n}\} \to 0.$$

To control the sum of variances, use the identical distributions together with the fact that $\mathbb{P}\xi_{n,i} = \mu_n/n = o(1/n)$ to deduce that

$$\textstyle\sum_i \text{var}(\xi_{n,i}) = \mathbb{P}X_1^2\{|X_1| \le \sqrt{n}\} - n(\mathbb{P}\xi_{n,1})^2 \to 1.$$

For the third moment bound use

$$\textstyle\sum_i \mathbb{P}|\xi_{n,i}|^3 \le \frac{n}{n^{3/2}}\mathbb{P}|X_1|^3\{|X_1| \le \sqrt{n}\} \le \mathbb{P}X_1^2\left(1 \wedge \frac{|X_1|}{\sqrt{n}}\right) \to 0.$$

It follows that $\sum_i \xi_{n,i} \rightsquigarrow N(0,1)$. Complete the proof via an appeal to Example <6>, using the inequality

$$\mathbb{P}\left\{\frac{\sum_{i\le n} X_i}{\sqrt{n}} \ne \textstyle\sum_{i\le n} \xi_{n,i}\right\} \le \textstyle\sum_{i\le n} \mathbb{P}\{|X_i| > \sqrt{n}\} \le n\mathbb{P}X_1^2\{|X_1| > \sqrt{n}\}/n \to 0$$

□ to show that $\sum_{i\le n} X_i/\sqrt{n} - \sum_i \xi_{n,i} \to 0$ in probability.

Similar truncation techniques can be applied to derive even more general forms of the central limit theorem from Theorem <19>. Often we have to deal with conditions expressible as $H_n(\epsilon) \to 0$ for each $\epsilon > 0$, for some sequence of functions H_n. It is convenient to be able to replace ϵ by a sequence $\{\epsilon_n\}$ that converges to zero and also have $H_n(\epsilon_n) \to 0$. Roughly speaking, if a condition holds for all fixed ϵ then it will also hold for sequences ϵ_n tending to zero slowly enough.

<22> **Lemma.** *Suppose $H_n(\epsilon) \to 0$ as $n \to \infty$, for each fixed $\epsilon > 0$. Then there exists a sequence $\epsilon_n \to 0$ such that $H_n(\epsilon_n) \to 0$.*

Proof. For each positive integer k there exists an n_k such that $|H_n(1/k)| < 1/k$ for $n \ge n_k$. Without loss of generality, assume $n_1 < n_2 < \ldots$. Define

$$\epsilon_n = \begin{cases} \text{arbitrary} & \text{for } n < n_1, \\ 1/k & \text{for } n_k \le n < n_{k+1}. \end{cases}$$

That is, for $n \ge n_1$, we have $\epsilon_n = 1/k_n$, where k_n is the positive integer k for which $n_k \le n < n_{k+1}$. Clearly $k_n \to \infty$ as $n \to \infty$. Also, for $n \ge n_1$, we have $|H_n(\epsilon_n)| < 1/k_n$, which converges to zero as $n \to \infty$. □

The form of the central limit theorem in the next Exercise is essentially due to Lindeberg (1922). The result actually includes Theorem <21> as a special case.

<23> **Exercise.** Let $\{X_{n,i}\}$ be a triangular array of random variables, independent within each row, such that:

(i) $\mathbb{P}X_{n,i} = 0$ for all n and i;

(ii) $\sum_i \mathbb{P}X_{n,i}^2 = 1$;

(iii) $L_n(\epsilon) := \sum_i \mathbb{P}X_{n,i}^2\{|X_{n,i}| > \epsilon\} \to 0$ for each $\epsilon > 0$ [Lindeberg's condition].

Show that $\sum_i X_{n,i} \rightsquigarrow N(0,1)$.

SOLUTION: Invoke Lemma <22> with $H_n(\epsilon) := L_n(\epsilon)/\epsilon^2$ to find ϵ_n tending to zero slowly enough to ensure that $L_n(\epsilon_n)/\epsilon_n^2 \to 0$. Define a triangular array of random variables $\xi_{n,i} := X_{n,i}\{|X_{n,i}| \le \epsilon_n\}$. Notice that

$$\mathbb{P}\{\xi_{n,i} \neq X_{n,i} \text{ for some } i\} \le \sum_i \mathbb{P}\{|X_{n,i}| > \epsilon_n\} \le L_n(\epsilon_n)/\epsilon_n^2 \to 0.$$

By Example <6> it suffices to show that $\sum_i \xi_{n,i} \rightsquigarrow N(0,1)$.

Check the conditions of Theorem <19>. For the first moments:

$$\left|\sum_i \mathbb{P}\xi_{n,i}\right| = \left|-\sum_i \mathbb{P}X_{n,i}\{|X_{n,i}| > \epsilon_n\}\right| \le L_n(\epsilon_n)/\epsilon_n \to 0.$$

For the variances, use the fact that $\mathbb{P}\xi_{n,i} = -\mathbb{P}X_{n,i}\{|X_{n,i}| > \epsilon_n\}$ to show that

$$\sum_i \text{var}(\xi_{n,i}) = \sum_i \mathbb{P}X_{n,i}^2\{|X_{n,i}| \le \epsilon_n\} - \sum_i \left(-\mathbb{P}X_{n,i}\{|X_{n,i}| > \epsilon_n\}\right)^2.$$

The first sum on the right-hand side equals $\sum_i \mathbb{P}X_{n,i}^2 - L_n(\epsilon_n)$, which tends to 1. The second sum is bounded above by $L_n(\epsilon_n)$. For the third moments:

$$\sum_i \mathbb{P}|\xi_{n,i}|^3 \le \epsilon_n \sum_i \mathbb{P}X_{n,i}^2 \to 0.$$

The central limit theorem follows. □

3. Multivariate limit theorems

The arguments for proving convergence in distribution of random $\mathbb{R}^k$ vectors—multivariate limit theorems—are similar to those for random variables. Indeed, with an occasional reinterpretation of a square as a squared length of a vector, and a product as an inner product, the arguments in Section 2 carry over to random vectors.

Perhaps the only subtlety in the multivariate analog of Theorem <19> arises from the factorization of quadratic terms like $\mathbb{P}\left(Y'\ddot{f}(X)Y\right)$, with Y independent of the $k \times k$ random matrix of second derivatives $\ddot{f}(X)$. We could resort to an explicit expansion into a sum of terms $\mathbb{P}\left(Y_i Y_j \ddot{f}(X)_{ij}\right)$, but it is more elegant to reinterpret the quadratic as the expected trace of a 1×1 matrix, then rearrange it as

$$\mathbb{P}\text{trace}(Y'\ddot{f}(X)Y) = \mathbb{P}\text{trace}(\ddot{f}(X)YY') = \text{trace}\left(\left(\mathbb{P}\ddot{f}(X)\right)\left(\mathbb{P}YY'\right)\right).$$

We are again in the position to approximate $\mathbb{P}f(\sum_i X_i)$ by means of a sequence of substitutions of variables with the same expected values and covariances.

The new variables should be chosen as multivariate normals. For each $\mu \in \mathbb{R}^k$ and each nonnegative definite matrix V, the $N(\mu, V)$ could be defined as the distribution of $\mu + RW$, with W a vector of independent $N(0, 1)$'s and R any $k \times k$ matrix for which $RR' = V$. It is easy to check that

$$\mathbb{P}(\mu + RW) = \mu + R\mathbb{P}W = \mu \qquad \text{and} \qquad \text{var}(\mu + RW) = R\text{var}(W)R' = V.$$

The Fourier tools from Chapter 8 offer the simplest method for showing that the distribution does not depend on the particular choice of R.

Problem [14] shows how to adapt the one-dimensional calculations to derive a multivariate analog of approximation <18>. The assertion of Theorem <19> holds if we reinterpret σ^2 to be a variance matrix. The derivations of other multivariate central limit theorems follow in much the same way as before.

<24> **Example.** If $X_1, X_2, \ldots$ are independent, identically distributed random vectors with $\mathbb{P}X_i = 0$ and $\mathbb{P}|X_i|^2 < \infty$, then $(X_1 + \ldots + X_n)/\sqrt{n} \rightsquigarrow N(0, V)$, where $V := \mathbb{P}(X_1 X_1')$. □

In short, the multivariate versions of the results from Section 2 present little extra challenge if we merely translate the methods into vector notation. Fans of the more traditional approach to the theory (based on pointwise convergence of distribution functions) might find the extensions via multivariate distribution functions more tedious. Textbooks seldom engage in such multivariate exercises, because there is a more pleasant alternative: By means of a simple Fourier device (to be discussed in Section 8.6), multivariate results can often be reduced directly to their univariate analogs. There is not much incentive to engage in multivariate proofs, except as a way of deriving explicit error bounds (see Section 10.4).

4. Stochastic order symbols

You are probably familiar with the $O(\cdot)$ and $o(\cdot)$ notation from real analysis. They allow one to avoid specifying constants in many arguments, thereby simplifying the notational load. For example, it is neater to write

$$f(x) = f(0) + xf'(0) + o(|x|) \qquad \text{near } 0,$$

than to write out the formal epsilon-delta details of the limit properties that define differentiability of f at 0.

The order symbols have stochastic analogs, which are almost indispensable in advanced asymptotic theory. As with any very concise notation, it is easy to conceal subtle errors if the symbols are not used carefully; but without them, all but the simplest arguments become notationally overwhelming.

<25> **Definition.** *For random vectors $\{X_n\}$ and nonnegative random variables $\{\alpha_n\}$ write $X_n = 0_p(\alpha_n)$ to mean: for each $\epsilon > 0$ there is a finite constant M_ϵ such that $\mathbb{P}\{|X_n| > M_\epsilon \alpha_n\} < \epsilon$ eventually. Write $X_n = o_p(\alpha_n)$ to mean: $\mathbb{P}\{|X_n| > \epsilon\alpha_n\} \to 0$ for each $\epsilon > 0$.*

Typically $\{\alpha_n\}$ is a sequence of constants, but occasionally random bounds are useful.

REMARK. The notation allows us to write things like $o_p(\alpha_n) = O_p(\alpha_n)$, but not $O_p(\alpha_n) = o_p(\alpha_n)$, meaning that if X_n is of order $o_p(\alpha_n)$ then it is also of order $O_p(\alpha_n)$ but not conversely. It might help to think of $O_p(\cdot)$ and $o_p(\cdot)$ as defining classes of sequences of random variables and perhaps even to write $X_n \in O_p(\alpha_n)$ or $X_n \in o_p(\alpha_n)$ instead of $X_n = O_p(\alpha_n)$ or $X_n = o_p(\alpha_n)$.

<26> **Example.** The assertion $X_n = o_p(1)$ means the same as $X_n \to 0$ in probability. When specialized to random vectors, the result in Example <6> becomes: if $X_n \rightsquigarrow P$ then $X_n + o_p(1) \rightsquigarrow P$. The $o_p(1)$ here replaces the sequence $X_n - X'_n$. □

<27> **Example.** If $X_n \rightsquigarrow P$ then $X_n = 0_p(1)$. From <9> with G as the open ball of radius M centered at the origin, $\liminf \mathbb{P}\{X_n \in G\} \geq PG$. If M is large enough, $PG > 1 - \epsilon$, which implies that $\mathbb{P}\{|X_n| \geq M\} < \epsilon$ eventually. □

<28> **Example.** Be careful with the interpretation of an assertion such as $O_p(1) + O_p(1) = O_p(1)$. The three $0_p(1)$ symbols do not refer to the same sequence of random vectors; it would be a major blunder to cancel out the $0_p(1)$ to conclude that $O_p(1) = 0$. The assertion is actually shorthand for: if $X_n = 0_p(1)$ and $Y_n = O_p(1)$ then $X_n + Y_n = 0_p(1)$.

The assertion is easy to verify. Given $\epsilon > 0$, choose a constant M so that $\mathbb{P}\{|X_n| > M\} < \epsilon$ and $\mathbb{P}\{|Y_n| > M\} < \epsilon$ eventually. Then, eventually,

$$\mathbb{P}\{|X_n + Y_n| > 2M\} \leq \mathbb{P}\{|X_n| > M\} + \mathbb{P}\{|Y_n| > M\} < 2\epsilon.$$

If you worry about the 2ϵ in the bound, replace ϵ by $\epsilon/2$ throughout the previous paragraph. □

<29> **Example.** If $\{X_n\}$ is a sequence of real random variables of order $O_p(1)$, what can be asserted about $\{1/X_n\}$? **Nothing.** The stochastic order symbol $O_p(\cdot)$ conveys no information about lower bounds. For example, if $X_n \equiv 1/n$ then $X_n = O_p(1)$ but $1/X_n = n \to \infty$. You should invent other examples.

The blunder of asserting $1/0_p(1) = 0_p(1)$ is quite common. Be warned. □

<30> **Example.** For a sequence of constants $\{\alpha_n\}$ that tends to zero, suppose $X_n = O_p(\alpha_n)$. Let g be a function, defined on the range space of the $\{X_n\}$, for which $g(x) = o(|x|)$ near 0. Then the random variables $g(X_n)$ are of order $o_p(\alpha_n)$. To prove the assertion, for given $\epsilon > 0$ and $\epsilon' > 0$, find M and then $\delta > 0$ such that

$\mathbb{P}\{|X_n| > M\alpha_n\} < \epsilon'$ eventually, and $|g(x)| \le \epsilon|x|/M$ when $|x| \le \delta$. When n is large enough, we have $M\alpha_n \le \delta$ and

$$\mathbb{P}\left\{|g(X_n)| > \frac{\epsilon}{M}M\alpha_n\right\} \le \mathbb{P}\{|X_n| > M\alpha_n\} < \epsilon'.$$

□ That is, $g(X_n) = o_p(\alpha_n)$, or, more cryptically, $o(O_p(\alpha_n)) = o_p(\alpha_n)$.

<31> **Example.** The so-called ***delta method*** gives a simple way to analyze smooth transformations of sequences of random vectors. Suppose x_0 is a fixed vector in $\mathbb{R}^k$, and $\{X_n\}$ is a sequence of random vectors for which $Z_n = \sqrt{n}(X_n - x_0) \rightsquigarrow Z$. Suppose g is a measurable function from $\mathbb{R}^k$ into $\mathbb{R}^l$ that is differentiable at x_0. That is, there exists an $l \times k$ matrix D such that

$$g(x_0 + \delta) = g(x_0) + D\delta + R(\delta),$$

where $|R(\delta)| = o(|\delta|)$ as $\delta \to 0$. If we replace δ by the random quantity $Z_n/\sqrt{n}$ we get $\sqrt{n}(g(X_n) - g(x_0)) = DZ_n + \sqrt{n}R(Z_n/\sqrt{n})$. From Example <30> we have $R(Z_n/\sqrt{n}) = o(O_p(1/\sqrt{n})) = o_p(1/\sqrt{n})$, from which it follows, via Example <26>,
□ that $\sqrt{n}(g(X_n) - g(x_0)) + o_p(1) \rightsquigarrow DZ$.

*5. Weakly convergent subsequences

In a compact metric space, each sequence has a convergent subsequence. Sequences of probability measures that concentrate most of their mass on compact subsets of a metric space have a similar property, a result that provides a powerful method for proving existence of probability measures.

<32> **Definition.** *A probability measure P on $\mathcal{B}(\mathcal{X})$ is said to be **tight** if to each positive ϵ there exists a compact set K_ϵ such that $PK_\epsilon > 1 - \epsilon$.*

For the purposes of weak convergence arguments it is more convenient to have tight measures identified with particular linear functionals on $BL(\mathcal{X})$, with $\mathcal{X}$ a metric space. The following characterization is a special case of a result proved in Section 6 of Appendix A.

<33> **Theorem.** *A linear functional $\lambda : BL(\mathcal{X})^+ \mapsto \mathbb{R}^+$ with $\lambda 1 = 1$ defines a tight probability measure if and only if it is functionally tight: to each positive ϵ there exists a compact set K_ϵ such that $\lambda\ell < \epsilon$ for every ℓ in $BL(\mathcal{X})^+$ for which $\ell \le K_\epsilon^c$.*

In order that a limit functional on $BL(\mathcal{X})^+$ inherit the functional tightness property from a convergent sequence, it suffices that an analogous property hold "uniformly along the sequence." It turns out that a property slightly weaker than requiring $\sup_n P_n K_\epsilon^c < \epsilon$ is enough.

<34> **Definition.** *Call a sequence of probability measures $\{P_n\}$ on the Borel sigma-field of a metric space **uniformly tight** if to each $\epsilon > 0$ there exists a compact set K_ϵ such that $\limsup_{n\to\infty} P_n G^c < \epsilon$ for every open set G containing K_ϵ.*

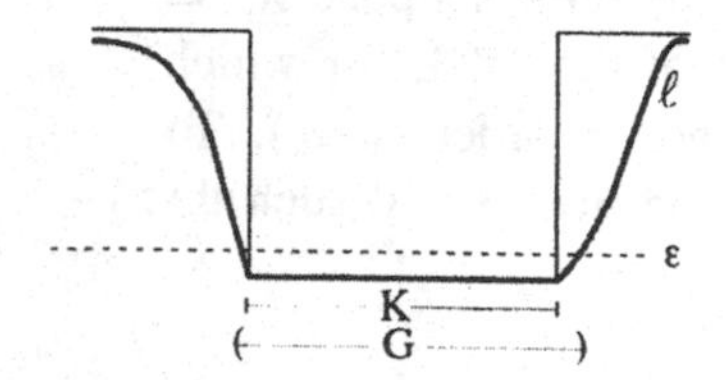

Uniform tightness implies (and, apart from a few inconsequential constant factors, is equivalent to) the assertion that for each $\epsilon > 0$ there is a compact set K_ϵ such that

$\limsup P_n \ell < 2\epsilon$ for every ℓ in $BL(\mathcal{X})^+$ for which $0 \le \ell \le K_\epsilon^c$: for such an ℓ, the open set $G_\epsilon := \{\ell < \epsilon\}$ contains K_ϵ and

<35> $$P_n \ell \le \epsilon + P_n G_\epsilon^c < 2\epsilon \qquad \text{eventually.}$$

This equivalent form of uniform tightness is better suited to the passage to the limit.

<36> **Theorem.** *(Prohorov 1956, Le Cam 1957) Every uniformly tight sequence of probability measures on the Borel sigma-field of a metric space has a subsequence that converges weakly to a tight probability measure.*

Construction of the limit distribution, in the form of a tight linear functional on $BL(\mathcal{X})^+$, will be achieved by a Cantor diagonalization argument, applied to a countable family of functions of the type descibed in the following Lemma.

<37> **Lemma.** *For each $\delta > 0$, $\epsilon > 0$, and each compact set K there exists a finite collection $\mathcal{G} = \{g_0, g_1, \ldots, g_k\} \subseteq BL(\mathcal{X})^+$ such that:*

(i) $g_0(x) + \ldots + g_k(x) = 1$ for each $x \in \mathcal{X}$;

(ii) the diameter of each set $\{g_i > 0\}$, for $i \ge 1$, is less than δ.

(iii) $g_0 < \epsilon$ on K.

REMARK. A finite collection of nonnegative, continuous functions that sum to one everywhere is called a ***continuous partition of unity***.

Proof. Let $x_1, \ldots, x_k$ be the centers of open balls of radius $\delta/4$ whose union covers the compact set K. Define functions $f_0 \equiv \epsilon/2$ and $f_i(x) := (1 - 2d(x, x_i)/\delta)^+$, for $i \ge 1$, in $BL(\mathcal{X})^+$. Notice that $f_i(x) = 0$ if $d(x, x_i) > \delta/2$, for $i \ge 1$. Thus the set $\{f_i > 0\}$ has diameter less than δ for $i \ge 1$.

The function $F(x) := \sum_{i=0}^k f_i(x)$ is everywhere greater than $\epsilon/2$, and it belongs to $BL(\mathcal{X})^+$. The nonnegative functions $g_i := f_i/F$ are bounded by 1 and satisfy a Lipschitz condition:

$$\begin{aligned} |g_i(x) - g_i(y)| &= \frac{|F(y)f_i(x) - F(x)f_i(y)|}{F(x)F(y)} \\ &\le \frac{|f_i(x) - f_i(y)|}{F(x)} + \frac{|F(y) - F(x)| f_i(y)}{F(x)F(y)} \\ &\le 2\left(\|f_i\|_{BL} + \|F\|_{BL}\right) d(x, y)/\epsilon. \end{aligned}$$

For each x in K there is an i for which $d(x, x_i) < \delta/4$. For that i we have $f_i(x) > 1/2$ and $g_0(x) \le f_0(x)/f_i(x) < \epsilon$. The g_i sum to 1 everywhere. They are the required functions. □

Proof of Theorem <36>. Write K_i for the compact set given by Definition <34> with $\epsilon := 1/i$. For each i in $\mathbb{N}$ write $\mathcal{G}_i$ for the finite collection of functions in $BL(\mathcal{X})^+$ constructed via Lemma <37> with $\delta := \epsilon := 1/i$ and K equal to K_i. The class $\mathcal{G} := \cup_{i \in \mathbb{N}} \mathcal{G}_i$ is countable.

For each g in $\mathcal{G}$ the sequence of real numbers $P_n g$ is bounded. It has a convergent subsequence. Via a Cantor diagonalization argument, we can construct a single subsequence $\mathbb{N}_1 \subseteq \mathbb{N}$ for which $\lim_{n \in \mathbb{N}_1} P_n g$ exists for every g in $\mathcal{G}$.

The approximation properties of $\mathcal{G}$ will ensure existence of the limit $\lambda \ell := \lim_{n \in \mathbb{N}_1} P_n \ell$ for every ℓ in $BL(\mathcal{X})^+$. With no loss of generality, suppose $\|\ell\|_{BL} \le 1$.

Given $\epsilon > 0$, choose an $i > 1/\epsilon$, then write $\mathcal{G}_i = \{g_0, g_1, \ldots, g_k\}$, with indexing as in Lemma <37>. The open set $G_i := \{g_0 < \epsilon\}$ contains K_i, which ensures that $\limsup_n P_n G_i^c < \epsilon$.

For each $j \geq 1$ let x_j be any point at which $g_j(x_j) > 0$. If x is any other point with $g_j(x) > 0$ we have $|\ell(x) - \ell(x_j)| \leq d(x, x_j) \leq \epsilon$. It follows that, for every x in $\mathcal{X}$,

$$\begin{aligned} |\ell(x) - \textstyle\sum_{j=1}^k \ell(x_j) g_j(x)| &\leq \ell(x) g_0(x) + \textstyle\sum_{j=1}^k |\ell(x) - \ell(x_j)| g_j(x) \\ &\leq \left(\epsilon + G_i^c\right) + \epsilon, \end{aligned}$$

which integrates to give

$$|P_n \ell - \textstyle\sum_{j=1}^k \ell(x_j) P_n g_j| \leq P_n G_i^c + 2\epsilon.$$

It then follows, via the existence of $\lim_{n \in \mathbb{N}_1} P_n g_j$, that $\limsup_{n \in \mathbb{N}_1} P_n \ell$ differs from $\liminf_{n \in \mathbb{N}_1} P_n \ell$ by at most 6ϵ. The limit $\lambda \ell := \lim_{n \in \mathbb{N}_1} P_n \ell$ exists.

The limit functional λ inherits linearity from $\mathbb{P}$. Clearly $\lambda 1 = 1$. It inherits tightness from the uniform tightness of the sequence, as in <35>. From Theorem <33>, the functional λ corresponds to a tight probability measure to which $\{P_n : n \in \mathbb{N}_1\}$ converges in distribution. □

REMARK. For readers who know more about general topology: The Cantor diagonalization argument could be replaced by an argument with ultrafilters, or universal subsets, for uniformly tight nets of probability measures on more general topological spaces. Lemma <37> was needed only to allow us to work with sequences; it could be avoided.

<38> **Example.** Let $\{X_n\}$ be a sequence of $\mathbb{R}^k$-valued random vectors of order $O_p(1)$. If $\mathbb{P}\{|X_n| > M\} < \epsilon$ eventually then $\limsup \mathbb{P}\{X_n \notin G\} \leq \epsilon$ for every open set G that contains the compact ball $\{x : |x| \leq M\}$. That is, $\{X_n\}$ is uniformly tight. It has a subsequence that converges in distribution to a probability measure on $\mathcal{B}(\mathbb{R}^k)$. □

6. Problems

[1] Let B be a subset of a metric space. For each pair of points x_1 and x_2 in $\mathcal{X}$, show that $\inf_{y \in B} d(x_1, y) \leq d(x_1, x_2) + \inf_{y \in B} d(x_2, y)$. Deduce that the function $f_B(x) := d(x, B)$ satisfies the Lipschitz condition $|f_B(x) - f_B(y)| \leq d(x, y)$.

[2] For real-valued functions f, g on $\mathcal{X}$, prove that $\|fg\|_{BL} \leq 2\|f\|_{BL}\|g\|_{BL}$.

[3] Let B be a subset of a metric space $\mathcal{X}$. Show that

$$\begin{aligned} \{x : d(x, B) = 0\} &= \bar{B} := \text{closure of } B, \\ \{x : d(x, B^c) > 0\} &= \mathring{B} := \text{interior of } B, \\ \{x : d(x, B) = 0 = d(x, B^c)\} &= \bar{B} \backslash \mathring{B} = \partial B = \text{boundary of } B. \end{aligned}$$

Hint: If $d(x, B) = 0$, there exists points $x_n \in B$ with $d(x, x_n) \to 0$.

[4] Let P be a probability measure on a separable metric space, and B be a Borel set for which $P(\partial B) > 0$.

(i) For each $\epsilon > 0$, show that there is a partition of ∂B into disjoint Borel sets $\{D_i : i \in \mathbb{N}\}$ each with diameter less than ϵ. Hint: Consider the union of balls of radius $\epsilon/3$ centered at the points of a countable dense subset of ∂B.

(ii) For each i, find points $x_i \in B$ and $y_i \in B^c$ such that $d(x_i, D_i) < \epsilon$ and $d(y_i, D_i) < \epsilon$. Define a probability measure P'_ϵ by replacing all the P mass in D_i by a point mass (of size PD_i) at x_i, for each i. Define P''_ϵ similarly, by concentrating the mass in each D_i at y_i. Show that $P'_\epsilon B = P\bar{B}$ and $P''_\epsilon B = P\mathring{B}$ for each $\epsilon > 0$.

(iii) Show that, for each ℓ with $\|\ell\|_{BL} := K < \infty$,
$$|P\ell - P'_\epsilon\ell| \le K\epsilon P(\partial B) \qquad \text{and} \qquad |P\ell - P''_\epsilon\ell| \le K\epsilon P(\partial B).$$
Deduce that $P'_\epsilon \rightsquigarrow P$ and $P''_\epsilon \rightsquigarrow P$ as $\epsilon \to 0$, even though we have $P'_\epsilon B \equiv P\bar{B} > PB$ and $P''_\epsilon B \equiv P\mathring{B} < PB$.

[5] For $\mathbf{x} = (x_1, x_2) \in \mathbb{R}^2$, define $Q_\mathbf{x} = \{(y_1, y_2) \in \mathbb{R}^2 : y_1 \le x_1, y_2 \le x_2\}$, the quadrant with vertex $\mathbf{x}$. Let $\{X_n\}$ be a sequence of random vectors and P be a probability measure such that $\mathbb{P}\{X_n \in Q_\mathbf{x}\} \to PQ_\mathbf{x}$ for all $\mathbf{x}$ such that $P(\partial Q_\mathbf{x}) = 0$. Show that $X_n \rightsquigarrow P$ by following these steps.

(i) Write $\mathcal{L}$ for the class of all lines parallel to a coordinate axis. Show that all except countably many lines in $\mathcal{L}$ have zero P measure. (Hint: How many horizontal lines can have P measure greater than $1/n$?)

(ii) Given $\ell \in BL(\mathbb{R}^2)$, with $0 \le \ell \le 1$, show that there exist disjoint rectangles $S_1, \dots, S_m$ with sides parallel to the coordinate axes, such that (a) $P(\partial S_i) = 0$ for each each i; (b) the set $B = \bigcup_i S_i$ has P-measure greater than $1 - \epsilon$; (c) the function ℓ oscillates by less than ϵ on each S_i.

(iii) Choose arbitrarily points $\mathbf{x}_i$ in S_i, for $i = 1, \dots, m$. Define functions $g_\epsilon(\mathbf{x}) = \sum_i \{\mathbf{x} \in S_i\}\ell(\mathbf{x}_i)$. Show that $|g_\epsilon(\mathbf{x}) - \ell(\mathbf{x})| \le \epsilon + \{\mathbf{x} \notin B\}$ for all $\mathbf{x}$.

(iv) Use a sandwiching argument to show that $\mathbb{P}\ell(X_n) \to P\ell$, then deduce that $X_n \rightsquigarrow P$.

[6] Let $\mathcal{X}$ be a metric space. Show that the map $\psi : (x, y) \to d(x, y)$ is continuous. Deduce that ψ is $\mathcal{B}(\mathcal{X}^2)\backslash\mathcal{B}(\mathbb{R})$-measurable. If $\mathcal{X}$ is separable, deduce that ψ is $\mathcal{B}(\mathcal{X}) \otimes \mathcal{B}(\mathcal{X})\backslash\mathcal{B}(\mathbb{R})$-measurable, and hence $\omega \mapsto d(X(\omega), Y(\omega))$ is measurable if X and Y are random elements of $\mathcal{X}$.

[7] If $P\ell = Q\ell$ for each ℓ in $BL(\mathcal{X})$, show that $P = Q$ as measures on $\mathcal{B}(\mathcal{X})$.

[8] Let $\mathcal{X}$ be a separable metric space. Show that $\Delta(P, Q) := \sup\{|P\ell - Q\ell| : \|\ell\|_{BL} \le 1\}$ is a metric for weak convergence. That is, show that $P_n \rightsquigarrow P$ if and only if $\Delta(P_n, P) \to 0$, by following these steps. (Read the proof of Lemma <37> for hints.)

(i) Show that Δ is a metric, and $|P\ell - Q\ell| \le \|\ell\|_{BL}\Delta(Q, P)$, for each ℓ. Deduce that $P_n\ell \to P\ell$, for each $\ell \in BL(\mathcal{X})$, if $D(P_n, P) \to 0$.

(ii) Let $\mathcal{X}_0 := \{x_i : i \in \mathbb{N}\}$ be dense in $\mathcal{X}$. For a fixed $\epsilon > 0$, define $f_0(x) \equiv \epsilon$ and $f_i(x) := (1 - d(x, x_i)/\epsilon)^+$. Define $G_k := \cup_{i=1}^k\{f_i > 1/2\}$. Choose k so that $PG_k^k < \epsilon$. Show that each function $\ell_i := f_i / \sum_{0 \le i \le k} f_i$ belongs to $BL(\mathcal{X})$. Show that $\ell_0(x) \le 2\epsilon$ for $x \in G_k$. Show that $\sum_{i=0}^k \ell_i \equiv 1$. Show that $\text{diam}\{\ell_i > 0\} \le 2\epsilon$ for $i \ge 1$.

(iii) For each $i \ge 1$, choose an x_i from $\{\ell_i > 0\}$. Show that

$$|\ell(x) - \sum_{i=1}^k \ell(x_i)\ell_i(x)| \le 4\epsilon + G_k^c \qquad \text{if } \|\ell\|_{BL} \le 1.$$

(iv) For each probability measure Q, and each ℓ with $\|\ell\|_{BL} \le 1$, show that

$$|Q\ell - P\ell| \le 8\epsilon + QG_k^c + PG_k^c + \sum_{i=1}^k |Q\ell_i - P\ell_i|.$$

(v) If $P_n \rightsquigarrow P$, deduce that $\Delta(P_n, P) < 10\epsilon$ eventually.

[9] Let $\mathcal{Y}$ and $\mathcal{Z}$ be metric spaces, such that $\mathcal{Y}$ is separable. Let d be the metric on $\mathcal{Y} \otimes \mathcal{Z}$ defined in Example <14>. Let P_0, P be probability measures on $\mathcal{B}(\mathcal{Y})$, and Q_0, Q be probability measure on $\mathcal{B}(\mathcal{Z})$. For a fixed ℓ in $BL(\mathcal{Y} \times \mathcal{Z})$, define $h_0(z) := P_0^y \ell(y, z)$ and $g_Q(y) := Q^z \ell(y, z)$.

(i) Show that $\|h_0\|_{BL} \le \|\ell\|_{BL}$ and $\|g_Q\|_{BL} \le \|\ell\|_{BL}$.

(ii) Let $\Delta_{\mathcal{Y}}$ and $\Delta_{\mathcal{Z}}$ denote the analogs of the metric from Problem [8]. Show that

$$\begin{aligned} |(P \otimes Q)\ell - (P_0 \otimes Q_0)\ell| &\le |Pg_Q - P_0 g_Q| + |Qh_0 - Q_0 h_0| \\ &\le \|\ell\|_{BL} \Delta_{\mathcal{Y}}(P, P_0) + |Qh_0 - Q_0 h_0|. \end{aligned}$$

(iii) Show that $\Delta_{\mathcal{Y} \times \mathcal{Z}}(P \otimes Q, P_0 \otimes Q_0) \le \Delta_{\mathcal{Y}}(P, P_0) + \Delta_{\mathcal{Z}}(Q, Q_0)$.

(iv) If $P_n \rightsquigarrow P_0$ and $Q_n \rightsquigarrow Q_0$, show that $P_n \otimes Q_n \rightsquigarrow P_0 \otimes Q_0$, even if $\mathcal{Z}$ is not separable. (For separable $\mathcal{Z}$, the result follows from (iii); otherwise use (ii).)

[10] Suppose $\{X_n\}$ are random elements of $\mathcal{X}$, and P is a probability measure P on $\mathcal{B}(\mathcal{X})$, for which $\mathbb{P}\{X_n \in B\} \to PB$ for every P-continuity set. Let f be a bounded measurable function on $\mathcal{X}$ (with no loss of generality assume $0 \le f \le 1$) that is continuous at all points except those of a P-negligible set $\mathcal{N}$.

(i) For each real t, show that the boundary of the set $\{f \ge t\}$ is contained in $\mathcal{N} \cup \{f = t\}$. Deduce that $\{f \ge t\}$ is a P-continuity set for almost all (Lebesgue measure) t. Hint: Consider sequences $x_n \to x$ and $y_n \to x$ with $f(x_n) \ge t > f(y_n)$.

(ii) Show that $\mathbb{P}f(X_n) = \int_0^1 \mathbb{P}\{f(X_n) \ge t\}\,dt \to Pf$.

[11] Suppose $\mu_n \to \mu$ and $\sigma_n^2 \to \sigma^2$, with both limits finite. Let Z have a $N(0, 1)$ distribution. Show that $|\mathbb{P}\ell(\mu_n+\sigma_n Z) - \mathbb{P}\ell(\mu+\sigma Z)| \le \|\ell\|_{BL} \mathbb{P}(1 \wedge (|\mu_n - \mu| + |\sigma_n - \sigma|\,|Z|))$. Deduce, by Dominated Convergence, that $N(\mu_n, \sigma_n^2) \rightsquigarrow N(\mu, \sigma^2)$.

[12] Suppose X_n has a $N(\mu_n, \sigma_n^2)$ distribution, and $X_n \rightsquigarrow P$.

(i) Show that $\mu := \lim \mu_n$ and $\sigma^2 := \lim \sigma_n^2$ must exist as finite limits, and that P must be the $N(\mu, \sigma^2)$ distribution. Hint: Choose M with $P\{M\} = 0 = P\{-M\}$ and $P[-M, M] > 3/4$. If $|\mu_n| > M$ or if σ_n is large enough, show that $\mathbb{P}\{|X_n| > M\} \ge 1/2$. Show that all convergent subsequences of (μ_n, σ_n) must converge to the same limit.

(ii) Extend the result to sequences of random vectors. Hint: Use part (i) to prove boundedness of $\{\mu_n\}$ and each diagonal element of $\{V_n\}$. Use Cauchy-Schwarz to deduce that all elements of $\{V_n\}$ are bounded.

[13] Suppose the random variables $\{X_n\}$ converge in distribution to X, and that $\{A_n\}$ and $\{B_n\}$ are sequences of constants with $A_n \to A$ and $B_n \to B$ (both limits finite). Show that $A_n X_n + B_n \rightsquigarrow AX + B$. Generalize to random vectors.

[14] Let Y be a random k-vector with $\mu := \mathbb{P}Y$ and $V := \text{var}(Y)$. Let V have the representation $V = L\Lambda^2 L'$, with L an orthogonal matrix and $\Lambda := \text{diag}(\lambda_1, \ldots, \lambda_k)$, each λ_i nonnegative. Define $R := L\Lambda$. Let W be a random k-vector of independent $N(0, 1)$ random variables.

(i) Show that $|\mu| \le \mathbb{P}|Y|$. Hint: For a unit vector u in the direction of μ, use the fact that $u \cdot Y \le |Y|$.

(ii) Show that $\mathbb{P}|RW|^3 = \mathbb{P}|\sum_i \lambda_i W_i|^3 \le (\text{trace } V)^{3/2}\mathbb{P}|N(0, 1)|^3$.

(iii) Show that $\mathbb{P}|\mu + RW|^3 \le 8\mathbb{P}|Y|^3 + 8\left(\mathbb{P}|Y|^2\right)^{3/2}\mathbb{P}|N(0, 1)|^3$.

[15] Let f be a bounded, measurable, real-valued function on the real line. Let k be a Lebesgue integrable function, with derivative k'. Suppose there exists a Lebesgue integrable function M with $|k(x + \delta) - k(x)| \le |\delta| M(x)$ for all $|\delta| \le 1$ and all x.

(i) Define $g(x) := \int f(x + y)k(y)\, dy = \int f(z)k(z - x)\, dz$. Use Dominated Convergence to justify differentiation under the integral sign to prove that g is differentiable, with derivative $g'(x) = -\int f(x + y)k'(y)\, dy$.

(ii) Let $k(x) := p(x)\exp(-x^2/2)$, with p a polynomial in x. Show that k and each of its derivatives satisfies the stated assumptions. Deduce that the corresponding g has bounded derivatives of all orders. Hint: Consider the case $p(x) := x^d$. Show that $|e^t - 1| \le |t|e^{|t|}$ for all real t.

(iii) For each $\sigma > 0$, show that the function $x \mapsto k(x/\sigma)/\sigma$ and each of its derivatives also satisfies the assumptions.

[16] Let $\{X_n\}$ be a sequence of random variables, all defined on the same probability space. If $X_n = o_p(1)$, we know from Chapter 2 that there is a subsequence $\{X_{n_i} : i \in \mathbb{N}\}$ for which $X_{n_i}(\omega) = o(1)$ for almost all ω. If, instead, $X_n = O_p(1)$, must there exist a subsequence for which $X_{n_i}(\omega) = O(1)$ for almost all ω? Hint: Let $\xi_i : i \in \mathbb{N}_0\}$ be a sequence of independent random variables, each distributed Uniform(0, 1). Consider $X_n := (\xi_0 - \xi_n)^{-1}$.

[17] Let ψ be a strictly increasing function on $\mathbb{R}^+$ with $\psi(0) = 0$ and $\psi(t) \to 1$ as $t \to \infty$. Show that a sequence of random vectors $\{X_n\}$ is of order $O_p(1)$ if and only if $\limsup \mathbb{P}\psi(|X_n|) < 1$.

[18] Let $\{X_n\}$ and $\{Y_n\}$ be sequences of random k-vectors, with X_n and Y_n defined on the same space and independent of each other. Suppose $X_n - Y_n = O_p(1)$. Show that there exists a sequence of nonrandom vectors $\{a_n\}$ for which $X_n - a_n = O_p(1)$. Hint: For probability measures P and Q, show that it $P^x Q^y \psi(x - y) \le M$ then $P^x \psi(x - y) \le M$ for at least one y. Use Problem [17].

[19] If X has a Poisson(λ) distribution, show that $\sqrt{X} - \sqrt{\lambda} \rightsquigarrow N(0, 1/4)$ as $\lambda \to \infty$.

[20] Let $\{X_{n,i}\}$ be a triangular array of random variables, independent within each row and satisfying

(a) $\sum_i \mathbb{P}\{|X_{n,i}| > \epsilon\} \to 0$ for each $\epsilon > 0$,

(b) $\sum_i \text{var}\big(X_{n,i}\{|X_{n,i}| \le \epsilon\}\big) \to 1$ for each $\epsilon > 0$.

Show that $\sum_i X_{n,i} - A_n \rightsquigarrow N(0,1)$, where $A_n := \sum_i \mathbb{P}X_{n,i}\{|X_{n,i}| \le 1\}$. Hint: Consider truncated variables $\eta_{n,i} := X_{n,i}\{|X_{n,i}| \le \epsilon_n\}$ and $\xi_{n,i} := \eta_{n,i} - \mathbb{P}\eta_{n,i}$, for a suitable $\{\epsilon_n\}$ sequence.

[21] Let $\{X_{n,i}\}$ be a triangular array of random variables, independent within each row and satisfying

(i) $\max_i |X_{n,i}| \to 0$ in probability,

(ii) $\sum_i \mathbb{P}X_{n,i}\{|X_{n,i}| \le \epsilon\} \to \mu$ for each $\epsilon > 0$,

(iii) $\sum_i \text{var}\big(X_{n,i}\{|X_{n,i}| \le \epsilon\}\big) \to \sigma^2 < \infty$ for each $\epsilon > 0$.

Show that $\sum_i X_{n,i} \rightsquigarrow N(\mu, \sigma^2)$. Hint: Define $\eta_{n,i} := X_{n,i}\{|X_{n,i}| \le \epsilon_n\}$ and $\xi_{n,i} := \eta_{n,i} - \mathbb{P}\eta_{n,i}$, where ϵ_n tends to zero slowly enough that: (a) $\mathbb{P}\{\max_i |X_{n,i}| > \epsilon_n\} \to 0$; (b) $\sum_i \mathbb{P}X_{n,i}\{|X_{n,i}| \le \epsilon_n\} \to \mu$; and (c) $\sum_i \text{var}\big(X_{n,i}\{|X_{n,i}| \le \epsilon_n\}\big) \to \sigma^2$.

[22] If each of the components $\{X_{ni}\}$, for $i = 1, \ldots, k$, of a sequence of random k-vectors $\{X_n\}$ is of order $O_p(1)$, show that $X_n = O_p(1)$.

[23] Suppose $f(x) = o(|x|)$ as $x \to 0$, and $g(x) = O(|x|^k)$ as $x \to 0$. Supppose $X_n = O_p(\alpha_n)$ for some sequence of constants α_n tending to zero. Derive a bound for the rate at which $f(g(X_n))$ tends to zero.

7. Notes

See Daw & Pearson (1972) and Stigler (1986a, Chapter 2) for discussion of De Moivre's 1733 derivation of the normal approximation to the binomial distribution (reproduced on pages 243–250 of the 1967 reprint of de Moivre 1718)—the first central limit theorem.

Theorem <19> is esentially due to Liapounoff (1900, 1901), although the method of proof is due to Lindeberg (1922). I adapted my exposition of Lindeberg's method, in Section 2, from Billingsley (1968, section 7), via Pollard (1984, Section III.4). The development of the CLT, from the simple idea described in Section 1 to formal limit theorems, has a long history, culminating in the work of several authors during the 1920's and 30's. For example, building on Lindeberg's ideas, Lévy (1931, Section 10) conjectured the form of the general necessary and sufficient condition for a sum of independent random variables to be approximately normally distributed, but established only the sufficiency. Apparently independently of each other, Lévy (1935) and Feller (1935) established necessary conditions for the CLT, under an assumption that individual summands satisfy a mild "asymptotic negligibility" condition. See the discussion by Le Cam (1986). Chapter 4 of Lévy (1937) and Chapter 3 of Lévy (1970) provide more insights into Lévy's thinking about the CLT and the role of the normal distribution. The idea of using truncation to obtain general CLT's from the Lindeberg version of the theorem runs through much of Lévy's work.

Later works, such as Gnedenko & Kolmogorov (1949/68, Chapter 5) and Petrov (1972/75, Section IV.4), treat the CLT as a special case of more general limit theorems for infinitely divisible distributions—compare, for example, the direct argument in Problem [20] with Theorem 3 in Section 25 of the former or with Theorem 16 in Chapter 4 of the latter.

Theorem <36> for complete separable metric spaces is due to Prohorov (1956). Independently, Le Cam (1957) proved similar results for more general topological spaces. The monograph of Billingsley (1968) is still an excellent reference for the theory of weak convergence on metric spaces. Together with the slightly more abstract account by Parthasarathy (1967), Billingsley's exposition stimulated widespread interest in weak convergence methods by probabilists and statisticians (including me) during the 1970's. See Dudley (1989, Chapter 11) for an elegant treatment that weaves in more recent ideas.

The stochastic order notation of Section 4 is due to Mann & Wald (1943). For further examples see the survey paper by Chernoff (1956) and Pratt (1959).

References

Alexandroff, A. D. (1940–43), 'Additive set functions in abstract spaces', *Mat. Sbornik*. Chapter 1: **50**(NS 8) 1940, 307–342; Chapters 2 and 3: **51**(NS 9) 1941, 563–621; Chapters 4 and 5: **55**(NS 13) 1943, 169–234.

Billingsley, P. (1968), *Convergence of Probability Measures*, Wiley, New York.

Chernoff, H. (1956), 'Large sample theory: parametric case', *Annals of Mathematical Statistics* **27**, 1–22.

Daw, R. H. & Pearson, E. S. (1972), 'Abraham De Moivre's 1733 derivation of the normal curve: a bibliographical note', *Biometrika* **59**, 677–680.

de Moivre, A. (1718), *The Doctrine of Chances*, first edn. Second edition 1738. Third edition, from 1756, reprinted in 1967 by Chelsea, New York.

Dudley, R. M. (1985), 'An extended Wichura theorem, definitions of Donsker classes, and weighted empirical distributions', *Springer Lecture Notes in Mathematics* **1153**, 141–178. Springer, New York.

Dudley, R. M. (1989), *Real Analysis and Probability*, Wadsworth, Belmont, Calif.

Feller, W. (1935), 'Über den Zentralen Grenzwertsatz der Wahrscheinlichskeitsrechnung', *Mathematische Zeitung* **40**, 521–559. Part II, same journal, **42** (1937), 301–312.

Gnedenko, B. V. & Kolmogorov, A. N. (1949/68), *Limit Theorems for Sums of Independent Random Variables*, Addison-wesley. English translation in 1968, of original Russian edition from 1949.

Hoffmann-Jørgensen, J. (1984), Stochastic Processes on Polish Spaces, Unpublished manuscript, Aarhus University, Denmark.

Le Cam, L. (1957), 'Convergence in distribution of stochastic processes', *University of California Publications in Statistics* **2**, 207–236.

Le Cam, L. (1986), 'The central limit theorem around 1935', *Statistical Science* **1**, 78–96.

Lévy, P. (1931), 'Sur les séries dont les termes sont des variables éventuelles indépendantes', *Studia Mathematica* **3**, 119–155.

Lévy, P. (1935), 'Propriétés asymptotique des sommes de variables aléatoires indépendantes ou enchaînées', *Journal de Math Pures Appl.* **14**, 347–402.

Lévy, P. (1937), *Théorie de l'addition des variables aléatoires*, Gauthier-Villars, Paris. Second edition, 1954.

Lévy, P. (1970), *Quelques Aspects de la Pensée d'un Mathématicien*, Blanchard, Paris.

Liapounoff, A. M. (1900), 'Sur une proposition de la théorie des probabilités', *Bulletin de l'Academie impériale des Sciences de St. Pétersbourg* **13**, 359–386.

Liapounoff, A. M. (1901), 'Nouvelle forme du théorème sur la limite de probabilités', *Mémoires de l'Academie impériale des Sciences de St. Pétersbourg*.

Lindeberg, J. W. (1922), 'Eine neue Herleitung des Exponentialgesetzes in der Wahrscheinlichkeitsrechnung', *Mathematische Zeitschrift* **15**, 211–225.

Mann, H. B. & Wald, A. (1943), 'On stochastic limit and order relationships', *Annals of Mathematical Statistics* **14**, 217–226.

Parthasarathy, K. R. (1967), *Probability Measures on Metric Spaces*, Academic, New York.

Petrov, V. V. (1972/75), *Sums of Independent Random Variables*, Springer-Verlag. Enlish translation in 1975, from 1972 Russian edition.

Pollard, D. (1984), *Convergence of Stochastic Processes*, Springer, New York.

Pratt, J. W. (1959), 'On a general concept of "in probability" ', *Annals of Mathematical Statistics* **30**, 549–558.

Prohorov, Y. V. (1956), 'Convergence of random processes and limit theorems in probability theory', *Theory Probability and Its Applications* **1**, 157–214.

Stigler, S. M. (1986a), *The History of Statistics: The Measurement of Uncertainty Before 1900*, Harvard University Press, Cambridge, Massachusetts.

Topsøe, F. (1970), *Topology and Measure*, Vol. 133 of *Springer Lecture Notes in Mathematics*, Springer-Verlag, New York.

Varadarajan, V. S. (1961), 'Measures on topological spaces', *Mat. Sbornik* **55**(97), 35–100. American Mathematical Society Translations **48** (1965), 161–228.

Chapter 8

Fourier transforms

SECTION 1 presents a few of the basic properties of Fourier transforms that make them such a valuable tool of probability theory.

SECTION 2 exploits a mysterious coincidence, involving the Fourier transform and the density function of the normal distribution, to establish inversion formulas for recovering distributions from Fourier transforms.

*SECTION *3 explains why the coincidence from Section 2 is not really so mysterious.*

SECTION 4 shows that the inversion formula from Section 2 has a continuity property, which explains why pointwise convergence of Fourier transforms implies convergence in distribution.

*SECTION *5 establishes a central limit theorem for triangular arrays of martingale differences.*

SECTION 6 extends the theory to multivariate distributions, pointing out how the calculations reduce to one-dimensional analogs for linear combinations of coordinate variables—the Cramér and Wold device.

*SECTION *7 provides a direct proof (no Fourier theory) of the fact that the family of (one-dimensional) distributions for all linear combinations of a random vector uniquely determines its multivariate distribution.*

*SECTION *8 illustrates the use of complex-variable methods to prove a remarkable property of the normal distribution—the Lévy-Cramér theorem.*

1. Definitions and basic properties

Some probabilistic calculations simplify when reexpressed in terms of suitable transformations, such as the probability generating function (especially for random variables taking only positive integer values), the Laplace transform (especially for random variables taking only nonnegative values), or the moment generating function (for random variables with rapidly decreasing tail probabilities). The Fourier transform shares many of the desirable properties of these transforms without the restrictions on the types of random variable to which it is best applied, but with the slight drawback that we must deal with random variables that can take complex values.

The integral of a complex-valued function, $f := g + ih$, is defined by splitting into real ($\Re f := g$) and imaginary ($\Im f := h$) parts, $\mu f := \mu g + i\mu h$. These integrals inherit linearity and the dominated convergence property from their real-valued counterparts. The increasing functional property becomes meaningless for complex integrals—the complex numbers are not ordered. The inequality $|\mu f| \le \mu|f|$ still holds if $|\cdot|$ is interpreted as the modulus of a complex number (Problem [1]).

The ***Fourier transform*** (which is often referred to as the ***characteristic function*** in the probability and statistics literature) of a probability measure P on $\mathcal{B}(\mathbb{R})$ is defined by

$$\psi_P(t) := P^x e^{ixt} \qquad \text{for } t \text{ in } \mathbb{R}.$$

Similarly, the Fourier transform of a real random variable X is defined by

$$\psi_X(t) := \mathbb{P}^\omega \exp(iX(\omega)t) \qquad \text{for } t \text{ in } \mathbb{R}.$$

That is, ψ_X is the Fourier transform of the distribution of X.

REMARK. Without the i in the exponent, we would be defining the moment generating function, Pe^{xt}, which might be infinite except at $t = 0$, as in the case of the Cauchy distribution (Problem [6]).

Fourier transforms are well defined for every probability measure on $\mathcal{B}(\mathcal{R})$, and

$$|\psi_P(t)| = |P^x \exp(ixt)| \le P^x|\exp(ixt)| = 1 \qquad \text{for all real } t.$$

They are uniformly continuous, because

$$|\psi_P(t+\delta) - \psi_P(t)| \le P^x|e^{ix(t+\delta)} - e^{ixt}| = P^x|e^{ix\delta} - 1|,$$

which tends to zero as $\delta \to 0$ by Dominated Convergence. As a map from $\mathbb{R}$ into the complex plane $\mathbb{C}$, the Fourier transform defines a curve that always lies within the unit disk. The curve touches the boundary of the disk at $1 + 0i$, corresponding to $t = 0$. If it also touches for some nonzero value of t, then P must concentrate on a regularly spaced, countable subset of $\mathbb{R}$ (Problem [2]). If P is absolutely continuous with respect to Lebesgue measure then $\psi_P(t) \to 0$ as $t \to \pm\infty$ (Problem [4]).

Fourier methods are particularly effective for dealing with sums of independent random variables, chiefly due to the following simplification.

<1> **Theorem.** *If $X_1, \ldots, X_n$ are independent then the Fourier transform of $X_1 + \ldots + X_n$ factorizes into $\psi_{X_1}(t) \ldots \psi_{X_n}(t)$, for all real t.*

Proof. Extend the factorization property of real functions of the X_j's to the complex functions $\exp(itX_j)$. □

REMARK. Be careful that you do not invent a false converse to the Theorem. If X has a Cauchy distribution and $Y \equiv X$ then $\psi_{X+Y}(t) = \exp(-2|t|) = \psi_X(t)\psi_Y(t)$, but X and Y are certainly not independent (Problem [6]).

There are various ways to extract information about a distribution from its Fourier transform. For example, the next Theorem shows that existence of finite moments of the distribution gives polynomial approximations to the Fourier transform, corresponding to the purely formal operation of taking expectations term-by-term in the Taylor expansion for $\exp(iXt)$.

<2> **Theorem.** *If $\mathbb{P}|X|^k < \infty$, for a positive integer k, then the Fourier transform has the approximation*

$$\psi_X(t) = 1 + it\mathbb{P}X + \frac{(it)^2}{2!}\mathbb{P}X^2 + \ldots + \frac{(it)^k}{k!}\mathbb{P}X^k + o(t^k)$$

for t near 0.

□ *Proof.* Apply Problem [5] to $\mathbb{P}\cos(Xt)$ and $\mathbb{P}\sin(Xt)$.

<3> **Example.** The Poisson(λ) distribution has Fourier transform

$$\psi(t) = \sum_{k=0}^{\infty} e^{ikt}\frac{\lambda^k e^{-\lambda}}{k!} = e^{-\lambda}\sum_{k=0}^{\infty}\frac{\left(\lambda e^{it}\right)^k}{k!} = \exp\left(\lambda e^{it} - \lambda\right).$$

An appeal to the central limit theorem and a suitable passage to the limit will lead us to the Fourier transform for the normal distribution.

Suppose $X_1, X_2, \ldots$ is a sequence of independent random variables, each distributed Poisson(1). By the central limit theorem for identically distributed summands, $Z_n := (X_1 + \ldots + X_n - n)/\sqrt{n} \rightsquigarrow N(0,1)$. For each fixed t the function e^{ixt} is bounded and continuous in x. Thus $\lim_{n\to\infty}\mathbb{P}\exp(itZ_n)$ is the Fourier transform of the $N(0,1)$ distribution. Evaluate the limit using Theorem <1>:

$$\begin{aligned}\mathbb{P}\exp(iZ_nt) &= \exp(-it\sqrt{n})\prod_{k=1}^{n}\mathbb{P}\exp(iX_kt/\sqrt{n}) \\ &= \exp(-it\sqrt{n} + ne^{it/\sqrt{n}} - n).\end{aligned}$$

The last exponent has the approximation

$$-it\sqrt{n} + n\left(1 + \frac{it}{\sqrt{n}} + \frac{1}{2!}\left(\frac{it}{\sqrt{n}}\right)^2 + O\left(\frac{it}{\sqrt{n}}\right)^3 - 1\right) = -\tfrac{1}{2}t^2 + O\left(\frac{1}{\sqrt{n}}\right).$$

Notice the way the error term behaves as a function of n for fixed t. In the limit we get $\exp(-t^2/2)$ as the Fourier transform of the $N(0,1)$ distribution. □

REMARK. If Z has a $N(0,1)$ distribution and s is real,

$$\begin{aligned}\mathbb{P}\exp(sZ) &= \frac{1}{\sqrt{2\pi}}\int_{-\infty}^{\infty}\exp\left(sz - \tfrac{1}{2}z^2\right)dz \\ &= \frac{1}{\sqrt{2\pi}}\int_{-\infty}^{\infty}\exp\left(\tfrac{1}{2}s^2 - \tfrac{1}{2}(z-s)^2\right)dz = \exp(s^2/2).\end{aligned}$$

Formally, we have only to replace s by it to get the Fourier transform for Z. A rigorous justification requires some complex-variable theory, such as the uniqueness of analytic continuations.

2. Inversion formula

Written out in terms of the $N(0,1)$ density, the final result from Example <3> becomes

$$\int_{-\infty}^{\infty}\frac{\exp(iyt - \frac{1}{2}y^2)}{\sqrt{2\pi}}\,dy = \exp(-\tfrac{1}{2}t^2) \qquad \text{for real } t.$$

The function on the right-hand side looks very like a normal density; it lacks only the standardizing constant. If we substitute t/σ for t, then make a change of variables $z = -y/\sigma$, we get an integral representation for the $N(0, \sigma^2)$ density ϕ_σ.

<4> $$\frac{1}{2\pi}\int_{-\infty}^{\infty} \exp\left(-izt - \tfrac{1}{2}\sigma^2 z^2\right) dz = \frac{\exp(-\frac{1}{2}t^2/\sigma^2)}{\sigma\sqrt{2\pi}} =: \phi_\sigma(t).$$

This equality is a special case of a general inversion formula that relates Fourier transforms to densities. Indeed, the general formula can be derived from the special case.

Suppose Z has a $N(0, 1)$ distribution, independently of a random variable X with Fourier transform $\psi(\cdot)$. From the convolution formula for densities with respect to Lebesgue measure (Section 4.4), for $\sigma > 0$ the sum $X + \sigma Z$ has a distribution with density $f^{(\sigma)}(y) := \mathbb{P}^\omega \phi_\sigma(y - X(\omega))$. Substituting for ϕ_σ from <4>, we have

$$f^{(\sigma)}(y) = \frac{1}{2\pi}\mathbb{P}^\omega \int_{-\infty}^{\infty} \exp\left(-iz(y - X(\omega)) - \tfrac{1}{2}\sigma^2 z^2\right) dz.$$

The integrand, which is bounded in absolute value by $\exp(-\sigma^2 z^2/2)$, is integrable (as a function of ω and z) with respect to the product of $\mathbb{P}$ with Lebesgue measure. Interchanging the order of integration, as justified by the Fubini theorem, we get the basic integral representation,

<5> $$f^{(\sigma)}(y) = \frac{1}{2\pi}\int_{-\infty}^{\infty} \psi(z)\exp\left(-izy - \tfrac{1}{2}\sigma^2 z^2\right) dz.$$

Notice that $\psi(t)\exp(-\sigma^2 t^2/2)$ is the Fourier transform of $X + \sigma Z$. If we write P for the distribution of X, the formula becomes

<6> $$\text{density of } P \star N(0, \sigma^2) = \frac{1}{2\pi}\int e^{-izy}(\text{Fourier transform of } P \star N(0, \sigma^2))\, dz,$$

a special case of the inversion formula to be proved in Theorem <10>.

Limiting arguments as σ tends to zero in the basic formula <5>, or <6>, lead to several important conclusions.

<7> **Theorem.** *The distribution of a random variable is uniquely determined by its Fourier transform.*

Proof. For h a bounded measurable function on $\mathbb{R}$ and $f^{(\sigma)}$ as defined above,

$$\mathbb{P}h(X + \sigma Z) = \int_{-\infty}^{\infty} h(y) f^{(\sigma)}(y)\, dy.$$

which shows, via formula <5>, that the Fourier transform of the random variable X uniquely determines the density for the distribution of $X + \sigma Z$. Specialize to h in $\mathcal{C}(\mathbb{R})$, the class of all bounded, continuous real functions on $\mathbb{R}$. By Dominated Convergence, $\mathbb{P}h(X + \sigma Z) \to \mathbb{P}h(X)$ as $\sigma \to 0$. Thus the Fourier transform uniquely determines all the expectations $\mathbb{P}h(X)$, with h ranging over $\mathcal{C}(\mathbb{R})$. A generating class argument shows that these expectations uniquely determine the distribution of X, as a probability measure on $\mathcal{B}(\mathbb{R})$. □

You might feel tempted to rearrange limit operations in the previous proof, to arrive at an assertion that

$$\mathbb{P}h(X) = \lim_{\sigma\to 0}\int_{-\infty}^{\infty} h(y) f^{(\sigma)}(y)\,dy \overset{?}{=} \int_{-\infty}^{\infty} h(y) \lim_{\sigma\to 0} f^{(\sigma)}(y)\,dy,$$

and thereby conclude that the distribution of X has density

<8>
$$f(y) \overset{?}{=} \lim_{\sigma\to 0} f^{(\sigma)}(y) \overset{?}{=} \frac{1}{2\pi}\int_{-\infty}^{\infty} \psi(z)\exp(-izy)\,dz$$

with respect to Lebesgue measure. Of course such temptation should be resisted. The migration of limits inside integrals typically requires some domination assumptions. Moreover, it would be exceedingly strange to derive densities for measures that are not absolutely continuous with respect to Lebesgue measure.

<9> **Example.** Let P denote the probability measure that puts mass $1/2$ at ± 1. It has Fourier transform $\psi(t) = \left(e^{it} + e^{-it}\right)/2$. The integral on the right-hand side of <8> does not exist for this Fourier transform. Application of formulas <5> then <4> gives

$$\begin{aligned} f^{(\sigma)}(y) &= \frac{1}{2\pi}\int_{-\infty}^{\infty} \frac{1}{2}\left(e^{iz} + e^{-iz}\right)\exp\left(-izy - \tfrac{1}{2}\sigma^2 z^2\right)dz \\ &= \frac{1}{2\sigma}\phi_\sigma(y-1) + \frac{1}{2\sigma}\phi_\sigma(y+1). \end{aligned}$$

That is, $f^{(\sigma)}$ is the density for the mixture $\tfrac{1}{2}N(-1,\sigma^2) + \tfrac{1}{2}N(+1,\sigma^2)$, a density that is trying to behave like two point masses. The limit of $f^{(\sigma)}$ does not exist in the ordinary sense. □

<10> **Theorem.** *If P has Fourier transform ψ for which $\int_{-\infty}^{\infty} |\psi(z)|\,dz < \infty$, then P is absolutely continuous with respect to Lebesgue measure, with a density*

$$f(y) := \frac{1}{2\pi}\int_{-\infty}^{\infty} \psi(z)\exp(-izy)\,dz$$

that is bounded and uniformly continuous.

Proof. The convolution density $f^{(\sigma)}(y)$, as in <5>, is bounded by the constant $C := \int |\psi|/2\pi$. Uniform continuity follows as for Fourier transforms. By Dominated Convergence $f^{(\sigma)}(y)$ converges pointwise to f. Thus $0 \le f \le C$. If h is continuous and vanishes outside a bounded interval $[-M, M]$, a second appeal to Dominated Convergence gives

$$Ph = \lim_{\sigma\to 0}\int_{-M}^{M} h(y) f^{(\sigma)}(y)\,dy = \int_{-M}^{M} h(y) f(y)\,dy.$$

That is, $Ph = \int hf$ for a large enough class of functions h to ensure (via a generating class argument) that P has density f. □

REMARK. The inversion formula for integrable Fourier transforms is the basis for a huge body of theory, including Edgeworth expansions, rates of convergence in the central limit theorem, density estimation, and more. See for example the monograph of Petrov (1972/75).

*3. A mystery?

I have always been troubled by the mysterious workings of Fourier transforms. For example, it seems an enormous stroke of luck that the Fourier transform of the normal distribution is proportional to its density function—the key to the inversion formula <5>. Perhaps it would be better to argue slightly less elegantly, to see what is really going on.

Start from the slightly less mysterious fact that there exists at least one random variable W whose Fourier transform ψ_0 is Lebesgue integrable. (See Problem [3] for one way to ensure integrability of the Fourier transform.) Symmetrize by means of an independent copy W' of W to get a random variable $W - W'$ with a real-valued Fourier transform $\psi = |\psi_0|^2$, which is also Lebesgue integrable. That is, there exists a probability distribution Q whose Fourier transform ψ is both nonnegative and Lebesgue integrable. For some finite constant c_0, the function $c_0\psi$ defines a density (with respect to Lebesgue measure $\mathfrak{m}$ on the real line) of a probability measure $\widetilde{Q}$ on $\mathcal{B}(\mathbb{R})$.

For h bounded and measurable,

$$P \otimes \widetilde{Q}\, h(x + \sigma y) = P^x \mathfrak{m}^y \left(c_0 \left(Q^z e^{iyz}\right) h(x + \sigma y)\right).$$

If h vanishes outside a bounded interval, Fubini lets us interchange the order of integration, make a change of variable $w = x + \sigma y$ in the Lebesgue integration, then change order of integration again, to reexpress the right-hand side as

$$\frac{c_0}{\sigma} Q^z \mathfrak{m}^w \left(h(w) P^x \exp\left(iz(w - x)/\sigma\right)\right),$$

a function of the Fourier transform of P. Again we have a special case of the inversion formula, from which all results flow.

The method in Section 2 corresponds to the case where both Q and $\widetilde{Q}$ are normal distributions, but that coincidence is not vital to the method.

4. Convergence in distribution

The representation <5> shows not only that the density $f^{(\sigma)}$ of $X + \sigma Z$ is uniquely determined by the Fourier transform ψ_X of X but also that it depends on ψ_X in a continuous way. The factor $\exp(-\sigma^2 z^2/2)$ ensures that small perturbations of ψ_X do not greatly affect the integral. The traditional way to make this continuity idea precise involves an assertion about pointwise convergence of Fourier transforms.

The proof makes use of a simple fact about smoothing and a simple consequence of Dominated Convergence known as Scheffé's Lemma:

Let $f, f_1, f_2, \ldots$ be nonnegative, μ-integrable functions for which $f_n \to f$ a.e. $[\mu]$ and $\mu f_n \to \mu f$. Then $\mu|f_n - f| \to 0$.

See Section 3.1 for the proof.

<11> **Lemma.** *Let $X, X_1, X_2, \ldots$ and Z be random variables for which $X_n + \sigma Z \rightsquigarrow X + \sigma Z$ for each $\sigma > 0$. Then $X_n \rightsquigarrow X$.*

Proof. For ℓ in $BL(\mathbb{R})$,

$$|\mathbb{P}\ell(X_n) - \mathbb{P}\ell(X)| \leq \mathbb{P}|\ell(X_n) - \ell(X_n + \sigma Z)| + |\mathbb{P}\ell(X_n + \sigma Z) - \mathbb{P}\ell(X + \sigma Z)| + \mathbb{P}|\ell(X + \sigma Z) - \ell(X)|.$$

On the right-hand side, the middle term tends to zero, by assumption. The other two terms are both bounded by $\|\ell\|_{BL}\mathbb{P}(1 \wedge \sigma|Z|)$, which tends to zero as $\sigma \to 0$. □

<12> **Continuity Theorem.** *Let $X, X_1, X_2, \ldots$ be random variables with Fourier transforms $\psi, \psi_1, \psi_2, \ldots$ for which $\psi_n(t) \to \psi(t)$ for each real t. Then $X_n \rightsquigarrow X$.*

Proof. Let Z have a $N(0,1)$ distribution independent of X and all the X_n. The random variables $X_n + \sigma Z$ have distributions with densities

$$f_n^{(\sigma)}(y) = \frac{1}{2\pi}\int_{-\infty}^{\infty} \psi_n(t)\exp\left(-ity - \tfrac{1}{2}\sigma^2 t^2\right) dt$$

with respect to Lebesgue measure, and $X + \sigma Z$ has a distribution with a similarly defined density $f^{(\sigma)}$. By Dominated convergence, $f_n^{(\sigma)}(y) \to f^{(\sigma)}(y)$ as $n \to \infty$, for each fixed y. If h is bounded and measurable,

$$\begin{aligned}
&|\mathbb{P}h(X_n + \sigma Z) - \mathbb{P}h(X + \sigma Z)| \\
&= |\int h(y) f_n^{(\sigma)}(y)\, dy - \int h(y) f^{(\sigma)}(y)\, dy| \\
&\leq M \int |f_n^{(\sigma)}(y) - f^{(\sigma)}(y)|\, dy \qquad \text{where } M = \sup|h| \\
&\to 0 \qquad \text{by Scheffé's Lemma.}
\end{aligned}$$

Thus $X_n + \sigma Z \rightsquigarrow X + \sigma Z$ for each $\sigma > 0$, and the asserted convergence in distribution follows via Lemma <11>. □

<13> **Example.** Suppose $X_1, X_2, \ldots$ are independent, identically distributed random variables with $\mathbb{P}X_k = 0$ and $\mathbb{P}X_k^2 = 1$. From Chapter 7 we know that

$$\frac{X_1 + \ldots + X_n}{\sqrt{n}} \rightsquigarrow N(0,1).$$

Here is the proof of the same result using Fourier transforms.

Let $\psi(\cdot)$ be the Fourier transform of X_1. From Theorem <2>,

$$\psi(t) = 1 - \tfrac{1}{2}t^2 + o(t^2) \qquad \text{as } t \to 0.$$

In particular, for fixed t,

$$\psi(t/\sqrt{n}) = 1 - \tfrac{1}{2}t^2/n + o(1/n) \qquad \text{as } n \to \infty.$$

The standardized sum has Fourier transform

$$\psi(t/\sqrt{n})^n = \left(1 - \tfrac{1}{2}t^2/n + o(1/n)\right)^n \to \exp\left(-\tfrac{1}{2}t^2\right).$$

The limit equals the $N(0,1)$ Fourier transform. By the Continuity Theorem, the asymptotic normality of the standardized sum follows. □

Certainly the calculations in the previous Example involve less work than the Lindeberg argument plus the truncations used in Chapter 7 to establish the same central limit theorem. I would point out, however, the amount of theory needed to establish the Continuity Theorem. Moreover, for the corresponding Fourier transform proofs of central limit theorems for more general triangular arrays, the calculations would parallel those for the Lindeberg method.

REMARK. A slightly stronger version of the Continuity Theorem (due to Cramér—see the Notes) also goes by the same name. The assumptions of the Theorem can be weakened (Problem [9]) to mere existence of a pointwise limit $\psi(t) := \lim_n \psi_n(t)$ with ψ continuous at the origin. Initially it need not be identified as the Fourier transform of some specified distribution. The stronger version of the Theorem asserts the existence of a distribution P for which ψ is the Fourier transform, such that $X_n \rightsquigarrow P$.

*5. A martingale central limit theorem

Fourier methods have some advantages over the methods of Chapter 7. For example, the proof of the following important theorem seems to depend in an essential way on the factorization properties of the exponential function.

REMARK. The Lindeberg method for independent summands, as explained in Section 7.2, can be extended to martingales differences under a natural assumption on the sum of conditional variances—see Lévy (1937, Section 67).

<14> **Theorem.** *(McLeish 1974) For each n in $\mathbb{N}$ let $\{\xi_{nj} : j = 0, \ldots, k_n\}$ be a martingale difference array, with respect to a filtration $\{\mathcal{F}_{nj}\}$, for which:*

(i) $\sum_j \xi_{nj}^2 \to 1$ in probability;

(ii) $\max_j |\xi_{nj}| \to 0$ in probability;

(iii) $\sup_n \mathbb{P} \max_j \xi_{nj}^2 < \infty$.

Then $\sum_j \xi_{nj} \rightsquigarrow N(0, 1)$ as $n \to \infty$.

Proof. Write S_n for $\sum_j \xi_{nj}$ and M_n for $\max_j |\xi_{nj}|$. Denote expectations conditional on $\mathcal{F}_{nj}$ by $\mathbb{P}_j(\cdot)$. The omission of the n subscript should cause no confusion, because all calculations will be carried out for a fixed n. Let me also abbreviate ξ_{nj} to ξ_j, and simplify notation by assuming $k_n = n$.

By the Continuity Theorem <12>, it suffices to show that $\mathbb{P}\exp(itS_n) \to \exp(-t^2/2)$, for each fixed t in $\mathbb{R}$. A Taylor expansion of $\log(1 + it\xi_j)$ when M_n is small gives $\exp\left(it\xi_j + t^2\xi_j^2/2\right) \approx 1 + it\xi_j$, and hence, via (i),

$$\exp(itS_n) \approx \exp\left(-t^2/2\right) \prod_{j \le n} \left(1 + it\xi_j\right).$$

Define $X_m := \prod_{j \le m}(1 + it\xi_j)$, for $m = 1, \ldots, n$, with $X_0 \equiv 1$. Each X_m has expected value 1, because $\mathbb{P}X_m = \mathbb{P}\left(X_{m-1}(1 + it\mathbb{P}_{m-1}\xi_m)\right) = \mathbb{P}X_{m-1} = \ldots = \mathbb{P}X_0$, which suggests $\mathbb{P}\exp(itS_n) \approx \exp(-t^2/2)\mathbb{P}X_n = \exp(-t^2/2)$.

For the formal proof, use the error bound

$$\log(1 + z) = z - z^2/2 + r(z) \qquad \text{with } |r(z)| \le |z|^3 \text{ for } |z| \le 1/2.$$

Temporarily write z_j for $it\xi_j$. Notice that $\sum_{j\le n} z_j^2 = -t^2 + o_p(1)$, by (i). Also, when $|t|M_n \le 1/2$, which happens with probability tending to 1,

$$\sum_{j\le n} |r(z_j)| \le |t|^3 \sum_{j\le n} |\xi_j|^3 \le |t|^3 M_n \sum_{j\le n} \xi_j^2 = o_p(1) \qquad \text{by (ii) and (i).}$$

Thus

$$\exp\left(itS_n + \tfrac{1}{2}t^2\right) = X_n \exp\left(\tfrac{1}{2}t^2 + \tfrac{1}{2}\sum_{j\le n} z_j^2 - \sum_{j\le n} r(z_j)\right) = \exp\left(o_p(1)\right),$$

and

$$Y_n := \exp\left(itS_n + \tfrac{1}{2}t^2\right) - X_n \to 0 \qquad \text{in probability.}$$

To strengthen the assertion to $\mathbb{P}Y_n \to 0$, it is enough to show that $\sup_n \mathbb{P}|Y_n|^2 < \infty$, for then

$$\mathbb{P}|Y_n| \le M^{-1}\mathbb{P}|Y_n|^2 + M\mathbb{P}\left(|Y_n|\{|Y_n| \le M\}\right) = O\left(M^{-1}\right) + o(1).$$

(Compare with uniform integrability, as defined in Section 2.8.) The contribution from S_n to Y_n is bounded in absolute value by a constant. We have only to control the contribution from X_n.

Consider first the obvious bound,

$$|X_n|^2 = \prod_{j\le n}\left(1 + |z_j|^2\right) \le \exp\left(\sum_{j\le n} |t\xi_j|^2\right).$$

By (ii), the expression on the right-hand side is of order $O_p(1)$, but it needn't be bounded everywhere by a constant. We can achieve something closer to uniform boundedness by means of a stopping time argument. Write $Q_n(m)$ for $\sum_{j\le m} \xi_{nj}^2$. Define stopping times $\tau_n := \inf\{m : Q_n(m) > 2\}$, with the usual convention that $\tau_n := n$ if the sum never exceeds 2. Redefine z_j as $it\xi_j\{\tau_n \le j\}$, a new sequence of martingale differences, for which $\mathbb{P}\{itS_n \ne \sum_{j\le n} z_j\} \le \mathbb{P}\{\tau_n < n\} \to 0$. We have only to prove that $\mathbb{P}\exp\sum_{j\le n}(z_j + t^2/2) \to 1$.

Repeat the argument from the previous paragraph but with X_n redefined using the new z_j's. We then have

$$|X_n|^2 = \prod_{j\le n}\left(1 + |t\xi_j|^2\{j \le \tau_n\}\right) \le \exp\left(|t|^2 \sum_j \xi_j^2\{j < \tau_n\}\right)\left(1 + |t|^2 M_n^2\right),$$

which gives $\sup_n \mathbb{P}|X_n|^2 \le \sup_n \exp(2t^2)(1 + t^2\mathbb{P}|M_n|^2) < \infty$. The asserted central limit theorem follows. □

REMARKS.

(i) Notice the role played by the stopping times τ_n. They ensure that properties of the increments that hold with probability tending to one can be made to hold everywhere, if we can ignore the effect of the increment corresponding to $\tau_n = j$. To control ξ_{n,τ_n} the Theorem had to impose constraints on $\max_j |\xi_{nj}|$.

(ii) The sum $\sum_j \xi_{nj}^2$ plays the same role as a sum of variances in the corresponding theory for independent variables. Martingale central limit theorems sometimes impose constraints on the sum of ***conditional variances*** $\mathbb{P}_{j-1}\xi_{nj}^2$. The sum of squared increments corresponds to the "square brackets" process $[X_n]$ of a martingale, and the sum of conditional variances corresponds to the "pointy brackets" process $\langle X_n\rangle$. These two processes are also called the ***quadratic variation process*** for X_n and the ***compensator*** for X_n^2, respectively.

6. Multivariate Fourier transforms

The (multivariate) Fourier transform of a random k-vector X is defined as $\psi_X(t) := \mathbb{P}\exp(it'X)$ for $t \in \mathbb{R}^k$. Many of the results for the one-dimensional Fourier transform carry over to the multidimensional setting with only notational changes. For example, if ψ_X is integrable with respect to Lebesgue measure $\mathfrak{m}_k$ on $\mathbb{R}^k$ then the distribution of X is absolutely continuous with respect to $\mathfrak{m}_k$ with density

$$f(y) := \frac{1}{(2\pi)^k}\mathfrak{m}_k^t\left(\psi_X(t)\exp(-it'y)\right).$$

Once again the Fourier transform uniquely determines the distribution of the random vector, and the pointwise convergence of Fourier transforms implies convergence in distribution. These two results have two highly useful consequences:

(i) The distribution of X is uniquely determined by the family of distributions of all linear combinations $t'X$, as t ranges over $\mathbb{R}^k$.

(ii) If $t'X_n \rightsquigarrow t'X$ for each t in $\mathbb{R}^k$ then $X_n \rightsquigarrow X$.

Both assertions follow from the trivial fact that $\mathbb{P}\exp(it'Y)$ is both the multivariate Fourier transform of the random vector Y, evaluated at t, and the Fourier transform of the random variable $t'Y$, evaluated at 1. The reduction to the one dimensional case via linear combinations is usually called the ***Cramér-Wold device***.

Consequence (ii) shows why one seldom bothers with direct proofs of multivariate limit theorems: They can usually be deduced from their one-dimensional analogues. For example, the multivariate central limit theorem of Section 7.3 is an immediate consequence of its univariate counterpart.

<15> **Example.** Suppose X is a random k-vector and Y is a random ℓ-vector for which

$$\mathbb{P}\exp(is'X + it'Y) = \mathbb{P}\exp(is'X)\mathbb{P}\exp(it'Y) \qquad \text{all } s \in \mathbb{R}^k, \text{ all } t \in \mathbb{R}^\ell.$$

Pass to image measures to deduce that the joint distribution $Q_{X,Y}$ of X and Y has the same Fourier transform as the product $Q_X \otimes Q_Y$ of the marginal distributions. By the uniqueness result (i) for Fourier transforms of distributions on $\mathbb{R}^{k+\ell}$, we must have $Q_{X,Y} = Q_X \otimes Q_Y$. That is, X and Y are independent.

You might regard this result as a sort of converse to Theorem <1>. Don't slip into the error of checking only that the factorization holds when s and t happen to be equal. □

<16> **Example.** A random n-vector X is said to have a ***multivariate normal distribution*** if each linear combination $t'X$, for $t \in \mathbb{R}^n$, has a normal distribution. In particular, each coordinate X_i has a normal distribution, with a finite mean and a finite variance. The random vector must have a well defined expectation $\mu := \mathbb{P}X$ and variance matrix $V := \mathbb{P}(X-\mu)(X-\mu)'$. The distribution of $t'X$ is therefore $N(t'\mu, t'Vt)$, which implies that X must have Fourier transform

$$\mathbb{P}\exp(it'X) = \exp(it'\mu - \tfrac{1}{2}t'Vt) \qquad \text{for all } t \text{ in } \mathbb{R}^n.$$

Write $N(\mu, V)$ for the probability distribution with this Fourier transform.

Every μ in $\mathbb{R}^n$ and nonnegative definite matrix V defines such a distribution. For if Z is a n-vector of independent $N(0,1)$ random variables and if we factorize

V as AA', with A an $n \times n$ matrix, then the random vector $X := \mu + AZ$ has Fourier transform $\psi_X(t) = \mathbb{P}\exp(it'(\mu + AZ)) = \exp(it'\mu)\psi_Z(A't)$. The random vector Z has Fourier transform $\psi_Z(s) = \prod_{j \le k} \psi_{Z_j}(s_j) = \exp\left(-\frac{1}{2}|s|^2\right)$, from which it follows that $\psi_Z(A't) = \exp\left(-\frac{1}{2}t'A'At\right) = \exp\left(-\frac{1}{2}t'Vt\right)$.

If V is nonsingular, the $N(\mu, V)$ distribution is absolutely continuous with respect to n-dimensional Lebesgue measure, with density

$$(2\pi|\det(V)|)^{-n/2} \exp\left(-\tfrac{1}{2}(x-\mu)'V^{-1}(x-\mu)\right).$$

This result follows via the Jacobian formula (Rudin 1974, Chapter 8) for change of variable from the density $(2\pi)^{-n/2}\exp(-|x|^2/2)$ for the $N(0, I_n)$ distribution. (In fact, the Jacobian formula for nonsingular linear transformations is easy to establish by means of invariance arguments.)

If V is singular, the $N(\mu, V)$ distribution concentrates on a translate of a subspace, $\{x \in \mathbb{R}^n : (x-\mu)'V(x-\mu) \neq 0\} = \{\mu + y : Ay = 0\}$, where $V = A'A$. □

*7. Cramér-Wold without Fourier transforms

Fourier methods were long regarded as essential underpinning for the Cramér-Wold device. Walther (1997) recently found a beautiful direct argument that avoids use of Fourier transforms altogether. I will describe only his method for showing that linear combinations characterize a distribution, which depends on two facts about the normal distribution:

(i) If Z is a vector of k independent random variables, each distributed $N(0,1)$, and if B is a vector of constants, then $B \cdot Z$ has a $N(0, |B|^2)$ distribution. (This fact can be established by a direct convolution argument, which makes no use of Fourier transforms.)

(ii) Write Φ for the standard normal distribution function. If $\theta_1, \dots, \theta_{2m+1}$ are distinct real numbers, then there exist real numbers $a_1, \dots, a_{2m+1}$ such that the function $g(t) := \sum_i a_i \Phi(\theta_i t)$, for $t \ge 0$, is of order $O\left(t^{2m+1}\right)$ near $t = 0$, and hence $g(t)/t^{2m+1}$ is Lebesgue integrable. Moreover, the $\{a_i\}$ can be chosen so that $\int_0^\infty g(t)/t^{2m+1}\,dt \neq 0$.

Walther gave an explicit construction for the constants $\{a_i\}$ for a particular choice of the $\{\theta_i\}$. (Actually, he also needed to add another well chosen constant a_0 to g to get the desired properties.) I will give a different proof of (ii) at the end of this Section.

Write $\mathfrak{m}$ for Lebesgue measure on $\mathbb{R}^{2m}$. The function g serves to define a $\mathfrak{m}$-integrable function F on $\mathbb{R}^{2m}$ by $F(u) := g(1/|u|)$, for which

$$\mathfrak{m}F = \mathfrak{m}^u g(1/|u|) = \int_0^\infty C_m r^{2m-1} g(1/r)\,dr = C_m \int_0^\infty g(t)/t^{2m+1}\,dt \neq 0,$$

where C_m denotes the surface area of the unit sphere in $\mathbb{R}^{2m}$.

Let h be any bounded, continuous, real function on $\mathbb{R}^{2m}$. Integrability of F justifies an appeal to Dominated Convergence to deduce that

$$\mathbb{P}h(X) = \lim_{\sigma \to 0} \mathbb{P}^\omega \mathfrak{m}^u \left(h(X(\omega) + \sigma u)F(u)\right)/\mathfrak{m}F.$$

A change of variable in the $\mathfrak{m}$ integral, followed by an appeal to Fubini, gives

$$\mathbb{P}^{\omega}\mathfrak{m}^{u}\left(h(X(\omega)+\sigma u)F(u)\right) = \sigma^{-2m}\mathfrak{m}^{y}\left(h(y)\mathbb{P}^{\omega}F\left(\frac{y-X(\omega)}{\sigma}\right)\right).$$

The last expectation is determined by the distributions of linear functions of X, as seen by an appeal to property (i), first conditioning on X, for a random normal vector Z that is independent of X:

$$\begin{aligned}\mathbb{P}F\left(\frac{y-X}{\sigma}\right) &= \sum_i a_i \mathbb{P}\Phi\left(\frac{\sigma\theta_i}{|y-X|}\right)\\ &= \sum_i a_i \mathbb{P}\{(y-X)\cdot Z \le \sigma\theta_i\}.\end{aligned}$$

Condition on Z to see that the last expression is uniquely determined by the distributions of $X \cdot z$ for z in $\mathbb{R}^m$.

Thus the distributions of the linear combinations $X \cdot z$, with z ranging over $\mathbb{R}^{2m}$, uniquely determines the expectation $\mathbb{P}h(X)$ for every bounded, continuous, real function h, which is enough to determine the distribution of X.

REMARK. The Cramér-Wold result for random vectors taking values in $\mathbb{R}^{2m-1}$ follows directly from the result for $\mathbb{R}^{2m}$: we have only to append one more coordinate variable to X. There is no loss in having a proof for only even dimensions.

Proof of assertion (ii)

The Taylor expansion of the exponential function,

$$e^{-x} = 1 - x + \frac{x^2}{2!} - \frac{x^3}{3!} + \ldots + \frac{(-x)^{m-1}}{(m-1)!} + \frac{(-x)^m}{m!}\exp(-x^*) \qquad \text{with } 0 < x^* < x,$$

shows that e^{-x} is the sum of a polynomial of degree $m-1$ plus a remainder term, $r(x)$, of order $O(x^m)$ near $x = 0$, such that $r(x) > 0$ for all $x > 0$ if m is even, and $r(x) < 0$ for all $x > 0$ if m is odd. Replace x by $x^2/2$, divide by $\sqrt{2\pi}$, then integrate from 0 to t to derive a corresponding expansion for the standard normal distribution function: $\Phi(t) = p(t) + R(t)$ for all $t \ge 0$, where $p(t) := \sum_{k=0}^{2m-1} \beta_k t^k$ and $R(t)$ is a remainder of order $O(t^{2m+1})$ near $t = 0$ that takes only positive (if m is even) or negative (if m is odd) values. Note that $\beta_0 = \Phi(0) = 1/2$. The constant $\kappa := \int_0^\infty R(t)t^{-2m-1}\,dt$ is finite and nonzero.

By construction,

$$g(t) := \sum_i a_i\Phi(\theta_i t) = \sum_i a_i R(\theta_i t) + \sum_{k=0}^{2m-1} \beta_k t^k \sum_i a_i\theta_i^k.$$

If we can choose the $\{a_i\}$ such that $\sum_i a_i\theta_i^k = 0$ for $k = 0, 1, \ldots, 2m-1$, then the contributions from the polynomials $p(\theta_i t)$ disappear, leaving

$$\int_0^\infty \frac{g(t)}{t^{2m+1}}\,dt = \sum_i a_i \int_0^\infty \frac{R(\theta_i t)}{t^{2m+1}}\,dt = \sum_i a_i\theta_i^{2m}\kappa$$

The integral is nonzero if we can also ensure that $\sum_i a_i\theta^{2m} \ne 0$.

A simple piece of linear algebra establishes existence of a suitable vector $a := (a_1, \ldots, a_{2m+1})$. Write U_k for $(\theta_1^k, \ldots, \theta_{2m+1}^k)$. The vector U_{2m} could not be a linear combination $\sum_{k=0}^{2m-1} \gamma_k U_k$, for otherwise the polynomial $\theta^{2m} - \sum_{k=0}^{2m-1} \gamma_k\theta^k$ of degree $2m$ would have $2m+1$ distinct roots, $\theta_1, \ldots, \theta_{2m+1}$: a contradiction. The component of U_{2m} that is orthogonal to all the other U_k vectors defines a suitable a.

*8. The Lévy-Cramér theorem

If X and Y are independent random variables, each normally distributed, it is easy to verify by direct calculation that $X+Y$ has a normal distribution. Surprisingly, there is a converse to this simple result.

<17> **Lévy-Cramér theorem.** *If X and Y are independent random variables with $X+Y$ normally distributed then both X and Y have normal distributions.*

> REMARK. The proof of the theorem makes use of several facts about analytic functions of a complex variable, such as existence of power series expansions. See Chapters 10 and 13 of the Rudin (1974) text for the required theory.

Proof. With no loss of generality we may assume $X+Y$ to have a standard normal distribution, and Y to have a zero median, that is, $\mathbb{P}\{Y \geq 0\} \geq \frac{1}{2}$ and $\mathbb{P}\{Y \leq 0\} \geq \frac{1}{2}$. Then, for each $x \geq 0$,

$$
\begin{aligned}
\mathbb{P}\{X \geq x\} &\leq 2\mathbb{P}\{Y \geq 0\}\mathbb{P}\{X \geq x\} \\
&= 2\mathbb{P}\{Y \geq 0,\ X \geq x\} && \text{independence} \\
&\leq 2\mathbb{P}\{X+Y \geq x\} \\
&\leq \exp(-x^2/2) && \text{normal tail bound from Section D.1.}
\end{aligned}
$$

A similar argument gives a similar bound for the lower tail. Thus

<18> $$\mathbb{P}\{|X| \geq x\} \leq 2\exp(-x^2/2) \qquad \text{for all } x \geq 0.$$

It follows (Problem [11]) that the function $g(z) := \mathbb{P}\exp(zX)$ is well defined and is an analytic function of z throughout the complex plane $\mathbb{C}$. It inherits a growth condition from the tail bound <18>:

$$
\begin{aligned}
|g(z)| \leq \mathbb{P}|\exp(zX)| &\leq \mathbb{P}\exp(|zX|) \\
&= 1 + |z|\int_0^\infty \exp(|z|x)\mathbb{P}\{|X| \geq x\}\,dx \\
&\leq 1 + 2|z|\int_{-\infty}^\infty \exp(|z|x - x^2/2)\,dx && \text{by <18>}
\end{aligned}
$$

<19> $$= \exp(C + |z|^2) \qquad \text{for some constant } C.$$

A similar argument shows that $h(z) := \mathbb{P}\exp(zY)$ is also well defined for every z. By independence, $g(z)h(z) = \exp(z^2/2)$ for all $z \in \mathbb{C}$, and hence $g(z) \neq 0$ for all z. It follows (Rudin 1974, Theorem 13.11) that there exists an analytic function $\gamma(\cdot)$ on $\mathbb{C}$ such that $g(z) = \exp(\gamma(z))$. We may choose γ so that $\gamma(0) = 0$, because $g(0) = 1$. (In effect, $\log g$ can be defined as a single-valued, analytic function on $\mathbb{C}$.) The analytic function γ has a power series expansion $\gamma(z) = \sum_{n=1}^\infty \gamma_n z^n$ that converges uniformly to γ on each bounded subset of $\mathbb{C}$.

Decompose $\gamma(re^{i\theta})$ into its real and imaginary parts $U(r,\theta)+iV(r,\theta)$. Then, from <19>, we have $\exp(U) = |\exp(U+iV)| \le \exp(C+r^2)$, which gives

<20> $$U(r,\theta) \le C + r^2 \qquad \text{for all } re^{i\theta} \in \mathbb{C}.$$

Uniform convergence of the power series expansion for γ on the circle $|z| = r$ lets us integrate term-by-term, giving

$$\int_0^{2\pi} \gamma(re^{i\theta}) \exp(-in\theta) = \begin{cases} 2\pi\gamma_n r^n & \text{for } n = 1, 2, 3, \dots \\ 0 & \text{for } n = 0, -1, -2, \dots \end{cases}$$

In particular, for $n = 1, 2, 3, \dots$, and real β,

$$\begin{aligned} &\int_0^{2\pi} \gamma(re^{i\theta})\Big(\exp(-in\theta - i\beta) + \exp(in\theta + i\beta) + 2\Big)\, d\theta \\ &\qquad = e^{-i\beta} \int_0^{2\pi} \gamma(re^{i\theta}) \exp(-in\theta)\, d\theta \\ &\qquad = 2\pi\gamma_n r^n e^{-i\beta}. \end{aligned}$$

Choose β so that $\gamma_n e^{-i\beta} = |\gamma_n|$, then equate real parts to deduce that

$$\int_0^{2\pi} U(r,\theta)\Big(2 + 2\cos(n\theta + \beta)\Big)\, d\theta = 2\pi|\gamma_n|r^n.$$

The integrand on the left-hand side is less than $4U(r,\theta) \le 4(C + r^2)$. Let r tend to infinity to deduce that $\gamma_n = 0$ for $n = 3, 4, 5, \dots$. That is,

$$\mathbb{P}\exp(zX) = \exp(\gamma_1 z + \gamma_2 z^2).$$

Problem [12] shows why γ_1 must be real valued and γ_2 must be nonnegative. That is, X has the Fourier transform that characterizes the normal distribution. □

9. Problems

[1] Let $f = (f_1, \dots, f_k)$ be a vector of μ-integrable, real-valued functions. Define μf as the vector $(\mu f_1, \dots, \mu f_k)$.

(i) Show that $|\mu f| \le \mu|f|$, where $|\cdot|$ denotes the usual Euclidean norm. Hint: Let α be a unit vector in the direction of μf. Note that $\alpha' f \le |f|$.

(ii) Let $f = f_1 + if_2$ be a complex-valued function, with μ-integrable real and imaginary parts f_1 and f_2. Show that $|\mu f| \le \mu|f|$, where $|\cdot|$ denotes the modulus of a complex number.

[2] Suppose a Fourier transform has $|\psi_P(t_0)| = 1$ for some nonzero t_0. That is, $\psi_P(t_0) = \exp(i\theta_0)$ for some real θ_0. Show that P concentrates on the lattice of points $\{(\theta_0 + 2n\pi)/t_0 : n \in \mathbb{Z}\}$. Hint: Show $\Re(1 - P^x \exp(it_0 x - i\theta_0)) = 0$. What do you know about $1 - \cos(t_0 x - \theta_0)$?

[3] Suppose is f is both integrable with respect to Lebesgue measure m on the real line and absolutely continuous, in the sense that $f(x) = \mathfrak{m}^x\big(\{t \le x\}\dot{f}(t)\big)$, for all x, for some integrable function $\dot{f}$.

(i) Show that $f(x) \to 0$ as $|x| \to \infty$. Hint: Show $\mathfrak{m}^t \dot{f}(t) = 0$ to handle $x \to \infty$.

(ii) Show that $\mathbb{m}^x\left(f(x)e^{ixt}\right) = -\mathbb{m}^x\left(e^{ixt}\dot{f}(x)\right)/(it)$ for $t \neq 0$. Hint: For safe Fubini, write the left-hand side as $\lim_{C\to\infty} \mathbb{m}^x\left(e^{ixt}\{|x| \le C\}\mathbb{m}^s\left(\dot{f}(s)\{s \le x\}\right)\right)$.

(iii) If a probability density has Lebesgue integrable derivatives up to kth order, prove that its Fourier transform is of order $O(|t|^{-k})$ as $|t| \to \infty$.

[4] Let $\mathbb{m}$ denote Lebesgue measure on $\mathcal{B}(\mathbb{R})$. For each f in $\mathcal{L}^1(\mathbb{m})$ show that $\mathbb{m}^x\left(f(x)e^{ixt}\right) \to 0$ as $t \to \pm\infty$. Hint: Check the result for f equal to the indicator function of a bounded interval. Show that the linear space spanned by such indicator functions is dense in $\mathcal{L}^1(\mathbb{m})$. Alternatively, approximate f by linear combinations of densities with integrable derivatives, then invoke Problem [3].

[5] Let $g(\cdot)$ be a bounded real function on the real line with bounded derivatives up to order $k+1$, and let X be a random variable for which $\mathbb{P}|X|^k < \infty$. Let $R_k(\cdot)$ be the remainder in the Taylor expansion up to kth power:

$$g(x) = g(0) + xg'(0) + \ldots + \frac{x^k}{k!}g^{(k)}(0) + R_k(x).$$

(i) Show that $|R_k(x)| \le C \min\left(|x|^k, |x|^{k+1}\right)$, for some constant C.

(ii) Invoke Dominated Convergence to show that

$$\mathbb{P}g(Xt) = g(0) + tg'(0)\mathbb{P}X + \ldots + \frac{t^k}{k!}g^{(k)}(0)\mathbb{P}X^k + o(t^k) \qquad \text{as } t \to 0.$$

[6] Let P denote the double-exponential distribution, given by the density $p(x) = \frac{1}{2}\exp(-|x|)$ with respect to Lebesgue measure on $\mathcal{B}(\mathbb{R})$.

(i) Show that $P^x e^{xs} = 1/(1-s^2)$ for real s with $|s| < 1$.

(ii) By a leap of faith (or by an appeal to analytic continuation), deduce that $\psi_P(t) = 1/(1+t^2)$ for $t \in \mathbb{R}$.

(iii) The Cauchy probability distribution is given by the density $q(x) = \pi^{-1}/(1+x^2)$. Apply the inversion formula from Theorem <10> to deduce that the Cauchy distribution has Fourier transform $\exp(-|t|)$.

[7] Suppose (X, Y) has a multivariate normal distribution with $\text{cov}(X, Y) = 0$. Show that X and Y are independent. (Hint: What is the Fourier transform of $\mathbb{P}_X \otimes \mathbb{P}_Y$?)

[8] Let $X_1, \ldots, X_n$ be independent, Uniform(0,1) distributed random variables. Calculate the logarithm of the Fourier transform of $n^{-1/2}(X_1 + \ldots + X_n - \frac{1}{2}n)$ up to a remainder term of order $O(n^{-2})$.

[9] Suppose X is a random variable with Fourier transform ψ_X.

(i) For each $\delta > 0$, show that

$$\frac{1}{\delta}\int_0^\delta (1 - \Re\psi_X(t))\, dt = \mathbb{P}\left(1 - \frac{\sin X\delta}{X\delta}\right) \ge c_0\mathbb{P}\{|X| \ge 1/\delta\},$$

where $1 - c_0 := \sup_{|x|\ge 1} |\sin x/x|$.

(ii) If ψ is complex-valued function, continuous at the origin, show that

$$\frac{1}{\delta}\int_0^\delta \Re\left(\psi(0) - \psi(t)\right) dt \to 0 \qquad \text{as } \delta \to 0.$$

(iii) Suppose $\{X_n\}$ is a sequence of random variables whose Fourier transforms ψ_n converge pointwise to a function ψ that is continuous at the origin. Show that X_n is of order $O_p(1)$.

(iv) Show that the ψ from part (iii) equals the Fourier transform of some random variable Z representing the limit in distribution of a subsequence of $\{X_n\}$. Deduce via Theorem <12> that $X_n \rightsquigarrow Z$.

[10] Suppose a random variable X has a Fourier transform ψ for which $\psi(t) = 1 + O(t^2)$ near $t = 0$.

(i) Show that there exists a finite constant C such that

$$C \geq \frac{\Re(1 - \psi(t))}{t^2} \geq \mathbb{P}\left(\frac{1 - \cos(Xt)}{(Xt)^2} X^2\{0 < |X| \leq M\}\right)$$

for all M. Deduce via Dominated Convergence that $2C \geq \mathbb{P}X^2$.

(ii) Show that $\mathbb{P}X = 0$.

[11] Suppose X is a random variable, and $f(z, X)$ is a jointly measurable function that is analytic in a neighborhood $\mathcal{N} := \{z \in \mathbb{C} : |z - z_0| < \delta\}$ of a point z_0 in the complex plane. Write $f'(z, X)$ for the derivative with respect to z. Suppose $|f'(z, X)| \leq M(X)$ for $|z - z_0| \leq \delta$, where $\mathbb{P}M(X) < \infty$. Show that $\mathbb{P}f(z, X)$ is analytic in $\mathcal{N}$, with derivative $\mathbb{P}f'(z, X)$. Hint: Reduce to the corresponding theorem for differentiation with respect to a real variable by defining $g(t, X) := f(z_0 + th, X)$ for $0 \leq t \leq 1$, with h fixed.

[12] Suppose a random variable X has Fourier transform $\phi(t) = \exp(ict - dt^2)$, for complex numbers $c := c_1 + ic_2$ and $d := d_1 + id_2$.

(i) Deduce from the facts that $|\phi(t)| \leq 1$ and $\phi(t) = 1 - c_2 t + ic_1 t + o(t)$ near $t = 0$ that $c_2 = 0$.

(ii) Show that $X - c_1$ has Fourier transform $\exp(-dt^2) = 1 + O(t^2)$ near $t = 0$. Deduce from Theorem <2> and Problem [10] that $d = \mathbb{P}|X - c_1|^2$, which is nonnegative.

10. Notes

Feller (1971, Chapters 15 and 16) is a good source for facts about Fourier transforms (characteristic functions). Much of my exposition in the first four Sections is based on his presentation, with help from Breiman (1968, Chapter 8).

The idea of generating functions is quite old—see the entries under *generating functions* in the index to Stigler (1986a), for descriptions of the contributions of De Moivre, Simpson, Lagrange, and Laplace.

Apparently Lévy borrowed the name *characteristic function* from Poincaré (who used it to refer to what is now known as the moment generating function), when rediscovering the usefulness of the Fourier transform for probability theory calculations, unaware of earlier contributions.

Lévy (1922) extended the classical Fourier inversion formula to general probability distributions on the real line. He also used an inversion formula to prove a form of the Continuity Theorem slightly weaker than Theorem <12>. (He required uniform convergence on bounded intervals, but his method of proof works just as well with pointwise convergence.) He noted that the theorem could be proved by reduction to the case of bounded densities, by convolution smoothing, offering the normal as a suitable source of smoothing—compare with Lévy (1925, page 197), where he used convolution with a uniform distribution for the same purpose. The slightly stronger version of the Continuity Theorem described in Problem [9] is due to Cramér (1937), albeit originally incorrectly stated (Cramér 1976, page 525).

The book by Hall & Heyde (1980) is one of the best references for martingale theory in discrete time. It contains a slightly stronger form of Theorem <14>.

The results in Section 6 concerning characterizations via linear combinations come from Cramér & Wold (1936). In the 1998 Addendum to his 1997 paper, Walther noted that Radon (1917) had proved similar results, also without the use of Fourier theory. I have not seen Radon's paper.

I borrowed the proof of the Lévy-Cramér theorem from Chow & Teicher (1978, Section 8.4). The result was conjectured by Lévy (1934, final paragraph) then proved by Cramér (1936)—see Cramér (1976, page 522) and Lévy (1970, page 111). The last part of the proof in Section 8 essentially establishes a special case of the result of Hadamard originally invoked by Cramér. See Le Cam (1986, page 80) and Loève (1973, page 3) for further discussion of why the result plays such a key role in the statement of necessary and sufficient conditions for the central limit theorem to hold.

References

Breiman, L. (1968), *Probability*, first edn, Addison-Wesley, Reading, Massachusets.

Chow, Y. S. & Teicher, H. (1978), *Probability Theory: Independence, Interchangeability, Martingales*, Springer, New York.

Cramér, H. (1936), 'Über eine Eigenschaft der normalen Verteilungsfinktion', *Mathematische Zeitung* **41**, 405–414.

Cramér, H. (1937), *Random Variables and Probability Distributions*, Cambridge University Press.

Cramér, H. (1976), 'Half a century with probability theory: some personal recollections', *Annals of Probability* **4**, 509–546.

Cramér, H. & Wold, H. (1936), 'Some theorems on distribution functions', *Journal of the London Mathematical Society* **11**, 290–294.

Feller, W. (1971), *An Introduction to Probability Theory and Its Applications*, Vol. 2, second edn, Wiley, New York.

Hall, P. & Heyde, C. C. (1980), *Martingale Limit Theory and Its Application*, Academic Press, New York, NY.

Le Cam, L. (1986), 'The central limit theorem around 1935', *Statistical Science* **1**, 78–96.

Lévy, P. (1922), 'Sur la determination des lois de probabilité par leurs fonctions caractéristiques', *Comptes Rendus de l'Academie des Sciences, Paris* **175**, 854–856.

Lévy, P. (1925), *Calcul des Probabilités*, Gauthier-Villars, Paris.

Lévy, P. (1934), 'Sur les intégrales dont les éléments sont des variables aléatoires indépendantes', *Ann. Ecole. Norm. Sup. Pisa(2)* **3**, 337–366.

Lévy, P. (1937), *Théorie de l'addition des variables aléatoires*, Gauthier-Villars, Paris. Page references from the 1954 second edition.

Lévy, P. (1970), *Quelques Aspects de la Pensée d'un Mathématicien*, Blanchard, Paris.

Loève, M. (1973), 'Paul Lévy, 1886-1971', *Annals of Probability* **1**, 1–18. Includes a list of Lévy's publications.

McLeish, D. L. (1974), 'Dependent central limit theorems and invariance principles', *Annals of Probability* **2**, 620–628.

Petrov, V. V. (1972/75), *Sums of Independent Random Variables*, Springer-Verlag. Enlish translation in 1975, from 1972 Russian edition.

Radon, J. (1917), 'Über die Bestimmung von Functionen durch ihre Integralwerte längs gewisser Mannigfaltigkeiten', *Ber. Verh. Sächs. Akad. Wiss. Leipzig, Math-Nat. Kl.* **69**, 262–277.

Rudin, W. (1974), *Real and Complex Analysis*, second edn, McGraw-Hill, New York.

Stigler, S. M. (1986a), *The History of Statistics: The Measurement of Uncertainty Before 1900*, Harvard University Press, Cambridge, Massachusetts.

Walther, G. (1997), 'On a conjecture concerning a theorem of Cramér and Wold', *Journal of Multivariate Analysis* **63**, 313–319. Addendum in same journal, **63**, 431.

Chapter 9

Brownian motion

SECTION 1 collects together some facts about stochastic processes and the normal distribution, for easier reference.

SECTION 2 defines Brownian motion as a Gaussian process indexed by a subinterval T of the real line. Existence of Brownian motions with and without continuous sample paths is discussed. Wiener measure is defined.

SECTION 3 constructs a Brownian motion with continuous sample paths, using an orthogonal series expansion of square integrable functions.

*SECTION *4 describes some of the finer properties—lack of differentiability, and a modulus of continuity—for Brownian motion sample paths.*

SECTION 5 establishes the strong Markov property for Brownian motion. Roughly speaking, the process starts afresh as a new Brownian motion after stopping times.

*SECTION *6 describes a family of martingales that can be built from a Brownian motion, then establishes Lévy's martingale characterization of Brownian motion with continuous sample paths.*

*SECTION *7 shows how square integrable functions of the whole Brownian motion path can be represented as limits of weighted sums of increments. The result is a thinly disguised version of a remarkable property of the isometric stochastic integral, which is mentioned briefly.*

*SECTION *8 explains how the result from Section 7 is the key to the determination of option prices in a popular model for changes in stock prices.*

1. Prerequisites

Broadly speaking, Brownian motion is to stochastic process theory as the normal distribution is to the theory for real random variables. They both arise as natural limits for sums of small, independent contributions; they both have rescaling and transformation properties that identify them amongst wider classes of possible limits; and they have both been studied in great detail. Every probabilist, and anyone dealing with continuous-time processes, should learn at least a little about Brownian motion, one of the most basic and most useful of all stochastic processes. This Chapter will define the process and explain a few of its properties.

The discussion will draw on a few basic ideas about stochastic processes, and a few facts about the normal distribution, which are summarized in this Section.

A ***stochastic process*** is just a family of random variables $\{X_t : t \in T\}$, all defined on the same probability space, say $(\Omega, \mathcal{F}, \mathbb{P})$. Throughout the Chapter, the index set T will always be $\mathbb{R}^+$ or a subinterval $[0, a]$ of $\mathbb{R}^+$. You should think of the parameter t as *time*, with the stochastic process evolving in time.

I will use the symbols $X_t(\omega)$ and $X(t, \omega)$ interchangeably. The latter notation suggests that we regard the whole process as a single function X on $T \times \Omega$, and use the single letter X to refer to the whole family of random variables. We can also treat $X(t, \omega)$ as a family of functions $X(\cdot, \omega)$ defined on T, one for each ω. Each of these functions, $t \mapsto X(t, \omega)$, is called a ***sample path*** of the process. Each viewpoint—a family of random variables, a single function of two arguments, and a family of sample paths—has its advantages.

As time passes, we learn more about the process and about other random variables defined on Ω, a situation represented (as in Chapter 6) by a ***filtration***: a family of sub-sigma-fields $\{\mathcal{F}_t : t \in T\}$ of $\mathcal{F}$ for which $\mathcal{F}_s \subseteq \mathcal{F}_t$ whenever $s < t$. A stochastic process $\{X_t : t \in T\}$ is said to be ***adapted*** to the filtration if X_t is $\mathcal{F}_t$-measurable for each t. On occasion it will be helpful to enlarge a filtration slightly, replacing $\mathcal{F}_t$ by the sigma-field $\mathcal{F}_t^*$ generated by $\mathcal{F}_t \cup \mathcal{N}$, with $\mathcal{N}$ the class of $\mathbb{P}$-negligible subsets of Ω. I will refer to $\{\mathcal{F}_t^* : t \in T\}$ as the ***completed filtration***.

The joint distributions of the subcollections $\{X_t : t \in S\}$, with S ranging over all the finite subsets of T, are called the ***finite dimensional distributions*** (or fidis) of the process. If all the fidis are multivariate normal, the process is said to be ***Gaussian***. If each X_t has zero expected value, the process is said to be ***centered***.

The striking behavior of Brownian motion will be largely determined by just a few properties of the normal distribution (see Section 8.6):

(a) A multivariate normal distribution is uniquely determined by its vector of means and its variance matrix. (The Fourier transform is a function of those two quantities.)

(b) If X and Y have a bivariate normal distribution, then X is independent of Y if and only if $\text{cov}(X, Y) = 0$. That is, under an assumption of joint normality, independence is equivalent to orthogonality of $X - \mu_X$ and $Y - \mu_Y$, in the $L^2(\mathbb{P})$ sense, where $\mu_X := \mathbb{P}X$ and $\mu_Y := \mathbb{P}Y$.

(c) If $Z_1, Z_2, \ldots Z_n$ are independent random variables, each with a $N(0, 1)$ distribution, then all linear combinations $\sum_i \alpha_i Z_i$ have normal distributions. The joint distributions of finite collections of linear combinations of $Z_1, \ldots, Z_n$ are multivariate normal.

(d) If $\{X_n\}$ is a sequence of random vectors with multivariate normal distributions that converges in distribution to a random vector X, then X has a multivariate normal distribution. The expected value $\mathbb{P}X_n$ must converge to $\mathbb{P}X$, and the covariance matrix $\text{var}(X_n)$ must converge to $\text{var}(X)$. Convergence in $L^2(\mathbb{P})$ implies convergence in distribution.

(e) If $Z_1, Z_2, \ldots Z_n$ are independent random variables, each with a $N(0, 1)$ distribution, then $\mathbb{P}\max_{i \le n} |Z_i| \le 2\sqrt{1 + \log n}$ and $\max_{i \le n} |Z_i| / \sqrt{2 \log n} \to 1$ almost surely as $n \to \infty$. (See Problems [1] and [2] for proofs.)

2. Brownian motion and Wiener measure

There are several closely related definitions for Brownian motion, each focusing on a different desirable property of the process. Let us start from a minimal definition, then build towards a more comfortable set of properties.

<1> **Definition.** *A stochastic process $B := \{B_t : t \in T\}$, adapted to a filtration $\{\mathcal{F}_t\}$, is said to be a **Brownian motion** (for that filtration) if its increments have the following two properties,*

(i) for all $s < t$ the increment $B_t - B_s$ is independent of $\mathcal{F}_s$,

(ii) for all $s < t$ the increment $B_t - B_s$ has a $N(0, t - s)$ distribution.

Equivalently,

(iii) $\mathbb{P}F \exp\left(i\theta(B_t - B_s) + \frac{1}{2}\theta^2(t - s)\right) = \mathbb{P}F$, for all $s < t$, all $\theta \in \mathbb{R}$, all $F \in \mathcal{F}_s$.

Proof of the equivalence. Necessity of (iii) follows from the independence and the fact that the $N(0, t - s)$ distribution has Fourier transform $\exp(-\theta^2(t - s)/2)$. For sufficiency, first take $F = \Omega$ to show that $B_t - B_s$ is $N(0, t - s)$ distributed. To establish independence, we have only to show that $\mathbb{P}(g(B_t - B_s)F) = (\mathbb{P}g(B_t - B_s))(\mathbb{P}F)$ for all bounded, measurable functions g. The assertion is trivial if $\mathbb{P}F = 0$. When $\mathbb{P}F \neq 0$, equality (iii) may be rewritten as

$$\mathbb{P}_F \exp(i\theta(B_t - B_s)) = \exp(-\theta^2(t - s)/2) = \mathbb{P}\exp(i\theta(B_t - B_s)),$$

where $\mathbb{P}_F(A) = \mathbb{P}(FA)/\mathbb{P}F$ for all A in $\mathcal{F}$. By the uniqueness theorem for Fourier transforms, $B_t - B_s$ has the same distribution under $\mathbb{P}_F$ as under $\mathbb{P}$. In particular, $\mathbb{P}_F g(B_t - B_s) = \mathbb{P}g(B_t - B_s)$, for all bounded, measurable functions g. □

Once we specify the distribution of B_0, the joint distribution of $B_0, B_{t_1}, \ldots, B_{t_k}$, for $0 < t_1 < \ldots < t_k$, is uniquely determined. That is, the fidis are uniquely detemined by Definition <1> and the distribution of B_0. If B_0 is integrable then so are all the B_t, and the process is a martingale. If B_0 has a normal distribution (possibly degenerate), then all the fidis are multivariate normal, which makes B a Gaussian process. If $B_0 \equiv x$, for a constant x, the process is said to ***start at*** x. In particular, when $B_0 \equiv 0$, the Brownian motion is a centered Gaussian process, whose fidis are uniquely determined by the covariances,

$$\text{cov}(B_s, B_t) = \text{cov}(B_s, B_s) + \text{cov}(B_s, B_t - B_s) = s + 0 \qquad \text{if } s \leq t.$$

More succinctly, $\text{cov}(B_s, B_t) = s \wedge t$ for all s, t in $\mathbb{R}^+$.

Often one speaks of a Brownian motion without explicit mention of the filtration, in which case it is implicit that $\mathcal{F}_t$ equals $\mathcal{F}_t^B := \sigma\{B_s : s \leq t\}$, the ***natural*** or ***Brownian filtration***. In that case, a simple generating class argument shows that property (i) is equivalent to the assertion:

(i)′ for all choices of $t_0 < t_1 < t_2 < \ldots < t_k$ from T, the random variables $\{B(t_j) - B(t_{j-1}) : j = 1, 2, \ldots, k\}$ and $B(t_0)$ are independent.

A centered Gaussian process with $\text{cov}(B_t, B_s) = t \wedge s$ for all s, t in T is a Brownian motion for the natural filtration.

A further property is usually added to the list of requirements for Brownian motion, namely that it have continuous sample paths:

(iv) For each fixed ω, the sample path $B(\cdot, \omega)$ is a continuous function on T.

Some authors give the mistaken impression that property (iv) follows from properties (i) and (ii). The proper assertion (Problem [3]) is that there exists a stochastic process that satisfies (i) and (ii), and has continous sample paths, or, more precisely, if $\{B_t : t \in T\}$ satisfies (i) and (ii), then there exists another process $\{B_t^* : t \in T\}$, defined on the same probability space, for which (i), (ii), and (iv) hold, and for which $\mathbb{P}\{B_t = B_t^*\} = 1$ at every t. The new process is called a ***version*** of the original process. Notice that B_t^* need not be $\mathcal{F}_t$-measurable if $\mathcal{F}_t$ does not contain enough negligible sets, but it is measurable with respect to the completed sigma-field $\mathcal{F}_t^*$. In fact, B^* is a Brownian motion for the completed filtration: the added negligible sets have no effect on the calculations required to prove that $B_t^* - B_s^*$ is independent of $\mathcal{F}_t^*$.

In truth, a Brownian motion that did not have all, or almost all, of its sample paths continuous would not be a very nice beast. Many of the beautiful properties of Brownian motion depend on the continuity of its sample paths.

<2> **Example.** Let $\{(B_t, \mathcal{F}_t) : 0 \le t \le 1\}$ be a Brownian motion with continuous sample paths. The quantity $M(\omega) := \sup_t |B(t, \omega)|$ is finite for each ω, because each continuous function is bounded on the compact set $[0, 1]$. It is an $\mathcal{F}_1$-measurable random variable, because $\{\omega : M(\omega) > x\} = \cup_{s \in S}\{|B_s(\omega)| > x\}$, for any countable, dense subset S of $[0, 1]$. A countable union of sets from $\mathcal{F}_1$ also belongs to $\mathcal{F}_1$.

What happens when the process does not have continuous sample paths, but (i) and (ii) are satisfied? To show you how bad it can get, I will perform pathwise surgery on B to create a version that behaves badly. As you will see, the issue is really one of managing uncountable families of negligible sets. Countable collections of negligible sets can be ignored, but uncountable collections are capable of causing real trouble.

From Problem [7], there is a partition of Ω into an uncountable union of disjoint, $\mathcal{F}_1$-measurable sets $\{\Omega_t : 0 \le t \le 1\}$, with $\mathbb{P}\Omega_t = 0$ for every t. Let $\beta(\omega)$ be an arbitrarily nasty, nonmeasurable, nonnegative function on Ω for which $\beta(\omega) > M(\omega)$ at each ω. Define $B^*(t, \omega) := B(t, \omega)\{\omega \notin \Omega_t\} + \beta(\omega)\{\omega \in \Omega_t\}$. By construction, $\mathbb{P}\{B_t^* = B_t\} = 1$ for every t. The joint distributions of finite, or countable, collections of B_t^* variables are the same as the joint distributions for the corresponding collections of B_t. In particular, B^* is a Brownian motion, with respect to the completed filtration.

The construction ensures that $|B^*(\cdot, \omega)|$ is maximized at the t for which $\omega \in \Omega_t$. and $\sup_s |B_s^*(\omega)| = \beta(\omega)$. We have built ourselves a process satisfying requirements (i) and (ii) of Definition <1> but by deliberately violating (iv), we have created a nasty nonmeasurablity. □

REMARK. The point of the Example is not that anyone might choose a bad Brownian motion, like B^*, in preference to one with continuous sample paths, but rather that there is nothing in requirements (i) and (ii) to exclude the bad version. Continuity of a path requires cooperation of uncountably many random variables, a cooperation that cannot be ensured by requirements expressed purely in terms of joint distributions of finite, or countable, subfamilies of random variables.

A Brownian motion that has continuous sample paths also defines a map, $\omega \mapsto B(\cdot, \omega)$ from Ω into the space $\mathbb{C}(T)$ of continuous, real valued functions on T. It becomes a random element of $\mathbb{C}(T)$ if we equip that space with its ***finite dimensional (or cylinder) sigma-field*** $\mathcal{C}(T)$, the sigma-field generated by the ***cylinder sets*** $\{x \in \mathbb{C}(T) : (x(t_1), \ldots, x(t_k)) \in A\}$, with $\{t_1, \ldots, t_k\}$ ranging over all finite subsets of T and A ranging over $\mathcal{B}(\mathbb{R}^k)$, for each finite k.

As an $\mathcal{F}\backslash\mathcal{C}(T)$-measurable map, $\omega \mapsto B(\cdot, \omega)$, from Ω into $\mathbb{C}(T)$, a Brownian motion induces a probability measure (its distribution or image measure) on $\mathcal{C}(T)$. The distribution is uniquely determined by the fidis, because the collection of cylinder sets is stable under finite intersections and it generates $\mathcal{C}(T)$. For the simplest case, where $B_0 \equiv 0$, the distribution is called ***Wiener measure*** on $\mathcal{C}(T)$, or, less precisely, Wiener measure on T. I will denote it by $\mathbb{W}$, relying on context to identify T.

REMARK. Each coordinate projection, $X_t(x) := x(t)$, defines a random variable on $\mathbb{C}(T)$. As a stochastic process on the probability space $(\mathbb{C}(T), \mathcal{C}(T), \mathbb{W})$, the family $\{X_t : t \in T\}$ is a Brownian motion with continuous paths, started at 0. For many purposes, the study of Brownian motion is just the study of $\mathbb{W}$.

For the remainder of the Chapter, you may assume that all Brownian motions satisfy requirements (i), (ii), and (iv), with $B_0 \equiv 0$. That is, unless explicitly warned otherwise, you may assume that all Brownian motions from now on are centered Gaussian processes with continuous sample paths and $\text{cov}(B_t, B_s) = t \wedge s$, a process that I will refer to as ***standard Brownian motion*** *(on T).*

3. Existence of Brownian motion

It takes some ingenuity to build a Brownian motion with continuous sample paths, a feat first achieved with mathematical rigor by Wiener (1923). This Section contains one construction, based on a few facts about Hilbert spaces, all of which are are established in Section 4 of Appendix B.

Suppose $\mathcal{H}$ is a Hilbert space with a countable orthonormal basis $\{\psi_i : i \in \mathbb{N}\}$. Let $\{\eta_i\}$ be a sequence of independent $N(0, 1)$ random variables, defined on some probability space $(\Omega, \mathcal{F}, \mathbb{P})$. For each fixed h in $\mathcal{H}$, the sequence of random variables $G_n(h) := \sum_{i=1}^n \langle h, \psi_i \rangle \eta_i$ converges in $L^2(\mathbb{P})$ to a limit, $G(h) := \sum_{i \in \mathbb{N}} \langle h, \psi_i \rangle \eta_i$, and by Parseval's identity,

$$\text{cov}\,(G_n(h_1), G_n(h_2)) = \sum\nolimits_{i=1}^n \langle h_1, \psi_i \rangle \langle h_2, \psi_i \rangle \to \sum\nolimits_{i=1}^\infty \langle h_1, \psi_i \rangle \langle h_2, \psi_i \rangle = \langle h_1, h_2 \rangle.$$

Note that $G(h)$ is uniquely determined as an element of $L^2(\mathbb{P})$, but it is only defined up to an almost sure equivalence as a random variable.

In particular, from the facts (c) and (d) in Section 1, the random variable $G(h)$ has a $N(0, \|h\|^2)$ distribution. Moreover, for each finite subset $\{h_1, \ldots, h_k\}$ of $\mathcal{H}$, the random vector $(G(h_1), \ldots, G(h_k))$ has a multivariate normal distribution with zero means and covariances given by $\mathbb{P}G(h_i)G(h_j) = \langle h_i, h_j \rangle$. That is, all the fidis of the process are centered multivariate normal. The family of random variables $\{G(h) : h \in \mathcal{H}\}$ is a Gaussian process that is sometimes called the ***isonormal process***, indexed by the Hilbert space $\mathcal{H}$ (compare with Dudley 1973, page 67).

REMARK. Notice that the map $h \mapsto G(h)$ is linear and continuous, as a function from $\mathcal{H}$ into $L^2(\mathbb{P})$. Thus $G(h)$ can be recovered, up to an almost sure equivalence, from the values $\{G(e_i) : i \in \mathbb{N}\}$ for any orthonormal basis $\{e_i : i \in \mathbb{N}\}$ for $\mathcal{H}$.

To build a Brownian motion indexed by $[0, 1]$, specialize to the case where $\mathcal{H} := L^2(\mathfrak{m})$, with $\mathfrak{m}$ equal to Lebesgue measure on $[0, 1]$. Write f_t for the indicator function of the interval $[0, t]$. The subset $\{f_t : t \in [0, 1]\}$ of $\mathcal{H}$ defines a centered Gaussian process $B_t := G(f_t)$ indexed by $[0, 1]$, with

$$\text{cov}(B_s, B_t) = \mathfrak{m}(f_s f_t) = \mathfrak{m}[0, s \wedge t] = s \wedge t.$$

That is, if we take $\mathcal{F}_t$ as $\sigma\{B_s : s \leq t\}$ then $\{(B_t, \mathcal{F}_t) : t \in [0, 1]\}$ is a centered Gaussian process with the covariances that identify it as a Brownian motion indexed by $[0, 1]$, in the sense that properties (i) and (ii) of Definition <1> hold. The question of sample path continuity is more delicate.

Each partial sum G_n defines a process with continuous paths, because

$$|G_n(f_s) - G_n(f_t)|^2 \leq \sum_{i=1}^n \langle f_s - f_t, \psi_i \rangle^2 \sum_{i=1}^n \eta_i^2$$

and $\sum_{i=1}^n \langle f_s - f_t, \psi_i \rangle^2 \leq \|f_s - f_t\|^2 = |s - t|$. We need to preserve continuity in limit as n tends to infinity. Convergence uniform in t would suffice.

Something slightly weaker than uniform convergence is easy to check if we work with the orthonormal basis of Haar functions on $[0, 1]$. It is most natural to specify this basis via a double indexing scheme. For $k = 0, 1, \ldots$ and $0 \leq i < 2^k$, define $H_{i,k}$ as a difference of indicator functions,

$$H_{i,k}(s) := \{i2^{-k} < s \leq (i + \tfrac{1}{2})2^{-k}\} - \{(i + \tfrac{1}{2})2^{-k} < s \leq (i + 1)2^{-k}\}.$$

Notice that $|H_{i,k}|$ is the indicator function of the interval $J_{i,k} := (i2^{-k}, (i+1)2^{-k}]$, and $H_{i,k} = J_{2i,k+1} - J_{2i+1,k+1}$. Thus $\mathfrak{m}H_{i,k}^2 = \mathfrak{m}J_{i,k} = 2^{-k}$. The functions $\psi_{i,k} := \sqrt{2^k} H_{i,k}$ are orthogonal, and each has $L^2(\mathfrak{m})$ norm 1. As shown in Section 3 of Appendix B, the collection of functions $\Psi := \{1\} \cup \{\psi_{i,k} : 0 \leq i < 2^k \text{ for } k = 0, 1, 2, \ldots\}$ is an orthonormal basis for $L^2(\mathfrak{m})$. The Brownian motion has the $L^2(\mathfrak{m})$ representation,

$$B_t := \bar{\eta}\langle f_t, 1 \rangle + \sum_{k=0}^{\infty} \sum_{0 \leq i < 2^k} \eta_{i,k} \langle f_t, \psi_{i,k} \rangle$$

<3>
$$= \bar{\eta}t + \sum_{k=0}^{\infty} 2^{k/2} X_k(t) \qquad \text{with } X_k(t) := \sum_{0 \leq i < 2^k} \eta_{i,k} \langle f_t, H_{i,k} \rangle,$$

where $\bar{\eta}$ and the $\{\eta_{i,k}\}$ are mutually independent $N(0, 1)$ random variables.

As a function of t, each $\langle f_t, H_{i,k} \rangle$ is nonzero only in the interval $J_{i,k}$, within which it is piecewise linear, achieving its maximum value of $2^{-(k+1)}$ at the midpoint, $(2i + 1)/2^{k+1}$:

<4>
$$\langle f_t, H_{i,k} \rangle = 2^{-(k+1)} \wedge \left(t - \frac{i}{2^k}\right)^+ - 2^{-(k+1)} \wedge \left(t - \frac{i + \frac{1}{2}}{2^k}\right)^+.$$

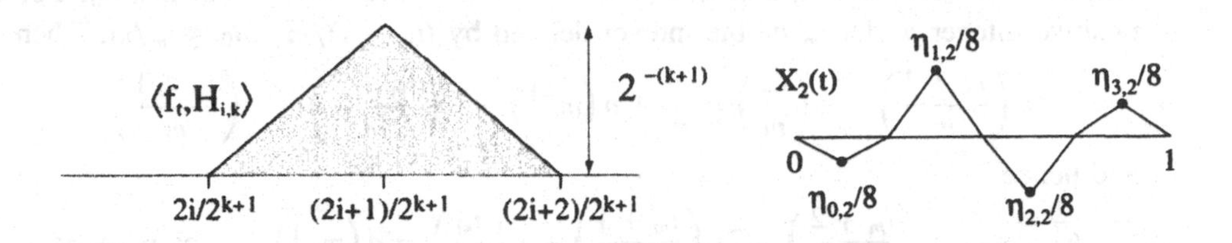

The process $X_k(t)$ has continuous, piecewise linear sample paths. It takes the value 0 at $t = i/2^k$ for $i = 0, 1, \ldots, 2^k$. It takes the value $\eta_{i,k}/2^{k+1}$ at the point $(2i+1)/2^{k+1}$. Thus $\sup_t |X_k(t)| = 2^{-(k+1)} \max_i |\eta_{i,k}|$. From property (e) in Section 1,

$$\sum\nolimits_{k=0}^{\infty} 2^{k/2}\mathbb{P} \sup_t |X_k(t)| \leq \sum\nolimits_{k=0}^{\infty} 2^{-k/2}\sqrt{1+\log(2^k)} < \infty,$$

which implies finiteness of $\sum_k 2^{k/2} \sup_t |X_k(t)|$ almost everywhere. With probability one, the random series <3> for B_t converges uniformly in $0 \leq t \leq 1$; almost all sample paths of B are continuous functions of t. If we redefine $B(t,\omega)$ to be identically zero for a negligible set of ω, we have a Brownian motion with all its sample paths continuous.

From the Brownian motion B indexed by $[0, 1]$ we could build a Brownian motion β indexed by $\mathbb{R}^+$ by defining

$$\beta_t := (1+t)B\left(\frac{t}{1+t}\right) - tB(1) \qquad \text{for } t \in \mathbb{R}^+.$$

Clearly $\{\beta_t : t \in \mathbb{R}^+\}$ is a centered Gaussian process with continuous paths and $\beta_0 = B_0 = 0$. You should check that $\text{cov}(\beta_s, \beta_t) = s \wedge t$, in order to complete the argument that it is a Brownian motion.

REMARK. We could also write Brownian motion indexed by $\mathbb{R}^+$ as a doubly infinite series,

$$B_t = \sum\nolimits_{k=-\infty}^{\infty} \sum\nolimits_{i=0}^{\infty} 2^{k/2}\eta_{i,k}\langle f_t, H_{i,k}\rangle,$$

which converges uniformly (almost surely) on each bounded subinterval. When we focus on $[0, 1]$, the terms for $k < 0$ contribute $t\sum_{k<0} 2^{k/2}\eta_{0,k}$, which corresponds to the $\bar{\eta}t$ in expansion <3>; and for $k \geq 0$, only the terms with $0 \leq i < 2^k$ contribute to B_t.

*4. Finer properties of sample paths

Brownian motion can be constructed to have all of its sample paths continuous, but, almost all of its paths must be nowhere differentiable. The heuristic explanation for this extreme irregularity is: existence of a derivative would imply approximate proportionality of successive small increments, implying a degree of cooperation that is highly unlikely for a succession of independent normals.

A formal proof can be built from the heuristic. Consider first a single continuous function x on $[0, 1]$, which happens to be differentiable at some point t_0 in $(0, 1)$, with a finite derivative v, that is, $x(t) = x(t_0) + (t - t_0)v + o(|t - t_0|)$ near t_0. For a positive integer m, let i_m be the integer defined by $(i_m - 1)/m < t_0 \le i_m/m$. Then

$$x\left(\frac{i_m+1}{m}\right) - x\left(\frac{i_m}{m}\right) = \frac{v}{m} + o\left(m^{-1}\right) = x\left(\frac{i_m+2}{m}\right) - x\left(\frac{i_m+1}{m}\right),$$

and hence

$$\Delta_{i_m,m}x := x\left(\frac{i_m+2}{m}\right) - 2x\left(\frac{i_m+1}{m}\right) + x\left(\frac{i_m}{m}\right) = o\left(m^{-1}\right) \qquad \text{as } m \to \infty.$$

Similarly, both $m\Delta_{i_m+2,m}x$ and $m\Delta_{i_m+4,m}x$ must also converge to zero. By considering successive second differences, we eliminate both t_0 and v, leaving the conclusion that if x is differentiable (with a finite derivative) at an unspecified point of $(0, 1)$ then, for all m large enough there must exist at least one i for which

$$m|\Delta_{i,m}x| \le 1 \qquad \text{and} \qquad m|\Delta_{i+2,m}x| \le 1 \qquad \text{and} \qquad m|\Delta_{i+4,m}x| \le 1.$$

Apply the same reasoning to each sample path of the Brownian motion $\{B_t : 0 \le t \le 1\}$ to see that the set of ω for which $B(\cdot, \omega)$ is somewhere differentiable on $(0, 1)$ is contained in

$$\liminf_{m\to\infty}\{\omega : m|\Delta_{i,m}B| \le 1,\ m|\Delta_{i+2,m}B| \le 1,\ m|\Delta_{i+4,m}B| \le 1 \text{ for some } i\},$$

where the $\Delta_{i,m}B$ are the second differences for the sample path $B(\cdot, \omega)$. Each of $\Delta_{0,m}B$, $\Delta_{2,m}B$, $\Delta_{4,m}B$, ... has a $N(0, 2/m)$ distribution—being a difference of two independent random variables, each $N(0, 1/m)$ distributed—and they are independent. By Fatou's lemma, the probability of the displayed event is smaller than the lim inf of the probabilities

$$\mathbb{P}\cup_{i=0}^{m-6}\{\omega : m|\Delta_{i,m}B| \le 1, m|\Delta_{i+2,m}B| \le 1,\ m|\Delta_{i+4,m}B| \le 1\}$$
$$\le m\left(\mathbb{P}\{|N(0, 2/m)| \le 1\}\right)^3 = O(m^{-1/2}).$$

Thus almost all Brownian motion sample paths are nowhere differentiable.

The nondifferentiability is also suggested by the fact that $(B_{t+\delta} - B_t)/\delta$ has a $N(0, 1/\delta)$ distribution, which could not settled down to a finite limit as δ tends to zero, for a fixed t. Indeed, the increment $B_{t+\delta} - B_t$ should be roughly of magnitude $\sqrt{\delta}$. The maximum such increment, $D(\delta) := \sup_{0\le t\le 1-\delta}|B_{t+\delta} - B_t|$, is even larger. For example, part (i) of Problem [2] shows, for small $\epsilon > 0$ and k large enough, that

$$\mathbb{P}\left\{\max_{0\le i<2^k} 2^{k/2}\left|B\left(\frac{i+1}{2^k}\right) - B\left(\frac{i}{2^k}\right)\right| \le (1-\epsilon)\sqrt{2\log 2^k}\right\} \le \exp\left(-C2^{k\theta}\right)$$

for constants $C > 0$ and $\theta > 0$ depending on ϵ. A Borel-Cantelli argument with $\delta_k := 2^{-k}$ then gives

$$\limsup_{\delta\to 0}\frac{D(\delta)}{\sqrt{2\delta\log(1/\delta)}} \ge \limsup_{k\to\infty}\max_i\frac{|B(i\delta_k + \delta_k) - B(i\delta_k)|}{\sqrt{2\delta_k\log(1/\delta_k)}} \ge 1 \tag{5}$$

almost surely. A similar argument (Lévy 1937, page 172; see McKean 1969, Section 1.6 for a concise presentation of the proof) leads to a sharper conclusion,

$$\limsup_{\delta\to 0}\frac{\Delta(\delta)}{\sqrt{2\delta\log(1/\delta)}} = 1 \qquad \text{almost surely.} \tag{6}$$

More broadly, $\sqrt{\delta \log(1/\delta)}$ gives a global bound for the magnitude of the increments—a ***modulus of continuity***—as shown in the next Theorem. To avoid trivial complications for δ near 1, it helps to increase the modulus slightly, by adding a term that becomes inconsequential for δ near zero.

<7> **Theorem.** *Define* $h(\delta) := \sqrt{\delta + \delta\log(1/\delta)}$ *for* $0 \le \delta \le 1$. *Then for almost all* ω *there is a finite constant* C_ω *such that*

$$|B_t(\omega) - B_s(\omega)| \le C_\omega h(|t-s|) \qquad \text{for all } s, t \text{ in } [0,1],$$

where B *is a standard Brownian motion indexed by* $[0, 1]$.

Proof. Consider a pair with $0 \le s < t \le 1$. Temporarily write δ for $t - s$. From the series representation <3>,

$$\begin{aligned} |B_t - B_s| &\le |(t-s)\bar{\eta}| + \sum\nolimits_{k=0}^{\infty} 2^{k/2}|X_k(t) - X_k(s)| \\ &\le \delta|\bar{\eta}| + \sum\nolimits_{k=0}^{\infty}\sum\nolimits_i 2^{k/2}\left|\langle f_t - f_s, H_{i,k}\rangle\right| M_k \qquad \text{where } M_k := \max_i |\eta_{i,k}|. \end{aligned}$$

For a fixed k, the function $f_t - f_s$ is orthogonal to all $H_{i,k}$ except possibly when t or s belongs to the support interval $J_{i,k}$. There are at most two such intervals, and for them we have the bound $|\langle f_t - f_s, H_{i,k}\rangle| \le \mathfrak{m}\left((s,t]\cap J_{i,k}\right) \le |s-t| \wedge 2^{-k}$. Thus

$$\text{<8>} \qquad \sup_{s\ne t}\frac{|B_t(\omega) - B_s(\omega)|}{h(|t-s|)} \le \sup_{0<\delta\le 1}\left(\delta|\bar{\eta}(\omega)| + 2\sum_{k=0}^{\infty} 2^{k/2}\left(\delta\wedge 2^{-k-1}\right)M_k(\omega)\right)/h(\delta)\,.$$

From Problem [2], for almost all ω there is a $k_0(\omega)$ such that $M_k(\omega) \le 2\sqrt{\log(1+2^k)}$ for $k \ge k_0(\omega)$. From Problem [5] we then have

$$\sum\nolimits_{k=k_0(\omega)}^{\infty} 2^{k/2}\left(\delta\wedge 2^{-k}\right)M_k(\omega) \le 2C_0 h(\delta)$$

for a finite constant C_0, which, together with the fact that $\delta \le h(\delta)$, gives

$$\sup_{s\ne t}\frac{|B_t(\omega) - B_s(\omega)|}{h(|t-s|)} \le |\bar{\eta}(\omega)| + 4C_0 + 2\sum\nolimits_{k=0}^{k_0(\omega)} 2^{k/2}M_k(\omega),$$

the desired bound. □

5. Strong Markov property

Let B be a standard Brownian motion indexed by $\mathbb{R}^+$. For each fixed t, define the ***restarted*** process $R^t B$ by moving the origin to the point (t, B_t). That is, $(R^t B)_s := B_{t+s} - B_t$ at each time s in $\mathbb{R}^+$.

For simplicity, let me temporarily write X_s instead of $(R^t B)_s$. As a process, $\{X_s : s \in \mathbb{R}^+\}$ is adapted to the filtration $\mathcal{G}_s := \mathcal{F}_{s+t}$ for $s \in \mathbb{R}^+$. Moreover, each increment $X_{s+\delta} - X_s$ has a $N(0,\delta)$ distribution, independent of $\mathcal{G}_s$. Thus X is also a standard Brownian motion. It has distribution $\mathbb{W}$, Wiener measure on $\mathcal{C}(\mathbb{R}^+)$. For each finite subset S of $\mathbb{R}^+$, the collection of random variables $\{X_s : s \in S\}$ is independent of $\mathcal{F}_t$. Via the usual sort of generating class argument it then follows that X, as a random element of $\mathcal{C}(\mathbb{R}^+)$, is independent of $\mathcal{F}_t$.

That is, $\mathbb{P}Fg(X) = (\mathbb{P}F)\mathbb{P}g(X)$ for each $F \in \mathcal{F}_t$ and (at least) for each bounded, $\mathcal{C}(\mathbb{R}^+)$-measurable function g on $C(\mathbb{R}^+)$. Equivalently,

$$\mathbb{P}(g(X) \mid \mathcal{F}_t) = \mathbb{W}g \qquad \text{almost surely.}$$

The fact that $R^t B$ is a Brownian motion independent of $\mathcal{F}_t$ is known as the ***Markov property*** of Brownian motion.

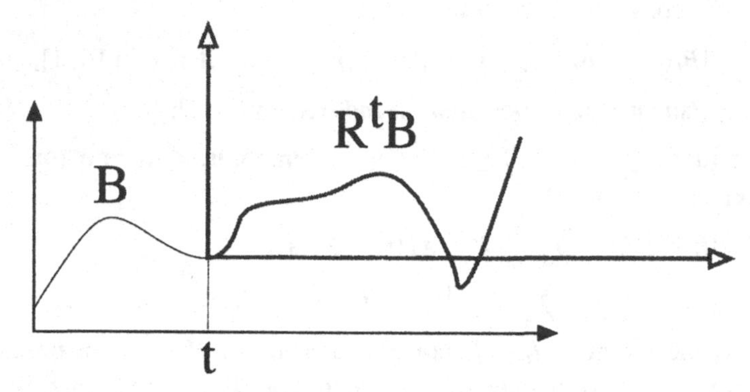

If we replace the fixed time t by a stopping time τ, we get a stronger assertion, known as the ***strong Markov property***. Roughly speaking, the restarted process $R^\tau B$ is a Brownian motion independent of the pre-τ sigma-field $\mathcal{F}_\tau$, which consists of all F for which $F\{\tau \le t\}$ is $\mathcal{F}_t$-measurable, for $0 \le t \le \infty$.

REMARK. Remember that $\mathcal{F}_\infty$ is defined, if not otherwise specified, as the sigma-field generated by $\cup\{\mathcal{F}_t : t \in \mathbb{R}^+\}$. We need to ensure that $\mathcal{F}_\tau \subseteq \mathcal{F}_\infty$ to avoid embarrassing ambiguities about the definition of $\mathcal{F}_\tau$ for the extreme case of a stopping time that is everywhere infinite.

We could write the stronger assertion as $\mathbb{P}(g(X) \mid \mathcal{F}_\tau) = \mathbb{W}g$ almost surely on the set $\{\tau < \infty\}$, but that equality does not quite capture everything we need, as you will discover when we consider the reflection principle, in Example <12>. In that Example we will meet a stopping time τ for which $B_\tau = \alpha$, a positive constant, whenever $\tau(\omega) < t$. We will need to make an assertion like

$$\mathbb{P}\{B_t \le \alpha \mid \mathcal{F}_\tau\} = \mathbb{P}\{B_t - B_\tau \le 0 \mid \mathcal{F}_\tau\} = \tfrac{1}{2} \qquad \text{on the set } \{\tau < t\}.$$

Intuitively speaking, the conditioning lets us treat the $\mathcal{F}_\tau$-measurable random variable τ as a constant, so that $B_t - B_\tau$ has a $N(0, t - \tau)$ conditional distribution, which is symmetric about 0. Unfortunately, this line of reasoning takes us beyond the properties of the abstract, Kolmogorov conditional expectation—if you were paying very careful attention while reading Chapter 5, you will recognize a covert appeal to existence of a conditional distribution. Fortunately, Fubini offers a way around the technicality.

Rewrite the Markov property as a pathwise decomposition of Brownian motion into two independent contributions: R^t shifted to the origin $(t, 0)$; and a killed process $K^t B$, defined by $(K^t B)_s := B_{s \wedge t}$ for $s \in \mathbb{R}^+$. More formally, define S^t as the operator that shifts functions to the right,

$$(S^t x)_s := \begin{cases} x(s-t) & \text{for } s \ge t \\ 0 & \text{for } 0 \le s < t. \end{cases}$$

Then $B = K^t B + S^t R^t B$. The Markov property lets us replace $R^t B$ by a new Brownian motion, independent of $\mathcal{F}_t$, without changing the distributional properties of the whole path. For example, for each $\mathcal{F}_t$-measurable Y and (at least) for each bounded, product measurable real function f, we have (via a generating class argument)

<9> $$\mathbb{P}f(Y, B) = \mathbb{P}^{\omega}\mathbb{W}^x f\left(Y(\omega), K^t B(\cdot, \omega) + S^t x\right).$$

The Y could take values in some arbitrary measurable space. We might take $Y = K^t B$, for instance.

REMARK. If Y takes values in $(\mathcal{X}, \alpha)$, we can define an $\mathcal{A}\otimes\mathcal{C}(\mathbb{R}^+)\otimes\mathcal{B}([0,\infty])$-measurable function g by $g(y, z, t) := \mathbb{W}^x f(y, K^t z + S^t x)$. Then the Markov property becomes $\mathbb{P}f(Y, B) = \mathbb{P}^{\omega} g\,(Y(\omega), B(\cdot, \omega), t)$, for $\mathcal{F}_t$-measurable Y. Multiple appeals to the Tonelli/Fubini theorem would establish the necessary measurability properties.

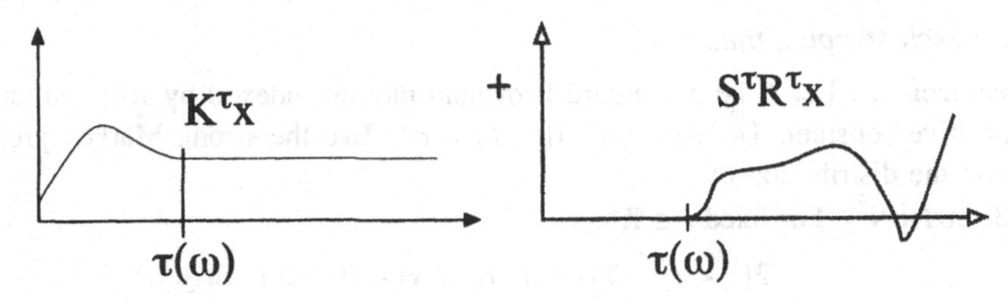

Now consider the effect of replacing the fixed t by a stopping time τ. For every sample path we have $B(\cdot, \omega) = K^{\tau(\omega)}B(\cdot, \omega) + S^{\tau(\omega)}R^{\tau(\omega)}B(\cdot, \omega)$. The decomposition even makes sense for those ω at which $\tau(\omega) = \infty$, because $K^{\infty}x \equiv x$ and S^{∞} shifts whatever we decide to define as $R^{\infty}x$ right out of the picture. For concreteness, perhaps we could take $S^{\infty}x \equiv 0$. Of course it would be a little embarrassing to assert that $R^{\tau}B$ is, conditionally, a Brownian motion at those ω.

<10> **Strong Markov property.** *Let B be a standard Brownian motion for a filtration $\{\mathcal{F}_t : t \in \mathbb{R}^+\}$, and let τ be a stopping time. Then for each $\mathcal{F}_{\tau}$-measurable random element Y, and (at least) for each bounded, product measurable function f,*

$$\mathbb{P}f(Y,B) = \mathbb{P}^{\omega}\mathbb{W}^x f(Y, K^{\tau(\omega)}B(\cdot, \omega) + S^{\tau(\omega)}x).$$

REMARK. In the notation from the previous Remark, the assertion becomes: $\mathbb{P}f(Y, B) = \mathbb{P}^{\omega} g\,(Y(\omega), B(\cdot, \omega), \tau(\omega))$, for $\mathcal{F}_{\tau}$-measurable Y.

Proof. A generating class argument reduces to the case where $f(y, z) := g(y)h(z)$, where g is bounded and measurable, and $h(z) := h_0\,(z(s_1), \ldots, z(s_k))$ with h_0 a bounded, continuous function on $\mathbb{R}^k$ and $s_1, \ldots, s_k$ fixed values in $\mathbb{R}^+$. Discretize the stopping time by rounding up to the next multiple of n^{-1},

$$\tau_n := 0\{\tau = 0\} + \sum_{i \in \mathbb{N}} \frac{i}{n}\left\{\frac{i-1}{n} < \tau \le \frac{i}{n}\right\} + \infty\{\tau = \infty\}.$$

For each n,

$$\mathbb{P}f(Y, B) = \mathbb{P}f(Y, B)\{\tau = \infty\} + \sum\nolimits_{i \in \mathbb{N}_0} \mathbb{P}\{\tau_n = i/n\}g(Y)h(K^{i/n}B + S^{i/n}R^{i/n}B).$$

The product $\{\tau_n = i/n\}g(Y)$ is $\mathcal{F}_{i/n}$-measurable, because Y is $\mathcal{F}_\tau$-measurable. By <9>, with $f(Y, B)$ replaced by $\big(\{\tau_n = i/n\}g(Y)\big)\, h(B)$, the ith summand equals

$$\mathbb{P}^\omega \mathbb{W}^x\{\tau_n = i/n\}g(Y)h(K^{i/n}B + S^{i/n}x) = \mathbb{P}^\omega \mathbb{W}^x\{\tau_n = i/n\}g(Y)h(K^{\tau_n}B + S^{\tau_n}x).$$

Sum over i to deduce that

$$\mathbb{P}f(Y, B) = \mathbb{P}^\omega \mathbb{W}^x g(Y)h(K^{\tau_n}B + S^{\tau_n}x).$$

As n tends to infinity, $\tau_n(\omega)$ converges to $\tau(\omega)$, and hence $K^{\tau_n}B + S^{\tau_n}x \to K^\tau B + S^\tau x$ pointwise (in particular, at each s_j), for each ω and each $x \in C(\mathbb{R}^+)$. Continuity of h_0 then gives $h(K^{\tau_n}B + S^{\tau_n}x) \to h(K^\tau B + S^\tau x)$. An appeal to Dominated Convergence completes the proof. □

<11> **Corollary.** *For each $\mathbb{W}$-integrable function f on $C(\mathbb{R}^+)$,*

$$\mathbb{P}\left(f(B) \mid \mathcal{F}_\tau\right) = \mathbb{W}^x f(K^\tau B + S^\tau x) \qquad \text{almost surely,}$$

for each stopping time τ.

<12> **Exercise.** Let B be a standard Brownian motion indexed by $\mathbb{R}^+$, and let α be a positive constant. Define $\tau := \inf\{t : B_t \geq \alpha\}$. Use the strong Markov property to find the distribution of τ.

SOLUTION: For fixed $t \in \mathbb{R}^+$,

$$\mathbb{P}\{\tau < t\} = \mathbb{P}\{\tau < t,\ B_t > \alpha\} + \mathbb{P}\{\tau < t,\ B_t \leq \alpha\}.$$

The first contribution on the right-hand side equals $\mathbb{P}\{B_t > \alpha\} = \mathbb{P}\{N(0, t) > \alpha\}$, because the inequality for τ is superfluous (by continuity) when $B_t > \alpha$. Invoke Theorem <10> to write the second term as

$$\mathbb{P}^\omega\left(\{\tau(\omega) < t\}\mathbb{W}^x\{x : B\left(t \wedge \tau(\omega)\right) + x\left(t - \tau(\omega)\right) \leq \alpha\}\right).$$

For each ω with $\tau(\omega) < t$ we have $B\left(t \wedge \tau(\omega)\right) = \alpha$ and $\mathbb{W}\{x : x(t - \tau(\omega)) \leq 0\} = 1/2$. By Fubini, the term equals $\frac{1}{2}\mathbb{P}\{\tau < t\}$. Thus

$$\mathbb{P}\{\tau < t\} = 2\mathbb{P}\{B_t > \alpha\} = 2\mathbb{P}\{N(0, 1) > \alpha/\sqrt{t}\} = \frac{2}{\sqrt{2\pi}}\int_{\alpha/\sqrt{t}}^{\infty} \exp(-y^2/2)\, dy.$$

If you differentiate you will discover that the distribution of τ has density $\alpha t^{-3/2}\exp(-\alpha^2/2t)/\sqrt{2\pi}$ with respect to Lebesgue measure on $\mathbb{R}^+$. □

*6. Martingale characterizations of Brownian motion

A centered Brownian motion is a martingale. This fact is just the simplest instance of a method for building martingales from polynomial functions of B_t and t.

<13> **Example.** If $\{(B_t, \mathcal{F}_t) : t \in T\}$ is a centered Brownian motion then a direct calculation shows that $\{B_t^2 - t\}$ is a martingale with respect to the same filtration: if $s < t$ and $F \in \mathcal{F}_s$, then

$$\begin{aligned}\mathbb{P}B_t^2 F &= \mathbb{P}(B_s^2 + 2B_s\Delta + \Delta^2)F \qquad \text{where } \Delta := B_t - B_s \\ &= \mathbb{P}B_s^2 F + 2\mathbb{P}(B_s F)\mathbb{P}\Delta + \left(\mathbb{P}\Delta^2\right)(\mathbb{P}F) \qquad \text{by independence.}\end{aligned}$$

After substitution of 0 for $\mathbb{P}\Delta$ and $t-s$ for $\mathbb{P}\Delta^2$, the equality rearranges to give $\mathbb{P}(B_t^2-t)F = \mathbb{P}(B_s^2-s)F$ for all F in $\mathcal{F}_s$, the asserted martingale property.

Similar arguments could be used for higher degree polynomials, but it is easier to code all the martingale assertions into a single identity. For each fixed complex θ, the process $M_t := \exp(\theta B_t - t\theta^2/2)$ is a (complex-valued) martingale: for F in $\mathcal{F}_s$,

$$\begin{aligned}\mathbb{P}M_tF &= \mathbb{P}\exp(\theta B_s + \theta\Delta - t\theta^2/2)F \\ &= \mathbb{P}\left(\exp(\theta B_s - s\theta^2/2)F\right)\mathbb{P}\exp(\theta\Delta - (t-s)\theta^2/2) \qquad \text{by independence} \\ &= \mathbb{P}M_sF \qquad \text{because } \Delta \text{ is } N(0,t-s) \text{ distributed.}\end{aligned}$$

A dominated convergence argument would justify the integration term-by-term to produce a power-series expansion,

$$\begin{aligned}\mathbb{P}M_tF &= \mathbb{P}\exp(\theta B_t)F\exp(-\theta^2t/2) \\ &= \left(1+\theta\mathbb{P}B_tF + \frac{\theta^2}{2}\mathbb{P}B_t^2F + \frac{\theta^3}{3!}\mathbb{P}B_t^3F+\ldots\right)\left(1-\frac{\theta^2t}{2}+\frac{\theta^4t^2}{8}+\ldots\right)\end{aligned}$$

with a similar expansion for $\mathbb{P}\exp(\theta B_s)$. The series converge for all complex θ. By equating coefficients of powers of θ, we obtain a sequence of equalities that establish the martingale property for $\{B_t\}$, $\{B_t^2-t\}$, ... As an exercise you might find the term involving B_t^3, then check the martingale property by direct calculation, as I did for B_t^2-t. □

Given that Fourier transforms determine distributions, it is not surprising that the martingale property of $\exp(\theta B_t - t\theta^2/2)$, for all θ, characterizes Brownian motion—essentially equivalence (iii) of Definition <1>. It is less obvious that the martingale property for the linear and quadratic polynomials alone should also characterize Brownian motion with continuous sample paths. This striking fact is actually just an elegant repackaging of a martingale central limit theorem. The continuity of the sample paths lets us express the process as a sum of many small martingale differences, whose individual contributions can be captured by Taylor expansions to quadratic terms.

<14> **Theorem.** *(Lévy) Suppose $\{(X_t,\mathcal{F}_t) : t\in\mathbb{R}^+\}$ is a martingale with continuous sample paths, and $X_0\equiv 0$. If $\{(X_t^2-t,\mathcal{F}_t) : t\in\mathbb{R}^+\}$ is also a martingale then the process is a standard Brownian motion.*

Proof. From equivalence (iii) of Definition <1>, it is enough to prove, for each real θ and each F in $\mathcal{F}_s$, that

<15> $$\mathbb{P}\exp\left(i\theta(X_t-X_s)+\theta_2(t-s)\right)F = \mathbb{P}F \qquad \text{where } \theta_2 := \tfrac{1}{2}\theta^2.$$

I will present the argument only for the notationally simplest case where $s=0$ and $t=1$, leaving to you the minor modifications needed for the more general case.

In principle, the method of proof is just Taylor's theorem. Break X_1 into a sum of increments $\sum_{k=1}^n \eta_k$, where $\eta_k := X(k/n) - X((k-1)/n)$. Write $\mathbb{P}_k$ for expectations conditional on $\mathcal{F}_{k/n}$. The two martingale assumptions give $\mathbb{P}_{k-1}\eta_k = 0$ and $v_{k-1} := \mathbb{P}_{k-1}\eta_k^2 = 1/n$. Notice that $\sum_{k=1}^n v_{k-1} = 1$. For fixed real θ, define

$$D_k := \exp\left(i\theta(\eta_1+\ldots+\eta_k)+\theta_2(v_0+\ldots+v_{k-1})\right),$$

so that $D_0 \equiv 1$ and $D_n = \exp(i\theta X_1 + \theta_2)$. Continuity of the paths should make all the η_k small, suggesting

$$\begin{aligned}\mathbb{P}_{k-1}D_k &= D_{k-1}\exp(\theta_2 v_{k-1})\mathbb{P}_{k-1}\left(1 + i\theta\eta_k - \theta_2\eta_k^2 + \ldots\right)\\ &\approx D_{k-1}\left(1 + \theta_2 v_{k-1} + \ldots\right)\left(1 - 0 + \theta_2 v_{k-1} + \ldots\right)\\ &\approx D_{k-1} \qquad \text{if } v_{k-1}^2 \text{ is small.}\end{aligned}$$

Averaging over an F in $\mathcal{F}_0$ we get $\mathbb{P}(D_k F) \approx \mathbb{P}(D_{k-1}F)$. Repeated appeals to this approximation give $\mathbb{P}D_n F \approx 1$, from which <15> would follow in the limit as $n \to \infty$.

For a rigorous proof we must pay attention to the remainder terms, and therein lies a small difficulty. Continuity of the sample paths does make all the increments η_k small *with high probability*, but we need slightly more control to ensure that the remainders cause no trouble when we take expectations. A stopping time trick will solve the problem. Here we need to make use of a result for martingales in continuous time:

If $\{(M_t, \mathcal{F}_t) : t \in T\}$ is a martingale with right continuous sample paths, and if τ is a stopping time for the filtration, then $\{(M_{t\wedge\tau}, \mathcal{F}_t) : t \in T\}$ is also a martingale.

The analogous result for discrete-time martingales was established as a Problem in Chapter 6. For continuous time, some extra measurability questions must be settled, and further approximation arguments are required. For details see Appendix E.

For Lévy's theorem, choose the stopping time so that the increments of the stopped process $X_{t\wedge\tau}$ are bounded by a small quantity. Fix an $\epsilon > 0$. For each n in $\mathbb{N}$, define

$$\tau_n := 1 \wedge \inf\{t \in \mathbb{R}^+ : |X_t - X_s| > \epsilon \text{ for some } s \text{ with } t - n^{-1} \le s \le t\}.$$

Each τ_n is a stopping time: by sample path continuity, the set $\{\tau_n \le t\}$ can be written as a countable union of $\mathcal{F}_t$-measurable sets, $\{|X_r - X_{r'}| > \epsilon\}$, with r and r' ranging over pairs of rational numbers from $[0, t]$ with $|r - r'| \le n^{-1}$. Sample path continuity (via uniform continuity on compact time intervals) also implies that, for each ω we have $\tau_n(\omega) = 1$ eventually, that is, for all $n \ge n_0(\omega)$.

For fixed n and ϵ, write Y_t for the martingale $X(t \wedge \tau_n)$. The martingale increments $\xi_k := Y(k/n) - Y((k-1)/n)$ are all bounded in absolute value by ϵ, and from the martingale property of $Y_t^2 - (t \wedge \tau_n)$,

$$V_{k-1} := \mathbb{P}_{k-1}\xi_k^2 = \mathbb{P}_{k-1}\left(\frac{k}{n} \wedge \tau_n - \frac{(k-1)}{n} \wedge \tau_n\right) \le \frac{1}{n}.$$

The conditional variance V_{k-1} is not equal to the constant $1/n$, but it is true that $V_0 + \ldots V_{n-1} \le 1$ and

$$\mathbb{P}(V_0 + \ldots V_{n-1}) = \sum_{i=1}^{n} \mathbb{P}\left(\frac{k}{n} \wedge \tau_n - \frac{k-1}{n} \wedge \tau_n\right) = \mathbb{P}(1 \wedge \tau_n) \to 1 \qquad \text{as } n \to \infty,$$

from which it follows that $1 - (V_0 + \ldots V_{n-1}) \to 0$ in probability.

Replace the D_k defined above by the analogous quantity for the Y process,

$$D_k := \exp\left(i\theta(\xi_1 + \ldots + \xi_k) + \theta_2(V_0 + \ldots + V_{k-1})\right).$$

To keep track of the remainder terms, use the bounds (compare with Problem [8])

$$e^x = 1 + x + A(x) \qquad \text{with } 0 \le A(x) \le x^2 \qquad \text{for } 0 \le x \le 1,$$
$$e^{iy} = 1 + iy - y^2/2 + B(y) \qquad \text{with } |B(y)| \le |y|^3 \text{ for all } y.$$

We still have $D_0 \equiv 1$, but now $\mathbb{P}D_nF = \mathbb{P}F\exp(i\theta Y_1 + \theta_2(V_0 + \ldots + V_{n-1}))$, which, by Dominated Convergence, converges to $\mathbb{P}F\exp(i\theta X_1 + \theta_2)$ as $n \to \infty$. With bounds on the remainder terms, the conditioning argument gives a more precise assertion,

$$\begin{aligned}
\mathbb{P}_{k-1}D_k &= D_{k-1}\exp(\theta_2 V_{k-1})\mathbb{P}_{k-1}\left(1 + i\theta\xi_k - \theta_2\xi_k^2 + B(\theta\xi_k)\right)\\
&= D_{k-1}\left(1 + \theta_2 V_{k-1} + A\left(\theta_2 V_{k-1}\right)\right)\left(1 - \theta_2 V_{k-1} + \mathbb{P}_{k-1}B(\theta\xi_k)\right)\\
&= D_{k-1} + R_k
\end{aligned}$$

where

$$\begin{aligned}
|R_k| &\le |D_{k-1}|\left(\theta_2^2 V_{k-1}^2 + A(\theta V_{k-1})\mathbb{P}_{k-1}|e^{i\theta\xi_k}| + (1 + \theta_2 V_{k-1})\mathbb{P}_{k-1}|B(\theta\xi_k)|\right)\\
&\le \exp(\theta_2)\left((\theta_2^2 + \theta^2)V_{k-1}^2 + (1 + \theta_2)|\theta|^3\mathbb{P}_{k-1}(\epsilon\xi_k^2)\right)\\
&\le C_\theta\left(n^{-1} + \epsilon\right)V_{k-1} \qquad \text{because } \mathbb{P}_{k-1}\xi_k^2 = V_{k-1} \le n^{-1},
\end{aligned}$$

for a constant C_θ that depends on θ. Averaging over an F in $\mathcal{F}_0$ we then get

$$|\mathbb{P}(FD_n) - \mathbb{P}F| \le \sum_{k=1}^n |\mathbb{P}(FD_k) - \mathbb{P}(FD_{k-1})| \le \sum_{k=1}^n \mathbb{P}|R_k| \le C_\theta\left(n^{-1} + \epsilon\right).$$

□ Let n tend to infinity then ϵ tend to zero to complete the argument.

REMARK. The Lévy characterization explains why Brownian motion plays a central role in the theory of Itô diffusions. Roughly speaking, such a diffusion is an adapted process $\{Z_t : t \in \mathbb{R}^+\}$ with continuous sample paths for which

$$\mathbb{P}\left(Z_{t+\delta} - Z_t \mid \mathcal{F}_t\right) \approx \mu(Z_t)\delta \qquad \text{and} \qquad \mathbb{P}\left((Z_{t+\delta} - Z_t)^2 \mid \mathcal{F}_t\right) \approx \sigma^2(Z_t)\delta$$

for small $\delta > 0$, where μ and σ are suitably smooth functions. If we break $[0, t]$ into a union of small intervals $[t_i, t_{i+1}]$, each of length δ, then the standardized increments

$$\Delta_i X := \left(Z(t_{i+1}) - Z(t_i) - \mu(Z_{t_i})\delta\right)/\sigma(Z_{t_i})$$

are martingale differences for which $\mathbb{P}\left((\Delta_i X)^2 \mid \mathcal{F}_{t_i}\right) \approx \delta$. The sum $X_t := \sum_i \Delta_i X$ is a discrete martingale for which $X_t^2 - t$ is also approximately a martingale. It is possible (see, for example, Stroock & Varadhan 1979, Section 4.5) to make these heuristics rigorous, by a formal passage to the limit as δ tends to zero, to build a Brownian motion X from Z, so that $Z_{t+\delta} - Z_t \approx \mu(Z_t)\delta + \sigma(Z_t)(X_{t+\delta} - X_t)$. By summing increments and formalizing another passage to the limit, we then represent Z as a solution to a stochastic integral equation,

$$Z_t = Z_0 + \int_0^t \mu(Z_s)\,ds + \int_0^t \sigma(Z_s)\,dX_s,$$

showing that the diffusion is driven by the Brownian motion X.

The probability theory needed to formalize the heuristics is mostly at the level of the current Chapter. However, if you wish to pursue the idea further it would be better to invest some time in studying systematically the methods of stochastic calculus, and the theory of the Itô stochastic integral, as developed by Stroock & Varadhan (1979), or (in the more general setting of stochastic integrals with respect to martingales) by Chung & Williams (1990). See also the comments at the end of Section 7 for more about stochastic integrals.

*7. Functionals of Brownian motion

Let B be a standard Brownian motion indexed by $[0, 1]$, with $\{\mathcal{F}_t^B : 0 \le t \le 1\}$ its natural filtration. How complicated can $\mathcal{F}_1^B$-measurable random variables be? Remember that these are the random variables expressible as measurable functionals of the whole sample path. The answer for square integrable random variables is quite surprising; and it has remarkable consequences, as you will see in Section 8.

<16> **Theorem.** *The Hilbert space* $\mathbb{H} := L^2(\Omega, \mathcal{F}_1^B, \mathbb{P})$ *equals the closure of the subspace* $\mathbb{H}_0$ *spanned by the constants together with the collection of all random variables of the form* $(B_t - B_s)h_s$, *where* h_s *ranges over all bounded,* $\mathcal{F}_s$*-measurable random variables, and* s, t *range over all pairs for which* $0 \le s < t \le 1$.

Proof. It is enough if we show that a random variable Z in $\mathbb{H}$ that is orthogonal to $\mathbb{H}_0$ is also orthogonal to every bounded random variable of the form $V := g(B_{t_1}, \dots, B_{t_k})$. For then a generating class argument would show that Z must be orthogonal to every random variable in $\mathbb{H}$, from which it would follow that $Z = 0$ almost surely. It even suffices to consider, in place of V, only random variables of the form $U := \prod_{j=1}^k \exp(i\theta_j B_{t_j})$, for real constants θ_j. For if

$$\mathbb{P}Z^+ \prod\nolimits_{j=1}^k \exp(i\theta_j B_{t_j}) = \mathbb{P}Z^- \prod\nolimits_{j=1}^k \exp(i\theta_j B_{t_j})$$

for all real $\{\theta_j\}$, the uniqueness theorem for Fourier transforms implies equality of the measures $\mathbb{Q}^\pm$, on $\sigma\{B_{t_1}, \dots, B_{t_k}\}$, with densities $Z^\pm$ with respect to $\mathbb{P}$, and hence $\mathbb{Q}^+ V = \mathbb{Q}^- V$ for all $V = g(B_{t_1}, \dots, B_{t_k})$.

Repeated application of the following equality (proved below),

<17>
$$\mathbb{P}ZYe^{i\theta B_t} = \mathbb{P}ZYe^{i\theta B_s} \exp\left(-\tfrac{1}{2}\theta^2(t-s)\right)$$

for all real θ, all $s < t$, and all bounded, $\mathcal{F}_s$-measurable random variables Y, will establish orthogonality of Z and U. The argument is easy. First put $Y_1 := \prod_{j=1}^{k-1} \exp(i\theta_j B_{t_j})$ and $\theta := \theta_k$ to deduce

$$\mathbb{P}ZY_1 \exp(i\theta_k B_{t_k}) = \mathbb{P}ZY_1 \exp(i\theta_k B_{t_{k-1}}) \exp\left(-\tfrac{1}{2}\theta_k^2(t_k - t_{k-1})\right).$$

Then repeat the exercise with $Y_2 := \prod_{j=1}^{k-2} \exp(i\theta_j B_{t_j})$ and $\theta := \theta_{k-1} + \theta_k$ to deduce

$$\mathbb{P}ZY_2 \exp(i\theta B_{t_{k-1}}) = \mathbb{P}ZY_2 \exp(i\theta B_{t_{k-2}}) \left(-\tfrac{1}{2}\theta^2(t_{k-1} - t_{k-2})\right).$$

And so on. In effect, we replace successive complex exponentials in B_{t_j} by a nonrandom factors, and replace θ_j by a sum $\sum_{\alpha \ge j} \theta_\alpha$. After k steps, we are left with a product of nonrandom factors multiplied by $\mathbb{P}Z$, which is zero because Z is orthogonal to the constant function 1.

To prove equality <17>, break the interval $[s, t]$ into n subintervals each of length $\delta := (t-s)/n$, and endpoints $s = s_0 < s_1 < \ldots < s_n = t$. Write X_j for $B(s_j)$ and Δ_j for $X_j - X_{j-1}$, which has a $N(0, \delta)$ distribution. Abbreviate the conditional expectation $\mathbb{P}(\cdot \mid \mathcal{F}_{s_j}^B)$ to $\mathbb{P}_j(\cdot)$.

Temporarily write $f(y)$ for $e^{i\theta y}$. Notice that $f'(y) = i\theta f(y)$ and $f''(y) = -2\theta_2 f(y)$ where $\theta_2 := \theta^2/2$. From Problem [8],

$$|f(x+h) - f(x) - hf'(x) - \tfrac{1}{2}h^2 f''(x)| \le |\theta h|^3/6.$$

Substitute $x := X_{j-1}$ and $h := \Delta_j$ to get

$$f(X_j) = f(X_{j-1}) + i\theta\Delta_j f(X_{j-1}) - \theta_2\Delta_j^2 f(X_{j-1}) + R_j \qquad \text{where } |R_j| \le |\theta\Delta_j|^3/6$$
$$= (1-\theta_2\delta)f(X_{j-1}) + i\theta\Delta_j f(X_{j-1}) - \theta_2(\Delta_j^2 - \delta)f(X_{j-1}) + R_j.$$

The random variable $\xi_j := -\theta_2(\Delta_j^2 - \delta)f(X_{j-1})$ is $\mathcal{F}_{s_j}^B$-measurable, with $\mathbb{P}_{j-1}\xi_j = 0$ and $\mathbb{P}|\xi_j|^2 \le C_1\delta^2$ for some constant C_1 depending on θ. Similarly, $\mathbb{P}|R_j|^2 \le C_2\delta^3$. Write γ for the constant $1-\theta_2\delta$. Notice that $\gamma^n = (1-\theta_2(t-s)/n)^n \to \exp(-\theta_2(t-s))$ as $n \to \infty$. Multiply the expansion of $f(X_j)$ by ZY then take expected values. The contribution from $\Delta_j f(X_{j-1})$ is zero because $Y\Delta_j f(X_{j-1}) \in \mathbb{H}_0$. Thus we have a recurrence formula,

$$\mathbb{P}ZYf(X_j) = \gamma\mathbb{P}ZYf(X_{j-1}) + \mathbb{P}ZY\left(\xi_j + R_j\right).$$

Repeated substitutions, starting with $j = n$, give an explicit formula,

$$\mathbb{P}ZYf(X_n) = \gamma^n\mathbb{P}ZYf(X_0) + \mathbb{P}ZY\left(\sum\nolimits_{j=1}^n \gamma^{n-j}(\xi_j + R_j)\right).$$

By Cauchy-Schwarz, the last term is bounded in absolute value by $\|ZY\|_2$ times

$$\|\sum\nolimits_{j=1}^n \gamma^{n-j}\xi_j\|_2 + \|\sum\nolimits_{j=1}^n \gamma^{n-j}R_j\|_2.$$

We may assume that δ is small enough to ensure that $|\gamma| < 1$. Orthogonality of the $\{\xi_j\}$ then gives

$$\|\sum\nolimits_{j=1}^n \gamma^{n-j}\xi_j\|_2^2 \le \sum\nolimits_{j=1}^n \mathbb{P}|\xi_j|^2 = O(n\delta^2) \to 0.$$

The other term is bounded by

$$\sum\nolimits_{j=1}^n \|R_j\|_2 = O(n\delta^{3/2}) \to 0.$$

□ In the limit, as n tends to infinity, we get the asserted equality <17>.

REMARK. Almost the same argument shows that $\mathbb{H}_0$ is dense in $\mathbb{L}_1 := L^1(\Omega, \mathcal{F}_1^B, \mathbb{P})$, under the L^1 norm. The proof rests on the fact that if some element X of $\mathbb{L}_1$ were not in the closure of $\mathbb{H}_0$, there would exist a bounded, $\mathcal{F}_1^B$-measurable random variable Z for which $\mathbb{P}(WZ) = 0$ for all W in $\mathbb{H}_0$ but $\mathbb{P}(XZ) = 1$. (Hahn-Banach plus $(L^1)^* = L^\infty$, for those of you who know some functional analysis.) The argument based on <17> would imply $\mathbb{P}(ZW) = 0$ for all bounded, $\mathcal{F}_1^B$-measurable random variables W, and, in particular, $\mathbb{P}Z^2 = 0$, a contradiction.

The Theorem tells us that for each X in $L^2(\Omega, \mathcal{F}_1^B, \mathbb{P})$ and each $\epsilon > 0$ there is a constant c_0 and grid points $0 = t_0 < t_1 < \ldots < t_{k+1} = 1$ such that

$$X(\omega) = c_0 + \sum\nolimits_{i=0}^k h_i(\omega)\Delta_i B + R_\epsilon(\omega) \qquad \text{where } \Delta_i B := B(t_{i+1}, \omega) - B(t_i, \omega),$$

with each h_i is bounded and $\mathcal{F}_{t_i}$-measurable, and $\mathbb{P}R_\epsilon^2 \le \epsilon^2$. Notice that $|\mathbb{P}X - c_0| = |\mathbb{P}R_\epsilon| \le \epsilon$, so we may as well absorb the difference $c_0 - \mathbb{P}X$ into the remainder, and assume $c_0 = \mathbb{P}X$.

REMARK. The representation takes an even neater form if we encode the $\{h_i\}$ into a single function on $[0,1] \times \Omega$, an ***elementary predictable process***, $H(t,\omega) := \sum_{i=0}^k h_i(\omega)\{t_i < t \le t_{i+1}\}$. The function H is measurable with respect to the predictable sigma-field $\mathcal{P}$, the sub-sigma-field of $\mathcal{B}[0,1] \otimes \mathcal{F}_1^B$ generated by the class of all adapted processes with left-continuous sample paths. Moreover, it is square integrable for the measure $\mu := \mathfrak{m} \otimes \mathbb{P}$, with $\mathfrak{m}$ equal to Lebesgue measure on $[0, 1]$,

$$\mu H^2 = \mathfrak{m}^t \mathbb{P}^\omega \sum_i h_i(\omega)^2 \{t_i < t \le t_{i+1}\} = \sum_i \mathbb{P} h_i^2 (t_{i+1} - t_i).$$

The stochastic integral of H with respect to B is defined as

$$\int_0^1 H\, dB := \sum_{i=0}^k h_i \Delta_i B.$$

The random variable defined in this way is also square integrable,

$$\mathbb{P}\left(\int_0^1 H\, dB\right)^2 = \sum_{i,j} \mathbb{P}\left(h_i h_j (\Delta_i B)(\Delta_j B)\right).$$

For $i < j$, the random variable $\Delta_j B$ is independent of the $\mathcal{F}_{t_{i+1}}$-measurable random variable $h_i h_j(\Delta_i B)$; and for $i = j$, the random variable $(\Delta_i B)^2$ is independent of h_i^2. The cross-product terms disappear, leaving

$$\mathbb{P}\left(\int_0^1 H\, dB\right)^2 = \sum_i \mathbb{P} h_i^2 \mathbb{P}(\Delta_i B)^2 = \mu H^2.$$

Thus the map $H \mapsto \int_0^1 H\, dB$ is an isometry from the space $\mathcal{H}_0$ of all elementary predictable processes into $\mathbb{L}_2 := L^2(\Omega, \mathcal{F}_1^B, \mathbb{P})$. It is not too difficult to show that $\mathcal{H}_0$ is dense in the space $\mathcal{H} := L^2\left([0, 1] \times \Omega, \mathcal{P}, \mu\right)$. The isometry therefore extends to a map from $\mathcal{H}$ into $\mathbb{L}_2$, which defines the stochastic integral $\int_0^1 H\, dB$ for all predictable processes that are square-integrable with respect to μ.

Theorem <16> implies that for each X in $\mathbb{L}_2$ there exists a sequence $\{H_n\}$ in $\mathcal{H}_0$ for which $\mathbb{P}\left|X - \mathbb{P}X - \int_0^1 H_n\, dB\right|^2 \to 0$. In consequence, $\{H_n\}$ is a Cauchy sequence,

$$\mu|H_n - H_m|^2 = \mathbb{P}\left|\int_0^1 (H_n - H_m)\, dB\right|^2 \to 0 \qquad \text{as } \min(m, n) \to \infty.$$

Completeness ensures existence of an H in $\mathcal{H}$ for which

$$\mathbb{P}\left|\int_0^1 H_n\, dB - \int_0^1 H\, dB\right|^2 = \mu|H_n - H|^2 \to 0.$$

In summary: $X = \mathbb{P}X + \int_0^1 H\, dB$ almost surely, a most elegant representation.

*8. Option pricing

The lognormal model of stock prices assumes that the price of a particular stock (standardized to take the value 1 at time 0) at time t is given by

$$S_t = \exp\left(\sigma B_t + \left(\mu - \tfrac{1}{2}\sigma^2\right) t\right) \qquad \text{for } 0 \le t \le 1, \tag{<18>}$$

where B is a standard Brownian motion. The drift parameter μ is unknown, but the volatility parameter σ is known (or is at least assumed to be well estimated). Notice that S is a martingale, for the filtration $\mathcal{F}_t := \sigma\{B_s : s \le t\} = \sigma\{S_s : s \le t\}$, if $\mu = 0$.

The strange form of the parametrization makes more sense if we consider relative increments in stock price over a small time interval,

$$\begin{aligned}\frac{S_{t+\delta} - S_t}{S_t} &= \exp\left(\left(\mu - \tfrac{1}{2}\sigma^2\right)\delta + \sigma\Delta B\right) - 1 \qquad \text{where } \Delta B := B_{t+\delta} - B_t \\ &= \left(\mu - \tfrac{1}{2}\sigma^2\right)\delta + \sigma\Delta B + \tfrac{1}{2}\left(\left(\mu - \tfrac{1}{2}\sigma^2\right)\delta + \sigma\Delta B\right)^2 + \dots \\ &= \mu\delta + \sigma\Delta B + \tfrac{1}{2}\sigma^2\left((\Delta B)^2 - \delta\right) + \dots\end{aligned}$$

The square $(\Delta B)^2$ has expected value δ. The term $-\sigma^2\delta/2$ centers it to have zero mean. As you will see, the centered variable eventually gets absorbed into a small error term, leaving $\mu\delta + \sigma\Delta B$ as the main contribution to the relative changes in stock price over short time intervals. The model is sometimes written in symbolic form as $dS_t = \mu S_t dt + \sigma S_t dB_t$.

An option may be thought of as a random variable Y that is, potentially, a function of the entire history $\{S_t : 0 \le t \le 1\}$ of the stock price: ones pays an amount y_0 at time 0 in return for the promise of a return Y at time 1 that depends on the performance of the stock. (That is, an option is a refined form of gambling on the stock market.)

The question whose answer makes investment bankers rich is: What is the appropriate price y_0? The elegant answer is: $y_0 = \mathbb{Q}Y$, where $\mathbb{Q}$ is a probability measure on $\mathcal{F}$ that makes the stock price a martingale, because (as you will soon learn) there are trading schemes whose net returns can be made as close to $Y - y_0$ as we please, in a probabilistic sense. If a trader offered the option for a price y smaller than y_0, one could buy the option and also engage in a trading scheme for a net return abitrarily close to $(Y - y) - (Y - y_0) = y_0 - y > 0$. A similar argument can be made against an asking price greater than y_0.

A trading scheme consists of a finite set of times $0 \le t_0 < t_1 < \ldots < t_{k+1} \le 1$ at which shares need to be traded: at time t_i buy a quantity K_i of the stock, at the cost $K_i S(t_i)$, then sell at time t_{i+1} for an amount $K_i S(t_{i+1})$, for a return of $K_i \Delta_i S$, where $\Delta_i S = S(t_{i+1}) - S(t_i)$ denotes the change in stock price per share over the time interval. The quantity K_i must be determined by the information available at time t_i, that is, K_i must be $\mathcal{F}_{t_i}$-measurable. We should also allow K_i to take negative values, a purchase of a negative quantity being a sale and a sale of a negative quantity being a purchase. The return from the trading scheme is $\sum_{i=0}^{k} K_i \Delta_i S$.

We could assume that the spacing of the times between trades is as small as we please. For example, purchase of K shares at time s followed by resale at time t has the same return as purchases of K at times $s + i\delta$ and sales of K at times $s + (i + 1)\delta$, for $i = 0, 1, \ldots, N - 1$, with $\delta := (t - s)/N$ for an arbitrarily large N. Conceptually, we could even pass to the mathematical limit of continuous trading, in which case the errors of approximation could be driven to zero (almost surely).

Existence of the trading scheme to nearly duplicate the return $Y - y_0$ will follow from the representation in the previous Section. Consider first the case where $\mu = 0$, which makes S a martingale. Let us assume that $\mathbb{P}Y^2$ is finite. Write y_0 for $\mathbb{P}Y$. For an arbitrarily small $\epsilon > 0$, Theorem <16> gives a finite collection of times $\{t_i\}$ and bounded, $\mathcal{F}_{t_i}$-measurable random variables h_i for which

$$\mathbb{P}\left|Y - y_0 - \sum_i h_i \Delta_i B\right|^2 < \epsilon^2 \qquad \text{where } \Delta_i B := B(t_{i+1}) - B(t_i).$$

If we could replace $\Delta_i B$ by $\Delta_i S / \left(\sigma S_{t_i}\right)$ for each i we would have a trading scheme $K_i := h_i/(\sigma S_{t_i})$ with the desired approximation property. If the time intervals $\delta_i := t_{i+1} - t_i$ are all small enough, a simple calculation will show that such a substitution increases the L^2 error only slightly. The error term

$$R_{i+1} := \frac{\Delta_i S}{S_{t_i}} - \sigma\Delta_i B = \exp\left(\sigma\Delta_i B - \tfrac{1}{2}\sigma^2\delta_i\right) - 1 - \sigma\Delta_i B$$

has zero expected value and it is independent of $\mathcal{F}_{t_i}$. You will see soon that it also has a small $L^2(\mathbb{P})$ norm. To simplify the algebra, temporarily write τ for $\sigma\sqrt{\delta}$ and Z for $\Delta_i B/\sqrt{\delta}$, which has a standard normal distribution. Then

$$\begin{aligned}
\mathbb{P}R_{i+1}^2 &= \mathbb{P}\left(\exp\left(\tau Z - \tfrac{1}{2}\tau^2\right) - 1 - \tau Z\right)^2 \\
&= \mathbb{P}\exp\left(2\tau Z - \tau^2\right) + 1 + \tau^2 - 2\mathbb{P}\left((1+\tau Z)\exp\left(\tau Z - \tfrac{1}{2}\tau^2\right)\right) \\
&= \exp(\tau^2) - 1 - \tau^2 \qquad \text{cf. } \tau e^{\tau^2/2} = \frac{\partial}{\partial\tau}e^{\tau^2/2} = \frac{\partial}{\partial\tau}\mathbb{P}e^{\tau Z} = \mathbb{P}\left(Ze^{\tau Z}\right) \\
&\le \tau^4 \qquad \text{when } \tau \le 1.
\end{aligned}$$

The approximation $\sum_i h_i\Delta_i B$ to $Y - y_0$ equals $\sum_i K_i\Delta_i S - \sum_i h_i R_{i+1}/\sigma$. The first sum represents the net return from a trading strategy. The other sum will be small in $L^2(\mathbb{P})$ when $\max_i \delta_i$ is small. Independence eliminates cross-product terms,

$$\mathbb{P}\left(h_i R_{i+1} h_j R_{j+1}\right) = \mathbb{P}\left(h_i R_{i+1} h_j\right)\left(\mathbb{P}R_{j+1}\right) = 0 \qquad \text{if } j > i,$$

and hence

$$\begin{aligned}
\mathbb{P}\left|\sum\nolimits_i h_i R_{i+1}\right|^2 &= \sum\nolimits_i \mathbb{P}\left|h_i R_{i+1}\right|^2 \\
&\le \sum\nolimits_i \left(\mathbb{P}h_i^2\right)\left(\sigma^4\delta_i^2\right) \qquad \text{if } \max_i \delta_i \le 1/\sigma^2 \\
&\le \sigma^4\left(\max\nolimits_i \delta_i\right)\sum\nolimits_i \left(\mathbb{P}h_i^2\right)\mathbb{P}(\Delta_i B)^2 \\
&\le \sigma^4\left(\max\nolimits_i \delta_i\right)\left\|\sum\nolimits_i h_i\Delta_i B\right\|_2^2 \to 0 \qquad \text{as } \max_i \delta_i \to 0.
\end{aligned}$$

The contribution from the R_i's can be absorbed into the other error of approximation, increasing the ϵ by an arbitrarily small amount, leaving the desired trading strategy approximation for $Y - y_0$.

Finally, what happens when μ is not zero? A change of measure will dispose of its effects. For a fixed constant α, let $\mathbb{Q}$ be the probability measure on $\mathcal{F}$ with density $\exp(\alpha B_1 - \alpha^2/2)$ with respect to $\mathbb{P}$. Problem [11] shows that the process $X_t := B_t - \alpha t$ is a Brownian motion under $\mathbb{Q}$. The stock price is also a function of X if we choose $\alpha = -\mu/\sigma$,

$$S_t = \exp\left(\sigma(X_t + \alpha t) + \left(\mu - \tfrac{1}{2}\sigma^2\right)t\right) = \exp\left(\sigma X_t - \tfrac{1}{2}\sigma^2 t\right) \qquad \text{for } 0 \le t \le 1.$$

If we replace $\mathbb{P}$ by $\mathbb{Q}$, and the Brownian motion B by the Brownian motion X, then a repeat of the argument for the case $\mu = 0$ shows that the appropriate price for the option is now $y_0 = \mathbb{Q}Y$.

REMARK. Of course there is a hidden assumption that Y is $\mathbb{Q}$ square-integrable, or just $\mathbb{Q}$-integrable if you accept the Remark following Theorem <16>.

9. Problems

[1] Let Z_i have a $N(0, \sigma_i^2)$ distribution, for $i = 1, 2, \ldots, n$. Prove that

$$\mathbb{P}\max_i |Z_i| \le \sqrt{\mathbb{P}\max_i |Z_i|^2} \le 2\sigma\sqrt{1 + \log n} \qquad \text{where } \sigma := \max_i \sigma_i,$$

Hint: Argue via Jensen's inequality that

$$\exp\left(\mathbb{P}\max_i |Z_i|^2/4\sigma^2\right) \le \mathbb{P}\exp\left(\max_i |Z_i|^2/4\sigma^2\right) \le \sum_i \mathbb{P}\exp\left(|Z_i|^2/4\sigma_i^2\right).$$

Bound the right-hand side by $n\sqrt{2}$, then take logarithms.

REMARK. For Problem [1] we do not need to assume that the $\{Z_i\}$ have a multivariate normal distribution. Nowhere in the argument do we need to know anything about the joint distribution. When the Z_i are independent, or nearly so, the bound is quite good. For the extreme case of n independent $N(0, \sigma^2)$ variables, the inequality from part (i) of Problem [2] shows that $\mathbb{P}\max_{i\le n} |Z_i| \ge C_0\sigma\sqrt{1+\log n}$ for a universal constant C_0. See Section 12.2 for a precise way—Sudakov's minoration—of capturing the idea of approximate independence when the $\{Z_i\}$ have a multivariate normal distribution. When the variables are highly dependent, the bound is not good: in the extreme cases where $Z_i \equiv Z_1$, which has a $N(0, \sigma^2)$ distribution, $\mathbb{P}\max_{i\le n} |Z_i| = \sigma\mathbb{P}|N(0,1)|$, which does not increase with n.

[2] Let $Z_1, Z_2, \ldots,$ be independent random variables, each distributed $N(0,1)$. Define $M_n := \max_{i\le n} |Z_i|$ and $\ell_n := \sqrt{2\log n}$ and $n(k) := 2^k$. Use the tail bound from Appendix D,

$$(2\pi)^{-1/2}\left(\frac{1}{x} - \frac{1}{x^3}\right)\exp\left(-\frac{x^2}{2}\right) \le \mathbb{P}\{Z_1 > x\} \le \tfrac{1}{2}\exp\left(-\frac{x^2}{2}\right) \qquad \text{for } x > 0,$$

to prove that $M_n/\sqrt{2\log n} \to 1$ almost surely. Argue as follows.

(i) For each small $\epsilon > 0$, show that there exist strictly positive constants C and θ for which $\mathbb{P}\{M_n \le (1-\epsilon)\ell_n\} = \left(1 - 2\mathbb{P}\{Z_1 > (1-\epsilon)\ell_n\}\right)^n \le \exp\left(-Cn^\theta\right)$, for all n large enough. Deduce that $\liminf_n M_n/\ell_n \ge 1$ almost surely.

(ii) For each $\epsilon > 0$, show that $\mathbb{P}\{M_n \ge (1+\epsilon)\ell_n\} \le n/n^{(1+\epsilon)^2}$. Deduce that $\limsup_k M_{n(k)}/\ell_{n(k)} \le 1$ almost surely.

(iii) Use the inequality $M_n/\ell_n \le M_{n(k+1)}/\ell_{n(k)}$ for $n(k) < n \le n(k+1)$, and the result from parts (i) and (ii), to conclude that $M_n/\ell_n \to 1$ almost surely.

[3] Let $\{(B_t, \mathcal{F}_t) : 0 \le t \le 1\}$ be a centered Brownian motion whose sample paths need not be continuous. Show that there exists a centered Brownian motion $\{B_t^* : 0 \le t \le 1\}$ with continuous sample paths, for which $\mathbb{P}\{B_t^* \ne B_t\} = 0$ for every t. Argue as follows. Let $S_n := \{i/2^n : i = 0, 1, \ldots, 2^n\}$ and $S := \cup_{n\in\mathbb{N}} S_n$. For each ω define

$$D_n(\omega) = \max_{0\le i<2^n} \left| B\left(\frac{i+1}{2^n}, \omega\right) - B\left(\frac{i}{2^n}, \omega\right) \right|$$

(i) Use Problem [1] to prove that $\sum_n \mathbb{P}D_n < \infty$. Deduce that there exists a negligible set N such that $\epsilon_n(\omega) := \sum_{i\ge n} D_i(\omega) \to 0$ as $n \to \infty$ for $\omega \notin N$.

(ii) Let s and t be points in S_n for which $|s - t| \le 2^{-m}$, where $n > m$. For each k, write s_k for the largest value in S_k for which $s_k \le s$, and define t_k similarly. (Thus $s = s_n$ and $t = t_n$.) Show that

$$|B(s) - B(t)| \le \sum_{i=m}^{n-1} |B(s_{i+1}) - B(s_i)| + |B(s_m) - B(t_m)| + \sum_{i=m}^{n-1} |B(t_{i+1}) - B(t_i)|.$$

(iii) Deduce that

$$\max\{|B(s,\omega) - B(t,\omega)| : s,t \in S_n \text{ and } |s-t| \le 2^{-m}\} \le 2\epsilon_m(\omega),$$

and hence

$$\sup\{|B(s,\omega) - B(t,\omega)| : s,t \in S \text{ and } |s-t| \le 2^{-m}\} \le 2\epsilon_m(\omega).$$

(iv) For each t in $[0,1]$ and each $\omega \in N^c$, define

$$B^*(t,\omega) = \lim_{n\to\infty} \sup\{B_s : s \in S \text{ and } t - 2^{-n} \le s \le t\}.$$

Define $B^*(t,\omega)$ to be zero if $\omega \in N$. (Notice that B_t^* is measurable with respect to the completion $\mathcal{F}_t^*$.) Show that, for all t, t' in $[0,1]$,

$$|B^*(t,\omega) - B^*(t',\omega)| \le 2\epsilon_m(\omega) \qquad \text{if } |t-t'| < 2^{-m}.$$

(v) Show that $B^*(t,\omega) = \lim_n B(t_n,\omega)$ almost surely, where $\{t_n\}$ is the sequence defined in step (ii). For each $\epsilon > 0$, deduce that

$$\mathbb{P}\{|B_t^* - B_t| > \epsilon\} \le \mathbb{P}\liminf_n\{|B(t_n) - B_t| > \epsilon\} \le \liminf_n \mathbb{P}\{|B(t_n) - B_t| > \epsilon\} = 0.$$

(Remember that $B(t) - B(t_n)$ has a $N(0, t-t_n)$ distribution.) Conclude that $B_t^* = B_t$ almost surely, for each t.

[4] Suppose $\{B_s : s \in S\}$ is a Brownian motion, with S the countable set of all dyadic rationals in $[0,1]$, as in the previous Problem. Modify the argument from that Problem to construct a standard Brownian motion indexed by $[0,1]$.

[5] Show that there exists a finite constant C_0 for which

$$\sum_{k=0}^{\infty} 2^{k/2}\left(\delta \wedge 2^{-k}\right)\sqrt{1+\log\left(2^k\right)} \le C_0\sqrt{\delta + \delta\log(1/\delta)} \qquad \text{for } 0 \le \delta \le 1.$$

Hint: For $2^{-m} \ge \delta \ge 2^{-m-1}$, bound the sum by

$$\delta\sqrt{1+\log\left(2^m\right)}\sum_{k=0}^{m} 2^{k/2} + \sum_{k>m} 2^{-k/2}\sqrt{1+\log\left(2^k\right)}.$$

The ratio of successive terms in the last sum converges to $1/\sqrt{2}$.

[6] Let B be a standard Brownian motion indexed by $[0,1]$. Show that the process $X_t = B_1 - B_{1-t}$, for $0 \le t \le 1$, is also a Brownian motion (with respect to own its natural filtration, not the filtration for B).

[7] Let $\{(B_t, \mathcal{F}_t) : 0 \le t \le 1\}$ be a Brownian motion with continuous sample paths. Define $\tau(\omega)$ as the smallest t at which the sample path $B(\cdot,\omega)$ achieves its maximum value, $M(\omega)$. (Note: τ is not a stopping time.) Follow these steps to show that the sets $\Omega_t := \{\tau = t\}$ for $0 \le t \le 1$ is a family suitable for the construction in Example <2>.

(i) Show that $\{\tau \le t\} = \{\sup_{s \le t} B_s(\omega) = M(\omega)\}$. Deduce that τ is $\mathcal{F}_1$-measurable. Hint: Work with the process at rational times.

(ii) For each t in $[0,1)$, show that $\{\tau = t\} \subseteq \{\sup_{t \le s \le 1}(B_s - B_t) \le 0\}$. Deduce via the result from Exercise <12> that $\mathbb{P}\{\tau = t\} = 0$.

(iii) Use Problem [6] to show that $\mathbb{P}\{\tau = 1\} = 0$.

[8] For all real y, show that $e^{iy} = 1 + iy + (iy)^2/2! + \ldots + (iy)^k/k! + B_k(y)$ with $|B_k(y)| \le |y|^{k+1}/(k+1)!$. Hint: for $y > 0$ use the fact that $i \int_0^y B_k(t)\,dt = B_{k+1}(y)$.

[9] Let $\mathcal{X} := C(\mathbb{R}^+)$, equipped with its metric d for uniform convergence on compacta.

(i) Show that $(\mathcal{X}, d)$ is separable. Hint: Consider the countable collection of piecewise linear functions obtained by interpolating between a finite number of "vertices" with rational coordinates.

(ii) Prove that the sigma-field $\mathcal{C}$ generated by the finite dimensional projections coincides with the Borel sigma-field $\mathcal{B}(\mathcal{X})$. Hint: For one inclusion use continuity of the projections. For the other inclusion replace sup-norm distances by suprema over subsets of rationals.

[10] Let $\{B_t : t \in \mathbb{R}^+\}$ be a standard Brownian motion. For fixed constants $\alpha > 0$ and $\beta > 0$ show that $\mathbb{P}\{B_t$ eventually hits the line $\alpha + \beta t\} = \exp(-2\alpha\beta)$, by following these steps. Write τ for the first hitting time on the linear barrier. For fixed real θ, let $X_\theta(t)$ denote the martingale $\exp(\theta B_t - \frac{1}{2}\theta^2 t)$.

(i) For each fixed t, show that $1 = \mathbb{P}X_\theta(t \wedge \tau)$. Hint: You might consult Appendix E if you wish to be completely rigorous.

(ii) If $\theta > 2\beta$, show that $0 \le X_\theta(t \wedge \tau) \le \exp(\theta\alpha)$.

(iii) If $\theta > 2\beta$, show that $X_\theta(t \wedge \tau) \to 0$ as $t \to \infty$ on the set $\{\tau = \infty\}$.

(iv) Deduce that $1 = \mathbb{P}\exp\left(\theta\alpha + (\theta\beta - \frac{1}{2}\theta^2)\tau\right)\{\tau < \infty\}$ for each $\theta > 2\beta$. Then let θ decrease to 2β to conclude the argument.

[11] Let $\{(B_t, \mathcal{F}_t) : 0 \le t \le 1\}$ be a Brownian motion defined on a probability space $(\Omega, \mathcal{F}, \mathbb{P})$. Let α be a positive real number.

(i) Show that the measure $\mathbb{Q}$ defined by $d\mathbb{Q}/d\mathbb{P} := \exp(\alpha B_1 - \frac{1}{2}\alpha^2)$ is a probability measure on $\mathcal{F}$.

(ii) Show that the process $X_t := B_t - \alpha t$ for $0 \le t \le 1$ is a Brownian motion under $\mathbb{Q}$. Hint: For fixed $s < t$, real θ, and F in $\mathcal{F}_s$, show that

$$\begin{aligned}\mathbb{Q}\exp\left(i\theta(X_t - X_s) + \tfrac{1}{2}\theta^2(t-s)\right)F \\ = \mathbb{P}\exp\left(\alpha(B_1 - B_t) - \tfrac{1}{2}\alpha^2(1-t)\right) \\ \times \mathbb{P}\exp\left((\alpha + i\theta)(B_t - B_s) - \tfrac{1}{2}(\alpha + i\theta)^2(t-s)\right) \\ \times \mathbb{P}\exp\left(\alpha B_s - \tfrac{1}{2}\alpha^2 s\right) F.\end{aligned}$$

Notice that right-hand side reduces to an expression that does not depend on θ, which identifies it as $\mathbb{Q}F$.

[12] Calculate the price of an option that allows you to purchase one unit of stock for a price K at time 1, if you wish. Hint: Interpret $(S_1 - K)^+$, or read a book about options.

10. Notes

The detailed study by Brush (1968) describes the history of Brownian motion: from the recognition by Brown, in 1828, that it represented a physical, rather than biological, phenomenon; through the mathematical theories of Einstein and Smoluchowski, and the experimental evidence of Perrin, in the first decade of the twentieth century.

Apparently Wiener (1923) was motivated to study the irregularity of the Brownian motion sample paths, in part, by remarks of Perrin regarding the haphazard motion of small particles, cited (page 133) in translation as "One realizes from such examples how near the mathematicians are to the truth in refusing, by a logical instinct, to admit the pretended geometrical demonstrations, which are regarded as experimental evidence for the existence of a tangent at each point of a curve." (Compare with Kac (1966, page 34), quoting from Wiener's autobiography.) Wiener constructed "Wiener measure" as a linear functional on the space of continuous functions, deriving the necessary countable additivity property from a property slightly weaker than the modulus condition of Theorem <7>.

The construction in Section 3 is essentially due to Lévy (1939, Section 6), who obtained the piecewise linear approximations to Brownian motion by an interpolation argument (compare with Theorem 1, Chapter 1 of Lévy 1948). He also referred to Lévy (1937, page 172) for the derivation of a Hölder condition <6> for the sample paths. The explicit construction via an orthogonal series expansion in the Haar basis is due to to Ciesielski (1961).

Theorem <14> in Section 6 is due to Lévy (1948, pages 77–78 of the second edition), who merely stated the result, with a reference to Theorem 67.3 of Lévy (1937), which established a central limit theorem for (discrete time) martingales under an assumption on the conditional variances analogous to the martingale property for $X_t^2 - t$. Doob (1953, Theorem 11.9, Chapter VII) provided a formal proof, similar to the proof in Section 6. The method could be streamlined slightly, with replacement of increments over deterministic time intervals by increments over random intervals. It could also be deduced from fancier results in stochastic calculus, whose derivations ultimately reduce to calculations with small increments of the process. I feel the direct method has pedagogic advantages, because it makes very clear the vital role of sample path continuity.

The proof of Theorem <16> is a disguised version of Itô's formula for stochastic integrals, applied to a particular function of Brownian motion—compare with Durrett (1984, Section 2.14). The result is due to Kunita & Watanabe (1967), although, as noted by Clark (1970), it follows easily from an expansion due to Itô (1950). See also Dudley (1977) for an extension to $\mathcal{F}_1^B$-measurable random variables that are not necessarily square integrable, verifying a result first asserted then retracted by Clark.

See Harrison & Pliska (1981) and Duffie (1992, Chapter 6) for a discussion of stochastic calculus and option trading. The book by Wilmott, Howison & Dewynne (1995) provides a gentler introduction to some of the finance ideas.

In the last three Sections of the Chapter, I was dabbling with ideas important in stochastic calculus, without introducing the formal machinery of the subject. Accordingly, there was some repetition of methods—chop the sample paths into small increments; make Taylor expansions; dispose of remainder terms by arguments reeking of martingale theory. As I remarked at the end of Section 6, if you wish to pursue the theory any further it would be a good idea to invest some time in studying the formal machinery. I found Chung & Williams (1990) a good place to start, with Métivier (1982) as a reliable backup, and Dellacherie & Meyer (1982) as a rigorous test of true understanding.

References

Brush, S. G. (1968), 'A history of random processes: I. Brownian movement from Brown to Perrin', *Archive for History of the Exact Sciences* pp. 1–36.

Chung, K. L. & Williams, R. J. (1990), *Introduction to Stochastic Integration*, Birkhäuser, Boston.

Ciesielski, Z. (1961), 'Holder condition for realization of Gaussian processes', *Transactions of the American Mathematical Society* **99**, 403–413.

Clark, J. M. C. (1970), 'The representation of functionals of Brownian motion by stochastic integrals', *Annals of Mathematical Statistics* **41**, 1282–1295. Correction, *ibid.* 42 (1971), 1778.

Dellacherie, C. & Meyer, P. A. (1982), *Probabilities and Potential B: Theory of Martingales*, North-Holland, Amsterdam.

Doob, J. L. (1953), *Stochastic Processes*, Wiley, New York.

Dudley, R. M. (1973), 'Sample functions of the Gaussian process', *Annals of Probability* **1**, 66–103.

Dudley, R. M. (1977), 'Wiener functionals as Itô integrals', *Annals of Probability* **5**, 140–141.

Duffie, D. (1992), *Dynamic Asset Pricing Theory*, Princeton Univeristy Press.

Durrett, R. (1984), *Brownian Motion and Martingales in Analysis*, Wadsworth, Belmont CA.

Harrison, J. M. & Pliska, S. R. (1981), 'Martingales and stochastic integrals in the theory of continuous trading', *Stochastic Processes and their Applications* **11**, 215–260.

Itô, K. (1950), 'Multiple Wiener integral', *J. Math. Society Japan* **3**, 158–169.

Kac, M. (1966), 'Wiener and integration in function spaces', *Bulletin of the American Mathematical Society* **72**, 52–68. One of several articles in a special issue of the journal, devoted to the life and work of Norbert Wiener.

Kunita, H. & Watanabe, S. (1967), 'On square integrable martingales', *Nagoya Math. J.* **30**, 209–245.

Lévy, P. (1937), *Théorie de l'addition des variables aléatoires*, Gauthier-Villars, Paris. References from the 1954 second edition.

Lévy, P. (1939), 'Sur certaines processus stochastiques homogènes', *Compositio mathematica* **7**, 283–339.

Lévy, P. (1948), *Processus stochastiques et mouvement brownien*, Gauthier-Villars, Paris. Second edition, 1965.

McKean, H. P. (1969), *Stochastic Integrals*, Academic Press.

Métivier, M. (1982), *Semimartingales: A Course on Stochastic Processes*, De Gruyter, Berlin.

Stroock, D. W. & Varadhan, S. R. S. (1979), *Multidimensional Diffusion Processes*, Springer, New York.

Wiener, N. (1923), 'Differential-space', *Journal of Mathematics and Physics* **2**, 131–174. Reprinted in *Selected papers of Norbert Wiener*, MIT Press, 1964.

Wilmott, P., Howison, S. & Dewynne, J. (1995), *The Mathematics of Financial Derivatives: a Student Introduction*, Cambridge University Press.

Chapter 10

Representations and couplings

SECTION 1 illustrates the usefulness of coupling, by means of three simple examples.

SECTION 2 describes how sequences of random elements of separable metric spaces that converge in distribution can be represented by sequences that converge almost surely.

*SECTION *3 establishes Strassen's Theorem, which translates the Prohorov distance between two probability measures into a coupling.*

*SECTION *4 establishes Yurinskii's coupling for sums of independent random vectors to normally distributed random vectors.*

SECTION 5 describes a deceptively simple example (Tusnády's Lemma) of a quantile coupling, between a symmetric Binomial distribution and its corresponding normal approximation.

SECTION 6 uses the Tusnády Lemma to couple the Haar coefficients for the expansions of an empirical process and a generalized Brownian Bridge.

SECTION 7 derives one of most striking results of modern probability theory, the KMT coupling of the uniform empirial process with the Brownian Bridge process.

1. What is coupling?

A coupling of two probability measures, P and Q, consists of a probability space $(\Omega, \mathcal{F}, \mathbb{P})$ supporting two random elements X and Y, such that X has distribution P and Y has distribution Q. Sometimes interesting relationships between P and Q can be coded in some simple way into the joint distribution for X and Y. Three examples should make the concept clearer.

<1> **Example.** Let P_α denote the Bin(n, α) distribution. As α gets larger, the distribution should "concentrate on bigger values." More precisely, for each fixed x, the tail probability $P_\alpha[x, n]$ should be an increasing function of α. A coupling argument will give an easy proof.

Consider a β larger than α. Suppose we construct a pair of random variables, X_α with distribution P_α and X_β with distribution P_β, such that $X_\alpha \le X_\beta$ almost surely. Then we will have $\{X_\alpha \ge x\} \le \{X_\beta \ge x\}$ almost surely, from which we would recover the desired inequality, $P_\alpha[x, n] \le P_\beta[x, n]$, by taking expectations with respect to $\mathbb{P}$.

How might we construct the coupling? Binomials count successes in independent trials. Couple the trials and we couple the counts. Build the trials from independent random variables U_i, each uniformly distributed on $(0, 1)$. That is, define $X_\alpha := \sum_{i\le n}\{U_i \le \alpha\}$ and $X_\beta := \sum_{i\le n}\{U_i \le \beta\}$. In fact, the construction couples all P_γ, for $0 \le \gamma \le 1$, simultaneously. □

<2> **Example.** Let P denote the $\text{Bin}(n, \alpha)$ distribution and Q denote the approximating $\text{Poisson}(n\alpha)$ distribution. A coupling argument will establish a total variation bound, $\sup_A |PA - QA| \le n\alpha^2$, an elegant means for expressing the Poisson approximation to the Binomial.

Start with the simplest case, where n equals 1. Find a probability measure $\mathbb{P}$ concentrated on $\{0, 1\} \times \mathbb{N}_0$ with marginal distributions $P := \text{Bin}(1, \alpha)$ and $Q := \text{Poisson}(\alpha)$. The strategy is simple: put as much mass as we can on the diagonal, $(0, 0) \cup (1, 1)$, then spread the remaining mass as needed to get the desired marginals. The atoms on the diagonal are constrained by the inequalities

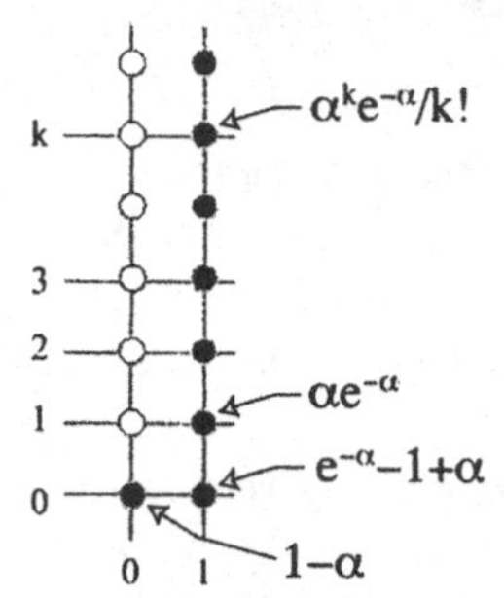

$$\mathbb{P}(0, 0) \le \min\left(P\{0\}, Q\{0\}\right) = \min\left(1 - \alpha, e^{-\alpha}\right),$$
$$\mathbb{P}(1, 1) \le \min\left(P\{1\}, Q\{1\}\right) = \min\left(\alpha, \alpha e^{-\alpha}\right).$$

To maximize, choose $\mathbb{P}(0, 0) := 1 - \alpha$ and $\mathbb{P}(1, 1) := \alpha e^{-\alpha}$. The rest is arithmetic. We need $\mathbb{P}(1, 0) := e^{-\alpha} - 1 + \alpha$ to attain the marginal probability $Q\{0\}$, and $\mathbb{P}(0, k) := 0$, for $k = 1, 2, \ldots$, to attain the marginal $P\{0\} = 1 - \alpha$. The choices $\mathbb{P}(1, k) := Q\{k\}$, for $k = 2, 3, \ldots$, are then forced. The total off-diagonal mass equals $\alpha - \alpha e^{-\alpha} \le \alpha^2$.

For the general case, take $\mathbb{P}$ to be the n-fold product of measures of the type constructed for $n = 1$. That is, construct n independent random vectors $(X_1, Y_1), \ldots, (X_n, Y_n)$ with each X_i distributed $\text{Bin}(1, \alpha)$, each Y_i distributed $\text{Poisson}(\alpha)$, and $\mathbb{P}\{X_i \ne Y_i\} \le \alpha^2$. The sums $X := \sum_i X_i$ and $Y := \sum_i Y_i$ then have the desired Binomial and Poisson distributions, and $\mathbb{P}\{X \ne Y\} \le \sum_i \mathbb{P}\{X_i \ne Y_i\} \le n\alpha^2$. The total variation bound follows from the inequality

$$|\mathbb{P}\{X \in A\} - \mathbb{P}\{Y \in A\}| = |\mathbb{P}\{X \in A, X \ne Y\} - \mathbb{P}\{Y \in A, X \ne Y\}| \le \mathbb{P}\{X \ne Y\},$$

for every subset A of integers. □

The first Example is an instance of a general method for coupling probability measures on the real line by means of quantile functions. Suppose P has distribution function F and Q has distribution function G, with corresponding quantile functions q_F and q_G. Remember from Section 2.9 that, for each $0 < u < 1$,

$$u \le F(x) \qquad \text{if and only if} \qquad q_F(u) \le x.$$

In particular, if U is uniformly distributed on $(0, 1)$ then

$$\mathbb{P}\{q_F(U) \le x\} = \mathbb{P}\{U \le F(x)\} = F(x),$$

so that $X := q_F(U)$ must have distribution P. We couple P with Q by using the same U to define the random variable $Y := q_G(U)$ with distribution Q.

A slight variation on the quantile coupling is available when G is one-to-one with range covering the whole of $(0, 1)$. In that case, q_G is a true inverse function for G, and $U = G(Y)$. The random variable $X := q_F G(Y)$ is then an increasing function of Y, a useful property. Section 5 will describe a spectacularly successful example of a quantile coupling expressed in this form.

<3> **Example.** Suppose $\{P_n\}$ is a sequence of probability measures on the real line, for which $P_n \rightsquigarrow P$. Write F_n and F for the corresponding distribution functions, and q_n and q for the quantile functions. From Section 7.1 we know that $F_n(x) \to F(x)$ at each x for which $P\{x\} = 0$, which implies (Problem [1]) that $q_n(u) \to q(u)$ at Lebesgue almost all u in $(0, 1)$. If we use a single U, distributed uniformly on $(0, 1)$, to construct the variables $X_n := q_n(U)$ and $X := q(U)$, then we have $X_n \to X$ almost surely. That is we have represented the weakly convergent sequence of measures by an almost surely convergent sequence of random variables.

REMARK. It might happen that the measures $\{P_n\}$ are the distributions of some other sequence of random variables, $\{Y_n\}$. Then, necessarily, $Y_n \rightsquigarrow P$; but the construction does *not* assert that Y_n converges almost surely. Indeed, we might even have the Y_n defined on different probability spaces, which would completely rule out any possible thought of almost sure convergence. The construction ensures that each X_n has marginal distribution P_n, the same as Y_n, but the joint distribution of the X_n's has nothing to do with the joint distribution of the Y_n's (which is only well defined if the Y_n all live on the same probability space). Indeed, that is the whole point of the construction: we have artificially manufactured the joint distribution for the X_n's in order that they converge, not just in the distributional sense, but also in the almost sure sense.

The representation lets us prove facts about weak convergence by means of the tools for almost sure convergence. For example, in the problems to Chapter 7, you were asked to show that $\Delta(P, Q) := \sup\{|P\ell - Q\ell| : \|\ell\|_{BL} \le 1\}$ defines a metric for weak convergence on the set of all Borel probability measures on a separable metric space. (Refer to Section 7.1 for the definition of the bounded Lipschitz norm.) If $\Delta(P_n, P) \to 0$ then $P_n f \to Pf$ for each f with $\|f\|_{BL} < \infty$, that is, $P_n \rightsquigarrow P$. Conversely, if $P_n \rightsquigarrow P$ and can we find X_n with distribution P_n and X with distribution P for which $X_n \to X$ almost surely (see Section 2 for the general case), then

$$\Delta(P_n, P) \le \sup_{\|\ell\|_{BL} \le 1} \mathbb{P}|\ell(X_n) - \ell(X)| \le \mathbb{P}\left(1 \wedge |X_n - X|\right) \to 0.$$

In effect, the general constructions of the representing variables subsume the specific calculations used in Chapter 7 to approximate $\{\ell : \|\ell\|_{BL} \le 1\}$ by a finite collection of functions. □

2. Almost sure representations

The representation from Example <3> has extensions to more general spaces. The result for separable metric spaces gives the flavor of the result without getting us caught up in too many measure theoretic details.

<4> **Theorem.** *For probability measures on the Borel sigma field of a separable metric space $\mathcal{X}$, if $P_n \rightsquigarrow P$ then there exist random elements X_n, with distributions P_n, and X, with distribution P, for which $X_n \to X$ almost surely.*

The main step in the proof involves construction of a joint distribution for X_n and X. To avoid a profusion of subscripts, it is best to isolate this part of the construction into a separate lemma. Once again, a single uniformly distributed U (that is, with distribution equal to Lebesgue measure m on $\mathcal{B}(0,1)$) will eventually provide the thread that ties together the various couplings into a single sequence converging almost surely. The construction builds the joint distribution via a probability kernel, K from $(0,1)\times\mathcal{X}$ into $\mathcal{X}$.

Recall, from Section 4.3, that such a kernel consists of a family of probability measures $\{K_{u,x}(\cdot) : u \in (0,1),\ x\in\mathcal{X}\}$ with $(u,x)\mapsto K_{u,x}B$ measurable for each fixed B in $\mathcal{B}(\mathcal{X})$. We define a measure on the product sigma-field of $(0,1)\times\mathcal{X}\times\mathcal{X}$ by

$$(\mathrm{m}\otimes P\otimes K)^{u,x,y} f(u,x,y) := \mathrm{m}^u\left(P^x K^y_{u,x} f(u,x,y)\right).$$

Less formally: we independently generate an observation u from the uniform distribution m and an observation x from P, then we generate a y from the corresponding $K_{u,x}$. The expression in parentheses on the right-hand side also defines a probability distribution, $(P\otimes K)_u$, on $\mathcal{X}\times\mathcal{X}$,

$$(P\otimes K)^{x,y}_u f(x,y) := P^x K^y_{u,x} f(x,y) \qquad \text{for each fixed } u.$$

In fact, $\{(P\otimes K)_u : u\in(0,1)\}$ is a probability kernel from $(0,1)$ to $\mathcal{X}\times\mathcal{X}$. Notice also that the marginal distribution $\mathrm{m}^u P^x K_{u,x}$ for y is a $\mathrm{m}\otimes P$ average of the $K_{u,x}$ probability measures on $\mathcal{B}(\mathcal{X})$. As an exercise in generating class methods, you might check all the measurability properties needed to make these assertions precise.

<5> **Lemma.** *Let P and Q be probability measures on the Borel sigma-field $\mathcal{B}(\mathcal{X})$. Suppose there is a partition of $\mathcal{X}$ into disjoint Borel sets $B_0, B_1, \dots, B_m$, and a positive constant ϵ, for which $QB_\alpha \ge (1-\epsilon)PB_\alpha$ for each α. Then there exists a probability kernel K from $(0,1)\times\mathcal{X}$ to $\mathcal{X}$ for which $Q = \mathrm{m}^u P^x K_{u,x}$ and for which $(P\otimes K)_u$ concentrates on $\cup_\alpha (B_\alpha\times B_\alpha)$ whenever $u\le 1-\epsilon$.*

Proof. Rewrite the assumption as $QB_\alpha = \delta_\alpha + (1-\epsilon)PB_\alpha$, where the nonnegative numbers δ_α must sum to ϵ because $\sum_\alpha QB_\alpha = \sum_\alpha PB_\alpha = 1$. Write $Q(\cdot\mid B_\alpha)$ for the conditional distribution, which can be taken as an arbitrary probability measure on B_α if $QB_\alpha = 0$. Partition the interval $(1-\epsilon, 1)$ into disjoint subintervals J_α with $\mathrm{m}J_\alpha = \delta_\alpha$. Define

$$K_{u,x}(\cdot) = \sum\nolimits_\alpha \left(\{u\in J_\alpha\} + \{u\le 1-\epsilon,\ x\in B_\alpha\}\right) Q(\cdot\mid B_\alpha).$$

When $u\le 1-\epsilon$ the recipe is: generate y from $Q(\cdot\mid B_\alpha)$ when $x\in B_\alpha$, which ensures that x and y then belong to the same B_α. Integrate over u and x to find the marginal probability that y lands in a Borel set A:

$$\mathrm{m}^u P^x K_{u,x}A = \sum\nolimits_\alpha \left(\delta_\alpha + (1-\epsilon)PB_\alpha\right) Q(A\mid B_\alpha) = \sum\nolimits_\alpha (QB_\alpha) Q(A\mid B_\alpha) = QA,$$

□ as asserted.

REMARK. Notice that the kernel K does nothing clever when $u\in J_\alpha$. If we were hoping for a result closer to the quantile coupling of Example <3>, we might instead try to select y from a B_β that is close to x, in some sense. Such refined behavior would require a more detailed knowledge of the partition.

Proof of Theorem <4>. The idea is simple. For each n we will construct an appropriate probability kernel $K^{(n)}_{u,x}$ from $(0,1)\times\mathcal{X}$ to $\mathcal{X}$, via an appeal to the

Lemma, with Q equal to the corresponding P_n and ϵ depending on n. We then independently generate $X_n(\omega)$ from $K_{u,x}^{(n)}$, for each n, with u an observation from $\mathfrak{m}$ independent of an observation $X(\omega) := x$ from P.

The inequality required by the Lemma would follow from convergence in distribution if each B_α were a P-continuity set (that is, if each boundary ∂B_α had zero P measure—see Section 7.1), for then we would have $P_n B_\alpha \to P B_\alpha$ as $n \to \infty$. Problem [4] shows how to construct such a partition $\pi := \{B_0, B_1, \ldots, B_m\}$ for an arbitrarily small $\epsilon > 0$, with two additional properties,

(i) $PB_0 \le \epsilon$

(ii) diameter$(B_\alpha) \le \epsilon$ for each $\alpha \ge 1$.

We shall need a a whole family of such partitions, $\pi_k := \{B_{\alpha,k} : \alpha = 0, 1, \ldots, m_k\}$, corresponding to values $\epsilon_k := 2^{-k}$ for each $k \in \mathbb{N}$.

To each k there exists an n_k for which $P_n B \ge (1 - \epsilon_k) P B$ for all B in π_k, when $n \ge n_k$. With no loss of generality we may assume that $1 < n_1 < n_2 < \ldots$, which ensures that for each n greater than n_1 there exists a unique $k := k(n)$ for which $n_k \le n < n_{k+1}$. Write $K_{u,x}^{(n)}$ for the probability kernel defined by Lemma <5> for $Q := P_n$ with $\epsilon := \epsilon_{k(n)}$, and $\pi_{k(n)}$ as the partition. Define $\mathbb{P}$ as the probability measure $\mathfrak{m} \otimes P \otimes \left(\otimes_{n \in \mathbb{N}} K_{u,x}^{(n)}\right)$ on the product sigma-field of $\Omega := (0, 1) \times \mathcal{X} \times \mathcal{X}^{\mathbb{N}}$. The generic point of Ω is a sequence $\omega := (u, x, y_1, y_2, \ldots)$. Define $X(\omega) := x$ and $X_n(\omega) := y_n$.

Why does X_n converge $\mathbb{P}$-almost surely to X? First note that $\sum_k P B_{0,k} < \infty$. Borel-Cantelli therefore ensures that, for almost all x and every u in $(0, 1)$, there exists a $k_0 = k_0(u, x)$ for which $u \le 1 - \epsilon_k$ and $x \notin B_{0,k}$ for all $k \ge k_0$. For such (u, x) and $k \ge k_0$ we have $(x, y_n) \in \cup_{\alpha \ge 1} B_{\alpha,k} \times B_{\alpha,k}$ for $n_k \le n < n_{k+1}$, by the concentration property of the kernels. That is, both $X(\omega)$ and $X_n(\omega)$ fall within the same $B_{\alpha,k}$ with $\alpha \ge 1$, a set with diameter less than ϵ_k. Think your way through that convoluted assertion and you will realize we have shown something even stronger than almost sure convergence. □

<6> **Example.** Suppose $P_n \rightsquigarrow P$ as probability measures on the Borel sigma-field of a separable metric space, and suppose that $\{T_n\}$ is a sequence of measurable maps into another metric space $\mathcal{Y}$. If P-almost all x have the property that $T_n(x_n) \to T(x)$ for every sequence $\{x_n\}$ converging to x, then the sequence of image measures also converges in distribution, $T_n P_n \rightsquigarrow TP$, as probability measures on the Borel sigma-field of $\mathcal{Y}$. The proof is easy is we represent $\{P_n\}$ by the sequence $\{X_n\}$, as in the Theorem. For each ℓ in $BL(\mathcal{Y})$, we have $\ell(T_n(X_n(\omega))) \to \ell(T(X(\omega)))$ for $\mathbb{P}$-almost all ω. Thus

$$(T_n P_n)\ell = \mathbb{P}\ell(T_n(X_n)) \to \mathbb{P}\ell(T(X)) = (TP)\ell,$$

by Dominated Convergence. □

I noted in Example <3> that if Y_n has distribution P_n, and if each Y_n is defined on a different probability space $(\Omega_n, \mathcal{F}_n, \mathbb{P}_n)$, then the convergence in distribution $Y_n \rightsquigarrow P$ cannot possibly imply almost sure convergence for Y_n. Nevertheless, using an argument similar to the proof of Theorem <4>, Dudley (1985) obtained something almost as good as almost sure convergence.

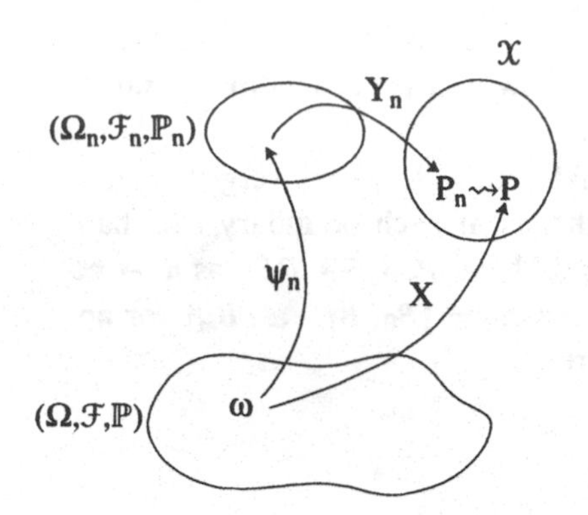

He built a single probability space $(\Omega, \mathcal{F}, \mathbb{P})$ supporting measurable maps ψ_n, into Ω_n, and X, into $\mathcal{X}$, with distributions $\mathbb{P}_n = \psi_n(\mathbb{P})$ and $P = X(\mathbb{P})$, for which $Y_n(\psi_n(\omega)) \to X(\omega)$ for $\mathbb{P}$ almost all ω. In effect, the ψ_n maps pull Y_n back to Ω, where the notions of pointwise and almost sure convergence make sense.

Actually, Dudley established a more delicate result, for Y_n that need not be measurable as maps into $\mathcal{X}$, a generalization needed to accommodate an application in the theory of abstract empirical processes. See Pollard (1990, Section 9) for a discussion of some of the conceptual and technical difficulties—such as the meaning of convergence in distribution for maps that don't have distributions in the usual sense—that are resolved by Dudley's construction. See Kim & Pollard (1990, Section 2) for an example of the subtle advantages of Dudley's form of the representation theorem.

*3. Strassen's Theorem

Once again let $(\mathcal{X}, d)$ be a separable metric space equipped with its Borel sigma-field $\mathcal{B}(\mathcal{X})$. For each subset A of $\mathcal{X}$, and each $\epsilon \geq 0$, define A^ϵ to be the closed set $\{x \in \mathcal{X} : d(x, A) \leq \epsilon\}$. The ***Prohorov distance*** between any P and Q from the set $\mathcal{P}$ of all probability measures on $\mathcal{B}(\mathcal{X})$ is defined as

$$\rho(P, Q) := \inf\{\epsilon > 0 : PB \leq QB^\epsilon + \epsilon \text{ for all } B \text{ in } \mathcal{B}(\mathcal{X})\}.$$

Despite the apparent lack of symmetry in the definition, ρ is a metric (Problem [3]) on $\mathcal{P}$.

REMARK. Separability of $\mathcal{X}$ is convenient, but not essential when dealing with the Prohorov metric. For example, it implies that $\mathcal{B}(\mathcal{X} \times \mathcal{X}) = \mathcal{B}(\mathcal{X}) \otimes \mathcal{B}(\mathcal{X})$, which ensures that $d(X, X')$ is measurable for each pair of random elements X and X'; and if $X_n \to X$ almost surely then $\mathbb{P}\{d(X_n, X) > \epsilon\} \to 0$ for each $\epsilon > 0$.

If $\rho(P_n, P) \to 0$ then, for each closed F we have $P_n F \leq PF^\epsilon + \epsilon$ eventually, and hence $\limsup_n P_n F \leq PF$, implying that $P_n \rightsquigarrow P$. Theorem <4> makes it easy to prove the converse. If X_n has distribution P_n and X has distribution P, and if $X_n \to X$ almost surely, then for each $\epsilon > 0$ there is an n_ϵ such that

$$\mathbb{P}\{d(X_n, X) > \epsilon\} < \epsilon \qquad \text{for } n \geq n_\epsilon.$$

For every Borel set B, when $n \geq n_\epsilon$ we have

<7> $$P_n B \leq \mathbb{P}\{X_n \in B, d(X_n, X) \leq \epsilon\} + \mathbb{P}\{d(X_n, X) > \epsilon\} \leq \mathbb{P}\{X \in B^\epsilon\} + \epsilon = PB^\epsilon + \epsilon.$$

Thus ρ is actually a metric for weak convergence of probability measures.

The Prohorov metric also has an elegant (and useful, as will be shown by Section 4) coupling interpretation, due to Strassen (1965). I will present a slightly restricted version of the result, by placing a tightness assumption on the probabilities, in order to simplify the statement of the Theorem. (Actually, the proof will establish a stronger result; the tightness will be used only at the very end, to tidy up.) Also, the role of ϵ is slightly easier to understand if we replace it by two separate constants.

<8> **Theorem.** *Let P and Q be tight probability measures on the Borel sigma field $\mathcal{B}$ of a separable metric space $\mathcal{X}$. Let ϵ and ϵ' be positive constants. There exists random elements X and Y of $\mathcal{X}$ with distributions P and Q such that $\mathbb{P}\{d(X,Y) > \epsilon\} \le \epsilon'$ if and only if $PB \le QB^\epsilon + \epsilon'$ for all Borel sets B.*

The argument for deducing the family of inequalities from existence of the coupling is virtually the same as <7>. For the other, more interesting direction, I follow an elegant idea of Dudley (1976, Lecture 18). By approximation arguments he reduced to the case where both P and Q concentrate on a finite set of atoms, and then existence of the coupling followed by an appeal to the classical Marriage Lemma (Problem [5]). I modify his argument to eliminate a few steps, by making an appeal to the following generalization (proved in Problem [6]) of that Lemma.

<9> **Lemma.** *Let ν be a finite measure on a finite set S and μ be a finite measure on a sigma-field $\mathcal{B}$ on a set T. Suppose $\{R_\alpha : \alpha \in S\}$ is a collection of measurable sets with the domination property that $\nu(A) \le \mu\left(\cup_{\alpha\in A} R_\alpha\right)$ for all $A \subseteq S$. Then there exists a probability kernel K from S to T with K_α concentrated on R_α for each α and $\sum_{\alpha\in S} \nu\{a\} K_\alpha \le \mu$.*

Proof of Theorem <8>. The measure $\mathbb{P}$ will live on $\mathcal{X} \times \mathcal{X}$, with X and Y as the coordinate maps. It will be the limit of a weakly convergent subsequence of a uniformly tight family $\{\mathbb{P}_\delta : \delta > 0\}$, obtained by an appeal to the Prohorov/Le Cam theorem from Section 7.5.

Construct $\mathbb{P}_\delta$ via a "discretization" of P, which brings the problem within the ambit of Lemma <9>. For a small, positive δ, which will eventually be sent to zero, partition $\mathcal{X}$ into finitely many disjoint, Borel sets $B_0, B_1, \ldots, B_m$ with $PB_0 < \delta$ and diameter$(B_\alpha) < \delta$ for $\alpha \ge 1$. (Compare with the construction in Problem [4].) Define a probability measure ν, concentrated on the finite set $S := \{0, 1, \ldots, m\}$, by $\nu\{\alpha\} := PB_\alpha$ for $\alpha = 0, \ldots, m$. Augment $\mathcal{X}$ by a point ∞. Extend Q to a measure μ on $T := \mathcal{X} \cup \{\infty\}$ by placing mass ϵ' at ∞. Define R_α as $B_\alpha^\epsilon \cup \{\infty\}$. With these definitions, the measures ν and μ satisfy the requirements of Lemma <9>: for each subset A of S,

$$\nu(A) = P\left(\cup_{\alpha\in A} B_\alpha\right) \le Q\left(\cup_{\alpha\in A} B_\alpha\right)^\epsilon + \epsilon' = Q\left(\cup_{\alpha\in A} B_\alpha^\epsilon\right) + \mu\{\infty\} = \mu\left(\cup_{\alpha\in A} R_\alpha\right).$$

The Lemma ensures existence of a probability kernel K, from S to T, with $K_\alpha B_\alpha^\epsilon + K_\alpha\{\infty\} = K_\alpha R_\alpha = 1$ for each α and $\sum_\alpha \nu\{\alpha\} K_\alpha A \le \mu A$ for every Borel subset A of T. In particular, $\sum_\alpha \nu\{\alpha\} K_\alpha B \le QB$ for all $B \in \mathcal{B}$. The nonnegative measure $Q - \sum_\alpha \nu\{\alpha\} K_\alpha\big|_{\mathcal{X}}$ on $\mathcal{B}$ has total mass

$$\tau := 1 - \textstyle\sum_\alpha \nu\{\alpha\} K_\alpha \mathcal{X} = \sum_\alpha \nu\{\alpha\} K_\alpha\{\infty\} \le \mu\{\infty\} = \epsilon'.$$

Write this measure as τQ_0, with Q_0 a probability measure on $\mathcal{B}$. (If $\tau = 0$, choose Q_0 arbitrarily.) We then have $Qh = \tau Q_0 h + \sum_\alpha \nu\{\alpha\} K_\alpha h$ for all $h \in \mathcal{M}^+(\mathcal{X})$.

Define a probability measure $\mathbb{P}_\delta$ on $\mathcal{B} \otimes \mathcal{B}$ by

$$\mathbb{P}_\delta f := P^x\left(\textstyle\sum_{\alpha=0}^m \{x \in B_\alpha\}\left(K_\alpha + K_\alpha\{\infty\} Q_0\right)^y f(x,y)\right) \qquad \text{for } f \in \mathcal{M}^+(\mathcal{X} \times \mathcal{X}).$$

REMARK. In effect, I have converted K to a probability kernel L from $\mathcal{X}$ to $\mathcal{X}$, by setting L_x equal to $K_\alpha|\mathcal{X} + K_\alpha\{\infty\}Q_0$ when $x \in B_\alpha$. The definition of $\mathbb{P}_\delta$ is equivalent to $\mathbb{P}_\delta := P \otimes L$, in the sense of Section 4.4.

The measure $\mathbb{P}_\delta$ has marginals P and Q because, for g and h in $\mathcal{M}^+(\mathcal{X})$,

$$\mathbb{P}_\delta^{x,y} g(x) = P^x\left(\sum\nolimits_\alpha \{x \in B_\alpha\}\left(K_\alpha\mathcal{X} + K_\alpha\{\infty\}\right) g(x)\right) = Pg,$$

$$\mathbb{P}_\delta^{x,y} h(y) = \sum\nolimits_\alpha P\{x \in B_\alpha\}\left(K_\alpha h + K_\alpha\{\infty\}Q_0 h\right) = \sum\nolimits_\alpha \nu\{\alpha\}K_\alpha h + \tau Q_0 h.$$

It concentrates most of its mass on the set $D := \cup_{\alpha=1}^m \left(B_\alpha \times B_\alpha^\epsilon\right)$,

$$\begin{aligned}\mathbb{P}_\delta D &\geq \sum\nolimits_{\alpha=1}^m P^x\left(\{x \in B_\alpha\}K_\alpha^y\{(x,y) \in D\}\right)\\ &= \sum\nolimits_{\alpha=1}^m P^x\left(\{x \in B_\alpha\}K_\alpha^y\{y \in B_\alpha^\epsilon\}\right)\\ &= \sum\nolimits_{\alpha=1}^m \nu\{\alpha\}K_\alpha\mathcal{X} = 1 - \tau - (PB_0)(K_0\mathcal{X}).\end{aligned}$$

When (x, y) belongs to D, we have $x \in B_\alpha$ and $d(y, B_\alpha) \leq \epsilon$ for some B_α with diameter$(B_\alpha) < \delta$, and hence $d(x, y) \leq \delta + \epsilon$. Thus $\mathbb{P}_\delta$ assigns measure at least $1 - \epsilon' - \delta$ to the closed set $F_{\delta+\epsilon} := \{(x, y) \in \mathcal{X} \times \mathcal{X} : d(x, y) \leq \delta + \epsilon\}$.

The tightness of both P and Q will let us eliminate δ, by passing to the limit along a subsequence. For each $\eta > 0$ there exists a compact set C_η for which $PC_\eta^c < \eta$ and $QC_\eta^c < \eta$. The probability measure $\mathbb{P}_\delta$, which has marginals P and Q, puts mass at most 2η outside the compact set $C_\eta \times C_\eta$. The family $\{\mathbb{P}_\delta : \delta > 0\}$ is uniformly tight, in the sense explained in Section 7.5. As shown in that Section, there is a sequence $\{\delta_i\}$ tending to zero for which $\mathbb{P}_{\delta_i} \rightsquigarrow \mathbb{P}$, with $\mathbb{P}$ a probability measure on $\mathcal{B}\otimes\mathcal{B}$. It is a very easy exercise to check that $\mathbb{P}$ has marginals P and Q. For each fixed $t > \epsilon$, the weak convergence implies

$$\mathbb{P}F_t \geq \limsup\nolimits_i \mathbb{P}_{\delta_i} F_t \geq \limsup\nolimits_i \mathbb{P}_{\delta_i} F_{\epsilon+\delta_i} \geq 1 - \epsilon'.$$

□ Let t decrease to ϵ to complete the proof.

*4. The Yurinskii coupling

The multivariate central limit theorem gives conditions under which a sum S of independent random vectors $\xi_1, \ldots, \xi_n$ has an approximate normal distribution. Theorem <4> would translate the corresponding distributional convergence into a coupling between the standardized sum and a random vector with the appropriate normal distribution. When the random vectors have finite third moments, Theorem <8> improves the result by giving a rate of convergence (albeit in probability).

<10> **Theorem.** *Let $\xi_1, \ldots, \xi_n$ be independent random k-vectors with $\mathbb{P}\xi_i = 0$ for each i and $\beta := \sum_i \mathbb{P}|\xi_i|^3$ finite. Let $S := \xi_1 + \ldots + \xi_n$. For each $\delta > 0$ there exists a random vector T with a $N(0, \text{var}(S))$ distribution such that*

$$\mathbb{P}\{|S - T| > 3\delta\} \leq C_0 B\left(1 + \frac{|\log(1/B)|}{k}\right) \qquad \text{where } B := \beta k \delta^{-3},$$

for some universal constant C_0.

REMARK. The result stated by Yurinskii (1977) took a slightly different form. I have followed Le Cam (1988, Theorem 1) in reworking the Yurinskii's methods. Both those authors developed bounds on the Prohorov distance, by making an explicit choice for δ. The Le Cam preprint is particularly helpful in its discussion of heuristics behind how one balances the effect of various parameters to get a good bound.

Proof. The existence of the asserted coupling (for a suitably rich probability space) will follow via Theorem <8> if we can show for each Borel subset A of $\mathbb{R}^k$ that

<11> $$\mathbb{P}\{S \in A\} \leq \mathbb{P}\{T \in A^{3\delta}\} + \text{ERROR},$$

with the ERROR equal to the upper bound stated in the Theorem. By choosing a smooth (bounded derivatives up to third order) function f that approximates the indicator function of A, in the sense that $f \approx 1$ on A and $f \approx 0$ outside $A^{3\delta}$, we will be able to deduce inequality <11> from the multivariate form of Lindeberg's method (Section 7.3), which gives a third moment bound for a difference in expectations,

<12> $$|\mathbb{P}f(S) - \mathbb{P}f(T)| \leq C\left(\mathbb{P}|\xi_1|^3 + \ldots + \mathbb{P}|\xi_k|^3\right) = C\beta.$$

More precisely, if the constant C_f is such that

<13> $$\left|f(x+y) - f(x) - y'\dot{f}(x) - \tfrac{1}{2}y'\ddot{f}(x)y\right| \leq C_f|y|^3 \qquad \text{for all } x \text{ and } y,$$

then we may take $C = \left(9 + 8\mathbb{P}|N(0,1)|^3\right)C_f \leq 15C_f$.

For a fixed Borel set A, Lemma <18> at the end of the Section will show how to construct a smooth function f for which approximation <13> holds with $C_f = (\sigma^2\delta)^{-1}$ and for which, if $\delta > \sigma\sqrt{k}$,

<14> $$(1-\epsilon)\{x \in A\} \leq f(x) \leq \epsilon + (1-\epsilon)\{x \in A^{3\delta}\} \qquad \text{where } \begin{cases} \epsilon := \left(\dfrac{1+\alpha}{e^\alpha}\right)^{k/2}, \\ 1+\alpha := \dfrac{\delta^2}{k\sigma^2}. \end{cases}$$

The Lindeberg bound <12>, with $C\beta = 15\beta/(\sigma^2\delta) = 15B(1+\alpha)$, then gives

$$\begin{aligned}
\mathbb{P}\{S \in A\} &\leq (1-\epsilon)^{-1}\mathbb{P}f(S) \\
&\leq (1-\epsilon)^{-1}\left(\mathbb{P}f(T) + 15B(1+\alpha)\right) \\
&\leq \mathbb{P}\{T \in A^{3\delta}\} + \epsilon' \qquad \text{where } \epsilon' := \frac{\epsilon + 15B(1+\alpha)}{(1-\epsilon)}.
\end{aligned} \tag{<15>}$$

We need to choose α, as a function of k and B, to make ϵ' small.

Clearly the bound <15> is useful only when ϵ is small, in which case the $(1-\epsilon)$ factor in the denominator contributes only an extra contant factor to the final bound. We should concentrate on the numerator. Similarly, the assertion of the Theorem is trivial if B is not small. Provided we make sure $C_0 \geq e$, we may assume $B \leq e^{-1}$, that is, $\log(1/B) \geq 1$.

To get within a factor 2 of minimizing a sum of two nonnegative functions, one increasing and the other decreasing, it suffices to equate the two contributions. This fact suggests we choose α to make

$$\alpha - \left(1 - \frac{2}{k}\right)\log(1+\alpha) \approx \frac{2}{k}\log(1/B) + O(k^{-1}).$$

If B is small then α will be large, which would make $\log(1+\alpha)$ small compared with α. If we make α slightly larger than $2k^{-1}\log(1/B)$ we should get close to equality. Actually, we can afford to have α a larger multiple of $\log(1/B)$, because extra multiplicative factors will just be absorbed into constant C_0. With these thoughts, it seems to me I cannot do much better than choose

$$\alpha := 3\left(1 + \frac{2}{k}\log(1/B)\right),$$

which at least has the virtue of giving a clean bound:

$$\log\epsilon \leq \frac{k}{2}\left(\log(1+\alpha) - \frac{2\alpha}{3}\right) - \frac{k\alpha}{6} \leq -\log(1/B) \leq -1.$$

and hence

$$\epsilon' = \frac{\epsilon + 15B(1+\alpha)}{(1-\epsilon)} \leq \frac{90}{1-e^{-1}}B\left(1 + \frac{\log(1/B)}{k}\right) \qquad \text{when } B \leq e^{-1}.$$

The proof is complete, except for the construction of the smooth function f satisfying <14>. □

Before moving on to the construction of f, let us see what we can do with the coupling from the Theorem in the case of identically distributed random vectors. For convenience of notation write $\mathcal{Y}_k(x)$ for the function $C_0 x\,(1 + |\log(1/x)|/k)$.

<16> **Example.** Let $\xi_1, \xi_2, \ldots$ be independent, identically distributed random k-vectors with $\mathbb{P}\xi_1 = 0$, $\text{var}(\xi_1) := V$, and $\mu_3 := \mathbb{P}|\xi_1|^3 < \infty$. Write S_n for $\xi_1 + \ldots \xi_n$. The central limit theorem asserts that $S_n/\sqrt{n} \rightsquigarrow N(0, V)$. Theorem <10>, asserts existence of a sequence of random vectors W_n, each distributed $N(0, V)$ for which

$$\mathbb{P}\left\{\left|\frac{S_n}{\sqrt{n}} - W_n\right| \geq 3\delta\right\} \leq \mathcal{Y}_k\left(\frac{kn\mu_3}{(\delta\sqrt{n})^3}\right).$$

For fixed k, we can make the right-hand side as small as we please by choosing δ as a large enough enough multiple of $n^{-1/6}$. Thus, with finite third moments,

$$\left|\frac{S_n}{\sqrt{n}} - W_n\right| = O_p(n^{-1/6}) \qquad \text{via the Yurinskii coupling.}$$

For $k = 1$, this coupling is not the best possible. For example, under an assumption of finite third moments, a theorem of Major (1976) gives a sequence of independent random variables $Y_1, Y_2, \ldots$, each distributed $N(0, V)$, for which

$$\left|\frac{S_n}{\sqrt{n}} - \frac{Y_1 + \ldots + Y_n}{\sqrt{n}}\right| = o_p(n^{-1/6}) \qquad \text{almost surely.}$$

Major's result has the correct joint distributions for the approximating normals, as n changes, as well as providing a slightly better rate. □

<17> **Example.** Yurinskii's coupling (and its refinements: see, for example, the discussion near Lemma 2.12 of Dudley & Philipp 1983) is better suited to situations where the dimension k can change with n.

Consider the case of a sequence of independent, identially distributed stochastic processes $\{X_i(t) : t \in T\}$. Suppose $\mathbb{P}X_1(t) = 0$ and $|X_1(t)| \leq 1$ for every t. Under suitable regularity conditions on the sample paths, we might try to show that the standardized partial sum processes, $Z_n(t) := (X_1(t) + \ldots + X_n(t))/\sqrt{n}$, behave like a centered Gaussian process $\{Z(t) : t \in T\}$, with the same covariance structure as X_1. We might even try to couple the processes in such a way that $\sup_t |Z_n(t) - Z(t)|$ is small in some probabilistic sense.

The obvious first step towards establishing a coupling of the processes is to consider behavior on large finite subsets $T(k) := \{t_1, \ldots, t_k\}$ of T, where k is allowed to increase with n. The question becomes: How rapidly can k tend to infinity?

For fixed k, write ξ_i for the random k-vector with components $X_i(t_j)$, for $j = 1, \ldots, k$. We seek to couple $(\xi_1 + \ldots + \xi_n)/\sqrt{n}$ with a random vector W_n, distributed like $\{Z(t_j) : j = 1, \ldots, k\}$. The bound is almost the same as in Example <16>, except for the fact that the third moment now has a dependence on k,

$$\mathbb{P}|\xi_1|^3 = k^{3/2}\mathbb{P}\left(\frac{1}{k}\sum_{j=1}^{k} X_1(t_j)^2\right)^{3/2} \le k^{3/2}\mathbb{P}\left(\frac{1}{k}\sum_{j=1}^{k} |X_1(t_j)|^3\right) \le k^{3/2}.$$

Via the general fact that $\max_j |x_j| \le \left(\sum_j x_j^2\right)^{1/2}$, the coupling bound becomes

$$\begin{aligned}\mathbb{P}\left\{\max_{j\le k} |Z_n(t_j) - W_{n,j}| \ge 3\delta\right\} &\le \mathbb{P}\left\{\left|\frac{\xi_1 + \ldots + \xi_n}{\sqrt{n}} - W_n\right| \ge 3\delta\right\} \\ &\le \mathcal{Y}_k\left(\frac{nk^{5/2}}{(\delta\sqrt{n})^3}\right) \\ &\to 0 \qquad \text{if } k = o\left(n^{1/5}\right) \text{ and } \delta \to 0 \text{ slowly enough.}\end{aligned}$$

□ That is, $\max_{j\le k}\left|Z_n(t_j) - W_{n,j}\right| = o_p(1)$ if k increases more slowly than $n^{1/5}$.

Smoothing of indicator functions

There are at least two methods for construction of a smooth approximation f to a set A. The first uses only the metric:

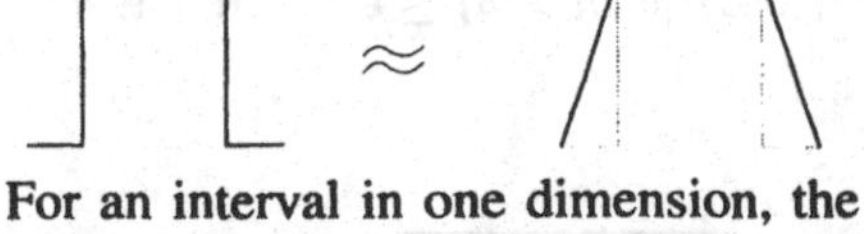

$$f(x) = (1 - d(x, A)/\delta)^+ .$$

For an interval in one dimension, the approximation has the effect of replacing the discontinuity at the boundary points by linear functions with slope $1/\delta$. The second method treats the indicator function of the set as an element of an $\mathcal{L}^1$ space, and constructs the approximation by means of convolution smoothing,

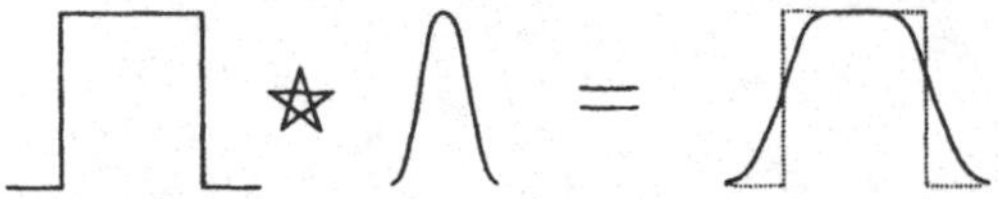

$$f(x) = \mathfrak{m}^w\left(\{w \in A\}\phi_\sigma(w - x)\right),$$

where ϕ_σ denotes the $N(0, \sigma^2 I_k)$ density and $\mathfrak{m}$ denotes Lebesgue measure on $\mathcal{B}(\mathbb{R}^k)$. (Any smooth density with rapidly decreasing tails would suffice.) A combination of the two methods of smoothing will give the best bound:

<18> **Lemma.** *Let A be a Borel subset of $\mathbb{R}^k$. let Z have a $N(0, I_k)$ distribution. For positive constants δ and σ define*

$$g(x) := \left(1 - \frac{d(x, A^\delta)}{\delta}\right)^+ \qquad \text{and} \qquad f(x) := \mathbb{P}g(x + \sigma Z) = \mathfrak{m}^w \left(g(w)\phi_\sigma(w - x)\right).$$

Then f satisfies <13> *with $C := (\sigma^2\delta)^{-1}$, and approximation* <14> *holds.*

Proof. The function f inherits some smoothness from g and some from the convolving standard normal density ϕ_σ, which has derivatives

$$\frac{\partial}{\partial z}\phi_\sigma(z) = -\frac{z}{\sigma^2}\phi_\sigma(z) \qquad \text{and} \qquad \frac{\partial^2}{\partial z^2}\phi_\sigma(z) = \left(\frac{zz'}{\sigma^4} - \frac{I_k}{\sigma^2}\right)\phi_\sigma(z).$$

For fixed x and y, the function $h(t) := f(x + ty)$, for $0 \le t \le 1$, has second derivative

$$\begin{aligned}\ddot{h}(t) &= \mathfrak{m}^w \left(g(w)\left(\frac{(y'(w - x - ty))^2}{\sigma^4} - \frac{|y|^2}{\sigma^2}\right)\phi_\sigma(w - x - ty)\right) \\ &= \sigma^{-2}\mathbb{P}\left(g(x + ty + \sigma Z)\left((y'Z)^2 - |y|^2\right)\right).\end{aligned}$$

The Lipschitz property $|g(x + ty + \sigma Z) - g(x + \sigma Z)| \le t|y|/\delta$ then implies

$$|\ddot{h}(t) - \ddot{h}(0)| \le \frac{t|y|}{\sigma^2\delta}\mathbb{P}\left((y'Z)^2 + |y|^2\right) \le \frac{2|y|^3}{\sigma^2\delta}.$$

The asserted inequality <13> then follows from a Taylor expansion,

$$|h(1) - h(0) - \dot{h}(0) - \tfrac{1}{2}\ddot{h}(0)| = \tfrac{1}{2}\left|\ddot{h}(t^*) - \ddot{h}(0)\right| \qquad \text{where } t^* \in (0, 1).$$

For approximation <14>, first note that $A^\delta \le g \le A^{2\delta}$ and $0 \le f \le 1$ everywhere. Also $\mathbb{P}\{|Z| > \delta/\sigma\} \le \epsilon$, from Problem [7]. Thus

$$f(x) \ge \mathbb{P}g(x + \sigma Z)\{|\sigma Z| \le \delta\} = \mathbb{P}\{|Z| \le \delta/\sigma\} \ge 1 - \epsilon \qquad \text{if } x \in A,$$

and

$$f(x) = \mathbb{P}g(x + \sigma Z)\{|\sigma Z| \le \delta\} + \mathbb{P}g(x + \sigma Z)\{|\sigma Z| > \delta\} \le \epsilon \qquad \text{if } x \notin A^{3\delta}.$$

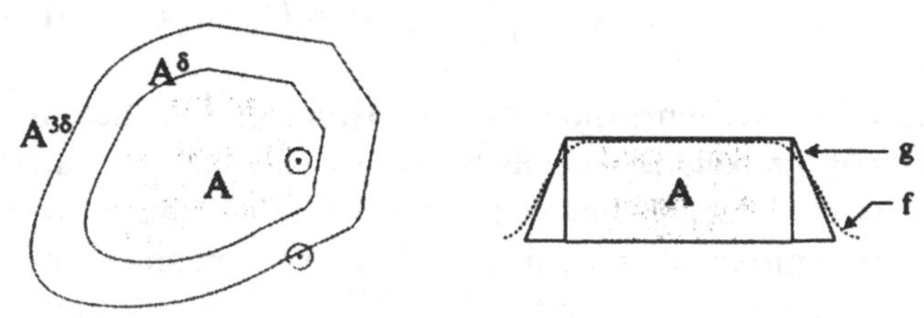

□

5. Quantile coupling of Binomial with normal

As noted in Section 1, if η is distributed $N(0, 1)$, with distribution function Φ, and if q denotes the Bin$(n, 1/2)$ quantile function, then the random variable $X := q(\Phi(\eta))$ has exactly a Bin$(n, 1/2)$ distribution. In a sense made precise by the following Lemma, X is very close to the random variable $Y := n/2 + \eta\sqrt{n/4}$, which has a $N(n/2, n/4)$ distribution. The coupling of the Bin$(n, 1/2)$ with its approximating $N(n/2, n/4)$ has been the starting point for a growing collection of striking approximation results, inspired by the publication of the fundamental paper of Komlós, Major & Tusnády (1975).

<19> **Tusnády's Lemma.** *For each positive integer n there exists a deterministic, increasing function $\tau(n,\cdot)$ such that the random variable $X := \tau(n,\eta)$ has a* Bin$(n, 1/2)$ *distribution whenever η has a $N(0,1)$ distribution. The random variable X satisfies the inequalities*

$$|X - Y| \le 1 + \frac{\eta^2}{8} \qquad \text{and} \qquad \left|X - \frac{n}{2}\right| \le 1 + \frac{\sqrt{n}|\eta|}{2},$$

where $Y := \frac{n}{2} + \eta\sqrt{\frac{n}{4}}$, *which has a* $N\left(\frac{n}{2}, \frac{n}{4}\right)$ *distribution.*

At first glance it is easy to underestimate the delicacy of these two inequalities. Both X and Y have mean $n/2$ and standard deviation of order $\sqrt{n}$. It would be no challenge to construct a coupling for which $|X - Y|$ is of order $\sqrt{n}$; the Lemma gives a coupling for which $|X - Y|$ is bounded by a quantity whose distribution does not even change with n.

The original proof (Tusnády 1977) of the Lemma is challenging. Appendix D contains an alternative derivation of similar inequalities. To simplify the argument, I have made no effort to derive the best constants for the bound. In fact, the precise constants appearing in the Lemma will have no importance for us. It will be enough for us to have a universal constant C_0 for which there exists couplings such that

<20>
$$|X - Y| \le C_0\left(1 + \eta^2\right) \qquad \text{and} \qquad \left|X - \frac{n}{2}\right| \le C_0\sqrt{n}\,(1 + |\eta|),$$

a weaker bound that follows easily from the inequalities in Appendix D.

6. Haar coupling—the Hungarian construction

Let $x_1, \dots, x_n$ be n independent observations from the uniform distribution P on $(0, 1]$. The ***empirical measure*** P_n is defined as the discrete distribution that puts mass $1/n$ at each of $x_1, \dots, x_n$. That is, $P_n f := \sum_{i=1}^n f(x_i)/n$, for each function f on $(0, 1]$. Notice that nP_nD has a Bin(n, PD) distribution for each Borel set D. The standardized measure $\nu_n := \sqrt{n}(P_n - P)$ is called the ***uniform empirical process***. For each square integrable function f,

$$\nu_n f = n^{-1/2}\sum\nolimits_{i=1}^n (f(x_i) - Pf) \rightsquigarrow N(0, \sigma_f^2) \qquad \text{where } \sigma_f^2 = Pf^2 - (Pf)^2.$$

More generally, for each finite set of square integrable functions $f_1, \dots, f_k$, the random vector $(\nu_n f_1, \dots, \nu_n f_k)$ has a limiting multivariate normal distribution with zero means and covariances $P(f_i f_j) - (Pf_i)(Pf_j)$. These finite dimensional distributions identify a Gaussian process that is closely related to the isonormal process $\{G(f) : f \in L^2(P)\}$ from Section 9.3.

Recall that G is a centered Gaussian process, defined on some probability space $(\Omega, \mathcal{F}, \mathbb{P})$, with $\text{cov}(G(f), G(g)) = \langle f, g\rangle = P(fg)$, the $L^2(P)$ inner product. The

Haar basis, $\Psi = \{1\} \cup \{\psi_{i,k} : 0 \le i < 2^k, k \in \mathbb{N}_0\}$, for $L^2(P)$ consists of rescaled differences of indicator functions of intervals $J_{i,k} := J(i,k) := (i2^{-k}, (i+1)2^{-k}]$,

$$\psi_{i,k} := 2^{k/2}\left(J_{2i,k+1} - J_{2i+1,k+1}\right) = 2^{k/2}\left(2J_{2i,k+1} - J_{i,k}\right) \qquad \text{for } 0 \le i < 2^k.$$

REMARK. For our current purposes, it is better to replace $L^2(P)$ by $\mathcal{L}^2(P)$, the space of square-integrable real functions whose P-equivalence classes define $L^2(P)$. It will not matter that each $G(f)$ is defined only up to a P-equivalence. We need to work with the individual functions to have $P_n f$ well defined. It need not be true that $P_n f = P_n g$ when f and g differ only on a P-negligible set.

Each function in $\mathcal{L}^2(P)$ has a series expansion,

$$f = (Pf) + \sum\nolimits_{k=0}^{\infty} \sum\nolimits_i \psi_{i,k} \langle f, \psi_{i,k} \rangle,$$

which converges in the $\mathcal{L}^2(P)$ sense. The random variables $\overline{\eta} := G(1)$ and $\eta_{i,k} := G(\psi_{i,k})$ are independent, each with a $N(0,1)$ distribution, and

$$G(f) = (Pf)\overline{\eta} + \sum\nolimits_{k=0}^{\infty} \sum\nolimits_i \eta_{i,k} \langle f, \psi_{i,k} \rangle,$$

with convergence in the $L^2(\mathbb{P})$ sense. If we center each function f to have zero expectation, we obtain a new Gaussian process, $\nu(f) := G(f - Pf) = G(f) - (Pf)G(1)$, indexed by $\mathcal{L}^2(P)$, whose covariances identify it as the limit process for ν_n. Notice that $\nu(\psi_{i,k}) = G(\psi_{i,k}) = \eta_{i,k}$ almost surely, because $P\psi_{i,k} = 0$. Thus we also have a series representation for ν,

<21> $$\nu(f) = G(f) - (Pf)\overline{\eta} = \sum\nolimits_{k=0}^{\infty} \sum\nolimits_i \eta_{i,k} \langle f, \psi_{i,k} \rangle = \sum\nolimits_{k=0}^{\infty} \sum\nolimits_i \nu(\psi_{i,k}) \langle f, \psi_{i,k} \rangle.$$

At least in a heuristic sense, we could attempt a similar series expansion of the empirical process,

<22> $$\nu_n(f) \stackrel{?}{=} \sum\nolimits_{k=0}^{\infty} \sum\nolimits_i \nu_n(\psi_{i,k}) \langle f, \psi_{i,k} \rangle.$$

REMARK. Don't worry about the niceties of convergence: when the heuristics are past I will be truncating the series at some finite k.

The expansion suggests a way of coupling the process ν_n and ν, namely, find a probability space on which $\nu_n(\psi_{i,k}) \approx \nu(\psi_{i,k}) = \eta_{i,k}$ for a large subset of the basis functions. Such a coupling would have several advantages. First, the peculiarities of each function f would be isolated in the behavior of the coefficients $\langle f, \psi_{i,k} \rangle$. Subject to control of those coefficients, we could derive simultaneous couplings for many different f's. Second, because the $\psi_{i,k}$ functions are rescaled differences of indicator functions of intervals, the $\nu_n(\psi_{i,k})$ are rescaled differences of Binomial counts. Tusnády's Lemma offers an excellent means for building Binomials from standard normals. With some rescaling, we can then build versions of the $\nu_n(\psi_{i,k})$ from the $\eta_{i,k}$.

The secret to success is a recursive argument, corresponding to the nesting of the $J_{i,k}$ intervals. Write node(i,k) for $(i + \frac{1}{2})/2^k$, the midpoint of $J_{i,k}$. Regard node$(2i, k+1)$ and node$(2i+1, k+1)$ as the children of node(i,k), corresponding to the decomposition of $J_{i,k}$ into the disjoint union of the two subintervals $J_{2i,k+1}$ and $J_{2i+1,k+1}$. The parent of node(i,k) is node$(\lfloor i/2 \rfloor, k-1)$.

For each integer i with $0 \le i < 2^k$ there is a path back through the tree,

$$\text{path}(i,k) := \{(i_0, 0), (i_1, 1), \ldots, (i_k, k)\} \qquad \text{where } i_0 = 0 \text{ and } i_k = i,$$

for which $J(i_k, k) \subset J(i_{k-1}, k-1) \subset \ldots \subset J(0,0) = (0,1]$. That is, the path traces through all the ancestors (parent, grandparent, ...) back to the root of the tree.

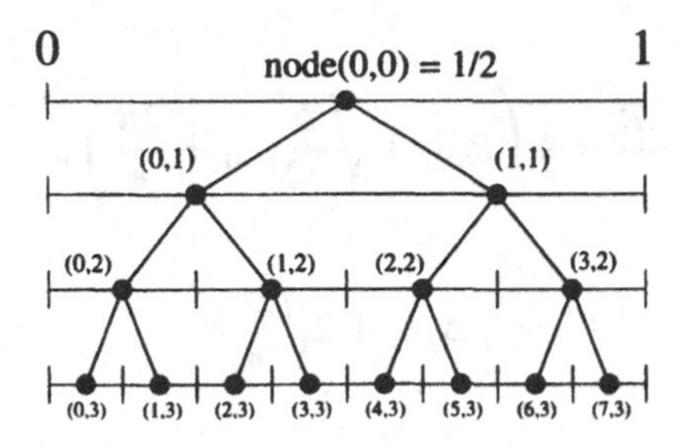

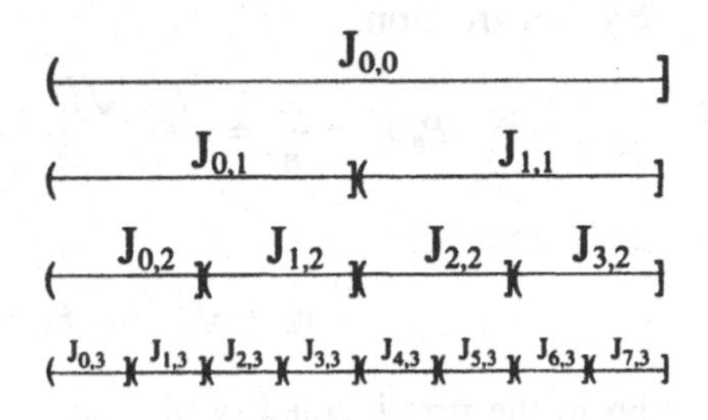

The recursive argument constructs successively refined approximations to P_n by assigning the numbers of observations $X_{i,k}$ amongst $x_1, x_2, \ldots, x_n$ that land in each interval $J_{i,k}$. Notice that, conditional on $X_{i,k} = N$, the two offspring counts must sum to N, with $X_{2i,k+1}$ having a conditional Bin$(N, 1/2)$ distribution. Via Lemma <19> define

$$X_{0,1} := \tau(n, \eta_{0,0}) =: n - X_{1,1},$$

$$X_{0,2} := \tau(X_{0,1}, \eta_{0,1}) =: X_{0,1} - X_{1,2}, \qquad X_{2,2} := \tau(X_{1,1}, \eta_{1,1}) =: X_{1,1} - X_{3,2},$$

and so on. That is, recursively divide the count $X_{i,k}$ at each node(i,k) between the two children of the node, using the normal variable $\eta_{i,k}$ to determine the Bin$(X_{i,k}, 1/2)$ count assigned to the child at node$(2i, k+1)$. The joint distribution for the $X_{i,k}$ variables is the same as the joint distribution for the empirical counts $nP_nJ_{i,k}$, because we have used the correct conditional distributions.

If we continued the process forever then, at least conceptually, we would identify the locations of the n observations, without labelling. Each point would be determined by a nested sequence of intervals. To avoid difficulties related to pointwise convergence of the Haar expansion, we need to stop at some finite level, say the mth, after which we could independently distribute the $X_{i,m}$ observations (if any) within $J_{i,m}$.

The recursive construction works well because Tusnády's Lemma, even in its weakened form <20>, provides us with a quadratic bound in the normal variables for the difference between $\nu_n(\psi_{i,k})$ and the corresponding $\eta_{i,k}$.

<23> **Lemma.** *There exists a universal constant C such that, for each k and $0 \le i_k < 2^k$,*

$$|\nu_n(\psi_{i,k}) - \eta_{i,k}| \le \frac{C}{\sqrt{n}} \sum_{j=0}^{k} 2^{j/2} \left(1 + \eta_{i_j,j}^2\right),$$

where $\{(i_j, j) : j = 0, 1, \ldots, k\}$ is a path from the root down to node(i_k, k).

Proof. Abbreviate $J(i_j, j)$ to J_j, and $\eta_{i_j,j}$ to η_j, and so on, for $j = 0, 1, \ldots, k$. Notice that the random variable P_nJ_j has expected value $PJ_j = 2^{-j}$, and a small variance, so we might hope that all of the random variables $\Delta_j := 2^j P_n J_j$ should be close to 1. Of course $\Delta_0 \equiv 1$.

Consider the effect of the split of J_j into its two subintervals, $J' := J(2i_j, j+1)$ and $J'' := J(2i_j+1, j+1)$. Write N for nP_nJ_j and X for nP_nJ', so that $\Delta_j = 2^jN/n$ and $\Delta' := 2^{j+1}P_nJ' = 2^{j+1}X/n$ and $\Delta'' := 2^{j+1}P_nJ'' = 2\Delta_j - \Delta'$. From inequality <20>, we have $X = N/2 + \sqrt{N}\eta_j/2 + R$, where

<24> $$|R| \le C_0(1+\eta_j^2) \qquad \text{and} \qquad |X - N/2| \le C_0\sqrt{N}\left(1+|\eta_j|\right).$$

By construction,

$$P_nJ' = \frac{X}{n} = \frac{N+\sqrt{N}\eta_j + 2R}{2n} = \tfrac{1}{2}\left(P_nJ_j + \sqrt{\frac{\Delta_j}{n2^j}}\,\eta_j + \frac{2R}{n}\right),$$

and hence

$$\nu_n\psi_j = \sqrt{n2^j}\,P_n\left(2J' - J_j\right) = \sqrt{\Delta_j}\,\eta_j + 2\sqrt{\frac{2^j}{n}}R.$$

From the first inequality in <24>,

<25> $$|\nu_n\psi_j - \eta_j| \le \left|\left(\sqrt{\Delta_j}-1\right)\eta_j\right| + 2C_0\sqrt{\frac{2^j}{n}}\left(1+\eta_j^2\right).$$

From the second inequality in <24>,

$$|\Delta'' - \Delta_j| = |\Delta' - \Delta_j| = \frac{2^{j+1}}{n}|X - N/2| \le C_0\sqrt{\frac{2^{j+2}}{n}}\sqrt{\Delta_j}\left(1+|\eta_j|\right).$$

Invoke the inequality $|\sqrt{a}-\sqrt{b}| \le |a-b|/\sqrt{b}$, for positive a and b, to deduce that

$$|\sqrt{\Delta_{j+1}} - \sqrt{\Delta_j}| \le \max\left(|\sqrt{\Delta'}-\sqrt{\Delta_j}|, |\sqrt{\Delta''}-\sqrt{\Delta_j}|\right) \le 2C_0 2^{j/2}\left(1+|\eta_j|\right)/\sqrt{n}.$$

From <25> with $j = k$, and the inequality from the previous line, deduce that

$$\begin{aligned}\sqrt{n}|\nu_n\psi_k - \eta_k| &\le 2C_0 2^{k/2}\left(1+\eta_k^2\right) + \sqrt{n}|\eta_k|\sum\nolimits_{j=0}^{k-1}|\sqrt{\Delta_{j+1}} - \sqrt{\Delta_j}| \\ &\le 2C_0 2^{k/2}\left(1+\eta_k^2\right) + 2C_0\sum\nolimits_{j=0}^{k-1}2^{j/2}\left(|\eta_k| + |\eta_k\eta_j|\right).\end{aligned}$$

Bound $|\eta_k| + |\eta_k\eta_j|$ by $1 + \eta_k^2 + \frac{1}{2}\eta_k^2 + \frac{1}{2}\eta_j^2$, then collect terms involving η_k^2, to complete the proof. □

7. The Komlós-Major-Tusnády coupling

The coupling method suggested by expansions <21> and <22> works particularly well when restricted to the set of indicator functions of intervals, $f_t(x) = \{0 < x \le t\}$, for $0 < t \le 1$. For that case, the limit process $\{\nu(0,t] : 0 \le t \le 1\}$, which can be chosen to have continuous sample paths, is called the ***Brownian Bridge***, or ***tied-down Brownian motion***, often written as $\{B^\circ(t) : 0 \le t \le 1\}$.

<26> **Theorem.** *(KMT coupling) There exists a Brownian Bridge $\{B^\circ(t) : 0 \le t \le 1\}$ with continuous sample paths, and a uniform empirical process ν_n, for which*

$$\mathbb{P}\left\{\sup_{0\le t\le 1}|\nu_n(0,t] - B^\circ(t)| \ge C_1\frac{x+\log n}{\sqrt{n}}\right\} \le C_0\exp(-x) \qquad \text{for all } x \ge 0,$$

with constants C_1 and C_0 that depend on neither n nor x.

REMARK. Notice that the exponent on the right-hand side is somewhat arbitrary; we could change it to any other positive multiple of $-x$ by changing the constant C_1 on the left-hand side. By the same reasoning, it would suffice to get a bound like $C_2\exp(-c_2x) + C_3\exp(-c_3x) + C_4\exp(-c_4x)$ for various positive constants C_i and c_i, for then we could recover the cleaner looking version by adjusting C_1 and C_0. In my opinion, the exact constants are unimportant; the form of the inequality is what counts. Similarly, it would suffice to consider only values of x bounded away from zero, such as $x \ge c_0$, because the asserted inequality is trivial for $x < c_0$ if $C_0 \ge e^{c_0}$.

It is easier to adjust constants at the end of an argument, to get a clean-looking inequality. When reading proofs in the literature, I sometimes find it frustrating to struggle with a collection of exquisitely defined constants at the start of a proof, eventually to discover that the author has merely been aiming for a tidy final bound.

Proof. We will build ν_n from B°, allocating counts down to intervals of length 2^{-m}, as described in Section 6. It will then remain only to control the behavior of both processes over small intervals. Let $T(m)$ denote the set of grid points $\{i/2^m : i = 0, 1, \ldots, 2^m\}$ in $[0, 1]$. For each t in $T(m)$, both series <21> and <22> terminate after $k = m$, because $[0, t]$ is orthogonal to each $\psi_{i,k}$ for $k > m$. That is, using the Hungarian construction we can determine $P_n J_{i,m}$ for each i, and then calculate

$$\nu_n(0, t] = \sum\nolimits_{k=0}^{m} \sum\nolimits_i \nu_n(\psi_{i,k})\langle f_t, \psi_{i,k}\rangle \qquad \text{for } t \text{ in } T(m),$$

which we need to show is close to

$$B^\circ(t) := \nu(0, t] = \sum\nolimits_{k=0}^{m} \sum\nolimits_i \eta_{i,k}\langle f_t, \psi_{i,k}\rangle \qquad \text{for } t \text{ in } T(m).$$

Notice that $B^\circ(0) = B^\circ(1) = 0 = \nu_n(0, 0] = \nu_n(0, 1]$. We need only consider t in $T(m)\backslash\{0, 1\}$. For each k, at most one coefficient $\langle f_t, \psi_{i,k}\rangle$ is nonzero, corresponding to the interval for which $t \in J_{i,k}$, and it is bounded in absolute value by $2^{-k/2}$. The corresponding nodes determine a path $(0, 0), \ldots, (i_j, j), \ldots, (i_m, m)$ down to the mth level. The difference between the processes at t is controlled by the quadratic function,

$$S_m(t) := \sum\nolimits_{j=0}^{m} \eta_{i_j,j}^2 \qquad \text{where } t \in J(i_j, j) \text{ for each } j,$$

of the normal variables at the nodes of this path:

$$\begin{aligned}
\sqrt{n}|\nu_n(0, t] - B^\circ(t)| &\le \sum\nolimits_{k=0}^{m} \sqrt{n}|\nu_n(\psi_{i_k,k}) - \eta_{i_k,k}|2^{-k/2} \\
&\le \sum_{j,k}\{0 \le j \le k \le m\}C2^{(j-k)/2}\left(1 + \eta_{i_j,j}^2\right) \qquad \text{by Lemma <23>} \\
&\le 4C\sum\nolimits_{j=0}^{m}\left(1 + \eta_{i_j,j}^2\right) \qquad \text{summing the geometric series} \\
&= 4C\left(m + 1 + S_m(t)\right). \qquad \text{<27>}
\end{aligned}$$

As t ranges over $T(m)$, or even over the whole of $(0, 1)$, the path defining $S_m(t)$ ranges over the set of all 2^m paths from the root down to the mth level. We bound the maximum difference between the two processes if we bound the maximum of $S_m(t)$. The maximum grows roughly linearly with m, the same rate as the contribution from a single t. More precisely

$$\mathbb{P}\{\max_t S_m(t) \ge 5m + x\} \le 2\exp(-x/4) \qquad \text{for each } x \ge 0. \qquad \text{<28>}$$

I postpone the proof of this result, in order not to break the flow of the main argument.

REMARK. The constants 5 and 4 are not magical. They could be replaced by any other pair of constants for which $\mathbb{P}\exp\left((N(0,1)^2 - c_1)/c_2\right) \le 1/2$.

From inequalities <27> and <28> we have

<29> $$\mathbb{P}\left\{\max_{t\in T(m)} |\nu_n(0,t] - B^\circ(t)| \ge 4C\frac{1+x+6m}{\sqrt{n}}\right\} \le \mathbb{P}\left\{\max_t S_m(t) \ge x+5m\right\} \le \exp(-x/4).$$

Provided we choose m smaller than a constant multiple of $x + \log n$, this term will cause us no trouble.

We now have the easy task of extrapolating from the grid $T(m)$ to the whole of $(0,1)$. We can make 2^{-m} exceedingly small by choosing m close to a large enough multiple of $x + \log n$. In fact, when $x \ge 2$, the choice of m such that

<30> $$2n^2e^x > 2^m \ge n^2e^x$$

will suffice. As an exercise, you might want to play around with other m and the various constants to get a neater statement for the Theorem.

We can afford to work with very crude estimates. For each s in $(0,1)$ write t_s for the point of $T(m)$ for which $t_s \le s < t_s + 2^{-m}$. Notice that

$$|\nu_n(0,s] - \nu_n(0,t_s]| \le \#\text{ points in } (t_s,s]/\sqrt{n} + \sqrt{n}2^{-m}.$$

The supremum over s is larger than $3/\sqrt{n}$ only when at least one $J_{i,m}$ interval, for $0 \le i < 2^m$ contains 2 or more observations, an event with probability less than

$$2^m\binom{n}{2}(2^{-m})^2 \le n^2 2^{-m} \le e^{-x} \qquad \text{for } m \text{ as in <30>}.$$

Similarly,

$$\begin{aligned}\sup_s |B^\circ(s) - B^\circ(t_s)| &\le \sup_s |G[0,s] - G[0,t_s]| + \sup_s |(s-t_s)\overline{\eta}| \\ &\le \max_{0\le i<2^m} \sup_{s\in J_{i,m}} |G[0,s] - G[0,i/2^m]| + 2^{-m}|\overline{\eta}|\end{aligned}$$

from which it follows that

$$\mathbb{P}\left\{\sup_s |B^\circ(s) - B^\circ(t_s)| \ge \frac{2x}{\sqrt{n}}\right\} \le 2^m\mathbb{P}\left\{\sup_{0\le s\le 2^{-m}} |B(s)| \ge \frac{x}{\sqrt{n}}\right\} + \mathbb{P}\left\{|N(0,1)| \ge \frac{2^m x}{\sqrt{n}}\right\}$$

where B is a Brownian motion. The second term on the right-hand side is less than $\exp\left(-4^m x^2/2n\right)$. By the reflection principle for Brownian motion (Section 9.5), the first term equals

$$2^m\mathbb{P}\left\{|B(2^{-m})| \ge \frac{x}{\sqrt{n}}\right\} = 2^{m+1}\mathbb{P}\left\{|N(0,1)| \ge \frac{2^{m/2}x}{\sqrt{n}}\right\} \le 2^{m+1}\exp\left(-\frac{2^m x^2}{2n}\right).$$

For $x \ge 2$ and m as in <30>, the sum of the two contributions from the Brownian Bridge is much smaller than e^{-x}.

From <29>, and the inequality

$$|\nu_n(0,s] - B^\circ(s)| \le |\nu_n(0,s] - \nu_n(0,t_s]| + |\nu_n(0,t_s] - B^\circ(t_s)| + |B^\circ(t_s) - B^\circ(s)|,$$

together with the bounds from the previous paragraph, you should be able to complete the argument. □

Proof of inequality <28>. Write R_m for $\max_t S_m(t)$. Think of the binary tree of depth m as two binary trees of depth $m-1$ rooted at node(0, 1) and node(1, 1), to see that that R_m has the same distribution as $\eta_{0,0}^2 + \max(T, T')$, where T and T' both have the same distribution as R_{m-1}, and $\eta_{0,0}$, T, and T' are independent. Write D_k for $\mathbb{P}\exp((R_k - 5k)/4)$. Notice that

$$e^{-5/4} D_0 = \mathbb{P}\exp\left(\tfrac{1}{4}\eta_{0,0}^2 - \tfrac{5}{4}\right) = \sqrt{2}\exp(-5/4) < 1/2.$$

For $m \geq 1$, independence lets us bound D_m by

$$\mathbb{P}\exp\left(\tfrac{1}{4}\eta_{0,0}^2 - \tfrac{5}{4}\right)\mathbb{P}\left(\tfrac{1}{4}\max\left(T - 5(m-1), T' - 5(m-1)\right)\right)$$
$$< \tfrac{1}{2}\left(\mathbb{P}\exp\left(\tfrac{1}{4}T - \tfrac{5}{4}(m-1)\right) + \mathbb{P}\exp\left(\tfrac{1}{4}T' - \tfrac{5}{4}(m-1)\right)\right) = D_{m-1}.$$

By induction, $\mathbb{P}\exp(R_m - 5m)/4 = D_m \leq D_0 = \sqrt{2}$. Thus

$$\mathbb{P}\{R_m \geq 5m + x\} \leq \mathbb{P}\exp\left((R_m - 5m)/4\right)\exp(-x/4) \leq \sqrt{2}\exp(-x/4),$$

as asserted. □

By means of the quantile transformation, Theorem <26> extends immediately to a bound for the empirical distribution function F_n generated from a sample $\xi_1, \ldots, \xi_n$ from a probability measure on the real line with distribution function F. Again writing q_F for the quantile function, and recalling that we can generate the sample as $\xi_i = q_F(x_i)$, we have

$$nF_n(t) = \sum\nolimits_{i \leq n}\{\xi_i \leq t\} = \sum\nolimits_{i \leq n}\{q_F(x_i) \leq t\} = \sum\nolimits_{i \leq n}\{x_i \leq F(t)\},$$

which implies $\sqrt{n}\,(F_n(t) - F(t)) = \nu_n\big(0, F(t)\big]$. Notice that $F(t)$ ranges over a subset of $[0, 1]$ as t ranges over $\mathbb{R}$; and when F has no discontinuities, the range covers all of $(0, 1)$. Theorem <26> therefore implies

$$\text{<31>}\qquad \mathbb{P}\left\{\sup_t |\sqrt{n}\,(F_n(t) - F(t)) - B^\circ(F(t))| \geq C_1 \frac{x + \log n}{\sqrt{n}}\right\} \leq C_0 e^{-x} \qquad \text{for } x \geq 0.$$

Put another way, we have an almost sure representation $F_n(t) = F(t) + n^{-1/2}B^\circ(F(t)) + R_n(t)$, where, for example, $\sup_t |R_n(t)| = O_p\left(n^{-1}\log n\right)$.

REMARK. From a given Brownian Bridge B° and a given n we have constructed a sample $x_1, \ldots, x_n$ from the uniform distribution. From the same B°, we could also generate a sample $x'_1, \ldots, x'_n, x'_{n+1}$ of size $n+1$. However, it is not true that $x_i = x'_i$ for $i \leq n$; it is not true that $x_1, \ldots, x_n, x'_{n+1}$ are mutually independent. If we wished to have the samples relate properly to each other we would have to change the Brownian Bridge with n. There is a version of KMT called the ***Kiefer*** coupling, which gets the correct joint distributions between the samples at the cost of a weaker error bound. See Csörgő & Révész (1981, Chapter 4) for further explanation.

Inequality <31> lets us deduce results about the empirical distribution function F_n from analogous results about the Brownian Bridge. For example, it implies $\sup_t \sqrt{n}|F_n(t) - F(t)| \rightsquigarrow \sup_t |B^\circ(F(t))|$. If F has no discontinuities, the limit distribution is the same as that of $\sup_s |B^\circ(s)|$. That is, we have an instant derivation of the Kolmogorov-Smirov theorem. The Csörgő & Révész book describes other consequences that make much better use of all the hard work that went into establishing the KMT inequality.

8. Problems

[1] Suppose F and F_n, for $n \in \mathbb{N}$ are distribution functions on the real line for which $F_n(x) \to F(x)$ for each x in a dense subset D of the real line. Show that the corresponding quantile functions Q_n converge pointwise to Q at all except (at worst) a countable subset of points in $(0, 1)$. Hint: Prove convergence at each continuity point u_0 of Q. Given points x', x'' in D with $x' < x_0 = Q(u_0) < x''$, find $\delta > 0$ such that $x' < Q(u_0 - \delta)$ and $Q(u_0 + \delta) \le x''$. Deduce that

$$F_n(x') < F(x') + \delta < u_0 \le F(x'') - \delta \le F_n(x'') \qquad \text{eventually,}$$

in which case $x' < Q_n(u_0) \le x''$.

[2] Let P and Q be two probability measures defined on the same sigma-field $\mathcal{A}$ of a set $\mathcal{X}$. The total variation distance $v = v(P, Q)$ is defined as $\sup_{A \in \mathcal{A}} |PA - QA|$.

(i) Suppose X are Y are random elements of $\mathcal{X}$, defined on the same probability space $(\Omega, \mathcal{F}, \mathbb{P})$, with distributions P and Q. Show that $\mathbb{P}^*\{X \ne Y\} \ge v(P, Q)$. Hint: Choose a measurable set $D \supseteq \{X \ne Y\}$ with $\mathbb{P}D = \mathbb{P}^*\{X \ne Y\}$. Note that $\mathbb{P}\{X \in A\} - \mathbb{P}\{Y \in A\} = \mathbb{P}\{X \in A\} \cap D - \mathbb{P}\{Y \in A\} \cap D$.

(ii) Suppose the diagonal $\Delta := \{(x, y) \in \mathcal{X} \times \mathcal{X} : x = y\}$ is product measurable. Recall from Section 3.3 that $v = 1 - (P \wedge Q)(\mathcal{X}) = (P - Q)^+(\mathcal{X}) = (Q - P)^+(\mathcal{X})$. Define a probability measure $\mathbb{P} = \frac{1}{v}(P - Q)^+ \otimes (Q - P)^+ + \lambda$, where λ is the image of $P \wedge Q$ under the map $x \mapsto (x, x)$. Let X and Y be the coordinate maps. Show that X has distribution P and Y has distribution Q, and $\mathbb{P}\{X \ne Y\} = v$.

[3] Show that the Prohorov distance is a metric. Hint: For the triangle inequality, use the inclusion $(B^\epsilon)^{\epsilon'} \subseteq B^{\epsilon + \epsilon'}$. For symmetry, consider $\rho(P, Q) < \delta < \epsilon$. Put $D^c = B^\epsilon$. Prove that $D^\delta \subseteq B^c$, then deduce that $1 - PB^\epsilon \le QD^\delta + \delta \le 1 - QB + \delta$.

[4] Let P be a Borel probability measure concentrated on the closure of a countable subset $S = \{x_i : i \in \mathbb{N}\}$ of a metric space $\mathcal{X}$. For fixed $\epsilon > 0$, follow these steps to show that there exist a partition of $\mathcal{X}$ into finitely many P-continuity sets $C_0, C_1, \ldots, C_m$ such that $PC_0 < \epsilon$ and diameter$(C_i) < \epsilon$ for $i \ge 1$.

(i) For each x in $\mathcal{X}$, show that there are at most countably many closed balls B centered at x with $P(\partial B) > 0$.

(ii) For each x_i in S, find a ball B_i centered at x_i with radius between $\epsilon/4$ and $\epsilon/2$ and $P(\partial B_i) = 0$.

(iii) Show that $\cup_{i \in \mathbb{N}} B_i$ contains the closure of S. Hint: Each point of the closure lies within $\epsilon/4$ of at least one x_i.

(iv) Show that $P\left(\cup_{i \le m} B_i\right) > 1 - \epsilon$ when m is large enough.

(v) Show that the sets $C_i := B_i \setminus \cup_{1 \le j < i} B_j$ and $C_0 := \left(\cup_{i \le m} B_i\right)^c$ have the desired properties.

[5] (Ye Olde Marriage Lemma) Suppose S is a finite set of princesses. Suppose each princess, σ, has a list, $K(\sigma)$, of frogs desirable for marriage. For each collection $A \subseteq S$, the combined list of frogs equals $K(A) = \bigcup\{K(\sigma) : \sigma \in A\}$. If

each princess is to find a frog on her list to marry, then clearly the "Desirable Frog Condition" (DFC), $\#K(A) \geq \#A$, for each $A \subseteq S$, must be satisfied. Show that DFC is also sufficient for happy princesses: under the DFC there exists a one-to-one map π from S into $K(S)$ such that $\pi(\sigma) \in K(\sigma)$ for every σ in S. Hint: Translate the following mathematical fairy tale into an inductive argument.

(i) Once upon a time there was a princess σ_0 who proposed to marry a frog τ_0 from her list. That would have left a collection $S\backslash\{\tau_0\}$ of princesses with lists $K(\sigma)\backslash\{\tau_0\}$ to choose from. If the analog of the DFC had held for those lists, an induction hypothesis would have made everyone happy.

(ii) Unfortunately, a collection $A_0 \subseteq S\backslash\{\sigma_0\}$ of princesses protested, on the grounds that $\#\mathcal{K}(A_0)\backslash\{\tau_0\} < \#A_0$; clearly not enough frogs to go around. They pointed out that the DFC held with equality for A_0, and that their happiness could be assured only if they had exclusive access to the frogs in $K(A_0)$.

(iii) Everyone agreed with the assertion of the A_0. They got their exclusive access, and, by induction, lived happily ever after.

(iv) The other princesses then got worried. Each collection B in $S\backslash A_0$ asked, "$\#K(B)\backslash K(A_0) \geq \#B$?" They were reassured, "Don't worry. Originally $\#K(B \cup A_0) \geq \#B + \#A_0$, and we all know that $\#K(A_0) = \#A_0$, so of course

$$\#K(B)\backslash K(A_0) = \#K(B \cup A_0) - \#K(A_0) \geq \#B.$$

You too can live happily ever after, by induction." And they did.

[6] Prove Lemma <9> by carrying out on the following steps. Write R_A for $\cup_{\alpha \in A} R_\alpha$. Argue by induction on the size of S. With no loss of generality, suppose $S = \{1, 2, \ldots, m\}$. Check the case $m = 1$. Work from the inductive hypothesis that the result is true for $\#S < m$.

(i) Suppose there exists a proper subset A_0 of S for which $\nu A_0 = \mu R_{A_0}$. Define $R'_\alpha = R_\alpha \backslash R_{A_0}$ for $\alpha \notin A_0$. Show that $\nu A \leq \mu R'_A$ for all $A \subseteq S\backslash A_0$. Construct K by invoking the inductive hypothesis separately for A_0 and $S\backslash A_0$. (Compare with part (iv) of Problem [5].)

Now suppose $\nu A < \mu R_A$ for all proper subsets A of S. Write L_α for the probability distribution $\mu(\cdot \mid R_\alpha)$, which concentrates on R_α.

(ii) Show that $\mu \geq \nu\{1\}L_1$. Hint: Show $\mu B \geq \nu\{1\}\mu(BR_1)/\mu R_1$ for all $B \subseteq R_1$.

Write ϵ_1 for the unit mass at 1. Let θ_0 be the largest value in $[0, \nu\{1\}]$ for which $(\mu - \theta_0 L_1)R_A \geq (\nu - \theta_0 \epsilon_1)A$ for every $A \subseteq S$.

(iii) If $\theta_0 = \nu\{1\}$, use the inductive hypothesis to find a probability kernel from $S\backslash\{1\}$ into T for which $(\mu - \nu\{1\}L_1) \geq \sum_{\alpha \geq 2} \nu\{\alpha\}K_\alpha$. Define $K_1 = L_1$.

(iv) If $\theta_0 < \nu\{1\}$, show that there exists an $A_0 \subseteq S$ for which $(\mu - \theta L_1)R_{A_0} < (\nu - \theta\epsilon_1)A_0$ when $\nu\{1\} \geq \theta > \theta_0$. Deduce that A_0 must be a proper subset of S for which $(\mu - \theta_0 L_1)R_{A_0} = (\nu - \theta_0\epsilon_1)A_0$. Invoke part (i) to find a probability kernel M for which $\mu - \theta_0 L_1 \geq (\nu\{1\} - \theta_0)M_1 + \sum_{\alpha \geq 2} \nu\{\alpha\}M_\alpha$. Define $K_1 := (\theta_0/\nu\{1\})L_1 + (1 - \theta_0/\nu\{1\})M_1$.

[7] Establish the bound $\mathbb{P}\{|N(0, I_k)| > \sqrt{kx}\} \le \left(xe^{1-x}\right)^{k/2}$, for $x > 1$, as needed (with $\sqrt{kx} = \delta/\sigma$) for the proof of Lemma <18>. Hint: Show that

$$\mathbb{P}\{|N(0, I_k)|^2 > kx\} \le \exp(-tkx)(1-2t)^{-k/2} \qquad \text{for } 0 < t < 1/2$$

which is minimized at $t = \frac{1}{2}(1 - x^{-1})$.

[8] Let F_m and G_n be empirical distribution functions, constructed from independent samples (of sizes m and n) from the same distribution function F on the real line. Show that

$$\sqrt{\frac{mn}{m+n}}\, \sup_t |F_m(t) - G_n(t)| \rightsquigarrow \sup_t |B^\circ(F(t))| \qquad \text{as } \min(m, n) \to \infty.$$

Hint: Use <31>. Show that $\alpha B_1^\circ(s) + \beta B_1^\circ(s)$ is a Brownian Bridge if $\alpha^2 + \beta^2 = 1$ and B_1°, B_1° are independent Brownian Bridges.

9. Notes

In increasing degrees of generality, representations as in Theorem <4> are due to Skorohod (1956), Dudley (1968), Wichura (1970), and Dudley (1985).

Prohorov (1956) defined his metric for probability measures on complete, separable metric spaces. Theorem <8> is due to Strassen (1965). I adapted the proof from Dudley (1976, Section 18), who used the Marriage Lemma (Problem [5]) to prove existence of the desired coupling in a special discrete case. Lemma <9> is a continuous analog of the Marriage Lemma, slightly extending the method of Pollard (1984, Lemma IV.24).

The discussion in Section 4 is adapted from an exposition of Yurinskii (1977)'s method by Le Cam (1988). I think the slightly weaker bound stated by Yurinskii may be the result of his choosing a slightly different tail bound for $|N(0, I_k)|$, with a correspondingly different choice for the smoothing parameter.

The idea for Example <17> comes from the construction used by Dudley & Philipp (1983) to build strong approximations for sums of independent random processes taking values in a Banach space. Massart (1989) refined the coupling technique, as applied to empiricial processes, using a Hungarian coupling in place of the Yurinskii coupling.

The proof of the KMT approximation in the original paper (Komlós et al. 1975) was based on the analog of the first inequality in <20>, for $|X - n/2|$ smaller than a tiny multiple of n. The proof of the elegant refinement in Lemma <19> appeared in a 1977 dissertation of Tusnády, in Hungarian. I have seen an annotated extract from the dissertation (courtesy of Sándor Csörgő). Csörgő & Révész (1981, page 133) remarked that Tusnády's proof is "elementary" but not "simple". I agree. Bretagnolle & Massart (1989, Appendix) published another proof, an exquisitely delicate exercise in elementary calculus and careful handling of Stirling's approximation. The method used in Appendix D resulted from a collaboration between Andrew Carter and me.

Lemma <23> repackages a construction from Komlós et al. (1975) that has been refined by several authors, most notably Bretagnolle & Massart (1989), Massart (1989), and Koltchinskii (1994).

References

Bretagnolle, J. & Massart, P. (1989), 'Hungarian constructions from the nonasymptotic viewpoint', *Annals of Probability* **17**, 239–256.

Csörgő, M. & Révész, P. (1981), *Strong Approximations in Probability and Statistics*, Academic Press, New York.

Dudley, R. M. (1968), 'Distances of probability measures and random variables', *Annals of Mathematical Statistics* **39**, 1563–1572.

Dudley, R. M. (1976), 'Convergence of laws on metric spaces, with a view to statistical testing'. Lecture Note Series No. 45, Matematisk Institut, Aarhus University.

Dudley, R. M. (1985), 'An extended Wichura theorem, definitions of Donsker classes, and weighted empirical distributions', *Springer Lecture Notes in Mathematics* **1153**, 141–178. Springer, New York.

Dudley, R. M. & Philipp, W. (1983), 'Invariance principles for sums of Banach space valued random elements and empirical processes', *Zeitschrift für Wahrscheinlichkeitstheorie und Verwandte Gebiete* **62**, 509–552.

Kim, J. & Pollard, D. (1990), 'Cube root asymptotics', *Annals of Statistics* **18**, 191–219.

Koltchinskii, V. I. (1994), 'Komlós-Major-Tusnády approximation for the general empirical process and Haar expansion of classes of functions', *Journal of Theoretical Probability* **7**, 73–118.

Komlós, J., Major, P. & Tusnády, G. (1975), 'An approximation of partial sums of independent rv-s, and the sample df. I', *Zeitschrift für Wahrscheinlichkeitstheorie und Verwandte Gebiete* **32**, 111–131.

Le Cam, L. (1988), On the Prohorov distance between the empirical process and the associated Gaussian bridge, Technical report, Department of Statistics, U.C. Berkeley. Technical report No. 170.

Major, P. (1976), 'The approximation of partial sums of independent rv's', *Zeitschrift für Wahrscheinlichkeitstheorie und Verwandte Gebiete* **35**, 213–220.

Massart, P. (1989), 'Strong approximation for multivariate empirical and related processes, via KMT constructions', *Annals of Probability* **17**, 266–291.

Pollard, D. (1984), *Convergence of Stochastic Processes*, Springer, New York.

Pollard, D. (1990), *Empirical Processes: Theory and Applications*, Vol. 2 of *NSF-CBMS Regional Conference Series in Probability and Statistics*, Institute of Mathematical Statistics, Hayward, CA.

Prohorov, Yu. V. (1956), 'Convergence of random processes and limit theorems in probability theory', *Theory Probability and Its Applications* **1**, 157–214.

Skorohod, A. V. (1956), 'Limit theorems for stochastic processes', *Theory Probability and Its Applications* **1**, 261–290.

Strassen, V. (1965), 'The existence of probability measures with given marginals', *Annals of Mathematical Statistics* **36**, 423–439.

Tusnády, G. (1977), A study of Statistical Hypotheses, PhD thesis, Hungarian Academy of Sciences, Budapest. In Hungarian.

Wichura, M. J. (1970), 'On the construction of almost uniformly convergent random variables with given weakly convergent image laws', *Annals of Mathematical Statistics* **41**, 284–291.

Yurinskii, V. V. (1977), 'On the error of the Gaussian approximation for convolutions', *Theory Probability and Its Applications* **2**, 236–247.

Chapter 11

Exponential tails and the law of the iterated logarithm

SECTION 1 introduces the law of the iterated logarithm (LIL) through the technically simplest case: independent standard normal summands.

SECTION 2 extends the results from Section 1 to sums of independent bounded random variables, by means of Bennett's exponential inequality. It is noted that the bounds on the variables could increase slowly without destroying the limit assertion, thereby pointing to the easy (upper) half of Kolmogorov's definitive LIL.

*SECTION *3 derives the very delicate exponential lower bound for bounded summands, needed to prove the companion lower half for Kolmogorov's LIL.*

*SECTION *4 shows how truncation arguments extend Kolmogorov's LIL to the case of independent, identically distributed summands with finite second moments.*

1. LIL for normal summands

Two important ideas run in tandem through this Chapter: the existence of exponential tail bounds for sums of independent random variables, and proofs of the law of the iterated logarithm (LIL) in various contexts. You could read the Chapter as either a study of exponential inequalities, with the LIL as a guiding application, or as a study of the LIL, with the exponential inequalities as the main technical tool.

The LIL's will all refer to partial sums $S_n := X_1 + \ldots + X_n$ for sequences of independent random variables $\{X_i\}$ with $\mathbb{P}X_i = 0$ and $\text{var}(X_i) := \sigma_i^2 < \infty$, for each i. The words *iterated logarithm* refer to the role played by function $L(x) := \sqrt{2x \log\log x}$. To avoid minor inconveniences (such as having to exclude cases involving logarithms or square roots of negative numbers), I arbitrarily define $L(x)$ as 1 for $x < e^e \approx 15.15$. Under various assumptions, we will be able to prove, with $V_n := \text{var}(S_n)$, that

<1> $$\limsup_{n\to\infty} S_n/L(V_n) = 1 \qquad \text{almost surely,}$$

together with analogous assertions about the lim inf and the almost sure behavior of the sequence $\{S_n/L(V_n)\}$. Equality <1> breaks naturally into a pair of assertions,

<2> $$\limsup_{n\to\infty} S_n/L(V_n) \le 1 \qquad \text{and} \qquad \limsup_{n\to\infty} S_n/L(V_n) \ge 1 \qquad \text{a.s.,}$$

inequalities that I will refer to as the ***upper*** and ***lower halves of the LIL***, or upper and lower LIL's, for short. In general, it will be easier to establish the upper half, because the exponential inequalities required for that case are easier to prove.

As you will see, several of the techniques used for proving LIL's are refinements of techniques used in Chapter 4 (appeals to the Borel-Cantelli lemma, truncation of summands, bounding of whole blocks of terms by means of maximal inequalities) for proving strong laws of large numbers (SLLN). Indeed, the LIL is sometimes described as providing a rate of convergence for the SLLN.

The theory is easiest to understand when specialized to the normal distribution, for which the following result holds.

<3> **Theorem.** *For the partial sums* $\{S_n\}$ *of a sequence of independent* $N(0,1)$ *random variables,*

(i) $\limsup_{n\to\infty} S_n/L_n = 1$ a.s.

(ii) $\liminf_{n\to\infty} S_n/L_n = -1$ a.s.

(iii) $S_n/L_n \in J$ *infinitely often, a.s., for every open subinterval* J *of* $[-1,1]$.

Proof. The key requirement for the proof of the upper LIL is an exponential tail bound, such as (see Appendix D for the proof)

<4> $$\mathbb{P}\{S_n \ge x\sqrt{n}\} \le \tfrac{1}{2}\exp(-x^2/2) \qquad \text{for } x \ge 0.$$

If we take $x_n := \gamma\sqrt{2\log n}$ for some fixed $\gamma > 1$, then

$$\sum\nolimits_n \mathbb{P}\{S_n \ge x_n\sqrt{n}\} \le \tfrac{1}{2}\sum\nolimits_n n^{-\gamma^2} < \infty,$$

which, by the Borel-Cantelli lemma, implies

$$\limsup_{n\to\infty} \frac{S_n(\omega)}{\sqrt{2n\log n}} \le \gamma \qquad \text{almost surely, for fixed } \gamma > 1.$$

Cast out a sequence of negligible sets, for a sequence of γ values decreasing to 1, to deduce that

<5> $$\limsup_{n\to\infty} \frac{S_n(\omega)}{\sqrt{2n\log n}} \le 1 \qquad \text{almost surely.}$$

This result is not quite what we need for the upper LIL. Somehow we must replace the $\sqrt{2n\log n}$ factor by $\sqrt{2n\log\log n}$, without disturbing the almost sure bound.

As with the proof of the SLLN in Section 4.6, the improvement is achieved by collecting the S_n into blocks, then applying Borel-Cantelli to a bound for a maximum over a block. To handle the contributions from within each block we need a maximal inequality, such as the following one-sided analog of the bound from Section 4.6.

<6> **Maximal Inequality.** *Let* $\xi_1, \dots, \xi_N$ *be independent random variables, and* x, ϵ, *and* β *be nonnegative constants such that* $\mathbb{P}\{\sum_{j=i}^N \xi_j \ge -\epsilon\} \ge 1/\beta$ *for* $2 \le i < N$. *Then* $\mathbb{P}\left\{\max_{i\le N}(\xi_1 + \ldots + \xi_i) \ge x + \epsilon\right\} \le \beta\mathbb{P}\{\xi_1 + \ldots + \xi_N \ge x\}$.

The proof is almost identical to the proof for the two-sided bound. For independent, standard normal summands, symmetry lets us take $\beta = 2$ with $\epsilon = 0$.

Define blocks $B_k := \{n : n_k < n \le n_{k+1}\}$, where $n_k/\rho^k \to 1$ for some constant $\rho > 1$ (depending on γ) that needs to be specified. For a fixed $\gamma > 1$,

$$\begin{aligned}
&\mathbb{P}\{S_n \ge \gamma L(n) \text{ for some } n \in B_k\} \\
&\quad \le \mathbb{P}\{\max_{n \in B_k} S_n \ge \gamma L(n_k)\} && \text{because } L(n) \text{ is increasing} \\
&\quad \le 2\mathbb{P}\{S_{n_{k+1}} \ge \gamma L(n_k)\} && \text{by <6> with } \beta = 2 \text{ and } \epsilon = 0 \\
&\quad \le \exp\left(-\tfrac{1}{2}\gamma^2 L(n_k)^2/n_{k+1}\right) && \text{by <4>.}
\end{aligned}$$

The expression in the exponent increases like

$$\frac{\gamma^2 \rho^k}{2\rho^{k+1}} 2 \log\log \rho^k = \frac{\gamma^2}{\rho}(\log k + \log\log \rho).$$

If $\rho < \gamma$, the bound for block B_k eventually decreases faster than $k^{-\alpha}$, for some $\alpha > 1$. The Borel-Cantelli lemma then implies that, with probability one, only finitely many of the events $\{S_n \ge \gamma L(n) \text{ for some } n \in B_k\}$ occur. As before, it then follows that $\limsup_{n\to\infty} S_n/L(n) \le 1$ almost surely, the upper half of the LIL. By symmetry, we also have $\limsup_n (-S_n/L(n)) \le 1$ almost surely, and hence

$$\limsup_n |S_n|/L(n) \le 1 \qquad \text{almost surely.} \tag{<7>}$$

The lower half of the LIL asserts that, with probability one, $S_n > \gamma L(n)$ infinitely often, for each fixed $\gamma < 1$. To prove this assertion, it is enough if we can find a sequence $\{n(k)\}$ along which $S_{n(k)} > \gamma L_{n(k)}$ infinitely often, with probability one. (I will write $n(k)$ and L_n, instead of n_k and $L(n)$, to avoid nearly invisible subscripts.) The proof uses the Borel-Cantelli lemma in the other direction, namely: if $\{A_k\}$ is a sequence of *independent* events for which $\sum_k \mathbb{P}A_k = \infty$ then the event $\{A_k$ occurs infinitely often$\}$ has probability one.

The sums $S_{n(k)}$ are not independent, because of shared summands. However the events $A_k := \{S_{n(k)} - S_{n(k-1)} > \gamma L_{n(k)}\}$ are independent. If we can choose $n(k)$ so that $\sum_k \mathbb{P}A_k = \infty$ then we will have

$$\limsup_k \frac{S_{n(k)} - S_{n(k-1)}}{L_{n(k)}} > \gamma \qquad \text{almost surely.}$$

If $n(k)$ increases rapidly enough to ensure that $L_{n(k-1)}/L_{n(k)} \to 0$, then <7> will force $S_{n(k-1)}/L_{n(k)} \to 0$ almost surely, and the lower LIL will follow.

We need a lower bound for $\mathbb{P}A_k$. The increment $S_{n(k)} - S_{n(k-1)}$ has a $N(0, m(k))$ distribution, where $m(k) := n(k) - n(k-1)$. From Appendix D,

$$\begin{aligned}
\mathbb{P}\{N(0,1) \ge x\} &\ge \left(\frac{1}{x} - \frac{1}{x^3}\right)\frac{\exp(-x^2/2)}{\sqrt{2\pi}} && \text{for } x > 1 \\
&\ge \exp\left(-\theta x^2/2\right) && \text{for all } x \text{ large enough, if } \theta > 1.
\end{aligned} \tag{<8>}$$

Thus, for fixed $\theta > 1$,

$$\mathbb{P}A_k \ge \exp\left(\frac{-\theta\gamma^2 2n(k)\log\log n(k)}{2m(k)}\right) \qquad \text{for } k \text{ large enough.}$$

The choice $n(k) := k^k$ ensures that both $n(k)/m(k) \to 1$ and $L_{n(k-1)}/L_{n(k)} \to 0$. With θ close enough to 1, the lower bound behaves like $(k \log k)^{-\alpha}$ for an $\alpha < 1$, making $\sum_k \mathbb{P}A_k$ diverge, and completing the proof of (i).

Assertion (ii) follows from (i) by symmetry of the normal distribution. For (iii) note that $\sum_n \mathbb{P}\{|S_n - S_{n-1}| > \sqrt{4\log n}\} \le \sum_n \exp(-2\log n) < \infty$. By Borel-Cantelli, $S_n - S_{n-1} = O\left(\sqrt{\log n}\right)$ a.s., which (after some algebra) implies $(S_n/L_n) - (S_{n-1}/L_{n-1}) \to 0$ a.s.. As $\{S_n/L_n\}$ oscillates between neighborhoods of $+1$ and -1 it must pass through each intervening J infinitely often. □

2. LIL for bounded summands

The upper LIL for normal variables relied on symmetry of the summands (for the appeal to the Maximal Inequality) and the exponential bound <4>. For sequences $\{S_n\}$ generated in other ways we will not have such a clean tail bound, and we need not have symmetry, but the arguments behind the LIL can be adapted in some cases.

<9> **Lemma.** *Let $T_n := \xi_1 + \ldots + \xi_n$ be a sum of independent random variables with $\mathbb{P}\xi_i = 0$ and $\sigma_i^2 := \text{var}(\xi_i) < \infty$. Suppose $\{W_n\}$ is an increasing sequence of constants with $\sigma_1^2 + \ldots + \sigma_n^2 \le W_n \to \infty$, and $\{n(k) : k \in \mathbb{N}\}$ is an increasing sequence for which $W_{n(k+1)}/W_{n(k)}$ is bounded. Then, for constants $\lambda > 1$ and $\delta > 0$,*

$$\begin{aligned} &\mathbb{P}\big\{T_n \ge (\lambda+\delta)L(W_n) \text{ for some } n \text{ with } n(k) \le n \le n(k+1)\big\} \\ &\qquad \le 2\mathbb{P}\{T_{n(k+1)} \ge \lambda L(W_{n(k)})\} \qquad \text{for all } k \text{ large enough.} \end{aligned}$$

Proof. Replace $L(W_n)$ by the lower bound $L(W_{n(k)})$, to put the inequality in the form amenable to an application of the Maximal Inequality <6>. Then argue, by Tchebychev's inequality, that for $n(k) < n \le n(k+1)$,

$$\mathbb{P}\{S_{n(k+1)} - S_n \ge -\delta L(W_{n(k)})\} \ge 1 - \frac{\text{var}\left(S_{n(k+1)} - S_n\right)}{\delta^2 L(W_{n(k)})} \ge 1 - \frac{W_{n(k+1)}}{2\delta^2 W_{n(k)} \log\log W_{n(k)}},$$

which tends to 1 as k tends to infinity. □

If we are to imitate the proof for normal summands, the other key requirement is existence of an exponential tail bound. For bounded summands there is a simple exponential inequality, which looks like <4> except for the appearance of an extra factor in the exponent, a factor involving the nonegative function

<10>
$$\psi(x) := \begin{cases} 2\left((1+x)\log(1+x) - x\right)/x^2 & \text{for } x \ge -1 \text{ and } x \ne 0 \\ 1 & \text{for } x = 0. \end{cases}$$

The function ψ is convex and decreasing (Appendix C). For the moment, it is enough to know that $\psi(x) \approx 1$ when $x \approx 0$, so that the inequalities look similar to the inequalities for normal tails when we focus on departures "not too far out into the tails."

<11> **Bennett's Inequality.** *Let $Y_1, \cdots, Y_n$ be independent random variables with*

(i) $\mathbb{P}Y_i = 0$ and $\sigma_i^2 := \mathbb{P}Y_i^2 < \infty$

(ii) $Y_i \le M$ for every i, for some finite constant M.

For each constant $W \ge \sigma_1^2 + \cdots + \sigma_n^2$,

$$\mathbb{P}\{Y_1 + \cdots + Y_n \ge x\} \le \exp\left(-\frac{x^2}{2W}\psi\left(\frac{Mx}{W}\right)\right) \qquad \text{for } x \ge 0.$$

Proof. For each $t > 0$,

$$\mathbb{P}\{Y_1 + \cdots + Y_n \geq x\} \leq e^{-xt} \prod\nolimits_{i \leq n} \mathbb{P}\exp(tY_i).$$

As shown in Appendix C, the function

<12>
$$\Delta(x) := 2(e^x - 1 - x)/x^2, \qquad \text{with } \Delta(0) = 1,$$

is nonnegative and increasing over the whole real line. Rewrite $\exp(tY_i)$ as

$$1 + tY_i + \tfrac{1}{2}(tY_i)^2\Delta(tY_i) \leq 1 + tY_i + \tfrac{1}{2}t^2Y_i^2\Delta(tM).$$

Take expectations, then invoke $1 + a \leq e^a$ to bound the tail probability by

$$\begin{aligned} e^{-xt}\prod\nolimits_{i\leq n}\left(1 + \tfrac{1}{2}t^2\sigma_i^2\Delta(tM)\right) &\leq \exp\left(-xt + \tfrac{1}{2}t^2\sum\nolimits_i \sigma_i^2\Delta(tM)\right) \\ &\leq \exp\left(-xt + \frac{W}{M^2}\left(e^{tM} - 1 - tM\right)\right). \end{aligned}$$

Minimize the exponent by putting Mt equal to $\log(1 + Mx/W)$, then rearrange to get the stated upper bound. □

The inequality $\psi(x) \geq (1 + x/3)^{-1}$, also established in Appendix C, gives a slight weakening of Bennett's inequality,

<13> **Corollary.** *(Bernstein's inequality) Under the conditions of <11>,*

<14>
$$\mathbb{P}\{Y_1 + \cdots + Y_n \geq x\} \leq \exp\left(-\frac{x^2}{2W + 2Mx/3}\right),$$

With Lemma <9> and Bennett's inequality, we have enough to establish an upper LIL, at least for summands X_i bounded in absolute value by a fixed constant M. Assume $\text{var}(S_n) := V_n \to \infty$ as $n \to \infty$. For a fixed $\rho > 1$ (depending on γ), define blocks by putting $n(k) := \max\{n : V_n \leq \rho^k\}$. The fact that $\sigma_{n(k)+1}^2 \leq M^2 = o(V_{n(k)})$ ensures that $V_{n(k)}/\rho^k \to 1$ as $k \to \infty$. From Lemma <9> with W_n equal to V_n,

$$\begin{aligned} &\mathbb{P}\{S_n \geq (\lambda + \delta)L(V_n) \text{ for some } n \text{ with } n(k) \leq n \leq n(k+1)\} \\ &\quad\leq 2\mathbb{P}\{S_{n(k+1)} \geq \lambda L(V_{n(k)})\} \\ &\quad\leq \exp\left(-\frac{\lambda^2 L(V_{n(k)})^2}{2V_{n(k+1)}}\psi\left(\frac{M\lambda L(V_{n(k)})}{V_{n(k+1)}}\right)\right). \end{aligned}$$

The expression in the exponent increases like $\lambda^2 2\rho^k \log\log(\rho^k)\psi\left(o(1)\right)/2\rho^{k+1}$. The $\psi(o(1))$ converges to 1, and therefore can be absorbed into other factors. If $\rho < \lambda$, the bound for block B_k again eventually decreases faster than $k^{-\alpha}$, for some $\alpha > 1$. The upper half of the LIL, as in first assertion of <2>, then follows via Borel-Cantelli and the casting out of a sequence of negligible sets.

Notice that uniform boundedness of the summands was needed twice:

(i) to show that $\sigma_{n(k)+1}^2 = o(V_{n(k)})$, thereby ensuring that $V_{n(k)}/\rho^k \to 1$ as $k \to \infty$;
(ii) to show that the argument of the ψ factor in the exponent tends to zero.

The same properties also hold for sequences with $|X_n| \leq M_n$, where $\{M_n\}$ is a slowly diverging sequence of constants. In fact, they hold if and

<15>
$$V_n \to \infty \qquad \text{and} \qquad M_n = o\left(\sqrt{V_n/\log\log V_n}\right) \qquad \text{as } n \to \infty,$$

a condition introduced by Kolmogorov (1929) to prove the LIL <1> for partial sums of independent random variables X_i with $\mathbb{P}X_i = 0$ and $|X_i| \le M_i$.

The proof of the upper LIL under <15> is essentially the same as the proof for uniformly bounded summands, as sketched above. The lower LIL requires an analog of the exponential lower bound <8>. The next Section establishes this lower bound, an extremely delicate exercise in Calculus. With this exponential bound, you could prove the corresponding lower LIL by modifying the analogous proof from Section 1 (or the proof of the lower LIL that will be sketched in Section 4).

*3. Kolmogorov's exponential lower bound

Let X_i, S_n and V_n be as before. Suppose also that $|X_i| \le \delta\sqrt{V_n}$ for $i = 1, 2, \ldots, n$, where δ is a small positive constant. Then for each constant $\theta > 1$ there exists an $x_0 > 0$ and a K (both depending only on θ) such that

<16>
$$\mathbb{P}\{S_n \ge x\sqrt{V_n}\} \ge \exp\left(-\tfrac{1}{2}\theta x^2\right) \qquad \text{for } x_0 \le x \le K/\delta.$$

Proof. The constants will be determined by a collection of requirements that emerge during the course of the argument. As with many proofs of this type, the requirements make little intuitive sense out of context, but it is useful to have them collected in one place, so that the dependences between the constants are clear. As you will soon see: the constant θ will determine a small $\epsilon > 0$, which in turn will determine an even smaller $\eta > 0$ (in fact, we will choose η slightly smaller than $\epsilon^2/2$), and a small, positive K depending on ϵ and η. Specifically, we will need K so small that

<17>
$$\psi(8K) \ge \max\left((1+\eta)^{-1}, \tfrac{1}{2}(1+\epsilon)\right) \qquad \text{and} \qquad \Delta(-2K)/\left(1+2K^2\right) > 1-\eta,$$

with ψ as in <10> and Δ as in <12>. To avoid an accumulation of many trivial constraints, assume (redundantly) that $0 < \eta < \epsilon < 1$. We will also need

$$\kappa := (1+\eta) - \frac{\epsilon^2}{(1+\eta)(1+\epsilon)^2} < 1-\eta$$

and

$$-\epsilon - (1+4\epsilon)(1+\epsilon) + \tfrac{1}{2}(1+\epsilon)^2(1-\eta) > -\theta/2.$$

The constant x_0 will need to be large enough that $\frac{1}{2} \ge \exp(-\epsilon x_0^2)$ and

$$2 + 3(1+\epsilon)x\exp\left(\tfrac{1}{2}x^2(1+\epsilon)^2\kappa\right) \le \tfrac{1}{2}\exp\left(\tfrac{1}{2}x^2(1+\epsilon)^2(1-\eta)\right) \qquad \text{for } x \ge x_0.$$

Now let the argument begin.

REMARK. As noted by Dudley (1989, page 379), it is notoriously difficult to manage the constants and ranges correctly for the Kolmogorov inequality. With the constraints made explicit, I hope that my errors will be easier to detect and repair.

With no loss of generality, assume $V_n = 1$ (equivalently, divide each X_i by $\sqrt{V_n}$), and consequently $\sigma_i^2 = \mathbb{P}X_i^2 \le \delta^2$ for each i. By almost the same reasoning as in the proof of Bennett's inequality, for $t > 0$,

$$\begin{aligned}
\mathbb{P}\exp(tS_n) &= \prod\nolimits_{i\le n} \mathbb{P}\left(1 + tX_i + \tfrac{1}{2}t^2X_i^2\Delta(tX_i)\right) \\
&\ge \prod\nolimits_{i\le n} \left(1 + \tfrac{1}{2}t^2\sigma_i^2\Delta(-t\delta)\right) \qquad \text{because } \Delta \text{ increasing, and } X_i \ge -\delta \\
&\ge \exp\left(\sum_{i\le n} \frac{\frac{1}{2}t^2\sigma_i^2\Delta(-t\delta)}{1 + \frac{1}{2}t^2\sigma_i^2\Delta(-t\delta)}\right) \qquad \text{via } \log(1+y) \ge \frac{y}{(1+y)}.
\end{aligned}$$

If $0 < t \le 2K/\delta$ then $\Delta(-2K) \le \Delta(-t\delta) \le \Delta(0) = 1$. We then have

<18>
$$\mathbb{P}\exp(tS_n) \ge \exp\left(\sum_{i\le n} \frac{\frac{1}{2}t^2\sigma_i^2\Delta(-2K)}{1+2K^2}\right) \ge \exp\left(\tfrac{1}{2}t^2(1-\eta)\right) \qquad \text{for } 0 < t\delta \le 2K,$$

the second equality coming from <17> and the fact that $\sum_i \sigma_i^2 = V_n = 1$. We also have an upper bound for the same quantity,

<19>
$$\mathbb{P}\exp(tS_n) = \mathbb{P}\int_{-\infty}^{\infty} te^{ty}\{S_n \ge y\}\, dy \le 1 + \int_0^{\infty} te^{ty}\mathbb{P}\{S_n \ge y\}\, dy.$$

The idea now is to choose t so that the last integrand is maximized somewhere in a small interval $J := [x, w]$, which contributes at most

<20>
$$\int_x^w te^{ty}\mathbb{P}\{S_n \ge x\}\, dy \le e^{tw}\mathbb{P}\{S_n \ge x\}$$

to the right-hand side of <19>. We need the other contributions to <19>, from y outside J, to be relatively small. For such y, we can use Bennett's inequality to bound the integrand by $t\exp\left(ty - \frac{1}{2}y^2\psi(y\delta)\right)$. If we were to ignore the ψ factor, the bound would be maximized at $y = t$, which suggests we make t slightly larger than x but smaller than w. Specifically, choose $t := (1+\epsilon)x$ and $w := (1+4\epsilon)x$, for a small ϵ that needs to be specified. (Note that $t \le 2x \le 2K/\delta$, as required for <18>.)

When y is large, the ψ factor has a substantial effect on the y^2. However, using the fact (Appendix C) that $y\psi(y)$ is an increasing function of y, and the constraint $x \le K/\delta$, we have $(y\delta)\psi(y\delta) \ge (8x\delta)\psi(8x\delta) \ge (8x\delta)\psi(8K)$ when $y \ge 8x$, hence

$$\tfrac{1}{2}y^2\psi(y\delta) \ge (\tfrac{1}{2}y)(8x)\psi(8K) = yt4\psi(8K)/(1+\epsilon) \ge 2yt \qquad \text{by <17>}.$$

The contribution from the region where $y \ge 8x$ is therefore small if $x \le K/\delta$:

<21>
$$\int_{8x}^{\infty} te^{ty}\mathbb{P}\{S_n \ge y\}\, dy \le \int_{8x}^{\infty} t\exp(ty - 2ty)\, dy = \exp(-8tx) \le 1.$$

Within the interval $[0, 8x]$ we have $\psi(y\delta) \ge \psi(8x\delta) \ge \psi(8K) \ge (1+\eta)^{-1}$ because $x \le K/\delta$, and by <17>, and the integrand $te^{ty}\mathbb{P}\{S_n \ge y\}$ is less than

$$t\exp\left(ty - \frac{y^2}{2(1+\eta)}\right) \le t\exp\left(\tfrac{1}{2}t^2(1+\eta) - \frac{|y-(1+\eta)t|^2}{2(1+\eta)}\right).$$

Notice that the exponent is maximized at $y = (1+\eta)t = (1+\eta)(1+\epsilon)x$, which lies in the interior of J, with

$$\min\left((1+\eta)t - x,\, w - (1+\eta)t\right) \ge \epsilon x \qquad \text{because } (1+\eta)(1+\epsilon) \le 1 + 3\epsilon.$$

If divided by $\sqrt{2\pi(1+\eta)}$, a constant smaller than 3, the factor contributed by the quadratic in y turns into the $N((1+\eta)t, (1+\eta))$ density. The contribution to the bound <19> from y in $[0, 8x]\backslash J$ is less than

<22> $$t\exp\left(\tfrac{1}{2}t^2(1+\eta)\right)3\mathbb{P}\{|N(0,(1+\eta))| \ge \epsilon x\} \le 3t\exp\left(\tfrac{1}{2}t^2(1+\eta) - \frac{\epsilon^2 x^2}{2(1+\eta)}\right).$$

Combining the inequalities from <18> and <19>, with the right-hand side of the latter broken into contributions bounded via <21>, <22>, and <20> then rewritten as functions of x, we have

$$\begin{aligned} 2+&3(1+\epsilon)x\exp\left(\tfrac{1}{2}x^2(1+\epsilon)^2\left((1+\eta) - \frac{\epsilon^2}{(1+\eta)(1+\epsilon)^2}\right)\right) \\ &+ \exp\left(x^2(1+4\epsilon)(1+\epsilon)\right)\mathbb{P}\{S_n \ge x\} \\ &\ge \exp\left(\tfrac{1}{2}x^2(1+\epsilon)^2(1-\eta)\right) \qquad \text{for } 0 < x \le K/\delta. \end{aligned} \tag{23}$$

We need to absorb the first two terms on the left-hand side into the right-hand side, which will happen for large enough x if we ensure that

$$(1+\eta) - \frac{\epsilon^2}{(1+\eta)(1+\epsilon)^2} < 1-\eta$$

Choose $\eta := \eta(\epsilon)$ to make this inequality hold (a value slightly smaller than $\epsilon^2/2$ will suffice), then find x_ϵ so that the sum of the first two terms in <23> is smaller than half the right-hand side when $x \ge x_\epsilon$. We may also assume that $\frac{1}{2} \ge \exp(-\epsilon x_\epsilon^2)$. Then we have

$$\mathbb{P}\{S_n \ge x\} \ge \exp\left(-\epsilon x^2 - x^2(1+4\epsilon)(1+\epsilon) + \tfrac{1}{2}x^2(1+\epsilon)^2(1-\eta)\right)$$

when $x_\epsilon \le x \le K/\delta$. Finally, we choose ϵ so small that

$$-\epsilon - (1+4\epsilon)(1+\epsilon) + \tfrac{1}{2}(1+\epsilon)^2(1-\eta) > -\theta/2,$$

which is possible because the left-hand side tends to $-1/2$ as ϵ decreases to zero. Put x_0 equal to the corresponding x_ϵ. □

*4. Identically distributed summands

Kolmogorov's LIL for bounded summands under the constraint <15> extends to identically distributed summands by means of a truncation argument, an idea due to Hartman & Wintner (1941). That is, the normality assumption can be dropped from Theorem <3>.

<24> **Theorem.** *For the sequence of partial sums $\{S_n\}$ of a sequence of independent, identically distributed random variables $\{X_i\}$ with $\mathbb{P}X_i = 0$ and $\text{var}(X_i) = 1$,*

(i) $\limsup_{n\to\infty} S_n/L_n = 1$ a.s.

(ii) $\liminf_{n\to\infty} S_n/L_n = -1$ a.s.

(iii) $S_n/L_n \in J$ *infinitely often, a.s., for every open subinterval* J *of* $[-1, 1]$.

Most of the ideas needed for the proof are contained in Sections 1 and 2. I will merely sketch the arguments needed to prove Theorem <24>, with emphasis on the way the new idea, truncation, fits neatly with the other techniques.

The truncated variables will satisfy an analog of <15> with $V_n := n$, except that the $o(\cdot)$ will be replaced by a fixed, small factor. The fragments discarded by the truncation will be controlled by the the following lemma, which formalizes an idea of DeAcosta (1983).

<25> **Lemma.** *The function*

$$g(x) := \sqrt{\frac{x}{2\log\log x}} = \frac{x}{L(x)} \qquad \text{for } x \geq e^e,$$

is strictly increasing and

$$\int_{e^e}^{g^{-1}(t)} \frac{1}{L(x)}\,dx \leq Ct \qquad \text{for } t \geq g(e^e),$$

for some constant C.

Proof. Differentiate.

$$2\frac{g'(x)}{g(x)} = \frac{1}{x} - \frac{1}{\log\log x} \cdot \frac{1}{\log x} \cdot \frac{1}{x} \geq \frac{1}{x}\left(1 - \frac{1}{e}\right) > 0 \qquad \text{for } x \geq e^e.$$

The inequality also implies that

$$\frac{1}{L(x)} = \frac{g(x)}{x} \leq Cg'(x) \qquad \text{where } C := \frac{2e}{e-1} \approx 3.16,$$

□ from which it follows, for $t \geq g(e^e)$, that $\int_{e^e}^{g^{-1}(t)} \frac{1}{L(x)}\,dx \leq \int_{e^e}^{g^{-1}(t)} Cg'(x)\,dx \leq Ct$.

Upper LIL

Consider first the steps needed to prove $\limsup (S_n/L_n) < \gamma$ a.s., for a fixed $\gamma > 1$. It helps to work with a smooth form of truncation, so that (Problem [4]) the variances of the truncated variables are necessarily smaller than the variances of the original variables. For each positive constant M define a function from $\mathbb{R}$ onto $[-M, M]$ by

<26> $$\tau(x, M) := -M\{x < -M\} + x\{|x| \leq M\} + M\{x > M\}.$$

For a fixed $\epsilon > 0$, which will depend on γ, define, for $i \geq 17 > 1 + e^e$,

$$Y_i := \tau(X_i, \epsilon g_i), \qquad Z_i := X_i - Y_i, \qquad \mu_i := \mathbb{P}Y_i, \qquad \xi_i := Y_i - \mu_i,$$

with $g_i := g(i)$, as defined by Lemma <25>. Note that $|\mu_i| \leq \mathbb{P}|Z_i|$, because $\mathbb{P}(Y_i + Z_i) = \mathbb{P}X_i = 0$.

REMARK. Notice the dependence of the truncation level on i. Compare with the truncation used to prove the SLLN under a first moment assumption in Section 4.7, and the truncations used to prove central limit theorems in Section 7.2. For almost sure convergence arguments it is common for each variable to have its own truncation level, because ultimately Borel-Cantelli assertions must depend on convergence of a single infinite series, rather than on convergence of changing sequences of partial sums to zero.

The partial sum S_n decomposes into $\sum_{i\leq n}\xi_i + \sum_{i\leq n}\mu_i + \sum_{i\leq n} Z_i$. The first sum will be handled essentially as in Section 2. The other two sums will be small relative to L_n, by virtue of the following bound,

$$\sum_{i=17}^{\infty}\frac{|\mu_i|}{L_i} \leq \sum_{i=17}^{\infty}\frac{\mathbb{P}|Z_i|}{L_i} \leq \sum_{i=17}^{\infty}\frac{\mathbb{P}|X_1|\{g^{-1}(|X_1|/\epsilon) > i\}}{L_i} \qquad \text{identical distributions}$$

$$\leq \mathbb{P}\left(|X_1|\sum_{i=17}^{\infty}\{g^{-1}(|X_1|/\epsilon) > i\}\int\frac{\{i-1\leq x < i\}}{L(x)}\,dx\right)$$

$$\leq \mathbb{P}\left(|X_1|\int\frac{\{e^e \leq x \leq g^{-1}(|X_1|/\epsilon)\}}{L(x)}\,dx\right)$$

$$\leq \mathbb{P}\left(|X_1| C|X_1|/\epsilon\right) \qquad \text{by Lemma <25>}$$

$$< \infty \qquad \text{because } \mathbb{P}X_1^2 < \infty.$$

The series started from $i = 1$ also converges. By Kronecker's lemma, $L_n^{-1}\sum_{i\leq n} L_i|\mu_i|/L_i \to 0$ as $n\to\infty$. Similarly, finiteness of the expected value of $\sum_i |Z_i|/L_i$ implies $\sum_{i\leq n} Z_i = o(L_n)$, almost surely.

To handle the contribution from the $\{\xi_i\}$, write T_n for $\sum_{i\leq n}\xi_i$ and V_n for $\text{var}(T_n)$. By Dominated Convergence, $\text{var}(\xi_i)\to 1$ as $i\to\infty$, implying $V_n/n \to 1$ as $n\to\infty$. Write γ as $1+2\delta$, with $\delta > 0$. We need to prove $\limsup(T_n/L_n) < 1+2\delta$ almost surely. As in Section 2, use blocks $B_k := \{n(k) < n \leq n(k+1)\}$, with $n(k)/\rho^k \to 1$, for a $\rho > 1$ to be specified. Invoke Lemma <9> with $W_n := n$ and $\lambda := 1+\delta$ to reduce to behavior along the geometrically spaced subsequence. Then invoke Bennett's inequality,

$$\mathbb{P}\{T_{n(k+1)} > \lambda L_{n(k)}\} \leq \exp\left(-\frac{\lambda^2 2n(k)\log\log n(k)}{2V_{n(k+1)}}\psi\left(\frac{2\epsilon g_{n(k+1)}\lambda L_{n(k)}}{V_{n(k+1)}}\right)\right).$$

The argument of the ψ factor behaves like

$$\frac{2\epsilon n(k+1)\lambda L_{n(k)}}{L_{n(k+1)}n(k+1)} < 2\epsilon\gamma.$$

For fixed γ, the ψ factor can be brought as close to 1 as we please, by choosing ϵ small enough. The other term in the exponent behaves like $(\lambda^2/\rho)\log\log\rho^k$. With appropriate choices for ρ and ϵ, we therefore have the tail probability decreasing faster than $k^{-\alpha}$, for some $\alpha > 1$, which leads to the desired upper LIL.

Lower LIL

For a fixed $\gamma < 1$ we need to show $\limsup(T_n/L_n) > \gamma$ almost surely. As in Section 1, look along a subsequence $n(k) := k^k$. Write T for $T_{n(k)} - T_{n(k-1)}$, and V for $\text{var}(T) = V_{n(k)} - V_{n(k-1)}$. We can make $V/n(k)$ as close to 1 as we please, by making k large enough. The summands contributing to T are bounded in abolute value by $\delta\sqrt{V}$, where $\delta := 2\epsilon g_{n(k)}/\sqrt{V}$. We need to bound $\mathbb{P}\{T > \gamma L_{n(k)}\}$ from below by a term of a divergent series. Fix a $\theta > 1$. Write x for $\gamma L_{n(k)}/\sqrt{V}$. Inequality <16> gives

$$\mathbb{P}\{T > \gamma L_{n(k)}\} = \mathbb{P}\{T \geq x\sqrt{V}\} \geq \exp\left(-\tfrac{1}{2}\theta x^2\right) = \exp\left(-\frac{\theta\gamma^2 2n(k)\log\log n(k)}{2V}\right)$$

provided $x_0 \le x \le K/\delta$, that is, provided

$$x_0 \le \gamma\sqrt{\frac{2n(k)\log\log n(k)}{n(k)(1+o(1))}} \le \frac{K}{2\epsilon}\sqrt{\frac{n(k)(1+o(1))2\log\log n(k)}{n(k)}}$$

With ϵ small enough, the range eventually contains the desired x value. The rest of the argument follows as in Section 1.

Cluster points

Assertion (iii) of Theorem <24> will follow from assertion (i), by means of an ingenious projection argument, borrowed from Finkelstein (1971). Construct new independent observations $\{\bar{X}_i\}$ with the same distribution as the $\{X_i\}$, and let $\bar{S}_n := \bar{X}_1 + \ldots + \bar{X}_n$. Write W_n for the random vector $(S_n, \bar{S}_n)/L_n$ and u_θ for the unit vector $(\cos\theta, \sin\theta)$. For each fixed θ, the random variables $W_n \cdot u_\theta = X_i \cos\theta + \bar{X}_i \sin\theta$ have mean zero, variance one, and they are identically distributed. From (i), $\limsup_{n\to\infty}(W_n \cdot u_\theta) = 1$ almost surely. Given $\epsilon > 0$, there exists a finite collection of halfspaces $\{(x, y) \cdot u_\theta \le 1 + \epsilon\}$ whose intersection lies inside the ball of radius $1 + 2\epsilon$ about the origin. It follows that $\limsup |W_n| \le 1 + 2\epsilon$ almost surely. The geometry of the circle then forces W_n to visit each neighborhood of the boundary point u_θ infinitely often, with probability one. The projection of such a neighborhood onto the horizontal axis gives a neighborhood of the point $\cos\theta$, which S_n/L_n must visit infinitely often. After a casting out of a countable sequence of negligible sets, for a countable collection of subintervals of $(-1, 1)$, we then deduce assertion (iii).

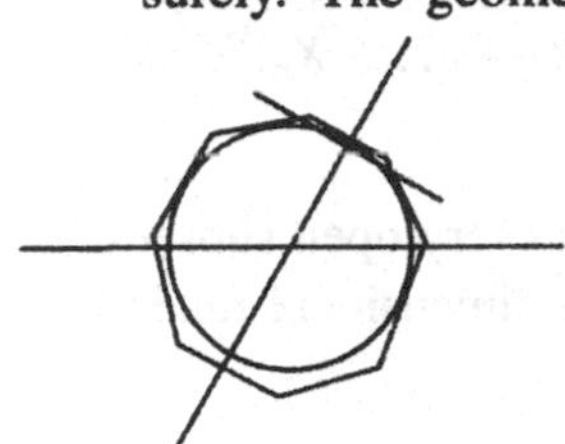

5. Problems

[1] Let X have a Bin(n, p) distribution. Define $q := 1 - p$. For $0 \le x \le nq$, show that

$$\mathbb{P}\{X \ge np + x\} \le \exp\left(-\frac{x^2}{2npq}\left(q\psi\left(\frac{x}{np}\right) + p\psi\left(\frac{-x}{nq}\right)\right)\right)$$
$$\le \exp\left(-\frac{x^2}{2npq}\psi\left(\frac{x(q-p)}{npq}\right)\right)$$

Hint: Bound the tail probability by $\exp\left(-t(np+x) + n\log(q + pe^t)\right)$ for $t \in \mathbb{R}^+$, then minimize the expression in the exponent by Calculus. For $0 < x < nq$ show that the minimum is achieved at $t := \log\left(\frac{1 + x/np}{1 - x/nq}\right)$. For the second bound, use convexity of ψ.

[2] Let $X_1, \ldots, X_n$ be independent random variables with $\mathbb{P}X_i := p_i$ and $0 \le X_i \le 1$ for each i. Let $p := (p_1 + \cdots + p_n)/n =: 1 - q$. Show that

$$\mathbb{P}\left\{\sum_{i\le n} X_i \ge np + x\right\} \le \exp\left(-\frac{x^2}{2npq}\psi\left(\frac{x(q-p)}{npq}\right)\right) \qquad \text{for } 0 \le x \le nq$$

Hint: For $t \in \mathbb{R}^+$, bound the tail probability by

$$\exp(-t(np+x))\mathbb{P}\exp(t\Sigma X_i) = \exp\left(-t(np+x) + \sum_{i\le n}\log(q_i + p_i e^t)\right).$$

Use concavity of the logarithm function to increase the bound to a form amenable to the method of Problem [1].

[3] Suppose X has a Poisson(λ) distribution.

(i) By direct minimization of $\exp(-t(\lambda+x))\mathbb{P}\exp(tX)$ over $\mathbb{R}^+$, prove that

$$\mathbb{P}\{X \ge \lambda + x\} \le \exp\left(-\frac{x^2}{2\lambda}\psi\left(\frac{x}{\lambda}\right)\right).$$

(ii) Derive the same tail bound by a passage to the limit in the binomial bound from Problem [1].

[4] For each random variable X with finite variance, and each constant M, show that $\text{var}(\tau(X,M)) \le \text{var}(X)$, with τ as defined in <26>. Hint: Let X' be an independent copy of X. Show that $2\text{var}(\tau(X,M)) = \mathbb{P}|\tau(X,M) - \tau(X',M)|^2$ and also that $|\tau(x,M) - \tau(x',M)| \le |x - x'|$ for all real x and x'.

[5] Let $\{X_n\}$ be a sequence of independent, identically distributed two-dimensional random vectors with $\mathbb{P}X_i = 0$ and $\text{var}(X_i) = I_2$. Define $S_n = X_1 + \ldots + X_n$.

(i) Show that $\limsup |S_n|/L_n \le 1$ almost surely.

(ii) Show that, with probability one, the sequence S_n/L_n visits every open subset of the unit ball $\{|x| < 1\}$ infinitely often. Hint: Project three-dimensional random vectors.

6. Notes

My understanding of the LIL began with the reading of Feller (1968, Section VIII.5), Lamperti (1966, Section 11) and Stout (1974, Chapter 5). I learned the idea of regarding the Bennett inequality as a slightly corrupted (by the presence of the ψ function in the exponent) analog the the normal tail bound from conversations with Galen Shorack and Jon Wellner. Shorack (1980) systematically exploited the idea to establish very sharp LIL results for the empirical distribution function. For a beautiful exposition of the many applications of the idea to the study of inequalities for the empirical distribution function and related processes see Shorack & Wellner (1986, Chapter 11).

The method used to establish the Bennett inequality, but not the form of the inequality, comes from Chow & Teicher (1978, page 338). They developed exponential inequalities suitable for derivation of LIL results. I am uncertain about the earlier history of the exponential bounds. Apparently (cf. Kolmogorov & Sarmanov 1960), Bernstein's inequality comes from a 1924 paper. The Bennett (1962) and Hoeffding (1963) papers contain other tail bounds for sums of random variables, with some references to further literature.

Apparently the first versions of the LIL with a log log bound are due to Khinchin (1923, 1924). For the early history of the LIL, leading up to the definitive version by Kolmogorov (1929), see Feller (1943). Hartman & Wintner (1941) extended to the case of independent identically distributed summands with finite second moments, by means of a truncation argument. (Actually, they did not assume identical distributions, but only a domination condition for the tails, which holds under the second moment condition for identically distributed summands.) My exposition in Section 4 draws from DeAcosta (1983), who gave an elegant alternative derivation (and extension) of the Hartman-Wintner version of the LIL.

I thank Jim Kuelbs for the proof of part (iii) of Theorem <3>.

References

Bennett, G. (1962), 'Probability inequalities for the sum of independent random variables', *Journal of the American Statistical Association* **57**, 33–45.

Chow, Y. S. & Teicher, H. (1978), *Probability Theory: Independence, Interchangeability, Martingales*, Springer, New York.

DeAcosta, A. (1983), 'A new proof of the Hartman-Wintner law of the iterated logarithm', *Annals of Probability* **11**, 270–276.

Dudley, R. M. (1989), *Real Analysis and Probability*, Wadsworth, Belmont, Calif.

Feller, W. (1943), 'The general form of the so-called law of the iterated logarithm', *Transactions of the American Mathematical Society* **54**, 373–402.

Feller, W. (1968), *An Introduction to Probability Theory and Its Applications*, Vol. 1, third edn, Wiley, New York.

Finkelstein, H. (1971), 'The law of the iterated logarithm for empirical distributions', *Annals of Mathematical Statistics* **42**, 607–615.

Hartman, P. & Wintner, A. (1941), 'On the law of the iterated logarithm', *American Journal of Mathematics* **63**, 169–176.

Hoeffding, W. (1963), 'Probability inequalities for sums of bounded random variables', *Journal of the American Statistical Association* **58**, 13–30.

Khinchin, A. Ya. (1923), 'Über dyadische Brüche', *Math. Zeit.* **18**, 109–116.

Khinchin, A. Ya. (1924), 'Über einen Satz der Wahrscheinlichkeitsrechnung', *Fundamenta Mathematicae* **6**, 9–20.

Kolmogorov, A. (1929), 'Über das Gesetz des Iterierten Logarithmus', *Mathematische Annalen* **101**, 126–135.

Kolmogorov, A. N. & Sarmanov, O. V. (1960), 'The work of S. N. Bernshtein on the theory of probability', *Theory Probability and Its Applications* **5**, 197–203.

Lamperti, J. (1966), *Probability: A Survey of the Mathematical Theory*, W. A. Benjamin, New York.

Shorack, G. R. (1980), 'Some law of the iterated logarithm type results for the empirical process', *Australian Journal of Statistics* **22**(1), 50–59.

Shorack, G. R. & Wellner, J. A. (1986), *Empirical Processes with Applications to Statistics*, Wiley, New York.

Stout, W. F. (1974), *Almost Sure Convergence*, Academic Press.

Chapter 12

Multivariate normal distributions

SECTION 1 explains why you will not learn from this Chapter everything there is to know about the multivariate normal distribution.
SECTION 2 introduces Fernique's inequality. As illustration, Sudakov's lower bound for the expected value of a maximum of correlated normals is derived.
*SECTION *3 proves Fernique's inequality.*
SECTION 4 introduces the Gaussian isoperimetric inequlity. As an application, Borell's tail bound for the distribution of the maximum of correlated normals is derived.
*SECTION *5 proves the Gaussian isoperimetric inequlity.*

1. Introduction

Of all the probability distributions on multidimensional Euclidean spaces the multivariate normal is the most studied and, in many ways, the most tractable. In years past, the statistical subject known as "Multivariate Analysis" was almost entirely devoted to the study of the multivariate normal. The literature on Gaussian processes—stochastic processes whose finite dimensional distributions are all multivariate normal—is vast. It is important to know a little about the multivariate normal.

As you saw in Section 8.6, the multivariate normal is uniquely determined by its vector of means and its matrix of covariances. In principle, everything that one might want to know about the distribution can be determined by calculation of means and covariances, but in practice it is not completely straightforward. In this Chapter you will see two elegant examples of what can be achieved: Fernique's (1975) inequality, which deduces important information about the spread in a multivariate normal distribution from its covariances; and Borell's (1975) Gaussian isoperimetric inequality, with a proof due to Ehrhard (1983a, 1983b). Both results are proved by careful Calculus.

The Chapter provides only a very brief glimpse of Multivariate Analysis and the theory of Gaussian processes, two topics that are covered in great detail in many specialized texts. I have chosen merely to present examples that give the flavor of some of the more modern theory. Both the Fernique and Borell inequalities have found numerous applications in the recent research literature.

2. Fernique's inequality

The larger σ^2, the more spread out is the $N(0, \sigma^2)$ distribution. Fernique (1975, page 18) proved a striking multivariate generalization of this simple fact.

<1> **Theorem.** *Suppose X and Y both have centered (zero means) multivariate normal distributions, with $\mathbb{P}|X_i - X_j|^2 \le \mathbb{P}|Y_i - Y_j|^2$ for all i, j. Then*

$$\mathbb{P} f(\max_i X_i - \min_i X_i) \le \mathbb{P} f(\max_i Y_i - \min_i Y_i)$$

for each increasing, convex function f on $\mathbb{R}^+$.

The theorem lets us deduce inequalties for multivariate normal distributions, with potentially complicated covariance structures, by making comparisons with simpler processes.

<2> **Example.** (Sudakov's minoration) Let $Y := (Y_1, Y_2, \ldots, Y_n)$ have a centered multivariate normal distribution, with $\mathbb{P}|Y_j - Y_k|^2 \ge \delta^2$ for all $j \ne k$. Fernique's inequality will show that $\mathbb{P} \max_{i \le n} Y_i \ge C\delta\sqrt{\log_2 n}$, where $C := \mathbb{P}|N(0,1)|/\sqrt{8}$.

The result is trivial for $n = 1$. For $n \ge 2$, let k be the largest integer for which $n \ge 2^k$. Note that $2k \ge k + 1 > \log_2 n \ge k$. Reindex the variables $\{Y_i : 1 \le i \le 2^k\}$ by the vectors α in the set $A = \{-1, +1\}^k$ of all k-tuples of ± 1 values. Write Y_α instead of Y_i. The precise correspondence between A and $\{i : 1 \le i \le 2^k\}$ is unimportant.

Build another centered multivariate normal family $\{X_\alpha : \alpha \in A\}$,

$$X_\alpha := \tfrac{1}{2}\delta k^{-1/2} \sum\nolimits_{i=1}^{k} \alpha_i W_i \qquad \text{where } W_1, \ldots, W_k \text{ are independent } N(0,1)\text{'s}.$$

For $\alpha \ne \beta$,

$$\mathbb{P}|X_\alpha - X_\beta|^2 = \tfrac{1}{4}\delta^2 k^{-1} \sum\nolimits_i (\alpha_i - \beta_i)^2 \le \delta^2 \le \mathbb{P}|Y_\alpha - Y_\beta|^2.$$

From Fernique's inequality with $f(t) := t$ we get

$$\mathbb{P}(\max_\alpha Y_\alpha - \min_\alpha Y_\alpha) \ge \mathbb{P}(\max_\alpha X_\alpha - \min_\alpha X_\alpha).$$

Symmetry of the multivariate normal implies that $\max_\alpha Y_\alpha$ has the same distribution as $\max_\alpha(-Y_\alpha) = -\min_\alpha Y_\alpha$, and similarly for the X's. The last inequality implies

$$\mathbb{P} \max_\alpha Y_\alpha \ge \mathbb{P} \max_\alpha X_\alpha = \tfrac{1}{2}\delta k^{-1/2} \mathbb{P}\left(\max_\alpha \sum\nolimits_{i=1}^{k} \alpha_i W_i\right).$$

For each realization $W := (W_1, \ldots, W_k)$, the maximum of $\sum_i \alpha_i W_i$ is achieved when each α_i takes the same sign as W_i. (Of course, the maximizing α depends on W.) The lower bound equals

$$\tfrac{1}{2}\delta k^{-1/2} \mathbb{P}\left(\sum\nolimits_{i=1}^{k} |W_i|\right) = \tfrac{1}{2}\delta k^{-1/2} k \mathbb{P}|N(0,1)| \ge \delta\sqrt{\log_2 n}\, \mathbb{P}|N(0,1)|/\sqrt{8},$$

□ as asserted.

REMARK. The lower bound is sharp within a constant, in the following sense. If $\mathbb{P}|Y_j - Y_k|^2 \le \delta^2$ for all $j \ne k$ then $\mathbb{P} \max_i Y_i = \mathbb{P} Y_1 + \mathbb{P} \max_i (Y_i - Y_1) = \mathbb{P} \max_i (Y_i - Y_1)$ and, by Jensen's inequality and monotonicity,

$$\exp\left(\mathbb{P} \max_i (Y_i - Y_1)/2\delta\right)^2 \le \mathbb{P} \max_i \exp\left((Y_i - Y_1)^2/4\delta^2\right) \le n \mathbb{P} \exp\left(N\left(0, \tfrac{1}{4}\right)^2\right).$$

Thus $\mathbb{P} \max_i Y_i$ is bounded above by $2\delta\sqrt{\log(\sqrt{2}n)}$.

*3. Proof of Fernique's inequality

A straightforward approximation argument (see Appendix C) reduces to the case where f has a bounded continuous second derivative with bounded support. Also we may assume that the covariance matrices $V_0 := \text{var}(X)$ and $V_1 := \text{var}(Y)$ are both nonsingular $n \times n$ matrices: We could prove the result for $X + \epsilon Z$ and $Y + \epsilon Z$, where Z is a distributed $N(0, I_n)$ independently of X and Y, then let ϵ tend to zero.

To simplify notation, I will write ∂_j for $\partial/\partial x_j$ and $\partial^2_{j,k}$ for $\partial^2/\partial x_j \partial x_k$.

For nonsingular V_0 and V_1, the covariance matrix $V_\theta := (1-\theta)V_0 + \theta V_1$ is nonsingular for $0 \le \theta \le 1$. The $N(0, V_\theta)$ distribution has a density g_θ with respect to Lebesgue measure $\mathfrak{M}$ on $\mathbb{R}^n$. A simple calculation (Problem [1]) based on the Fourier inversion formula for the $N(0, V_\theta)$ density gives

<3> $$\frac{\partial g_\theta}{\partial \theta}(x) = \tfrac{1}{2} \sum\nolimits_{j,k} \Delta_{j,k} \partial^2_{j,k} g_\theta(x) \qquad \text{where } \Delta := V_1 - V_0.$$

The Theorem will be proved if we can show that the function

$$H(\theta) := \mathfrak{M}^x \left(f\left(\max_i x_i - \min_i x_i\right) g_\theta(x) \right)$$

is increasing in θ. The assumptions on f justify differentation under the $\mathfrak{M}$ to get

$$H'(\theta) = \mathfrak{M}^x f\left(\max_i x_i - \min_i x_i\right) \frac{\partial}{\partial \theta} g_\theta(x)$$

<4> $$= \tfrac{1}{2} \sum\nolimits_{j,k} \Delta_{j,k} \mathfrak{M}^x \left(f\left(\max_i x_i - \min_i x_i\right) \partial^2_{j,k} g_\theta \right) \qquad \text{by <3>}.$$

Two integrations by parts will replace the $\mathfrak{M}$-integrals by integrals involving the nonnegative functions f' and f'', reducing the expression for $H'(\theta)$ to a sum of the form $\sum_{j<k} \left(\Delta_{jj} + \Delta_{kk} - 2\Delta_{jk}\right)$ (something nonnegative). To establish such a representation, we need to keep track of contributions from subregions of $\mathbb{R}^n$ defined by inequalities involving the functions

$$L(x) := \max_i x_i \qquad \text{and} \qquad S(x) := \min_i x_i$$
$$L_j(x) := \max_i \{i \neq j\} x_i \qquad \text{and} \qquad S_j(x) := \min_i \{i \neq j\} x_i.$$

Notice that $L(x) = L_j(x) \vee x_j$ and $S(x) = S_j(x) \wedge x_j$, for each j.

Let $\mathfrak{m}_j$ denote Lebesgue measure on the jth coordinate space $\mathbb{R}$, and $\mathfrak{M}_j$ denote $(n-1)$-dimensional Lebesgue measures on the product of the remaining coordinate subspaces. That is, $\mathfrak{m}_j$ integrates over x_j and $\mathfrak{M}_j$ integrates over the remaining $n-1$ coordinate variables. The product of $\mathfrak{m}_j \otimes \mathfrak{M}_j$ equals $\mathfrak{M}$, Lebesgue measure on $\mathbb{R}^n$.

The function $x_j \mapsto fL - S)$ is absolutely continuous with almost sure derivative $f'(x_j - S_j)\{x_j > L_j\} - f'(L_j - x_j)\{x_j < S_j\} = f'(L - S)\left(\{x_j = L\} - \{x_j = S\}\right)$. Here, and subsequently, I ignore the $\mathfrak{M}$-negligible set of x for which there is a tie for maximum or minimum. The function $\partial_k g_\theta$ decreases to zero exponentially fast as $\max_i |x_i| \to \infty$. An integration by parts with respect to x_j gives

$$\begin{aligned}\mathfrak{M}^x f(L-S)\, \partial^2_{j,k} g_\theta &= \mathfrak{M}_j \left(\mathfrak{m}_j f(x_j \vee L_j - x_j \wedge S_j) \partial_j (\partial_k g_\theta) \right) \\ &= -\mathfrak{M}_j \left(\mathfrak{m}_j f'(L-S) \left(\{x_j = L\} - \{x_j = S\}\right) \partial_k g_\theta \right)\end{aligned}$$

<5> $$= -\mathfrak{M} \left(f'(L-S) \left(\{x_j = L\} - \{x_j = S\}\right) \partial_k g_\theta \right).$$

The second integration-by-parts proceeds slightly differently for $j = k$ or $j \neq k$. To simplify notation, I will temporarily assume $j = 1$ and either $k = 1$ or $k = 2$; and I will replace $x_3, \ldots, x_n$ or $x_2, x_3, \ldots, x_n$ by a long dash (—) to keep attention focussed on the variables actively involved in the calculation.

Mixed partial derivative

With $j := 1$ and $k := 2$, consider the final integrand of <5> as a function of x_2. Rewrite the difference of indicator functions as $\{x_2 < x_1 = L_2\} - \{x_2 > x_1 = S_2\}$. Integration by parts, with respect to x_2, gives

$$\begin{aligned}
&\mathfrak{m}_2\left(f'(L-S)\left(\{x_1 = L\} - \{x_1 = S\}\right)\partial_2 g_\theta\right) \\
&\quad = f'(L-S)\{x_1 = L_2\}g_\theta(x)\Big|_{x_2=-\infty}^{x_1} - f'(L-S)\{x_1 = S_2\}g_\theta(x)\Big|_{x_2=x_1}^{\infty} \\
&\qquad - \mathfrak{m}_2\left(f''(L-S)\left(\{x_2 = L\} - \{x_2 = S\}\right)\left(\{x_1 = L\} - \{x_1 = S\}\right)g_\theta\right) \\
&\quad = f'(L_2 - S_2)g_\theta(x_1, x_1, —)\left(\{x_1 = L_2\} + \{x_1 = S_2\}\right) \\
&\qquad + \mathfrak{m}_2\left(f''(L-S)\left(\{x_1 = S,\, x_2 = L\} + \{x_1 = L,\, x_2 = S\}\right)g_\theta\right) \qquad \text{a.e. } [\mathfrak{M}_2].
\end{aligned}$$

The $\mathfrak{M}_2$ negligible set allows for x where x_1 and x_2 tie for maximum or minimum. Integrate with respect to $\mathfrak{M}_2$ to get an expression for the mixed partial derivative.

$$\begin{aligned}
\mathfrak{M}f(L-S)\partial_{1,2}^2 g_\theta &= -\mathfrak{M}_2\left(f'(L_2 - S_2)\left(\{x_1 = S_2\} + \{x_1 = L_2\}\right)g_\theta(x_1, x_1, —)\right) \\
&\quad - \mathfrak{M}\left(f''(L-S)\left(\{x_1 = S,\, x_2 = L\} + \{x_1 = L,\, x_2 = S\}\right)g_\theta(x)\right) \\
&:= -A_{1,2} - B_{1,2}. \qquad \text{<6>}
\end{aligned}$$

Notice the curious form of the integral for $A_{1,2}$. It runs over $n-1$ variables (x_2 omitted) ranging over the sets where either x_1 is the smallest or the largest of those variables. The second argument of g_θ, previously occupied by x_2, is now occupied by the extremal value x_1. We would get exactly the same integral if we interchanged the roles of x_1 and x_2. Write $A_{j,k}$ for the analogous integral with x_j taking over the role of x_1 and x_k taking over the role of x_2. Nonnegativity of f', and the symmetry of the roles of the two variables, ensure that $A_{j,k} = A_{k,j} \geq 0$ for all $j \neq k$. Similarly, write $B_{j,k}$ for the analog of $B_{1,2}$ with x_j taking over the role of x_1 and x_k taking over the role of x_2. Nonegativity of f'' implies that $B_{j,k} = B_{k,j} \geq 0$ for all $j \neq k$.

Repeated partial derivative

The calculations for $\partial_{1,1}^2 g_\theta$ are similar. Integrate first with respect to x_1.

$$\begin{aligned}
&\mathfrak{m}_1\left(f'(L-S)\left(\{x_1 = L\} - \{x_1 = S\}\right)\partial_1 g_\theta\right) \\
&\quad = f'(L-S)g_\theta(x)\Big|_{x_1=L_1}^{\infty} - f'(L-S)g_\theta(x)\Big|_{x_1=-\infty}^{S_1} \\
&\qquad - \mathfrak{m}_1\left(f''(L-S)\left(\{x_1 = L\} - \{x_1 = S\}\right)^2 g_\theta(x)\right) \\
&\quad = -f'(L_1 - S_1)\left(g_\theta(S_1, —) + g_\theta(L_1, —)\right) \\
&\qquad - \mathfrak{m}f''(L-S)g_\theta(x)\left(\{x_1 = S\} + \{x_1 = L\}\right)
\end{aligned}$$

Then integrate over the remaining $n-1$ variables, to conclude that

$$\begin{aligned}
\mathfrak{M}\left(f(L-S)\partial_{1,1}^2 g_\theta\right) &= \mathfrak{M}_1 f'(L_1 - S_1)\left(g_\theta(S_1, —) + g_\theta(L_1, —)\right) \\
&\quad + \mathfrak{M}\left(f''(L-S)g_\theta(x)\left(\{x_1 = S\} + \{x_1 = L\}\right)\right).
\end{aligned}$$

When split according to which of the variables $x_2, \ldots, x_n$ achieves the extremal value S_1 or L_1, the first integral breaks into the sum $\sum_{k\geq 2} A_{1,k}$, and the second into $\sum_{k\geq 2} B_{1,k}$. The other repeated derivatives contribute similar expressions.

Collection of terms

Substitute the results from the two integrations by parts into <4>.

$$\begin{aligned} H'(\theta) &= \tfrac{1}{2}\sum\nolimits_j \Delta_{j,j}\mathfrak{M}^x f(L-S)\,\partial^2_{j,j}g_\theta + \sum\nolimits_{j<k}\Delta_{j,k}\mathfrak{M}^x f(L-S)\,\partial^2_{j,k}g_\theta \\ &= \tfrac{1}{2}\sum_j \Delta_{j,j}\sum_k\{k\neq j\}\left(A_{j,k}+B_{j,k}\right) - \sum\nolimits_{j<k}\Delta_{j,k}\left(A_{j,k}+B_{j,k}\right) \\ &= \tfrac{1}{2}\sum\nolimits_{j<k}\left(\Delta_{j,j}+\Delta_{k,k}-2\Delta_{j,k}\right)\left(A_{j,k}+B_{j,k}\right). \end{aligned}$$

The assumption of the Theorem tells us that

$$\begin{aligned} \Delta_{jj}+\Delta_{kk}-2\Delta_{jk} &= \mathbb{P}Y_j^2 - \mathbb{P}X_j^2 + \mathbb{P}Y_k^2 - \mathbb{P}X_k^2 - 2(\mathbb{P}Y_jJ_k - \mathbb{P}X_jX_k) \\ &= \mathbb{P}|Y_j - Y_k|^2 - \mathbb{P}|X_j - X_k|^2 \geq 0. \end{aligned}$$

□ Thus $H'(\theta) \geq 0$, and Fernique's inequality follows.

4. Gaussian isoperimetric inequality

Let γ_k denote the standard normal, $N(0, I_k)$, distribution on $\mathbb{R}^k$. Write Φ for the one-dimensional $N(0,1)$ distribution function, and write $\phi(x) := (2\pi)^{-1/2}\exp(-x^2/2)$ for its density with respect to Lebesgue measure.

For each Borel subset A of $\mathbb{R}^k$ and each $r > 0$ define $A^r := \{x : d(x, A) \leq r\}$, a set that I will call a neighborhood of A. The isoperimetric problem requires minimization of $\gamma_k A^r$ over all A with a fixed value of $\gamma_k A$. Borell (1975) showed that the minimizing choice is a closed half-space H. The neighborhood H^r is another half-space. By rotational symmetry of the standard normal, the calculation of $\gamma_k H^r$ reduces to a one-dimensional problem: if $\gamma_k H := \Phi(\alpha)$ then $\gamma_k H^r = \Phi(r+\alpha)$.

The term *isoperimetric* comes from an analogy with the classical isoperimetric inequality for minimization of surface area of a set with fixed Lebesgue measure. If one avoids the tricky problem of defining the surface area of a general Borel set A by substituting the Lebesgue measure of the thin shell $A^r\backslash A$, for very small r, the problem becomes one of minimizing the Lebesgue measure $\mathfrak{m}A^r$ for a fixed value of $\mathfrak{m}A$. Replace Lebesgue measure by γ_k and we have the Gaussian analog of the modified isoperimetric problem.

<7> **Theorem.** *The Gaussian measure of the neighbourhood A^r is minimized, for a given value of $\gamma_k A$, by choosing A as a closed half-space. More generally, $\gamma_k A^r \geq \Phi(r+\alpha)$, for every Borel subset A of $\mathbb{R}^k$ with $\gamma_k A \geq \Phi(\alpha)$.*

It is the reduction from a k-dimensional problem, with k arbitrarily large, to a one-dimensional calculation for the lower bound that makes Borell's result so powerful, as shown by the inequalities in the next Example.

<8> **Example.** Recall that a median of a (real valued) random variable X is any constant m for which $\mathbb{P}\{X \geq m\} \geq 1/2$ and $\mathbb{P}\{X \leq m\} \geq 1/2$. Such an m always exists, but it need not be unique.

Suppose $Y_1, \ldots, Y_n$ have a multivariate normal distribution. Define $S := \max_{i \leq k} |Y_i|$, and let M be a median of S. (In fact, the inequalities will force M to be unique.) Define $\sigma^2 := \max_i \operatorname{var}(X_i)$. Borell's concentration inequality asserts that $\mathbb{P}\{S > M + r\} \leq \mathbb{P}\{N(0, \sigma^2) \geq r\}$ and $\mathbb{P}\{S < M - r\} \leq \mathbb{P}\{N(0, \sigma^2) \leq -r\}$, for each $r \geq 0$. More succinctly,

<9> $$\mathbb{P}\{|\max|Y_i| - M)| > r\} \leq \mathbb{P}\{|N(0, \sigma^2))| > r\} \qquad \text{for each } r \geq 0.$$

That is, the spread in the distribution of $\max|Y_i|$ about its median is no worse than the spread in the distribution of the Y_i with largest variance.

In special cases, such as independent variables (Problem [3]), one can get tighter bounds, but Borell's inequality has two great virtues: it is impervious to the effects of possible dependence between the Y_i, and it does not depend on n. In consequence, it implies similar concentration inequalities for general Gaussian processes, with surprising consequences.

Each half of the Borell inequality follows easily from Theorem <7> if we represent the Y_i as linear functions $Y_i(x) := \mu_i + \theta_i'x$ on $\mathbb{R}^n$ equipped with the measure γ_n. (That is, regard the vector of Y_i's as a linear transformation of a vector of independent $N(0, 1)$'s.) The assumption about the variances becomes $\operatorname{var}(Y_i) = |\theta_i|^2 \leq \sigma^2$ for each i. By definition of the median M, the set

$$A := \{x \in \mathbb{R}^n : \max_{i \leq n} |Y_i(x)| \leq M\}$$

has $\gamma_n A \geq 1/2 = \Phi(0)$. If a point x lies within a distance r of a point x_0 in A, then $|\theta_i'x - \theta_i'x_0| \leq |\theta_i|\,|x - x_0| \leq \sigma r$, for each i. Thus the neighborhood A^r is contained within $\{x : \max_{i \leq n} |Y_i(x)| \leq M + \sigma r\}$, and

$$\mathbb{P}\{\max_{i \leq k} |Y_i| \leq M + \sigma r\} \geq \gamma_n A^r \geq \Phi(0 + r),$$

as asserted by the upper half of the Borell inequality. The derivation for the companion inequality is similar. □

<10> **Example.** Suppose $X_1, X_2, \ldots$ is a Gaussian sequence of random variables. The supremum $S := \sup_i |X_i|$ might take infinite values, but if $\mathbb{P}\{S < \infty\} > 0$ then Borell's inequality <9> will show that the distribution of S must have an upper tail that decreases like that of a normal distribution. More precisely, the constant $\sigma^2 := \sup_i \operatorname{var}(X_i)$ must be finite, and there must exist some finite constant M such that

<11> $$\mathbb{P}\{S > M + \sigma r\} \leq \bar{\Phi}(r) = \mathbb{P}\{N(0, 1) > r\} \qquad \text{for all } r \geq 0.$$

Of course there is no comparable result needed for the lower tail, because S is nonnegative.

The tail bound <11> ensures that S has finite moments of all orders, and in fact $\mathbb{P}\exp(\alpha S^2) < \infty$ for all $\alpha < (2\sigma^2)^{-1}$. See Problem [8].

We can establish <11> by reducing the problem to finite sets of random variables, to which Borell's inequality applies. Write S_n for $\max_{i\le n}|X_i|$ and M_n for its median. Define $\sigma_n^2 := \max_{i\le n}\operatorname{var}(X_i)$. From <9>,

$$\mathbb{P}\{|S_n - M_n| > \sigma_n r\} \le \bar{\Phi}(r). \tag{<12>}$$

The assumption on S ensures existence of a finite constant C and an $\epsilon > 0$ such $\mathbb{P}\{S \le C\} \ge \epsilon$, which implies that $\mathbb{P}\{|X_i| \le C\} \ge \mathbb{P}\{S_n \le C\} \ge \epsilon$ for all n and $i \le n$. These inequalities place an upper bound of $2C^2/\pi\epsilon^2$ on $\operatorname{var}(X_i)$, and thereby also on σ^2, because $\mathbb{P}\{|N(\mu,\tau^2)| \le C\} \le 2C/\tau\sqrt{2\pi}$. Choose an r_0 for which $\bar{\Phi}(r_0) < \epsilon$. From <12> we have $\mathbb{P}\{S_n < M_n - \sigma_n r_0\} < \epsilon$, which excludes the possibility that $M_n - \sigma_n r_0$ might be larger than C. The constant $M := \sup_n M_n$ is therefore bounded above by $C + \sigma r_0$. From <12> we then get $\mathbb{P}\{S_n > M + \sigma r\} \le \bar{\Phi}(r)$ for all n and all $r \ge 0$, an assertion stronger than <11>. □

*5. Proof of the isoperimetric inequality

First note that we may assume A is closed, because $\bar{A}^r = A^r$ and $\gamma_k\bar{A} \ge \gamma_k A$. As a convenient abbreviation, I will call a closed set B an ***improvement over*** A if $\gamma_k B \ge \gamma_k A$ and $\gamma_k B^r \le \gamma_k A^r$. Following Ehrhard (1983a, 1983b), I prove the Theorem in three steps:

(i) Establish the Theorem for the one-dimensional case, which Problem [6] shows can be reduced by a simple approximation argument to the case where A is a finite union of disjoint closed intervals $J_i := [x_i^-, x_i^+]$, where $-\infty < x_1^- < x_1^+ < x_2^- < \ldots < x_n^+ < +\infty$. The method of proof depends only on the fact that the logarithm of the standard normal density ϕ is a concave function. It works by showing that we can improve A by successively fusing each J_i with its neighboring interval on the left or the right, until eventually we are left with either a single semi-infinite interval or a union of two such intervals. A further convexity argument disposes of the two-interval case.

(ii) Establish the two-dimension version of the Theorem by an analog of the classical Steiner symmetrization method (Billingsley 1986, Section 19), which draws on the one-dimensional result. For a Borel subset A of $\mathbb{R}^2$, Fubini's Theorem asserts that the y-section $A_y := \{x \in \mathbb{R} : (x,y) \in A\}$ is Borel measurable, and that $\gamma_1 A_y$ is a Borel measurable function. Define a function $g(y)$ to satisfy the equality $\Phi(g(y)) := \gamma_1 A_y$. Then the y-sections B_y of the set $B := \{(x,y) : y \le g(x)\}$ have the same γ_1 measure as the corresponding A_y.

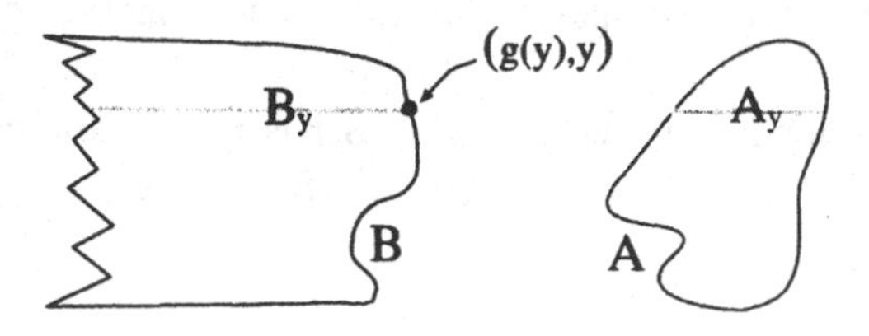

Intuitively speaking, the set B is obtained from A by sliding and stretching each of its x-sections into a semi-infinite interval with the same γ_1 measure. (The jagged left edge for B in the picture is supposed to suggest that the set extends off to $-\infty$.) I will call the operation that transforms A into B a **1-*shift***.

Problem [5] shows that B is closed if A is closed. Fubini's theorem ensures that $\gamma_2 B = \gamma_2 A$. The one-dimensional version of the Theorem will then ensure that B is an improvement over A.

The same 1-shift idea works for every direction, not just for slices parallel to the x-axis. I will write S_u for the 1-shift operator in the direction of a unit vector u. (The picture corresponds to the case where u is the unit vector $u := (-1, 0)$ that points back along the x-axis.) By means of a sequence of such shifts, we can rearrange A into a set arbitrarily close to a half space, with an improvement at each step. A formal limit argument then establishes the two-dimensional version of the Theorem.

(iii) Establish the k-dimensional version of the Theorem by induction on the dimension. It will turn out that the two dimensional case involves most of the hard work. For example, two applications of the result for two dimensions will give the result for three dimensions.

Details of the proof

(i) One dimension

We have to show that a half-line is an improvement over $A := \cup_{i \leq n} J_i$, a finite union of disjoint closed intervals.

If $x_1^+ \geq x_2^- - 2r$ we may replace J_1 and J_2 by the single interval $[x_1^-, x_2^+]$ without changing $\gamma_1 A^r$. Thus, with no loss of generality, we may suppose J_1 and the set $J = \cup_{i \geq 2} J_i$ are a distance at least $2r$ apart, in which case $\gamma_1 A^r = \gamma_1 J_1^r + \gamma_1 J^r$.

Define $2\delta := \gamma_1 J_1 = \Phi(x_1^+) - \Phi(x_1^-)$ and $2t_1 := \Phi(x_1^+) + \Phi(x_1^-)$, so that J_1 has endpoints $\Phi^{-1}(t_1 \pm \delta)$. Consider the effect of replacing J_1 by another interval $I_t := [x^-, x^+]$. If we take $x^- := \Phi^{-1}(t - \delta)$ and $x^+ := \Phi^{-1}(t + \delta)$, with $\delta \leq t \leq t* = \Phi(x_2^-) - \delta$, then $\gamma_1 I_t = \gamma_1 J_1$. When $t = \delta$, the interval I_t is semi-infinite, $[-\infty, x^+]$; when $t = t^*$, the intervals I_t and J_2 touch at $x^+ = x_2^-$. The sets $B_t := I_t \cup J$ and A have the same γ_1 measure.

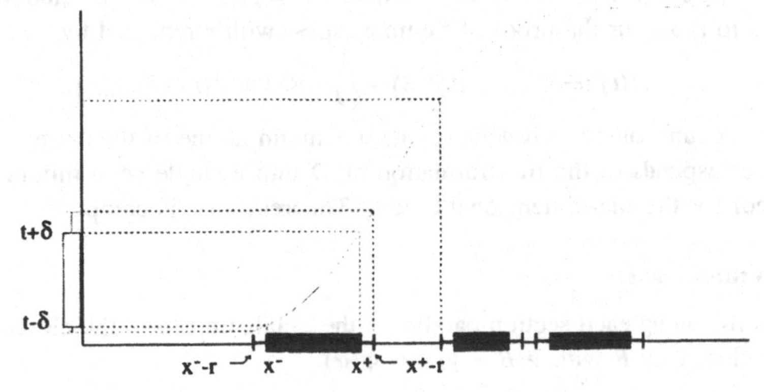

When $t < \Phi(x_2^- - 2r) - \delta$ the neighborhoods I_t^r and J^r are disjoint, and $\gamma_1 B_t^r = \gamma_1 I_t^r + \gamma_1 J^r$. For larger t, the two neighborhoods overlap, and $\gamma_1 B_t^r \leq \gamma_1 I_t^r + \gamma_1 J^r$. If we choose t with $\gamma_1 I_t^r \leq \gamma_1 J_1$ then B_t is an improvement over A. The following Lemma shows that we get the most improvement by pushing t to one of the extreme positions, because a concave function on $[\delta, t^*]$ achieves its minimum at one of the endpoints.

<13> **Lemma.** *For each fixed $r > 0$ the function $G(t) := \gamma_1 I_t^r$ is a concave function of t on $[\delta, 1-\delta]$.*

Proof. It is enough if we show that the derivative G' is a decreasing function on $(\delta, 1-\delta)$. By direct differentiation of $\Phi(x^+ + r) - \Phi(x^- - r)$ as a function of t we have

$$G'(t) = \frac{\phi(x^+ + r)}{\phi(x^+)} - \frac{\phi(x^- - r)}{\phi(x^-)}.$$

Concavity of $\log\phi$ implies that $\phi'(x)/\phi(x)$ is a decreasing function of x. Thus $h(x) := \phi(x+r)/\phi(x)$ is a decreasing function of x, because

$$h'(x) = \left(\frac{\phi'(x+r)}{\phi(x+r)} - \frac{\phi'(x)}{\phi(x)}\right)\frac{\phi(x+r)}{\phi(x)} \leq 0.$$

REMARK. Actually, $\log h(x) = -xr - r^2/2$, which is clearly decreasing. I wrote the argument using concavity because I suspect there might be a more general version of the isoperimetric inequality provable by similar methods (cf. Bobkov 1996).

Both x^+ and x^- are increasing functions of t. Thus G' equals a decreasing function of t minus the reciprocal of another decreasing function of t, which makes G' decreasing, and G concave. □

With the appropriate t, the improvement B_t is also a finite union of disjoint intervals, $\cup_{i=1}^n J_i'$, with either $J_1' := \emptyset$ or $J_1' := [-\infty, x^+]$.

Now repeat the argument, replacing J_2' by an interval that abuts either J_1' or J_3'. And so on. After at most n such operations, A is replaced by either a single semi-infinite interval (the desired minimizing set—it doesn't matter whether the interval extends off to $-\infty$ or to $+\infty$) or a union $D := [-\infty, z^-] \cup [z^+, \infty]$, for which $\gamma D = \gamma A$ and $\gamma D^r \leq \gamma A^r$.

For the second possibility we may assume that $z^+ - z^- > 2r$ to avoid the trivial case where $D^r = \mathbb{R}$. The complement of D is an interval of the form $(\Phi^{-1}(t-\delta), \Phi^{-1}(t+\delta))$ for some t, where $2\delta = \gamma_1(z^-, z^+)$. Calculations almost identical to those for the proof of Lemma <13>, with r replaced by $-r$, show that

$$H(t) := \Phi\left(\Phi^{-1}(t+\delta) - r\right) - \Phi\left(\Phi^{-1}(t-\delta) + r\right)$$

is a convex function of t. It achieves its maximum at one of the extreme positions, which corresponds to the transformation of D into a single semi-infinite interval. The proof for the one-dimensional case of Theorem <7> is complete.

(ii) Two-dimensions

The 1-shifts along each section parallel to the x-axis transform the closed set A into another closed set B with $\gamma_2 B = \gamma_2 A = \Phi(\alpha)$.

<14> **Lemma.** *The set B is an improvement over A.*

Proof. We need to show that $\gamma_2 B^r \le \gamma_2 A^r$. By Fubini, it is enough if we can show that $\gamma_1 (B^r)_y \le \gamma_1 (A^r)_y$, for all y.

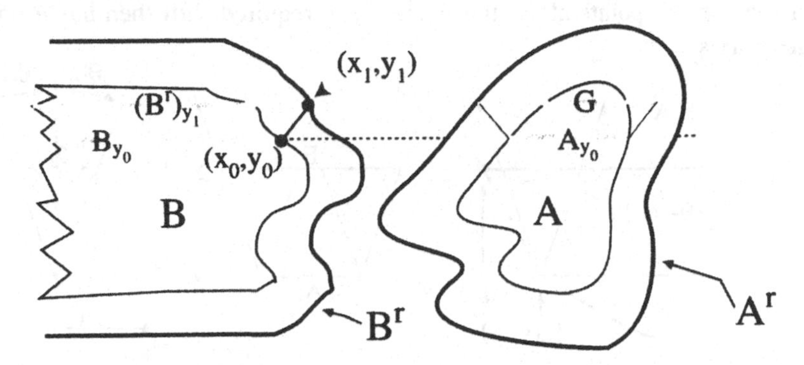

The section $(B^r)_{y_1}$ through B^r at a fixed y_1 is a closed set. Consider the boundary point (x_1, y_1) in $(B^r)_{y_1}$, with x_1 as large as possible. By definition, there exists a point (x_0, y_0) in B that lies within a distance r of (x_1, y_1): if $|x_1 - x_0| := \epsilon$ and $|y_1 - y_0| := \delta$ then $\epsilon^2 + \delta^2 \le r^2$. (Both ϵ and δ depend on y_1, of course.) For each $t \ge 0$ the point $(x_1 - t, y_1)$ lies within a distance r of $(x_0 - t, y_0)$, which also belongs to B. Thus B^r is of the form $\{(x, y) \in \mathbb{R}^2 : x \le g_r(y)\}$, for some function $g_r(\cdot)$.

Write $B_{y_0}[\epsilon]$ and $A_{y_0}[\epsilon]$ for the one-dimensional ϵ-neighborhoods of the y_0-sections B_{y_0} and A_{y_0}. Notice that $(A^r)_{y_1}$ contains the set $G = A_{y_0}[\epsilon] \otimes \{y_1\}$, because all points of G lie within a distance $(\epsilon^2 + \delta^2)^{1/2} \le r$ of A. Thus

$$
\begin{aligned}
\gamma_1 (A^r)_{y_1} &\ge \gamma_1 A_{y_0}[\epsilon] && \text{because } (A^r)_{y_1} \supseteq G \\
&\ge \Phi(g(y_0) + \epsilon) && \text{from part (i), because } \gamma_1 A_{y_0} = \Phi(g(y_0)) \\
&\ge \Phi(g_r(y_1)) && \text{because } g_r(y_1) = x_1 \le x_0 + \epsilon \le g(y_0) + \epsilon \\
&= \gamma_1 (B^r)_{y_1} && \text{definition of } g_r.
\end{aligned}
$$

□ It follows that B is an improvement over A.

To make precise the idea that a sequence of shifts can make A look more like a halfspace, we need some way of measuring how close a set is to being a half-space.

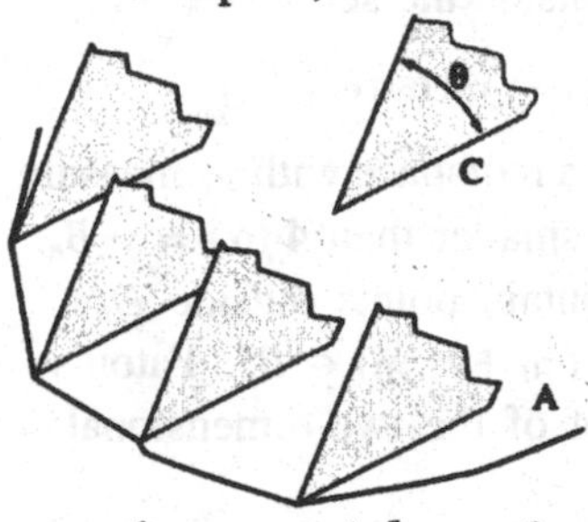

The picture suggests a method. The idea is that there should be a cone C of directions that we can follow from each point of A without leaving A. (The jagged edges on A and the cone C are meant to suggest that both sets should be extended off to infinity—there is not enough room on the page to display the whole sets.) If the vertex angle θ of the cone were a full 180°, the set A would be a half-space (or the whole of $\mathbb{R}^2$). More formally, let us say that a set A ***has a θ spread*** if there exists a cone C with vertex angle θ such that $\{x + y : x \in A,\ y \in C\} \subseteq A$. Call C the ***spreading cone***.

For example, the set B produced by Lemma <14> has spread of at least zero, with spreading cone $C := \{(x, 0) : x \le 0\}$.

<15> **Lemma.** *If a closed set A has θ spread, for some θ less than π, there exists a shift S_u such that $S_u A$ has $(\pi+\theta)/2$ spread.*

Proof. To make the picture easier to draw I will assume that the axis of symmetry of the cone C points along the y-axis. The required shift then has u pointing along the x-axis.

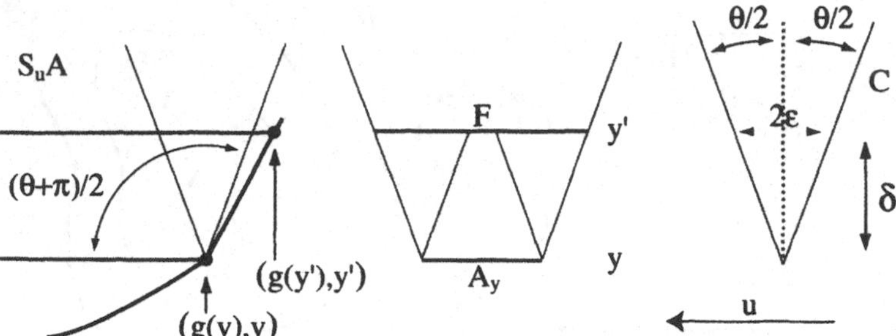

Consider the sections through $S_u A$ at heights y and $y' = y + \delta$, with $\delta > 0$. Both sections are intervals, $(-\infty, g(y)]$ and $(-\infty, g(y')]$, where $\gamma_1 A_y := \Phi(g(y))$ and $\gamma_1 A_{y'} := \Phi(g(y'))$. Define $\epsilon := \delta \tan(\theta/2)$. If $x \in A_y$ then $(x+t, y+\delta) \in A$ for all t with $|t| \le \epsilon$, because A has θ spread. That is, the section $A_{y'}$ contains a one-dimension neighborhood $F := A_y[\epsilon]$. The cross section at y' has γ_1 measure greater than $\gamma_1 A_y[\epsilon]$, which by part (i) is greater than $\Phi(\gamma_1 A_y + \epsilon)$. That is,

$$\Phi(g(y')) = \gamma_1 A_{y'} \ge \gamma_1 A_y[\epsilon] \ge \Phi(g(y) + \epsilon),$$

whence $g(y') \ge g(y) + \epsilon$. It follows that $S_u A$ has spread at least $(\pi+\theta)/2$. □

Repeated application of Lemma <15>, starting from A produces improvements $B_1, B_2, B_3, B_4, \ldots$ with spreads at least 0, $\pi/2$, $3\pi/4$, $7\pi/8$, and so on. Each B_i has γ_2 measure equal to $\Phi(\alpha) = \gamma_2 A$, and $\gamma_2 A^r \ge \gamma_2 B_1^r \ge \gamma_2 B_2^r \ge \ldots$. We may even rotate each B_n so that its spreading cone has axis parallel to the x axis.

The fact that B_n has spread $\pi - 2\epsilon_n$, with $\epsilon_n \to 0$, and the fact that $\gamma_2 B_n = \Phi(\alpha)$ together force B_n^r to lie close to the halfspace $H := \{(x, y) \in \mathbb{R}^2 : x < \alpha + r\}$ eventually. More precisely, it forces $\liminf B_n^r \ge H$, in the sense that each point of H must belong to B_n^r for all n large enough.

For if a point $(x_0 + r, y_0)$ of H were not in B_n^r then the point (x_0, y_0) could not lie in B_n, which would ensure that no points of B_n lie outside the set

$$D_n := \{(x, y) : x \le |y - y_0| \tan \epsilon_n\}.$$

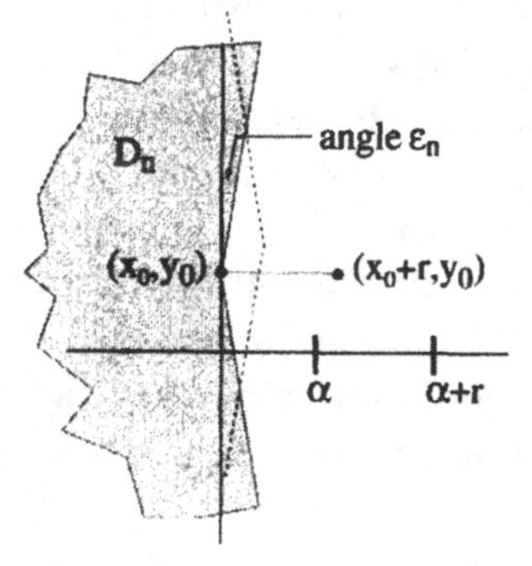

However, the set D_n converges to a halfspace with γ_2 measure equal to $\Phi(x_0)$, which is strictly smaller than $\Phi(\alpha) = \gamma_2 B_n$. Eventually B_n must therefore contain points outside D_n, in which case $(x_0, y_0) \in B_n$ and $(x_0 + r, y_0) \in B_n^r$. Fatou's Lemma then completes the proof of the two-dimensional isoperimetric assertion:

$$\gamma_2 A^r \ge \liminf \gamma_2 B_n^r \ge \gamma_2 \liminf B_n^r \ge \gamma_2 H = \Phi(\alpha + r).$$

(iii) More than two dimensions

A formal proof uses induction on the dimension, invoking the result from part (ii) to reduce for $\mathbb{R}^k$ to the result for $\mathbb{R}^{k-1}$. Then another application of part (ii) reduces to $\mathbb{R}^{k-2}$. And so on.

For simplicity of notation, I will explain only the reduction from $\mathbb{R}^3$ to $\mathbb{R}^2$. Consider a closed subset A of $\mathbb{R}^3$ with $\gamma_2 A := \alpha$. Write A_y for its y-section, so that $A = \cup_z A_y \otimes \{y\}$. Define a function g by the equality $\Phi(g(y)) := \gamma_2 A_y$. The closed set $B := \{(x, y, z) : x \leq g(y)\}$ has $\gamma_2 B_y = \gamma_2 A_y$ for every y, and hence (Fubini) $\gamma_3 B = \gamma_3 A$. Call B a ***2-shift*** of A.

The set B has all its z-sections equal to $C = \{(x, y) \in \mathbb{R}^2 : x \leq g(y)\}$. That is, $B = C \otimes \{z \in \mathbb{R}\}$. The closed set B^r has all its x-sections equal to the two-dimensional neighborhood $C[r]$.

The proof that B improves upon A is almost identical to the proof of Lemma <14>. Indeed the picture from the proof can be reinterpreted as a z-section of the (three-dimensional) picture for the present proof. Only small changes in wording are needed. In fact, here is a repeat of the argument, with changes indicated in boldface:

We need to show that $\boldsymbol{\gamma_3} B^r \leq \boldsymbol{\gamma_3} A^r$. *By Fubini, it is enough if we can show that* $\boldsymbol{\gamma_2}(B^r)_y \leq \boldsymbol{\gamma_2}(A^r)_y$, *for all y.*

The section $(B^r)_{y_1}$ *through* B^r *at a fixed* y_1 *is a closed set. Consider the boundary point* $(x_1, y_1, \mathbf{z_1})$ *in* $(B^r)_{y_1}$ *with* $\mathbf{z_1}$ *as large as possible. By definition, there exists a point* $(x_0, y_0, \mathbf{z_0})$, **with** $\mathbf{z_0} = \mathbf{z_1}$, *in B that lies within a distance r of* $(x_1, y_1, \mathbf{z_1})$: *if* $|x_1 - x_0| = \epsilon$ *and* $|y_1 - y_0| = \delta$ *then* $\epsilon^2 + \delta^2 \leq r^2$. *(Both* ϵ *and* δ *depend on* $\mathbf{z_1}$, *of course.) For each* $t \geq 0$ *the point* $(x_1 - t, y_1, \mathbf{z_0})$ *lies within a distance r of* $(x_0 - t, y_0, \mathbf{z_0})$, *which also belongs to B. Thus* B^r *is of the form* $\{(x, y, \mathbf{z}) \in \mathbb{R}^{\mathbf{3}} : x \leq g_r(y), -\infty < \mathbf{z} < \infty\}$, *for some function* $g_r(\cdot)$.

Write $B_{y_0}[\epsilon]$ *and* $A_{y_0}[\epsilon]$ *for the* **two-dimensional** ϵ*-neighborhoods of the* y_0*-sections* B_{y_0} *and* A_{y_0}. ...

And so on.

The set $B := C \otimes \{z \in \mathbb{R}\}$ is not a halfspace, but it can be transformed into one by means of a second 2-shift, with sections taken orthogonal to the y axis. The isoperimetric theorem for $\mathbb{R}^3$ then follows.

The argument for higher dimensions is similar.

6. Problems

[1] Let V_0 and V_1 be positive definition matrices. For $0 \leq \theta \leq 1$ let g_θ denote the $N(0, (1 - \theta)V_0 + \theta V_1)$ density on $\mathbb{R}^n$. Show that

$$\frac{\partial g_\theta}{\partial \theta} = \tfrac{1}{2} \sum\nolimits_{j,k} \Delta_{j,k} \frac{\partial^2 g_\theta}{\partial z_j \partial z_k} \qquad \text{where } \Delta := V_1 - V_0,$$

by following these steps.

(i) Use Fourier inversion to show that $g_\theta(x) = (2\pi)^{-n} \int \exp\left(-ix't - \frac{1}{2}t'V_\theta t\right) dt$.

(ii) Justify differentiation under the integral sign to show that

$$\frac{\partial g_\theta(x)}{\partial \theta} = (2\pi)^{-n} \int \frac{\partial}{\partial \theta} \exp\left(-ix't - \tfrac{1}{2}t'(\theta\Delta + V_0)t\right) dt$$
$$= (2\pi)^{-n} \int -\tfrac{1}{2}t'\Delta t \exp\left(-ix't - \tfrac{1}{2}t'(\theta\Delta + V_0)t\right) dt$$

and

$$\frac{\partial^2 g_\theta(x)}{\partial x_j \partial x_k} = (2\pi)^{-n} \int \frac{\partial^2}{\partial x_j \partial x_k} \exp\left(-ix't - \tfrac{1}{2}t'(\theta\Delta + V_0)t\right) dt$$
$$= (2\pi)^{-n} \int (-it_j)(-it_k) \exp\left(-ix't - \tfrac{1}{2}t'(\theta\Delta + V_0)t\right) dt.$$

(iii) Collect terms.

[2] Suppose X_n has a multivariate normal distribution, and $X_n \rightsquigarrow X$. Show that X also has a multivariate normal distribution. Hint: Reduce to the one-dimensional case by working with linear combinations. Also show that if $N(\mu_n, \sigma_n^2)$ converges then both $\{\mu_n\}$ and $\{\sigma_n\}$ must be bounded. Argue along subsequences.

[3] Let $Z_1, Z_2, \ldots$ be independent random variables, each distributed $N(0,1)$. Show that the distribution of $M_n := \max_{i \le n} Z_i$ concentrates largely in a range of order $(\log n)^{-1/2}$ by following these steps. Define $\alpha_n := (2\log n)^{1/2}$ and $L_n = \frac{1}{2}\log\log n$.

(i) Remember (Appendix D) that the tail function $\bar{\Phi}(x) := \mathbb{P}\{N(0,1) > x\}$ decreases like $\phi(x)/x$ as $x \to \infty$, in the sense that the ratio of the two functions tends to 1. For a fixed constant C, define $x_n := \alpha_n - (C + L_n)/\alpha_n$. Show that $\log\left(n\bar{\Phi}(x_n)\right)$ converges to $C_0 := C - \frac{1}{2}\log(4\pi)$ as $n \to \infty$.

(ii) Deduce that $\log \mathbb{P}\{M_n \le x_n\} = n \log\left(1 - \bar{\Phi}(x_n)\right) \to e^{-C_0}$.

(iii) Deduce that $\alpha_n(M_n - \alpha_n) + L_n$ converges in distribution as $n \to \infty$.

[4] Let X and Y be random variables for which both $\mathbb{P}\{X > x\} \le \mathbb{P}\{Z > x\}$ and $\mathbb{P}\{X \le -x\} \le \mathbb{P}\{Z \le -x\}$, for all $x \ge 0$. Show that $\mathbb{P}\exp(tX) \le \mathbb{P}\exp(tY)$ for all nonnegative t. Hint: Consider $\mathbb{P}\int_{-\infty}^{\infty} t\exp(tx)\{X > x\}\,dx$.

[5] Let A be a closed subset of $\mathbb{R}^2$, with sections $A_y := \{x \in \mathbb{R} : (x,y) \in A\}$ having Gaussian measure $g(y) := \gamma_1 A_y$. Let B denote the 1-shift, as in Lemma <14>.

(i) If $y_n \to y$, show that $\limsup A_{y_n} \le A_y$, in the sense of pointwise convergence of indicator functions. (If $x \in A_{y_n}$ infinitely often, deduce that $(x,y) \in A$.)

(ii) If $y_n \to y$, use Fatou's lemma to prove that $\limsup g(y_n) \le g(y)$.

(iii) If $(x_n, y_n) \in B$ and $(x_n, y_n) \to (x, y)$, show that $x \le g(y)$, that is, $(x,y) \in B$.

[6] Reduce the one-dimensional version of Theorem <7> to the case where A is a finite union of intervals, by following these steps. Let $\Phi(\alpha) := \gamma_1 A$.

(i) Show that it is enough prove that $\gamma A^{r+\delta} \ge \Phi(r+\alpha)$ for each $\delta > 0$, because $A^{r+\delta} \downarrow A^r$ as δ decreases to zero.

(ii) Define an open set $G := \{x : d(x, A) < \delta\}$. Show that $\gamma G > \alpha$. Hint: The set $G \backslash \bar{A}$ is open.

(iii) Show that there exists a countable family $\{J_i\}$ of disjoint closed intervals for which $\gamma_1 (G \backslash \cup_i J_i) = 0$.

(iv) Choose N so that the closed set $A_N := \cup_{i \le N} J_i$ has γ_1 measure greater than $\Phi(\alpha)$. Show that the assertion of the Theorem for A_N implies that $\gamma A^{r+\delta} \ge \Phi(r + \alpha)$.

[7] (Slepian's (1962) inequality) Let $X = (X_1, \ldots, X_n)$ and $Y = (Y_1, \ldots, Y_n)$ both have centered multivariate normal distributions with $\text{var}(X_j) = \text{var}(Y_j)$ for each j and $\mathbb{P}X_j X_k \le \mathbb{P}Y_j Y_k$ for all $j \ne k$. Prove that $\mathbb{P} \cup_j \{X_j > \alpha_j\} \ge \mathbb{P} \cup_j \{Y_j > \alpha_j\}$ for all real numbers $\alpha_1, \ldots, \alpha_n$. Hint: Use equality <3> to show that $\mu^x \prod_{i \le n} \{x_i \le \alpha_i\} g_\theta(x)$ is a decreasing function of θ.

[8] Let S be a nonnegative random variable with $\mathbb{P}\{S > M + r\} \le C_0 \exp\left(-\frac{1}{2} r^2/\sigma^2\right)$ for all $r \ge 0$, for positive constants C_0, M, and σ^2.

(i) Show that $\mathbb{P} \exp(\alpha S^2) = 1 + \int_0^\infty 2y\alpha \exp(\alpha y^2) \mathbb{P}\{S > y\}\, dy$.

(ii) For $\alpha < 1/(2\sigma^2)$, prove that $\mathbb{P} \exp(\alpha S^2) < \infty$.

7. Notes

There is a huge amount of literature on Gaussian process theory, with which I am only partially acquainted. I took the proof of the Sudakov minoration (Example <2>) from Fernique (1975, page 27). According to Dudley (1999, notes to Section 2.3), a stronger, but incorrect, result was first stated by Sudakov (1971). The result in Example <10> is due to Marcus & Shepp (1972), who sharpened earlier results of Landau & Shepp (1970) and Fernique.

See Dudley (1973) for a detailed discussion of sample path properties of Gaussian process, particularly regarding entropy characterizations of boundedness or continuity of paths. Dudley (1999, Chapter 2) contains much interesting information about Gaussian processes and their recent history. I also found the books by Jain & Marcus (1978), Adler (1990, Section II), and Lifshits (1995) useful references.

The Lifshits book contains an exposition of Ehrhard's method, similar to the one in Section 5, and proofs (Section 14) of the Fernique and Slepian inequalities. See also the notes in that section of his book for a discussion of related work of Schläfli and the contributions of Sudakov.

Borell's isoperimetric inequality was, apparently, also proved by similar methods by Tsirel'son and Sudakov in a 1974 paper which I have not seen. Tsirel'son (1975) mentioned the result and the method. The notes of Ledoux (1996, Section 4) discuss the isoperimetric inequality in great detail.

References

Adler, R. J. (1990), *An introduction to continuity, Extrema, and Related Topics for General Gaussian Processes*, Vol. 12 of *Lecture Notes–Monograph series*, Institute of Mathematical Statistics, Hayward, CA.

Billingsley, P. (1986), *Probability and Measure*, second edn, Wiley, New York.

Bobkov, S. (1996), 'Extremal properties of half-spaces for log-concave distributions', *Annals of Probability* **24**, 35–48.

Borell, C. (1975), 'The Brunn-Minkowski inequality in Gauss space', *Inventiones Math.* **30**, 207–216.

Dudley, R. M. (1973), 'Sample functions of the Gaussian process', *Annals of Probability* **1**, 66–103.

Dudley, R. M. (1999), *Uniform Central Limit Theorems*, Cambridge University Press.

Ehrhard, A. (1983a), 'Un principe de symétrisation dans les espaces de Gauss', *Springer Lecture Notes in Mathematics* **990**, 92–101.

Ehrhard, A. (1983b), 'Symétrisation dans l'espace de Gauss', *Mathematica Scandinavica* **53**, 281–301.

Fernique, X. (1975), 'Regularité des trajectoires des fonctions aléatoires gaussiennes', *Springer Lecture Notes in Mathematics* **480**, 1–97.

Jain, N. C. & Marcus, M. B. (1978), Advances in probability, *in* J. Kuelbs, ed., 'Probability in Banach Spaces', Vol. 4, Dekker, New York, pp. 81–196.

Landau, H. J. & Shepp, L. A. (1970), 'On the supremum of a Gaussian process', *Sankhyā: The Indian Journal of Statistics, Series A* **32**, 369–378.

Ledoux, M. (1996), 'Isoperimetry and Gaussian analysis', *Springer Lecture Notes in Mathematics* **1648**, 165–294.

Lifshits, M. A. (1995), *Gaussian Random Functions*, Kluwer.

Marcus, M. B. & Shepp, L. A. (1972), 'Sample behavior of Gaussian processes', *Proceedings of the Sixth Berkeley Symposium on Mathematical Statistics and Probability* **2**, 423–441.

Slepian, D. (1962), 'The one-sided barrier problem for Gaussian noise', *Bell System Technical Journal* **41**, 463–501.

Sudakov, V. N. (1971), 'Gaussian random processes and measures of solid angles in Hilbert space', *Soviet Math. Doklady* **12**, 412–415.

Tsirel'son, V. S. (1975), 'The density of the distribution of the maximum of a Gaussian process', *Theory Probability and Its Applications* **20**, 847–856.

Appendix A

Measures and integrals

SECTION 1 introduces a method for constructing a measure by inner approximation, starting from a set function defined on a lattice of sets.

SECTION 2 defines a "tightness" property, which ensures that a set function has an extension to a finitely additive measure on a field determined by the class of approximating sets.

SECTION 3 defines a "sigma-smoothness" property, which ensures that a tight set function has an extension to a countably additive measure on a sigma-field.

SECTION 4 shows how to extend a tight, sigma-smooth set function from a lattice to its closure under countable intersections.

SECTION 5 constructs Lebesgue measure on Euclidean space.

SECTION 6 proves a general form of the Riesz representation theorem, which expresses linear functionals on cones of functions as integrals with respect to countably additive measures.

1. Measures and inner measure

Recall the definition of a ***countably additive measure*** on ***sigma-field***. A sigma-field $\mathcal{A}$ on a set $\mathcal{X}$ is a class of subsets of $\mathcal{X}$ with the following properties.

(SF_1) *The empty set $\emptyset$ and the whole space $\mathcal{X}$ both belong to $\mathcal{A}$.*

(SF_2) *If A belongs to $\mathcal{A}$ then so does its complement A^c.*

(SF_3) *For countable $\{A_i : i \in \mathbb{N}\} \subseteq \mathcal{A}$, both $\cup_i A_i$ and $\cap_i A_i$ are also in $\mathcal{A}$.*

A function μ defined on the sigma-field $\mathcal{A}$ is called a countably additive (nonnegative) measure if it has the following properties.

(M_1) $\mu\emptyset = 0 \le \mu A \le \infty$ *for each A in $\mathcal{A}$.*

(M_2) $\mu\left(\cup_i A_i\right) = \sum_i \mu A_i$ *for sequences $\{A_i : i \in \mathbb{N}\}$ of pairwise disjoint sets from $\mathcal{A}$.*

If property SF_3 is weakened to require stability only under finite unions and intersections, the class is called a ***field***. If property M_2 is weakened to hold only for disjoint unions of finitely many sets from $\mathcal{A}$, the set function is called a ***finitely additive measure***.

Where do measures come from? Typically one starts from a nonnegative real-valued set-function μ defined on a small class of sets $\mathcal{K}_0$, then extends to a sigma-field $\mathcal{A}$ containing $\mathcal{K}_0$. One must at least assume "measure-like" properties for μ on $\mathcal{K}_0$ if such an extension is to be possible. At a bare minimum,

(M_0) *μ is an increasing map from $\mathcal{K}_0$ into $\mathbb{R}^+$ for which $\mu\emptyset = 0$.*

Note that we need $\mathcal{K}_0$ to contain $\emptyset$ for M_0 to make sense. I will assume that M_0 holds thoughout this Appendix. As a convenient reminder, I will also reserve the name ***set function on*** $\mathcal{K}_0$ for those μ that satisfy M_0.

The extension can proceed by various approximation arguments. In the first three Sections of this Appendix, I will describe only the method based on approximation of sets from inside. Although not entirely traditional, the method has the advantage that it leads to measures with a useful approximation property called $\mathcal{K}_0$***-regularity***:

$$\mu A = \sup\{\mu K : A \supseteq K \in \mathcal{K}_0\} \qquad \text{for each } A \text{ in } \mathcal{A}.$$

REMARK. When $\mathcal{K}$ consists of compact sets, a measure with the inner regularity property is often called a ***Radon measure***.

The desired regularity property makes it clear how the extension of μ must be constructed, namely, by means of the ***inner measure*** μ_*, defined for *every* subset A of $\mathcal{X}$ by $\mu_* A := \sup\{\mu K : A \supseteq K \in \mathcal{K}_0\}$.

In the business of building measures it pays to start small, imposing as few conditions on the initial domain $\mathcal{K}_0$ as possible. The conditions are neatly expressed by means of some picturesque terminology. Think of $\mathcal{X}$ as a large expanse of muddy lawn, and think of subsets of $\mathcal{X}$ as paving stones to lay on the ground, with overlaps permitted. Then a collection of subsets of $\mathcal{X}$ would be a ***paving*** for $\mathcal{X}$. The analogy might seem far-fetched, but it gives a concise way to describe properties of various classes of subsets. For example, a field is nothing but a $(\emptyset, \cup f, \cap f, {}^c)$ paving, meaning that it contains the empty set and is stable under the formation of finite unions $(\cup f)$, finite intersections $(\cap f)$, and complements $({}^c)$. A $(\emptyset, \cup c, \cap c, {}^c)$ paving is just another name for a sigma-field—the $\cup c$ and $\cap c$ denote countable unions and intersections. With inner approximations the natural assumption is that $\mathcal{K}_0$ be at least a $(\emptyset, \cup f, \cap f)$ paving—a ***lattice*** of subsets.

REMARK. Note well. A lattice is not assumed to be stable under differences or the taking of complements. Keep in mind the prime example, where $\mathcal{K}_0$ denotes the class of compact subsets of a (Hausdorff) topological space, such as the real line. Inner approximation by compact sets has turned out to be a *good thing* for probability theory.

For a general lattice $\mathcal{K}_0$, the role of the closed sets (remember f for fermé) is played by the class $\mathcal{F}(\mathcal{K}_0)$ of all subsets F for which $FK \in \mathcal{K}_0$ for every K in $\mathcal{K}_0$. (Of course, $\mathcal{K}_0 \subseteq \mathcal{F}(\mathcal{K}_0)$. The inclusion is proper if $\mathcal{X} \notin \mathcal{K}_0$.) The sigma-field $\mathcal{B}(\mathcal{K}_0)$ generated by $\mathcal{F}(\mathcal{K}_0)$ will play the role of the Borel sigma-field.

The first difficulty along the path leading to countably additive measures lies in the choice of the sigma-field $\mathcal{A}$, in order that the restriction of μ_* to $\mathcal{A}$ has the desired countable additivity properties. The Carathéodory splitting method identifies a suitable class of sets by means of an apparently weak substitute for the finite additivity property. Define $\mathcal{S}_0$ as the class of all subsets S of $\mathcal{X}$ for which

<1> $$\mu_* A = \mu_*(AS) + \mu_*(AS^c) \qquad \text{for all subsets } A \text{ of } \mathcal{X}.$$

If $A \in \mathcal{S}_0$ then μ_* adds the measures of the disjoint sets AS and AS^c correctly. As far as μ_* is concerned, S splits the set A "properly."

<2> **Lemma.** *The class $\mathcal{S}_0$ of all subsets S with the property <1> is a field. The restriction of μ_* to $\mathcal{S}_0$ is a finitely additive measure.*

Proof. Trivially $\mathcal{S}_0$ contains the empty set (because $\mu_*\emptyset = 0$) and it is stable under the formation of complements. To establish the field property it suffices to show that $\mathcal{S}_0$ is stable under finite intersections.

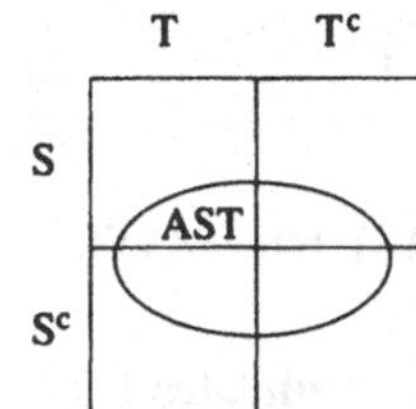

Suppose S and T belong to $\mathcal{S}_0$. Let A be an arbitrary subset of $\mathcal{X}$. Split A into two pieces using S, then split each of those two pieces using T. From the defining property of $\mathcal{S}_0$,

$$\begin{aligned}\mu_*A &= \mu_*(AS) + \mu_*(AS^c)\\ &= \mu_*(AST) + \mu_*(AST^c) + \mu_*(AS^cT) + \mu_*(AS^cT^c).\end{aligned}$$

Decompose $A(ST)^c$ similarly to see that the last three terms sum to $\mu_*A(ST)^c$. The intersection ST splits A correctly; the class $\mathcal{S}_0$ contains ST; the class is a field. If $ST = 0$, choose $A := S \cup T$ to show that the restriction of μ_* to $\mathcal{S}_0$ is finitely additive. □

At the moment there is no guarantee that $\mathcal{S}_0$ includes all the members of $\mathcal{K}_0$, let alone all the members of $\mathcal{B}(\mathcal{K}_0)$. In fact, the Lemma has nothing to do with the choice of μ and $\mathcal{K}_0$ beyond the fact that $\mu_*(\emptyset) = 0$. To ensure that $\mathcal{S}_0 \supseteq \mathcal{K}_0$ we must assume that μ has a property called ***$\mathcal{K}_0$-tightness***, an analog of finite additivity that compensates for the fact that the difference of two $\mathcal{K}_0$ sets need not belong to $\mathcal{K}_0$. Section 2 explains $\mathcal{K}_0$-tightness. Section 3 adds the assumptions needed to make the restriction of μ_* to $\mathcal{S}_0$ a countable additivity measure.

2. Tightness

If $\mathcal{S}_0$ is to contain every member of $\mathcal{K}_0$, every set $K \in \mathcal{K}_0$ must split every set $K_1 \in \mathcal{K}_0$ properly, in the sense of Definition <1>,

$$\mu_*(K_1) = \mu_*(K_1K) + \mu_*(K_1\backslash K).$$

Writing K_0 for K_1K, we then have the following property as a necessary condition for $\mathcal{K}_0 \subseteq \mathcal{S}_0$. It will turn out that the property is also sufficient.

<3> **Definition.** *Say that a set function μ on $\mathcal{K}_0$ is $\mathcal{K}_0$-tight if $\mu K_1 = \mu K_0 + \mu_*(K_1\backslash K_0)$ for all pairs of sets in $\mathcal{K}_0$ with $K_1 \supseteq K_0$.*

The intuition is that there exists a set $K \in \mathcal{K}_0$ that almost fills out $K_1\backslash K_0$, in the sense that $\mu K \approx \mu K_1 - \mu K_0$. More formally, for each $\epsilon > 0$ there exists a $K_\epsilon \in \mathcal{K}_0$ with $K_\epsilon \subseteq K_1\backslash K_0$ and $\mu K_\epsilon > \mu K_1 - \mu K_0 - \epsilon$. As a convenient abbreviation, I will say that such a K_ϵ fills out the difference $K_1\backslash K_0$ within an ϵ.

Tightness is as close as we come to having $\mathcal{K}_0$ stable under proper differences. It implies a weak additivity property: if K and H are disjoint members of $\mathcal{K}_0$ then $\mu(H \cup K) = \mu H + \mu K$, because the supremum in the definition of $\mu_*\big((H \cup K)\backslash K\big)$ is achieved by H. Additivity for disjoint $\mathcal{K}_0$-sets implies superadditivity for the inner measure,

<4> $$\mu_*(A \cup B) \geq \mu_* A + \mu_* B \qquad \text{for all disjoint } A \text{ and } B,$$

because the union of each inner approximating H for A and each inner approximating K for B is an inner approximating set for $A \cup B$. Tightness also gives us a way to relate $\mathcal{S}_0$ to $\mathcal{K}_0$.

<5> **Lemma.** *Let $\mathcal{K}_0$ be a $(\emptyset, \cup f, \cap f)$ paving, and μ be $\mathcal{K}_0$-tight set function. Then*

(i) *$S \in \mathcal{S}_0$ if and only if $\mu K \leq \mu_* (KS) + \mu_* (K \backslash S)$ for all K in $\mathcal{K}_0$;*

(ii) *the field $\mathcal{S}_0$ contains the field generated by $\mathcal{F}(\mathcal{K}_0)$.*

Proof. Take a supremum in (i) over all $K \subseteq A$ to get $\mu_* A \leq \mu_* (AS) + \mu_* (A \backslash S)$. The superadditivity property <4> gives the reverse inequality.

If $S \in \mathcal{F}(\mathcal{K}_0)$ and $K \in \mathcal{K}_0$, the pair $K_1 := K$ and $K_0 := KS$ are candidates for the tightness equality, $\mu K = \mu (KS) + \mu_* (K \backslash S)$, implying the inequality in (i). □

3. Countable additivity

Countable additivity ensures that measures are well behaved under countable limit operations. To fit with the lattice properties of $\mathcal{K}_0$, it is most convenient to insert countable additivity into the construction of measures via a limit requirement that has been called ***σ-smoothness*** in the literature. I will stick with that term, rather than invent a more descriptive term (such as σ-continuity from above), even though I feel that it conveys not quite the right image for a set function.

<6> **Definition.** *Say that μ is σ-smooth (along $\mathcal{K}_0$) at a set K in $\mathcal{K}_0$ if $\mu K_n \downarrow \mu K$ for every decreasing sequence of sets $\{K_n\}$ in $\mathcal{K}_0$ with intersection K.*

REMARK. It is important that μ takes only (finite) real values for sets in $\mathcal{K}_0$. If λ is a countably additive measure on a sigma-field $\mathcal{A}$, and $A_n \downarrow A_\infty$ with all A_i in $\mathcal{A}$, then we need not have $\lambda A_n \downarrow \lambda A_\infty$ unless $\lambda A_n < \infty$ for some n, as shown by the example of Lebesgue measure with $A_n = [n, \infty)$ and $A_\infty = \emptyset$.

Notice that the definition concerns only those decreasing sequences in $\mathcal{K}_0$ for which $\bigcap_{n \in \mathbb{N}} K_n \in \mathcal{K}_0$. At the moment, there is no presumption that $\mathcal{K}_0$ be stable under countable intersections. As usual, the σ is to remind us of the restriction to *countable* families. There is a related property called τ-smoothness, which relaxes the assumption that there are only countably many K_n sets—see Problem [1].

Tightness simplifies the task of checking for σ-smoothness. The next proof is a good illustration of how one makes use of $\mathcal{K}_0$-tightness and the fact that μ_* has already been proven finitely additive on the field $\mathcal{S}_0$.

<7> **Lemma.** *If a $\mathcal{K}_0$-tight set function on a $(\emptyset, \cup f, \cap f)$ paving $\mathcal{K}_0$ is σ-smooth at $\emptyset$ then it is σ-smooth at every set in $\mathcal{K}_0$.*

Proof. Suppose $K_n \downarrow K_\infty$, with all K_i in $\mathcal{K}_0$. Find an $H \in \mathcal{K}_0$ that fills out the difference $K_1 \backslash K_\infty$ within ϵ. Write L for $H \cup K_\infty$. Finite additivity of μ_* on $\mathcal{S}_0$ lets us break μK_n into the sum

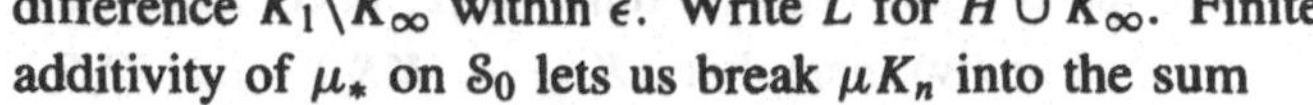

$$\mu K_\infty + \mu(K_n H) + \mu_*(K_n \backslash L).$$

The middle term decreases to zero as $n \to \infty$ because $K_n H \downarrow K_\infty H = \emptyset$. The last term is less than

$$\mu_*(K_1 \backslash L) = \mu K_1 - \mu K_\infty - \mu H = \mu_*(K_1 \backslash K_\infty) - \mu H,$$

which is less than ϵ, by construction. □

If $\mathcal{K}_0$ is a stable under countable intersections, the σ-smoothness property translates easily into countable additivity for μ_* as a set function on $\mathcal{S}_0$.

<8> **Theorem.** *Let $\mathcal{K}_0$ be a lattice of subsets of $\mathcal{X}$ that is stable under countable intersections, that is, a $(\emptyset, \cup f, \cap c)$ paving. Let μ be a $\mathcal{K}_0$-tight set function on a $\mathcal{K}_0$, with associated inner measure $\mu_* A := \sup\{\mu K : A \supseteq K \in \mathcal{K}_0\}$. Suppose μ is σ-smooth at $\emptyset$ (along $\mathcal{K}_0$). Then*

(i) *the class*

$$\mathcal{S}_0 := \{S \subseteq \mathcal{X} : \mu K \le \mu_*(KS) + \mu_*(K \backslash S) \text{ for all } K \text{ in } \mathcal{K}_0\}$$

is a sigma-field on $\mathcal{X}$;

(ii) *$\mathcal{S}_0 \supseteq \mathcal{B}(\mathcal{K}_0)$, the sigma-field generated by $\mathcal{F}(\mathcal{K}_0)$;*

(iii) *the restriction of μ_* to $\mathcal{S}_0$ is a $\mathcal{K}_0$-regular, countably additive measure;*

(iv) *$\mathcal{S}_0$ is complete: if $S_1 \supseteq B \supseteq S_0$ with $S_i \in \mathcal{S}_0$ and $\mu_*(S_1 \backslash S_0) = 0$ then $B \in \mathcal{S}_0$.*

Proof. From Lemma <5>, we know that $\mathcal{S}_0$ is a field that contains $\mathcal{F}(\mathcal{K}_0)$. To prove (i) and (ii), it suffices to show that the union $S := \cup_{i \in \mathbb{N}} T_i$ of a sequence of sets in $\mathcal{S}_0$ also belongs to $\mathcal{S}_0$, by establishing the inequality $\mu K \le \mu_*(KS) + \mu_*(K \backslash S)$, for each choice of K in $\mathcal{K}_0$.

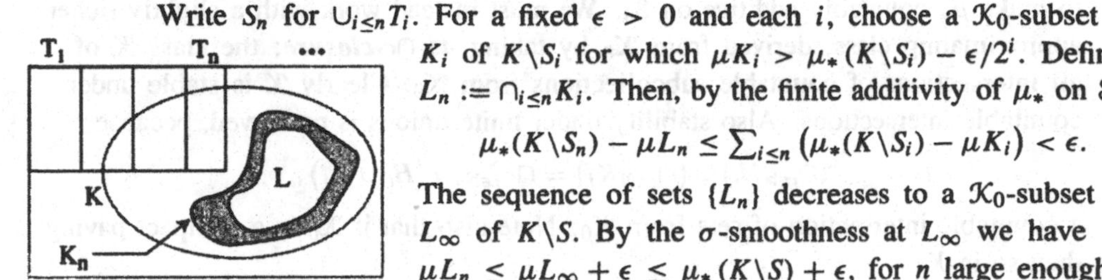

Write S_n for $\cup_{i \le n} T_i$. For a fixed $\epsilon > 0$ and each i, choose a $\mathcal{K}_0$-subset K_i of $K \backslash S_i$ for which $\mu K_i > \mu_*(K \backslash S_i) - \epsilon/2^i$. Define $L_n := \cap_{i \le n} K_i$. Then, by the finite additivity of μ_* on $\mathcal{S}_0$,

$$\mu_*(K \backslash S_n) - \mu L_n \le \sum_{i \le n} \left(\mu_*(K \backslash S_i) - \mu K_i\right) < \epsilon.$$

The sequence of sets $\{L_n\}$ decreases to a $\mathcal{K}_0$-subset L_∞ of $K \backslash S$. By the σ-smoothness at L_∞ we have $\mu L_n \le \mu L_\infty + \epsilon \le \mu_*(K \backslash S) + \epsilon$, for n large enough, which gives $\mu_*(K \backslash S_n) \le \mu L_n + \epsilon \le \mu_*(K \backslash S) + 2\epsilon$, whence

$$\begin{aligned} \mu K &= \mu_*(KS_n) + \mu_*(K \backslash S_n) \qquad \text{because } S_n \in \mathcal{S}_0 \\ &\le \mu_*(KS) + \mu_*(K \backslash S) + 2\epsilon. \end{aligned}$$

It follows that $S \in \mathcal{S}_0$.

When $K \subseteq S$, the inequality $\mu K \le \mu_*(KS_n) + \mu_*(K \backslash S) + 2\epsilon$ and the finite additivity of μ_* on $\mathcal{S}_0$ imply $\mu K \le \sum_{i \le n} \mu_*(KT_i) + 2\epsilon$. Take the supremum over all $\mathcal{K}_0$-subsets of S, let n tend to infinity, then ϵ tend to zero, to deduce that $\mu_* S \le \sum_{i \in \mathbb{N}} \mu_* T_i$. The reverse inequality follows from the superadditivity property <4>. The set function μ_* is countably additive on the the sigma-field $\mathcal{S}_0$.

For (iv), note that $\mu K = \mu_* (K S_0) + \mu_* (K \backslash S_0)$, which is smaller than

$$\mu_* (K B) + \mu_* (K \backslash S_1) + \mu_*(K S_1 S_0^c) \le \mu_* (K B) + \mu_* (K \backslash B) + 0,$$

□ for every K in $\mathcal{K}_0$.

In one particularly important case we get σ-smoothness for free, without any extra assumptions on the set function μ. A paving $\mathcal{K}_0$ is said to be ***compact*** (in the sense of Marczewski 1953) if: to each countable collection $\{K_i : i \in \mathbb{N}\}$ of sets from $\mathcal{K}_0$ with $\cap_{i \in \mathbb{N}} K_i = \emptyset$ there is some finite n for which $\cap_{i \le n} K_i = \emptyset$. In particular, if $K_i \downarrow \emptyset$ then $K_n = \emptyset$ for some n. For such a paving, the σ-smoothness property places no constraint on μ beyond the standard assumption that $\mu\emptyset = 0$.

<9> **Example.** Let $\mathcal{K}_0$ be a collection of closed, compact subsets of a topological space $\mathcal{X}$. Suppose $\{K_\alpha : \alpha \in A\}$ is a subcollection of $\mathcal{K}_0$ for which $\cap_{\alpha \in A} K_\alpha = \emptyset$. Arbitrarily choose an α_0 from A. The collection of open sets $G_\alpha := K_\alpha^c$ for $\alpha \in A$ covers the compact set K_{α_0}. By the definition of compactness, there exists a finite subcover. That is, for some $\alpha_1, \dots, \alpha_m$ we have $K_{\alpha_0} \subseteq \cup_{i=1}^m G_{\alpha_i} = \left(\cap_{i=1}^m K_{\alpha_i}\right)^c$. Thus
□ $\cap_{i=0}^m K_{\alpha_i} = \emptyset$. In particular, $\mathcal{K}_0$ is also compact in the Marczewski sense.

REMARK. Notice that the Marczewski concept involves only countable subcollections of $\mathcal{K}_0$, whereas the topological analog from Example <9> applies to arbitrary subcollections. The stronger property turns out to be useful for proving τ-smoothness, a property stronger than σ-smoothness. See Problem [1] for the definition of τ-smoothness.

4. Extension to the $\cap c$-closure

If $\mathcal{K}_0$ is not stable under countable intersections, σ-smoothness is not quite enough to make μ_* countably additive on $\mathcal{S}_0$. We must instead work with a slightly richer approximating class, derived from $\mathcal{K}_0$ by taking its ***$\cap c$-closure***: the class $\mathcal{K}$ of all intersections of countable subcollections from $\mathcal{K}_0$. Clearly $\mathcal{K}$ is stable under countable intersections. Also stability under finite unions is preserved, because

$$\left(\cap_{i \in \mathbb{N}} H_i\right) \cup \left(\cap_{j \in \mathbb{N}} K_j\right) = \cap_{i,j \in \mathbb{N} \times \mathbb{N}} \left(H_i \cup K_j\right),$$

a countable intersection of sets from $\mathcal{K}_0$. Note also that if $\mathcal{K}_0$ is a compact paving then so is $\mathcal{K}$.

The next Lemma shows that the natural extension of μ to a set function on $\mathcal{K}$ inherits the desirable σ-smoothness and tightness properties.

<10> **Lemma.** *Let μ be a $\mathcal{K}_0$-tight set function on a $(\emptyset, \cup f, \cap f)$ paving $\mathcal{K}_0$, which is σ-smooth along $\mathcal{K}_0$ at $\emptyset$. Then the extension $\tilde{\mu}$ of μ to the $\cap c$-closure $\mathcal{K}$, defined by*

<11>
$$\tilde{\mu} H := \inf\{\mu K : H \subseteq K \in \mathcal{K}_0\} \qquad \text{for } H \in \mathcal{K},$$

is $\mathcal{K}$-tight and σ-smooth (along $\mathcal{K}$) at $\emptyset$.

Proof. Lemma <7> gives us a simpler expression for $\tilde{\mu}$. If $\{K_n : n \in \mathbb{N}\} \subseteq \mathcal{K}_0$ and $K_n \downarrow L \in \mathcal{K}$, then $\tilde{\mu}L = \inf_n \mu K_n$ because, for each $\mathcal{K}_0$-subset K with $K \supseteq L$,

$$\tilde{\mu}K_n \leq \mu(K_n \cup K) \downarrow \mu K \qquad \text{by } \sigma\text{-smoothness of } \mu \text{ at } K.$$

The σ-smoothness at $\emptyset$ is easy. If $H_n := \bigcap_{j \in \mathbb{N}} K_{nj} \in \mathcal{K}$ and $H_n \downarrow \emptyset$ then the sets $K_n := \bigcap_{i \leq n, j \leq n} K_{ij}$ belong to $\mathcal{K}_0$, and $H_n = H_1 H_2 \ldots H_n \subseteq K_n \downarrow \emptyset$. It follows that $\tilde{\mu}H_n \leq \mu K_n \downarrow 0$.

The $\mathcal{K}$-tightness is slightly trickier. Suppose $H_1 \supseteq H_0$, with both sets in $\mathcal{K}$. Let $\{K_n\}$ be a decreasing sequence of sets in $\mathcal{K}_0$ with intersection H_1. For a fixed $\epsilon > 0$, choose a K in $\mathcal{K}_0$ with $K \supseteq H_0$ and $\mu K < \tilde{\mu}H_0 + \epsilon$. With no loss of generality we may assume that $K \subseteq K_1$. Invoke $\mathcal{K}_0$-tightness to find a $\mathcal{K}_0$-subset L of $K_1 \backslash K$ for which $\mu L > \mu K_1 - \mu K - \epsilon > \mu K_1 - \tilde{\mu}H_0 - 2\epsilon$. Notice that the sequence $\{LK_n\}$ decreases to LH_1, a $\mathcal{K}$-subset of $H_1 \backslash K \subseteq H_1 \backslash H_0$. The finite additivity of μ_*, when restricted to $\mathcal{S}_0$, gives

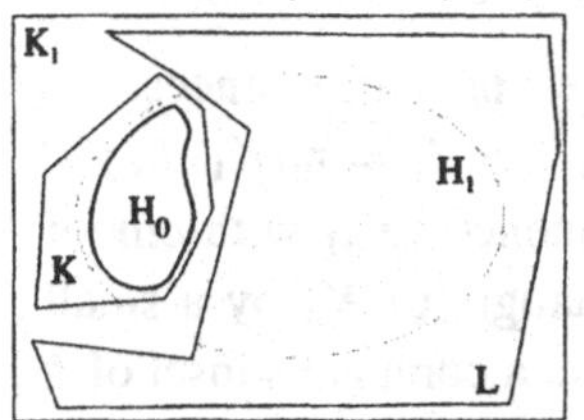

$$\begin{aligned} \mu(LK_n) &= \mu L + \mu K_n - \mu(L \cup K_n) \\ &> (\mu K_1 - \tilde{\mu}H_0 - 2\epsilon) + \mu K_n - \mu K_1 \\ &\to -\tilde{\mu}H_0 - 2\epsilon + \tilde{\mu}H_1 \qquad \text{as } n \to \infty. \end{aligned}$$

Thus $\tilde{\mu}(LH_1) \geq \tilde{\mu}H_1 - \tilde{\mu}H_0 - 2\epsilon$, as required for $\mathcal{K}$-tightness. □

REMARK. It is helpful, but not essential, to have a different symbol for the extension of μ to a larger domain while we are establishing properties for that extension. For example, it reminds us not to assume that $\tilde{\mu}$ has the same properties as μ before we have proved as much. Once the result is proven, the $\tilde{\mu}$ has served its purpose, and it can then safely be replaced by μ.

A similar argument might be made about the distinction between $\mathcal{K}_0$ and $\mathcal{K}$, but there is some virtue in retaining the subscript as a reminder than $\mathcal{K}_0$ is assumed stable only under finite intersections.

Together, Theorem <8> and Lemma <10> give a highly useful extension theorem for set functions defined initially on a lattice of subsets.

<12> **Theorem.** *Let $\mathcal{K}_0$ be a $(\emptyset, \cup f, \cap f)$ paving of subsets of $\mathcal{X}$, and let $\mathcal{K}$ denote its $\cap c$-closure. Let $\mu : \mathcal{K}_0 \to \mathbb{R}^+$ be a $\mathcal{K}_0$-tight set function that is sigma-smooth along $\mathcal{K}_0$ at $\emptyset$. Then μ has a unique extension to a complete, $\mathcal{K}$-regular, countably additive measure on a sigma-field $\mathcal{S}$, defined by*

$$\mu K := \inf\{\mu K_0 : K \subseteq K_0 \in \mathcal{K}_0\} \qquad \text{for } K \in \mathcal{K},$$

$$\mu S := \sup\{\mu K : S \supseteq K \in \mathcal{K}\} \qquad \text{for } S \in \mathcal{S}.$$

The sigma-field $\mathcal{S}$ contains all sets F for which $FK \in \mathcal{K}$ for all K in $\mathcal{K}$. In particular, $\mathcal{S} \supseteq \mathcal{K} \supseteq \mathcal{K}_0$.

REMARK. Remember: σ-smoothness is automatic if $\mathcal{K}_0$ is a compact paving.

5. Lebesgue measure

There are several ways in which to construct Lebesgue measure on $\mathbb{R}^k$. The following method for $\mathbb{R}^2$ is easily extended to other dimensions.

Take $\mathcal{K}_0$ to consist of all finite unions of semi-open rectangles $(\alpha_1, \beta_1]\otimes(\alpha_2, \beta_2]$. Each difference of two semi-open rectangles can be written as a disjoint union of

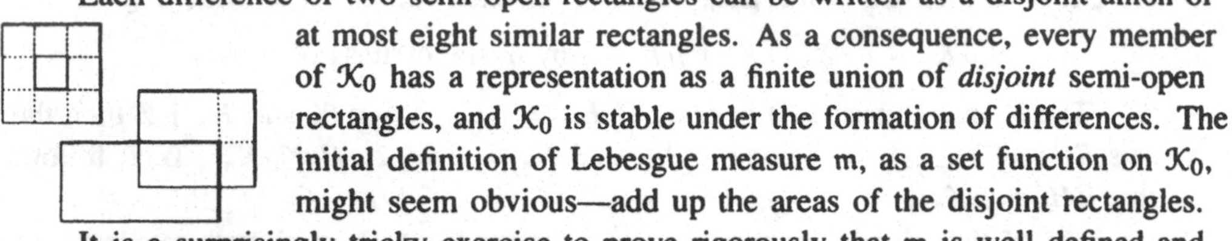

at most eight similar rectangles. As a consequence, every member of $\mathcal{K}_0$ has a representation as a finite union of *disjoint* semi-open rectangles, and $\mathcal{K}_0$ is stable under the formation of differences. The initial definition of Lebesgue measure $\mathfrak{m}$, as a set function on $\mathcal{K}_0$, might seem obvious—add up the areas of the disjoint rectangles. It is a surprisingly tricky exercise to prove rigorously that $\mathfrak{m}$ is well defined and finitely additive on $\mathcal{K}_0$.

REMARK. The corresponding argument is much easier in one dimension. It is, perhaps, simpler to consider only that case, then obtain Lebesgue measure in higher dimensions as a completion of products of one-dimensional Lebesgue measures.

The $\mathcal{K}_0$-tightness of $\mathfrak{m}$ is trivial, because $\mathcal{K}_0$ is stable under differences: if $K_1 \supseteq K_0$, with both sets in $\mathcal{K}_0$, then $K_1\backslash K_0 \in \mathcal{K}_0$ and $\mathfrak{m}K_1 - \mathfrak{m}K_0 = \mathfrak{m}(K_1\backslash K_0)$.

To establish σ-smoothness, consider a decreasing sequence $\{K_n\}$ with empty intersection. Fix $\epsilon > 0$. If we shrink each component rectangle of K_n by a small enough amount we obtain a set L_n in $\mathcal{K}_0$ whose closure $\bar{L}_n$ is a compact subset of K_n and for which $\mathfrak{m}(K_n\backslash L_n) < \epsilon/2^n$. The family of compact sets $\{\bar{L}_n : n = 1, 2, \dots\}$ has empty intersection. For some finite N we must have $\cap_{i\le N}\bar{L}_i = \emptyset$, so that

$$\mathfrak{m}K_N \le \mathfrak{m}\left(\cap_{i\le N} L_i\right) + \textstyle\sum_{i\le N}\mathfrak{m}(K_i\backslash L_i) \le 0 + \sum_{i\le N}\epsilon/2^i.$$

It follows that $\mathfrak{m}K_n$ tends to zero as n tends to infinity. The finitely additive measure $\mathfrak{m}$ is $\mathcal{K}_0$-smooth at $\emptyset$. By Theorem <12>, it extends to a $\mathcal{K}$-regular, countably additive measure on $\mathcal{S}$, a sigma-field that contains all the sets in $\mathcal{F}(\mathcal{K})$.

You should convince yourself that $\mathcal{K}$, the $\cap c$-closure of $\mathcal{K}_0$, contains all compact subsets of $\mathbb{R}^2$, and $\mathcal{F}(\mathcal{K})$ contains all closed subsets. The sigma-field $\mathcal{S}$ is complete and contains the Borel sigma-field $\mathcal{B}(\mathbb{R}^2)$. In fact $\mathcal{S}$ is the Lebesgue sigma-field, the closure of the Borel sigma-field.

6. Integral representations

Throughout the book I have made heavy use of the fact that there is a one-to-one correspondence (via integrals) between measures and increasing linear functionals on $\mathcal{M}^+$ with the Monotone Convergence property. Occasionally (as in Sections 4.8 and 7.5), I needed an analogous correspondence for functionals on a subcone of $\mathcal{M}^+$. The methods from Sections 1, 2, and 3 can be used to construct measures representing such functionals if the subcone is stable under lattice-like operations.

<13> **Definition.** *Call a collection $\mathcal{H}^+$ of nonnegative real functions on a set $\mathcal{X}$ a* ***lattice cone*** *if it has the following properties. For h, h_1 and h_2 in $\mathcal{H}^+$, and α_1 and α_2 in $\mathbb{R}^+$:*

(H_1) *$\alpha_1 h_1 + \alpha_2 h_2$ belongs to $\mathcal{H}^+$;*

(H_2) *$h_1\backslash h_2 := (h_1 - h_2)^+$ belongs to $\mathcal{H}^+$;*

(H_3) *the pointwise minimum $h_1 \wedge h_2$ and maximum $h_1 \vee h_2$ belong to $\mathcal{H}^+$;*

(H_4) *$h \wedge 1$ belongs to $\mathcal{H}^+$.*

The best example of a lattice cone to keep in mind is the class $\mathcal{C}_0^+(\mathbb{R}^k)$ of all nonnegative, continuous functions with compact support on some Euclidean space $\mathbb{R}^k$.

REMARK. By taking the positive part of the difference in H_2, we keep the function nonnegative. Properties H_1 and H_2 are what one would get by taking the collection of all positive parts of members of a vector space of functions. Property H_4 is sometimes called ***Stone's condition***. It is slightly weaker than an assumption that the constant function 1 should belong to $\mathcal{H}^+$. Notice that the cone $\mathcal{C}_0^+(\mathbb{R}^k)$ satisfies H_4, but it does not contain nonzero constants. Nevertheless, if $h \in \mathcal{H}^+$ and α is a positive constant then the function $(h-\alpha)^+ = (h - \alpha(1 \wedge h/\alpha))^+$ belongs to $\mathcal{H}^+$.

<14> **Definition.** *Say that a map $T : \mathcal{H}^+ \mapsto \mathbb{R}^+$ is an **increasing linear functional** if, for h_1, h_2 in $\mathcal{H}^+$, and α_1, α_2 in $\mathbb{R}^+$:*

(T_1) $T(\alpha_1 h_1 + \alpha_2 h_2) = \alpha_1 T h_1 + \alpha_2 T h_2$;

(T_2) $Th_1 \le Th_2$ *if* $h_1 \le h_2$ *pointwise.*

*Call the functional **σ-smooth at** 0 if*

(T_3) $Th_n \downarrow 0$ *whenever the sequence $\{h_n\}$ in $\mathcal{H}^+$ decreases pointwise to zero.*

*Say that T has the **truncation property** if*

(T_4) $T(h \wedge n) \to Th$ *as $n \to \infty$, for each h in $\mathcal{H}^+$.*

REMARK. For an increasing linear functional, T_3 is equivalent to an apparently stronger property,

(T_3') if $h_n \downarrow h_\infty$ with all h_i in $\mathcal{H}^+$ then $Th_n \downarrow Th_\infty$,

because $Th_n \le Th_\infty + T(h_n \backslash h_\infty) \downarrow Th_\infty + 0$. Property T_4 will allow us to reduce the representation of arbitrary members of $\mathcal{H}^+$ as integrals to the representation for bounded functions in $\mathcal{H}^+$.

If μ is a countably additive measure on a sigma-field $\mathcal{A}$, and all the functions in $\mathcal{H}^+$ are μ-integrable, then the $Th := \mu h$ defines a functional on $\mathcal{H}^+$ satisfying T_1 through T_4. The converse problem—find a μ to represent a given functional T—is called the ***integral representation problem***. Theorem <8> will provide a solution to the problem in some generality.

Let $\mathcal{K}_0$ denote the class of all sets K for which there exists a countable subfamily of $\mathcal{H}^+$ with pointwise infimum equal to (the indicator function of) K. Equivalently, by virtue of H_3, there is a decreasing sequence in $\mathcal{H}^+$ converging pointwise to K. It is easy to show that $\mathcal{K}_0$ is a $(\emptyset, \cup f, \cap c)$-paving of subsets for $\mathcal{X}$. Moreover, as the next Lemma shows, the functions in $\mathcal{H}^+$ are related to $\mathcal{K}_0$ and $\mathcal{F}(\mathcal{K}_0)$ in much the same way that nonnegative, continuous functions with compact support in $\mathbb{R}^k$ are related to compact and closed sets.

<15> **Lemma.** *For each h in $\mathcal{H}^+$ and each nonnegative constant α,*

(i) $\{h \ge \alpha\} \in \mathcal{K}_0$ *if* $\alpha > 0$, and *(ii)* $\{h \le \alpha\} \in \mathcal{F}(\mathcal{K}_0)$.

Proof. For (i), note that $\{h \ge \alpha\} = \inf_{n \in \mathbb{N}} \left(1 \wedge n\left(h - \alpha + n^{-1}\right)^+\right)$, a pointwise infimum of a sequence of functions in $\mathcal{H}^+$. For (ii), for a given K in $\mathcal{K}_0$, find a sequence $\{h_n : n \in \mathbb{N}\} \subseteq \mathcal{H}^+$ that decreases to K. Then note that $K\{h \le \alpha\} = \inf_n h_n \backslash (nh - n\alpha)^+$, a set that must therefore belong to $\mathcal{K}_0$. □

<16> **Theorem.** *Let $\mathcal{H}^+$ be a lattice cone of functions, satisfying requirements H_1 through H_4, and T be an increasing, linear functional on $\mathcal{H}^+$ satisfying conditions T_1 through T_4. Then the set function defined on $\mathcal{K}_0$ by $\mu K := \inf\{Th : K \le h \in \mathcal{H}^+\}$ is $\mathcal{K}_0$-tight and σ-smooth along $\mathcal{K}_0$ at $\emptyset$. Its extension to a $\mathcal{K}_0$-regular measure on $\mathcal{B}(\mathcal{K}_0)$ represents the functional, that is, $Th = \mu h$ for all h in $\mathcal{H}^+$. There is only one $\mathcal{K}_0$-regular measure on $\mathcal{B}(\mathcal{K}_0)$ whose integral represents T.*

REMARK. Notice that we can replace the infimum in the definition of μ by an infimum along any decreasing sequence $\{h_n\}$ in $\mathcal{H}^+$ with pointwise limit K. For if $K \le h \in \mathcal{H}^+$, then $\inf_n Th_n \le \inf_n T(h_n \vee h) = Th$, by T_2 and T_3'.

Proof. We must prove that μ is σ-smooth along $\mathcal{K}_0$ at $\emptyset$ and $\mathcal{K}_0$-tight; and then prove that $Th \ge \mu h$ and $Th \le \mu h$ for every h in $\mathcal{H}^+$.

σ-smoothness: Suppose $K_n \in \mathcal{K}_0$ and $K_n \downarrow \emptyset$. Express K_n as a pointwise infimum of functions $\{h_{n,i}\}$ in $\mathcal{H}^+$. Write h_n for $\inf_{m\le n, i\le n} h_{m,i}$. Then $K_n \le h_n \downarrow 0$, and hence $\mu K_n \le Th_n \downarrow 0$ by the σ-smoothness for T and the definition of μ.

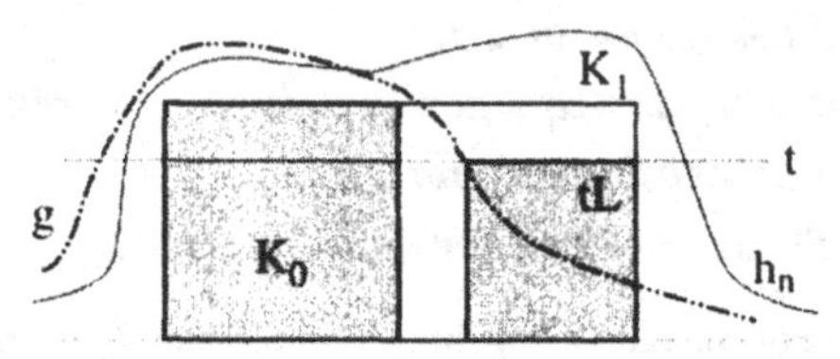

$\mathcal{K}_0$-tightness: Consider sets $K_1 \supseteq K_0$ in $\mathcal{K}_0$. Choose $\mathcal{H}^+$ functions $g \ge K_0$ and $h_n \downarrow K_1$ and fix a positive constant $t < 1$. The $\mathcal{H}^+$-function $g_n := (h_n - n(g\backslash t))^+$ decreases pointwise to the set $L := K_1\{g \le t\} \subseteq K_1\backslash K_0$. Also, it is trivially true that $g \ge tK_1\{g > t\}$. From the inequality $g_n + g \ge tK_1$ we get $\mu K_1 \le T(g_n + g)/t$, because $(g_n + g)/t$ is one of the $\mathcal{H}^+$-functions that enters into the definition of μK_1. Let n tend to infinity, take an infimum over all $g \ge K_0$, then let t increase to 1, to deduce that $\mu K_1 \le \mu L + \mu K_0$, as required for $\mathcal{K}_0$-tightness.

By Theorem <8>, the set function μ extends to a $\mathcal{K}_0$-regular measure on $\mathcal{B}(\mathcal{K}_0)$

Inequality $Th \ge \mu h$: Suppose $h \ge u := \sum_{j=1}^k \alpha_j A_j \in \mathcal{M}^+_{\text{simple}}$. We need to show that $Th \ge \mu u := \sum_j \alpha_j \mu A_j$. We may assume that the $\mathcal{A}$-measurable sets A_j are disjoint. Choose $\mathcal{K}_0$ sets $K_j \subseteq A_j$, thereby defining another simple function $v := \sum_{j=1}^k \alpha_j K_j \le u$. Find sequences h_{nj} from $\mathcal{H}^+$ with $h_{nj} \downarrow \alpha_j K_j$, so that $\sum_j Th_{nj} \downarrow \sum_j \alpha_j \mu K_j = \mu v$. With no loss of generality, assume $h \ge h_{nj}$ for all n and j. Then we have a pointwise bound, $\sum_j h_{nj} \le h + \sum_{i<j} h_{ni} \wedge h_{nj}$, because $\max_j h_{nj} \le h$ and each of the smaller h_{nj} summands must appear in the last sum. Thus

$$\mu v := \sum_j \alpha_j \mu K_j \le \sum_j Th_{nj} \le Th + \sum_{i<j} T\left(h_{ni} \wedge h_{nj}\right).$$

As n tends to infinity, $h_{ni} \wedge h_{nj} \downarrow K_i K_j = \emptyset$. By σ-smoothness of T, the right-hand side decreases to Th, leaving $\mu v \le Th$. Take the supremum over all $K_j \subseteq A_j$, then take the supremum over all $u \le h$, to deduce that $\mu h \le Th$.

Inequality $Th \le \mu h$: Invoke property T_4 to reduce to the case of a bounded h. For a fixed $\epsilon > 0$, approximate h by a simple function $s_\epsilon := \epsilon \sum_{i=1}^N \{h \ge i\epsilon\}$, with steps of size ϵ. Here N is a fixed value large enough to make $N\epsilon$ an upper bound for h. Notice that $\{h \geq i\epsilon\} \in \mathcal{K}_0$, by Lemma <15>. Find sequences h_{ni} from $\mathcal{H}^+$ with $h_{ni} \downarrow \{h \geq i\epsilon\}$. Then we have

$$s_\epsilon \leq h \leq (h \wedge \epsilon) + s_\epsilon \leq (h \wedge \epsilon) + \epsilon \sum_{i=1}^N h_{ni},$$

from which it follows that

$$\begin{aligned} Th &\leq T(h \wedge \epsilon) + \epsilon \sum_{i=1}^N T h_{ni} \\ &\to T(h \wedge \epsilon) + \epsilon \sum_{i=1}^N \mu\{h \geq i\epsilon\} \qquad \text{as } n \to \infty \\ &\leq T(h \wedge \epsilon) + \mu h \qquad \text{because } \epsilon \sum_{i=1}^N \{h \geq i\epsilon\} = s_\epsilon \leq h \\ &\to \mu h \qquad \text{as } \epsilon \to 0, \text{ by } \sigma\text{-smoothness of } T. \end{aligned}$$

Uniqueness: Let ν be another $\mathcal{K}_0$-regular representing measure. If $h_n \downarrow K \in \mathcal{K}_0$, and $h_n \in \mathcal{H}^+$, then $\mu K = \lim_n \mu h_n = \lim_n T h_n = \lim_n \nu h_n = \nu K$. Regularity extends the equality to all sets in $\mathcal{B}(\mathcal{K}_0)$. □

<17> **Example.** Let $\mathcal{H}^+$ equal $\mathcal{C}_0^+(\mathcal{X})$, the cone of all nonnegative, continuous functions with compact support on a locally compact, Hausdorff space $\mathcal{X}$. For example, $\mathcal{X}$ might be $\mathbb{R}^k$. Let T be an increasing linear functional on $\mathcal{C}_0^+(\mathcal{X})$.

Property T_4 holds for the trivial reason that each member of $\mathcal{C}_0^+(\mathcal{X})$ is bounded. Property T_3 is automatic, for a less trivial reason. Suppose $h_n \downarrow 0$. Without loss of generality, $K \geq h_1$ for some compact K. Choose h in $\mathcal{C}_0^+(\mathcal{X})$ with $h \geq K$. For fixed $\epsilon > 0$, the union of the open sets $\{h_n < \epsilon\}$ covers K. For some finite N, the set $\{h_N < \epsilon\}$ contains K, in which case $h_N \leq \epsilon K \leq \epsilon h$, and $T h_N \leq \epsilon T h$. The σ-smoothness follows.

The functional T has a representation $Th = \mu h$ on $\mathcal{C}_0^+(\mathcal{X})$, for a $\mathcal{K}_0$-regular measure μ. The domain of μ need not contain all the Borel sets. However, by an analog of Lemma <10> outlined in Problem [1], it could be extended to a Borel measure without disturbing the representation. □

<18> **Example.** Let $\mathcal{H}^+$ be a lattice cone of bounded continuous functions on a topological space, and let $T : \mathcal{H}^+ \to \mathbb{R}^+$ be a linear functional (necessarily increasing) with the property that to each $\epsilon > 0$ there exists a compact set K_ϵ for which $Th \leq \epsilon$ if $0 \leq h \leq K_\epsilon^c$. (In Section 7.5, such a functional was called functionally tight.)

Suppose $1 \in \mathcal{H}^+$. The functional is automatically σ-smooth: if $1 \geq h_i \downarrow 0$ then eventually $K_\epsilon \subseteq \{h_i < \epsilon\}$, in which case $T h_i \leq T\left((h_i - \epsilon)^+ + \epsilon\right) \leq \epsilon + \epsilon T(1)$. In fact, the same argument shows that the functional is also τ-smooth, in the sense of Problem [2].

The functional T is represented by a measure μ on the sigma-field generated by $\mathcal{H}^+$. Suppose there exists a sequence $\{h_i\} \subseteq \mathcal{H}^+$ for which $1 \geq h_i \downarrow K_\epsilon$. (The version of the representation theorem for τ-smooth functionals, as described by Problem [2], shows that it is even enough to have $\mathcal{H}^+$ generate the underlying topology.) Then $\mu K_\epsilon = \lim_i T h_i = T(1) - \lim_i T(1 - h_i) \geq T(1) - \epsilon$. That is, μ is a tight measure, in the sense that it concentrates most of its mass on a compact set. It is inner regular with respect to approximation by the paving of compact sets. □

7. Problems

[1] A family of sets $\mathcal{U}$ is said to be downward filtering if to each pair U_1, U_2 in $\mathcal{U}$ there exists a U_3 in $\mathcal{U}$ with $U_1 \cap U_2 \supseteq U_3$. A set function $\mu : \mathcal{K}_0 \to \mathbb{R}^+$ is said to be ***τ-smooth*** if $\inf\{\mu K : K \in \mathcal{U}\} = \mu(\cap\mathcal{U})$ for every downward filtering family $\mathcal{U} \subseteq \mathcal{K}_0$. Write $\mathcal{K}$ for the $\cap a$-closure of a $(\emptyset, \cup f, \cap f)$ paving $\mathcal{K}_0$, the collection of all possible intersections of subclasses of $\mathcal{K}_0$.

(i) Show that $\mathcal{K}$ is a $(\emptyset, \cup f, \cap a)$ paving (stable under arbitrary intersections).

(ii) Show that a $\mathcal{K}_0$-tight set function that is τ-smooth at $\emptyset$ has a $\mathcal{K}$-tight, τ-additive extension to $\mathcal{K}$.

[2] Say that an increasing functional T on $\mathcal{H}^+$ is ***τ-smooth at zero*** if $\inf\{Th : h \in \mathcal{V}\}$ for each subfamily $\mathcal{V}$ of $\mathcal{H}^+$ that is downward filtering to the zero function. (That is, to each h_1 and h_2 in $\mathcal{V}$ there is an h_3 in $\mathcal{V}$ with $h_1 \wedge h_2 \geq h_3$ and the pointwise infimum of all functions in $\mathcal{V}$ is everywhere zero.) Extend Theorem <16> to τ-smooth functionals by constructing a $\mathcal{K}$-regular representing measure from the class $\mathcal{K}$ of sets representable as pointwise infima of subclasses of $\mathcal{H}^+$.

8. Notes

The construction via $\mathcal{K}$-tight inner measures is a reworking of ideas from Topsøe (1970). The application to integral representations is a special case of results proved by Pollard & Topsøe (1975).

The book by Fremlin (1974) contains an extensive treatment of the relationship between measures and linear functionals. The book by König (1997) develops the theory of measure and integration with a heavy emphasis on inner regularity.

See Pfanzagl & Pierlo (1969) for an exposition of the properties of pavings compact in the sense of Marczewski.

References

Fremlin, D. H. (1974), *Topological Riesz Spaces and Measure Theory*, Cambridge University Press.

König, H. (1997), *Measure and Integration: An Advanced Course in Basic Procedures and Applications*, Springer-Verlag.

Marczewski, E. (1953), 'On compact measures', *Fundamenta Mathematicae* pp. 113–124.

Pfanzagl, J. & Pierlo, N. (1969), *Compact systems of sets*, Vol. 16 of *Springer Lecture Notes in Mathematics*, Springer-Verlag, New York.

Pollard, D. & Topsøe, F. (1975), 'A unified approach to Riesz type representation theorems', *Studia Mathematica* **54**, 173–190.

Topsøe, F. (1970), *Topology and Measure*, Vol. 133 of *Springer Lecture Notes in Mathematics*, Springer-Verlag, New York.

Appendix B

Hilbert spaces

SECTION 1 defines Hilbert space and presents two basic inequalities.

SECTION 2 establishes the existence of orthogonal projections onto closed subspaces of a Hilbert space.

SECTION 3 defines orthonormal bases of Hilbert spaces. Vectors in the space have representations as infinite linear combinations (convergent series) of basis vectors.

SECTION 4 shows how to construct a random process from an orthonormal sequence of random variables and an orthonormal basis.

1. Definitions

Hilbert space is an infinite dimensional generalization of ordinary Euclidean space. Arguments involving Hilbert spaces look similar to their analogs for Euclidean space, with the addition of occasional precautions against possible difficulties with infinite dimensionality.

<1> **Definition.** *A Hilbert space is a vector space $\mathcal{H}$ equipped with an* ***inner product*** *$\langle \cdot, \cdot \rangle$ (a map from $\mathcal{H} \otimes \mathcal{H}$ into $\mathbb{R}$) which satisfies the following requirements.*

(a) $\langle \alpha f + \beta g, h \rangle = \alpha \langle f, h \rangle + \beta \langle g, h \rangle$ for all real α, β all f, g, h in $\mathcal{H}$.

(b) $\langle f, g \rangle = \langle g, f \rangle$ for all f, g in $\mathcal{H}$.

(c) $\langle f, f \rangle \geq 0$ with equality if and only if $f = 0$.

(d) $\mathcal{H}$ is complete for the norm defined by $\|f\| := \sqrt{\langle f, f \rangle}$. That is, if $\{f_n\}$ is a ***Cauchy sequence*** *in $\mathcal{H}$, meaning $\|f_n - f_m\| \to 0$ as $\min(m, n) \to \infty$, then there exists an f in $\mathcal{H}$ for which $\|f_n - f\| \to 0$.*

Two elements f and g of $\mathcal{H}$ are said to be ***orthogonal***, written $f \perp g$, if $\langle f, g \rangle = 0$. An element f is said to be orthogonal to a subset G of $\mathcal{H}$, written $f \perp G$, if $f \perp g$ for every g in G.

The prime examples of Hilbert spaces are ordinary Euclidean space and $L^2(\mu)$, the set of equivalence classes of measurable real-valued functions whose squares are μ-integrable, for a fixed measure μ. See Section 2.7 for discussion of why we need to work with μ-equivalence classes to get property (c).

Hilbert space shares several properties with ordinary Euclidean space.

Cauchy-Schwarz inequality: $|\langle f, g\rangle| \leq \|f\| \, \|g\|$ for all f, g in $\mathcal{H}$.

The inequality is trivial if either $\|f\| = 0$ or $\|g\| = 0$. Otherwise it follows immediately from an expansion of the left-hand side of the inequality

$$\left\| \frac{f}{\|f\|} \pm \frac{g}{\|g\|} \right\|^2 \geq 0.$$

Triangle Inequality: $\|f + g\| \leq \|f\| + \|g\|$ for all f, g in $\mathcal{H}$.

The square of the left-hand side equals $\|f\|^2 + 2\langle f, g\rangle + \|g\|^2$, which is less than $\|f\|^2 + 2\|f\| \, \|g\| + \|g\|^2 = (\|f\| + \|g\|)^2$, by Cauchy-Schwarz.

2. Orthogonal projections

Many proofs for $\mathbb{R}^k$ that rely only on completeness carry over to Hilbert spaces. For example, if $\mathcal{H}_0$ is a subspace of $\mathbb{R}^k$, then to each vector x there exists a vector x_0 in $\mathcal{H}_0$ that is closest to x. The vector x_0 is characterized by the property that $x - x_0$ is orthogonal to $\mathcal{H}_0$. The vector x_0 is called the ***orthogonal projection*** of x onto $\mathcal{H}_0$, or the component of x in the subspace $\mathcal{H}_0$; the vector $x - x_0$ is called the component of x orthogonal to $\mathcal{H}_0$.

Projections also exist for ***closed*** subspaces of a Hilbert space $\mathcal{H}$. (Recall that a subset $\mathcal{G}$ of $\mathcal{H}$ is said to be closed if it contains all its limit points: if $\{g_n\} \subseteq \mathcal{G}$ and $\|g_n - f\| \to 0$ then $f \in \mathcal{G}$.) Every finite dimensional subspace is automatically closed (Problem [2]); infinite dimensional subspaces need not be closed (Problem [3]).

<2> **Theorem.** *Let $\mathcal{H}_0$ be a closed subspace of a Hilbert space $\mathcal{H}$. For each f in $\mathcal{H}$ there is a unique f_0 in $\mathcal{H}_0$, the orthogonal projection of f onto $\mathcal{H}_0$, for which $f - f_0$ is orthogonal to $\mathcal{H}_0$. The point f_0 minimizes $\|f - h\|$ over all h in $\mathcal{H}_0$.*

Proof. For a fixed f in $\mathcal{H}$, define $\delta := \inf\{\|f - h\| : h \in \mathcal{H}_0\}$. Choose $\{h_n\}$ in $\mathcal{H}_0$ such that $\|f - h_n\| \to \delta$. For arbitrary g, g' in $\mathcal{H}$, cancellation of cross-product terms leads to the identity

$$\|g + g'\|^2 + \|g - g'\|^2 = 2\|g\|^2 + 2\|g'\|^2.$$

Put $g := f - h_n$ and $g' := f - h_m$ to get

$$4\|f - (h_n + h_m)/2\|^2 + \|h_m - h_n\|^2 = 2\|f - h_n\|^2 + 2\|f - h_m\|^2.$$

The first term on the left-hand side must be $\geq 4\delta^2$ because $(h_n + h_m)/2$ belongs to $\mathcal{H}_0$. Both terms on the right-hand side converge to $2\delta^2$ as $\min(m, n) \to \infty$. Thus $\|h_m - h_n\| \to 0$ as $\min(m, n) \to \infty$. That is, $\{h_n\}$ is a Cauchy sequence.

Completeness of $\mathcal{H}$ ensures that h_n converges to some f_0 in $\mathcal{H}$. As $\mathcal{H}_0$ is closed and each h_n belongs to $\mathcal{H}_0$, we deduce that $f_0 \in \mathcal{H}_0$. The infimum δ is achieved at f_0, because $\|f - f_0\| \leq \|f - h_n\| + \|h_n - f_0\| \to \delta$.

To prove the orthogonality assertion, for a fixed g in $\mathcal{H}_0$ consider the squared distance from f to $f_0 + tg$,

$$\|f - (f_0 + tg)\|^2 = \|f - f_0\|^2 + 2t\langle f - f_0, g\rangle + t^2\|g\|^2,$$

as a function of the real variable t. The vector $f_0 + tg$ belongs to $\mathcal{H}_0$. It is one of those vectors in the range of the the infimum that defines δ. It follows that

$$\delta^2 + 2t\langle f - f_0, g\rangle + t^2\|g\|^2 \geq \delta^2 \qquad \text{for all real } t.$$

For such an inequality to hold, the coefficient, $\langle f - f_0, g\rangle$, of the linear term in t must be zero. (Otherwise what would happen for t close to zero?)

To prove uniqueness, suppose both f_0 and f_1 have the projection property. Then $f_1 - f_0$ in $\mathcal{H}_0$ would be orthogonal to both $f - f_0$ and $f - f_1$, from which it follows that $f_1 - f_0$ is orthogonal to itself, whence $f_1 = f_2$. □

<3> **Corollary.** *(Riesz-Fréchet Theorem) To each continuous linear map T from a Hilbert space $\mathcal{H}$ into $\mathbb{R}$ there exists a unique element h_0 in $\mathcal{H}$ such that $Th = \langle h, h_0\rangle$ for all h in $\mathcal{H}$.*

Proof. Uniqueness is easy to prove: if $\langle h, h_0\rangle = \langle h, h_1\rangle$ for $h := h_0 - h_1$ then $\|h_0 - h_1\| = 0$.

Existence is also easy when $Th \equiv 0 = \langle h, 0\rangle$. So let us assume that there exists an h_2 with $Th_2 \neq 0$. Let h_3 denote the component of h_2 that is orthogonal to the closed (by continuity of T) linear subspace $\mathcal{H}_0 := \{h \in \mathcal{H} : Th = 0\}$ of $\mathcal{H}$. Note that $Th_3 = Th_2 - T(h_2 - h_3) = Th_2 \neq 0$.

For h in $\mathcal{H}$, define $C_h := Th/Th_3$. The difference $h - C_h h_3$ belongs to $\mathcal{H}_0$, because $T(h - C_h h_3) = 0$, and therefore $0 = \langle h - C_h h_3, h_3\rangle = \langle h, h_3\rangle - (Th/Th_3)\|h_3\|^2$. The choice $h_0 := (Th_3/\|h_3\|^2)h_3$ gives the desired representation. □

3. Orthonormal bases

A family of vectors $\Psi = \{\psi_i : i \in I\}$ in a Hilbert space $\mathcal{H}$ is said to be ***orthonormal*** if $\langle\psi_i, \psi_i\rangle = 1$ for all i and $\langle\psi_i, \psi_j\rangle = 0$ for all $i \neq j$. The subspace $\mathcal{H}_0$ spanned by Ψ consists of all linear combinations $\sum_{i\in J} \alpha_i \psi$, with J ranging over all finite subsets of I. If $\mathcal{H}_0$ is dense in $\mathcal{H}$ (that is, if the closure $\overline{\mathcal{H}_0}$ of $\mathcal{H}_0$ equals $\mathcal{H}$) then the family ψ is called an ***orthonormal basis*** for $\mathcal{H}$.

<4> **Lemma.** *Let Ψ be an orthonormal basis for $\mathcal{H}$.*

(i) For each h in $\mathcal{H}$, the set $I_h := \{i \in I : \langle h, \psi_i\rangle \neq 0\}$ is countable, and for every enumeration $\{i(1), i(2), \ldots\}$ of I_h, the sum $\sum_{k=1}^{n}\langle h, \psi_{i(k)}\rangle\psi_{i(k)}$ converges in norm to h as $n \to \infty$.

*(ii) $\langle g, h\rangle = \sum_{i\in I}\langle g, \psi_i\rangle\langle h, \psi_i\rangle$ **(Parseval's identity)** for all $g, h \in \mathcal{H}$. In particular, $\|h\|^2 = \sum_{i\in I} |\langle h, \psi_i\rangle|^2$.*

REMARK. The sum in assertion (ii) actually runs over only a countable subset of I. The assertion should be understood as convergence of partial sums for all possible enumerations of that countable subset.

Proof. For each finite subset J of I, the subspace $\mathcal{H}_J$ spanned by the finite set $\{\psi_i : i \in J\}$ is closed—see Problem [2]. The projection of h onto $\mathcal{H}_J$ equals

$h_J := \sum_{i\in J}\langle h, \psi_i\rangle\psi_i$, because $\langle h - h_J, \psi_i\rangle = \langle h, \psi_i\rangle - \langle h, \psi_i\rangle = 0$ for each i in J. The orthogonality implies

$$\|h\|^2 = \|h - h_J\| + \|h_J\|^2 = \|h - h_J\|^2 + \sum\nolimits_{i\in J}\langle h, \psi_i\rangle^2.$$

For each $\epsilon > 0$, the set $I_h(\epsilon) := \{i \in I : |\langle h, \psi_i\rangle| \geq \epsilon\}$ has cardinality N_ϵ no greater than $\|h\|^2/\epsilon^2$, because $\|h\|^2 \geq \sum_i\{i \in I_h(\epsilon)\}\langle h, \psi_i\rangle^2 \geq \epsilon^2 N_\epsilon$. It follows that the set $I_h = \cup_{k\in\mathbb{N}} I_h(1/k)$ is countable, as asserted.

For a fixed h let $\{i(1), i(2), \ldots\}$ be an enumeration of I_h. Write $J(n)$ for the initial segment $\{i(1), \ldots, i(n)\}$. Denseness of $\mathcal{H}_0$ means that to each $\epsilon > 0$ there is some finite linear combination $h_\epsilon = \sum_{i\in J}\alpha_i\psi_i$ such that $\|h - h_\epsilon\| < \epsilon$. By definition of I_h, the vector h is orthogonal to ψ_i for each i in $J\backslash I_h$. It follows that the vector $h - \sum_i\{i \in J \cap I_h\}\alpha_i\psi_i$ is orthogonal to $\{\psi_i : i \in J\backslash I_h\}$, whence

$$\|h - h_\epsilon\|^2 = \|h - \sum\nolimits_i\{i \in J \cap I\}\alpha_i\psi_i\|^2 + \|\sum\nolimits_i\{i \in J\backslash I\}\alpha_i\psi_i\|^2.$$

We reduce the distance between h and h_ϵ if we discard from J those i not in I_h. Without loss of generality we may therefore assume that J is a subset of I_h.

When n is large enough that $J \subseteq J(n)$ we have $h - h_{J(n)}$ orthogonal to the subspace $\mathcal{H}_{J(n)}$, which contains both $h_{J(n)}$ and h_ϵ. For such n we have

$$\begin{aligned} \epsilon^2 &> \|h - h_{J(n)} + h_{J(n)} - h_\epsilon\|^2 && \text{definition of } h_\epsilon \\ &= \|h - h_{J(n)}\|^2 + \|h_{J(n)} - h_\epsilon\|^2 && \text{by orthogonality} \\ &\geq \|h - h_{J(n)}\|^2. \end{aligned}$$

That is, for each $\epsilon > 0$ we have $\|h - h_{J(n)}\| < \epsilon$ for all n large enough—precisely what it means to say that $\sum_{k=1}^\infty\langle h, \psi_i(k)\rangle\psi_{i(k)}$ converges (in norm) to h. The representation $\|h\|^2 = \sum_{k=1}^\infty\langle h, \psi_{i(k)}\rangle^2$ then follows via continuity (Problem [1]) of the map $g \mapsto \|g\|^2$.

For Parseval's identity, let $\{j(1), j(2), \ldots\}$ be an enumeration of $I_g \cup I_h$. Then, from the special case just established,

$$\begin{aligned} \|g \pm h\|^2 &= \sum\nolimits_{k=1}^\infty\langle g \pm h, \psi_{j(k)}\rangle^2 \\ &= \sum\nolimits_{k=1}^\infty\left(\langle g, \psi_{j(k)}\rangle^2 \pm 2\langle g, \psi_{j(k)}\rangle\langle h, \psi_{j(k)}\rangle + \langle h, \psi_{j(k)}\rangle^2\right). \end{aligned}$$

□ The series for $4\langle g, h\rangle = \|g + h\|^2 - \|g - h\|^2$ is obtained by subtraction.

<5> **Example.** (Haar basis) Let m denote Lebesgue measure on the Borel sigma-field of $(0, 1]$. For $k = 0, 1, 2, \ldots$, partition $(0, 1]$ into subintervals $J_{i,k} := \left(i2^{-k}, (1 + 1)2^{-k}\right]$, for $i = 0, 1, \ldots, 2^k - 1$. Define functions $H_{i,k} := J_{2i,k+1} - J_{2i+1,k+1}$ and $\psi_{i,k} := \sqrt{2^k}H_{i,k}$, for $0 \leq i < 2^k$, and $k = 0, 1, 2, \ldots$.

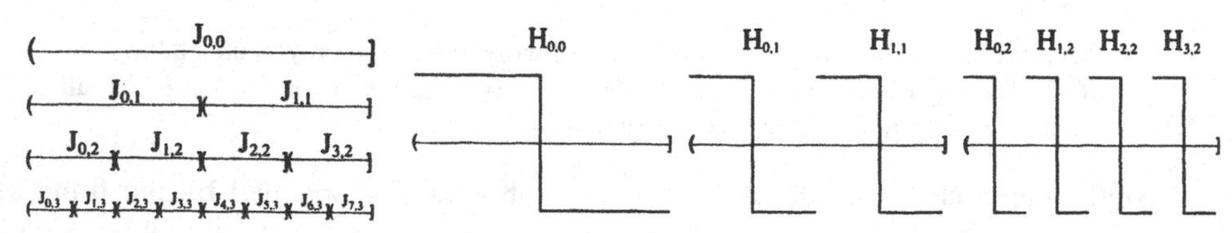

The collection of functions $\Psi := \{1\} \cup \{\psi_{i,k} : k \in \mathbb{N}_0;\ 0 \leq i < 2^k\}$ is an orthonormal family in $L^2(\mathrm{m})$. A generating class argument will show that Ψ is an orthonormal basis.

Each $J_{i,k}$ belongs to the subspace $\mathcal{H}_0$ spanned by Ψ:

$$J_{0,1} = \tfrac{1}{2}\left(1 + H_{0,0}\right) = 1 - J_{1,1}$$
$$J_{0,2} = \tfrac{1}{2}\left(J_{0,1} + H_{0,1}\right) = J_{0,1} - J_{1,2}, \qquad J_{2,2} = \tfrac{1}{2}\left(J_{1,1} + H_{1,1}\right) = J_{1,1} - J_{3,2}$$

and so on. Take finite sums to deduce that $\mathcal{H}_0$ contains the class $\mathcal{E}$ of all indicator functions of dyadic intervals $(i/2^k, j/2^k]$. Note that $\mathcal{E}$ is stable under finite intersections and that it generates the Borel sigma-field. The class $\mathcal{D}$ of all Borel sets whose indicators belong to the closure $\overline{\mathcal{H}}_0$ is a λ-system (easy proof), which contains $\mathcal{E}$. Thus $\overline{\mathcal{H}}_0$ contains $\sigma(\mathcal{E})$, the Borel sigma-field on $(0, 1]$.

If $\overline{\mathcal{H}}_0$ were not equal to the whole of $L^2(\mathfrak{m})$ there would exist a square-integrable function f orthogonal to $\overline{\mathcal{H}}_0$. In particular, f would be orthogonal to both $\{f > 0\}$ and $\{f < 0\}$, which would force $\mathfrak{m}f^+ = \mathfrak{m}f^- = 0$. That is, $f = 0$ a.e. $[\mathfrak{m}]$.

The component of a function h in the subspace spanned by 1 is just the constant function $\mathfrak{m}h$. Each function h in $L^2(\mathfrak{m})$ has a series expansion,

$$h = \mathfrak{m}h + \sum_{k \in \mathbb{N}_0} \sum_{i < 2^k} \psi_{i,k} \langle h, \psi_{i,k} \rangle,$$

with convergence in the $L^2(\mathfrak{m})$ sense. □

4. Series expansions of random processes

Suppose $\mathcal{H}$ is a Hilbert space (necessarily separable—see Problem [6]) with a countable orthonormal basis $\Psi = \{\psi_i : i \in \mathbb{N}\}$. Let $\{\xi_i : i \in \mathbb{N}\}$ be a sequence of random variables, on some probability space $(\Omega, \mathcal{F}, \mathbb{P})$, that is orthonormal for the $L^2(\mathbb{P})$ inner product: $\mathbb{P}\xi_i \xi_j = 1$ if $i = j$, zero otherwise. For each h in $\mathcal{H}$, the sequence of random variables

$$X_n(h) := \sum_{i=1}^{n} \langle h, \psi_i \rangle \xi_i$$

is a Cauchy sequence in $L^2(\mathbb{P})$, because convergence of $\sum_i \langle h, \psi_i \rangle^2$ implies

$$\mathbb{P} \left| \sum_{n \le i \le m} \langle h, \psi_i \rangle \xi_i \right|^2 = \sum_{n \le i \le m} \langle h, \psi_i \rangle^2 \to 0 \qquad \text{as } n \to \infty.$$

Write $X(h)$ for the limit $\sum_{i=1}^{\infty} \langle h, \psi_i \rangle \xi_i$, which is defined up to an almost sure equivalence as a square integrable random variable.

By Problem [1] and the Parseval identity, for g and h in $\mathcal{H}$ we have

$$\mathbb{P}X(g)X(h) = \lim_{n \to \infty} \mathbb{P}X_n(g)X_n(h) = \lim_{n \to \infty} \sum_{i=1}^{n} \langle g, \psi_i \rangle \langle h, \psi_i \rangle = \langle g, h \rangle.$$

In particular, $\mathbb{P}X(h)^2 = \|h\|^2$. The map $h \mapsto X(h)$ is a ***linear isometry*** between $\mathcal{H}$ and a linear subspace of $L^2(\mathbb{P})$: the map preserves inner products and distances.

The most important example of a series expansion arises when all the ξ_i have independent, standard normal distributions. The family of random variables $\{X(h) : h \in \mathcal{H}\}$ is then called the ***isonormal process*** indexed by $\mathcal{H}$. The particular case known as ***Brownian motion*** is discussed in Chapter 9.

5. Problems

[1] Suppose $g_n \to g$ and $h_n \to h$, as elements of a Hilbert space. Show that $\langle g_n, h_n \rangle \to \langle g, h \rangle$. Hint: Use Cauchy-Schwarz to bound terms like $\langle g_n - g, h_n \rangle$.

[2] Let $\mathcal{F}$ be a finite subset of a Hilbert space $\mathcal{H}$. Show that the subspace generated by $\mathcal{F}$ is a closed subset of $\mathcal{H}$. Hint: Without loss of generality assume the elements $\{f_1, \ldots, f_k\}$ of $\mathcal{F}$ are linearly independent. For each i, find a vector ψ_i such that $\langle f_i, \psi_i \rangle = 1$ but $\langle f_i, \psi_j \rangle = 0$ for $j \neq i$. If $h_n := \sum_i \alpha_i(n) f_i$ converges to some h, deduce that $\{\alpha_i(n)\}$ converges for each i.

[3] Let λ be a finite measure on $\mathcal{B}[0, 1]$. Let $\mathcal{H}$ denote the collection of λ-equivalence classes $\{ [h] : h \text{ is continuous } \}$. Show that $\mathcal{H}$ is not a closed subspace of $L^2(\lambda)$ if λ equals Lebesgue measure. Could it be a closed subspace for some other choice of λ?

[4] Let $\mathcal{K}$ be a closed, convex subset of a Hilbert space $\mathcal{H}$.

(i) Show that to each f in $\mathcal{H}$ there is a unique f_0 in $\mathcal{K}$ for which $\|f - f_0\| = \inf\{\|f - h\| : h \in \mathcal{K}\}$. Hint: Mimic the proof of Theorem <2>.

(ii) Show that $\langle f - f_0, g - f_0 \rangle \leq 0$ for all g in $\mathcal{K}$. Hint: Consider the distance from f to $(1 - t) f_0 + t g$ for $0 \leq t \leq 1$.

(iii) Give a (finite-dimensional) example where $\langle f - f_0, g - f_0 \rangle < 0$ for all g in $\mathcal{K} \backslash \{f_0\}$.

[5] Use Zorn's Lemma to prove that every Hilbert space has at least one orthonormal basis. Hint: Order orthonormal bases by inclusion. If Ψ is maximal for this ordering, show that there can be no nonzero element orthogonal to every member of Ψ.

[6] Let $\mathcal{H}$ be a Hilbert space with an orthonormal basis $\Psi := \{\psi_i : i \in I\}$. Show that I is countable if and only if $\mathcal{H}$ is separable (that is, it contains a countable, dense subset). Hint: if I is countable, consider finite linear combinations $\sum_{i \in J} \alpha_i \psi_i$ with the α_i rational. Conversely, if $\{h_1, h_2, \ldots\}$ is dense, construct an orthonormal basis inductively by defining $g_i := h_i - \sum_j \{j < i\} \langle h_i, \psi_j \rangle$ and $\psi_i := g_i / \|g_i\|$ when $g_i \neq 0$.

6. Notes

Halmos (1957, Chapter I) is an excellent source for basic facts about Hilbert space. See Dudley (1973) and Dudley (1989, page 378) for the isonormal process.

References

Dudley, R. M. (1973), 'Sample functions of the Gaussian process', *Annals of Probability* **1**, 66–103.

Dudley, R. M. (1989), *Real Analysis and Probability*, Wadsworth, Belmont, Calif.

Halmos, P. R. (1957), *Introduction to Hilbert space*, second edn, Chelsea.

Appendix C

Convexity

SECTION 1 defines convex sets and functions.
SECTION 2 shows that convex functions defined on subintervals of the real line have left- and right-hand derivatives everywhere.
SECTION 3 shows that convex functions on the real line can be recovered as integrals of their one-sided derivatives.
SECTION 4 shows that convex subsets of Euclidean spaces have nonempty relative interiors.
SECTION 5 derives various facts about separation of convex sets by linear functions.

1. Convex sets and functions

A subset C of a vector space is said to be convex if it contains all the line segments joining pairs of its points, that is,

$$\alpha x_1 + (1-\alpha)x_2 \in C \qquad \text{for all } x_1, x_2 \in C \text{ and all } 0 < \alpha < 1.$$

A real-valued function f defined on a convex subset C (of a vector space $\mathcal{V}$) is said to be convex if

$$f(\alpha x_1 + (1-\alpha)x_2) \le \alpha f(x_1) + (1-\alpha) f(x_2) \qquad \text{for all } x_1, x_2 \in C \text{ and } 0 < \alpha < 1.$$

Equivalently, the ***epigraph*** of the function,

$$\text{epi}(f) := \{(x,t) \in C \times \mathbb{R} : t \ge f(x)\},$$

is a convex subset of $C \times \mathbb{R}$. Some authors (such as Rockafellar 1970) define $f(x)$ to equal $+\infty$ for $x \in \mathcal{V}\backslash C$, so that the function is convex on the whole of $\mathcal{V}$, and epi(f) is a convex subset of $\mathcal{V} \times \mathbb{R}$.

This Appendix will establish several facts about convex functions and sets, mostly for Euclidean spaces. In particular, the facts include the following results as special cases.

(i) For a convex function f defined at least on an open interval of the real line (possibly the whole real line), there exists a countable collection of linear functions for which $f(x) = \sup_{i\in\mathbb{N}} (\alpha_i + \beta_i x)$ on that interval.

(ii) If a real-valued function f has an increasing, real-valued right-hand derivative at each point of an open interval, then f is convex on that interval. In particular, if f is twice differentiable, with $f'' \ge 0$, then f is convex.

(iii) If a convex function f on a convex subset $C \subseteq \mathbb{R}^n$ has a local minimum at a point x_0, that is, if $f(x) \geq f(x_0)$ for all x in a neighborhood of x_0, then $f(w) \geq f(x_0)$ for all w in C.

(iv) If C_1 and C_2 are disjoint convex subsets of $\mathbb{R}^n$ then there exists a nonzero ℓ in $\mathbb{R}^n$ for which $\sup_{x \in C_1} x \cdot \ell \leq \inf_{x \in C_2} x \cdot \ell$. That is, the linear functional $x \mapsto x \cdot \ell$ ***separates*** the two convex sets.

2. One-sided derivatives

Let f be a convex function, defined and real-valued at least on an interval J of the real line.

Consider any three points $x_1 < x_2 < x_3$, all in J. (For the moment, ignore the point x_0 shown in the picture.) Write α for $(x_2 - x_1)/(x_3 - x_1)$, so that $x_2 = \alpha x_3 + (1 - \alpha)x_1$. By convexity, $y_2 := \alpha f(x_3) + (1 - \alpha) f(x_1) \geq f(x_2)$. Write $S(x_i, x_j)$ for $\left(f(x_j) - f(x_i)\right)/(x_j - x_i)$, the slope of the chord joining the points $(x_i, f(x_i))$ and $(x_j, f(x_j))$. Then

$$
\begin{aligned}
S(x_2, x_3) &= \frac{f(x_3) - f(x_2)}{x_3 - x_2} \\
&\geq \frac{f(x_3) - y_2}{x_3 - x_2} = S(x_1, x_3) = \frac{y_2 - f(x_1)}{x_2 - x_1} \\
&\qquad\qquad\qquad\qquad\qquad \geq \frac{f(x_2) - f(x_1)}{x_2 - x_1} = S(x_1, x_2).
\end{aligned}
$$

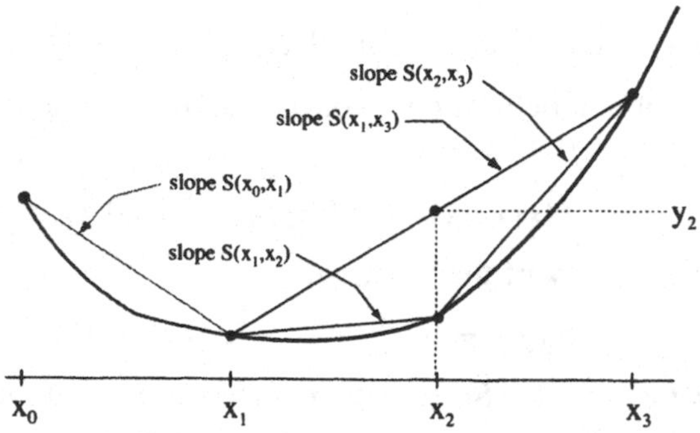

From the second inequality it follows that $S(x_1, x)$ decreases as x decreases to x_1. That is, f has right-hand derivative $D_+(x_1)$ at x_1, if there are points of J that are larger than x_1. The limit might equal $-\infty$, as in the case of the function $f(x) = -\sqrt{x}$ defined on $\mathbb{R}^+$, with $x_1 = 0$. However, if there is at least one point x_0 of J for which $x_0 < x_1$ then the limit $D_+(x_1)$ must be finite: Replacing $\{x_1, x_2, x_3\}$ in the argument just made by $\{x_0, x_1, x_2\}$, we have $S(x_0, x_1) \leq S(x_1, x_2)$, implying that $-\infty < S(x_0, x_1) \leq D_+(x_1)$.

The inequality $S(x_1, x) \leq S(x_1, x_2) \leq S(x_2, x')$ if $x_1 < x < x_2 < x'$, leads to the conclusion that D_+ is an increasing function. Moreover, it is continuous from the right, because

$$D_+(x_2) \le S(x_2, x_3) \to S(x_1, x_3) \qquad \text{as } x_2 \downarrow x_1, \text{ for fixed } x_3$$
$$\to D_+(x_1) \qquad \text{as } x_3 \downarrow x_1.$$

Analogous arguments show that $S(x_0, x_1)$ increases to a limit $D_-(x_1)$ as x_0 increases to x_1. That is, f has left-hand derivative $D_1(x_1)$ at x_1, if there are points of J that are smaller than x_1.

If x_1 is an interior point of J then both left-hand and right-hand derivatives exist, and $D_-(x_1) \le D_+(x_1)$. The inequality may be strict, as in the case where $f(x) = |x|$ with $x_1 = 0$. The left-hand derivative has properties analogous to those of the right-hand derivative. The following Theorem summarizes.

<1> **Theorem.** *Let f be a convex, real-valued function defined (at least) on a bounded interval $[a, b]$ of the real line. The following properties hold.*

(i) *The right-hand derivative $D_+(x)$ exists,*

$$\frac{f(y) - f(x)}{y - x} \downarrow D_+(x) \qquad \text{as } y \downarrow x,$$

for each x in $[a, b)$. The function $D_+(x)$ is increasing and right-continuous on $[a, b)$. It is finite for $a < x < b$, but $D_+(a)$ might possibly equal $-\infty$.

(ii) *The left-hand derivative $D_-(x)$ exists,*

$$\frac{f(x) - f(z)}{x - z} \uparrow D_-(x) \qquad \text{as } z \uparrow x,$$

for each x in $(a, b]$. The function $D_-(x)$ is increasing and left-continuous function on $(a, b]$. It is finite for $a < x < b$, but $D_-(b)$ might possibly equal $+\infty$.

(iii) *For $a \le x < y \le b$,*

$$D_+(x) \le \frac{f(y) - f(x)}{y - x} \le D_-(y).$$

(iv) *$D_-(x) \le D_+(x)$ for each x in (a, b), and*

$$f(w) \ge f(x) + c(w - x) \qquad \text{for all } w \text{ in } [a, b],$$

for each real c with $D_-(x) \le c \le D_+(x)$.

Proof. Only the second part of assertion (iv) remains to be proved. For $w > x$ use

$$\frac{f(w) - f(x)}{w - x} = S(x, w) \ge D_+(x) \ge c;$$

for $w < x$ use

$$\frac{f(x) - f(w)}{x - w} = S(w, x) \le D_-(x) \le c,$$

where $S(\cdot, \cdot)$ denotes the slope function, as above. □

<2> **Corollary.** *If a convex function f on a convex subset $C \subseteq \mathbb{R}^n$ has a local minimum at a point x_0, that is, if $f(x) \ge f(x_0)$ for all x in a neighborhood of x_0, then $f(w) \ge f(x_0)$ for all w in C.*

Proof. Consider first the case $n = 1$. Suppose $w \in C$ with $w > x_0$. The right-hand derivative $D_+(x_0) = \lim_{y \downarrow x_0} (f(y) - f(x_0))/(y - x_0)$ must be nonnegative, because $f(y) \geq f(x_0)$ for y near x_0. Assertion (iv) of the Theorem then gives

$$f(w) \geq f(x_0) + (w - x_0)D_+(x_0) \geq f(x_0).$$

The argument for $w < x_0$ is similar.

For general $\mathbb{R}^n$, apply the result for $\mathbb{R}$ along each straight line through x_0. □

Existence of finite left-hand and right-hand derivatives ensures that f is continuous at each point of the open interval (a, b). It might not be continuous at the endpoints, as shown by the example

$$f(x) = \begin{cases} -\sqrt{x} & \text{for } x > 0 \\ 1 & \text{for } x = 0. \end{cases}$$

Of course, we could recover continuity by redefining $f(0)$ to equal 0, the value of the limit $f(0+) := \lim_{w \downarrow 0} f(w)$.

<3> **Corollary.** *Let f be a convex, real-valued function on an interval $[a, b]$. There exists a countable collection of linear functions $d_i + c_i w$, for which the convex function $\psi(w) := \sup_{i \in \mathbb{N}} (d_i + c_i w)$ is everywhere $\leq f(w)$, with equality except possibly at the endpoints $w = a$ or $w = b$, where $\psi(a) = f(a+)$ and $\psi(b) = f(b-)$.*

Proof. Let $\mathcal{X}_0 := \{x_i : i \in \mathbb{N}\}$ be a countable dense subset of (a, b). Define $c_i := D_+(x_i)$ and $d_i := f(x_i) - c_i x_i$. By assertion (iv) of the Theorem, $f(w) \geq d_i + c_i w$ for $a \leq w \leq b$ for each i, and hence $f(w) \geq \psi(w)$.

If $a < w < b$ then (iv) also implies that $f(x_i) \geq f(w) + (x_i - w)D_+(w)$, and hence

$$\psi(w) \geq f(x_i) + c_i(w - x_i) \geq f(w) - (x_i - w)\left(D_+(x_i) - D_+(w)\right) \qquad \text{for all } x_i.$$

Let x_i decrease to w (through $\mathcal{X}_0$) to conclude, via right-continuity of D_+ at w, that $\psi(w) \geq f(w)$.

If $D_+(a) > -\infty$ then f is continuous at a, and

$$f(a) \geq \psi(a) \geq \limsup_{x_i \downarrow a} (f(x_i) + (a - x_i)c_i) = f(a+) = f(a).$$

If $D_+(a) = -\infty$ then f must be decreasing in some neighborhood $\mathcal{N}$ of a, with $c_i < 0$ when $x_i \in \mathcal{N}$, and

$$\psi(a) \geq \sup_{x_i \in \mathcal{N}} (f(x_i) + (a - x_i)c_i) \geq \sup_{x_i \in \mathcal{N}} f(x_i) = f(a+).$$

If $\psi(a)$ were strictly greater than $f(a+)$, the open set

$$\{w : \psi(w) > f(a+)\} = \cup_i \{w : d_i + c_i w > f(a+)\}$$

would contain a neighborhood of a, which would imply existence of points w in $\mathcal{N}\backslash\{a\}$ for which $\psi(w) > f(a+) \geq f(w)$, contradicting the inequality $\psi(w) \leq f(w)$. A similar argument works at the other endpoint. □

3. Integral representations

Convex functions on the real line are expressible as integrals of one-sided derivatives.

<4> **Theorem.** *If f is real-valued and convex on $[a, b]$, with $f(a) = f(a+)$ and $f(b) = f(b-)$, then both $D_+(x)$ and $D_-(x)$ are integrable with respect to Lebesgue measure on $[a, b]$, and*

$$f(x) = f(a) + \int_a^x D_+(t)\,dt = f(a) + \int_a^x D_-(t)\,dt \qquad \text{for } a \le x \le b.$$

Proof. Choose α and β with $a < \alpha < \beta < x$. For a positive integer n, define $\delta := (\beta - \alpha)/n$ and $x_i := \alpha + i\delta$ for $i = 0, 1, \ldots, n$. Both D_+ and D_- are bounded on $[\alpha, \beta]$. For $i = 2, \ldots, n-1$, part (iii) of Theorem <1> and monotonicity of both one-sdied derivatives gives

$$\int_{x_{i-2}}^{x_{i-1}} D_+(t)\,dt \le \delta D_+(x_{i-1}) \le f(x_i) - f(x_{i-1}) \le \delta D_-(x_i) \le \int_{x_i}^{x_{i+1}} D_-(t)\,dt,$$

which sums to give

$$\int_\alpha^{x_{n-2}} D_+(t)\,dt \le f(x_{n-1}) - f(x_1) \le \int_{x_2}^{\beta} D_-(t)\,dt.$$

Let n tend to infinity, invoking Dominated Convergence and continuity of f, to deduce that $\int_\alpha^\beta D_+(t)\,dt \le f(\beta) - f(\alpha) \le \int_\alpha^\beta D_-(t)\,dt$. Both inequalities must actually be equalities, because $D_-(t) \le D_+(t)$ for all t in (a, b).

Let α decrease to a. Monotone Convergence—the functions $D_\pm$ are bounded above by $D_+(\beta)$ on $(a, \beta]$—and continuity of f at a give $f(\beta) - f(a) = \int_a^\beta D_+(t)\,dt = \int_a^\beta D_-(t)\,dt$. In particular, the negative parts of both $D_\pm$ are integrable. Then let β increase to x to deduce, via a similar argument, the asserted integral expressions for $f(x) - f(a)$, and the integrability of $D_\pm$ on $[a, b]$. □

Conversely, suppose f is a continuous function defined on an interval $[a, b]$, with an increasing, real-valued right-hand derivative $D_+(t)$ existing at each point of $[a, b)$. On each closed proper subinterval $[a, x]$, the function D_+ is bounded, and hence Lebesgue integrable. From Section 3.4, $f(x) = \int_a^x D_+(t)\,dt$ for all $a \le x < b$. Equality for $x = b$ also follows, by continuity and Monotone Convergence. A simple argument will show that f is then convex on $[a, b]$.

More generally, suppose D is an increasing, real-valued function defined (at least) on $[a, b)$. Define $g(x) := \int_a^x D(t)\,dt$, for $a \le x \le b$. (Possibly $g(b) = \infty$.) Then g is convex. For if $a \le x_0 < x_1 \le b$ and $0 < \alpha < 1$ and $x_\alpha := (1-\alpha)x_0 + \alpha x_1$, then

$$\begin{aligned}
&(1-\alpha)g(x_0) + \alpha g(x_1) - g(x_\alpha) \\
&\quad = \int_a^b \big((1-\alpha)\{t \le x_0\} + \alpha\{t \le x_1\} - \{t \le x_\alpha\}\big)\,D(t)\,dt \\
&\quad = \int_a^b \big(\alpha\{x_\alpha < t \le x_1\} - (1-\alpha)\{x_0 < t \le x_\alpha\}\big)\,D(t)\,dt \\
&\quad \ge \big(\alpha(x_1 - x_\alpha) - (1-\alpha)(x_\alpha - x_0)\big)\,D(x_\alpha) = 0.
\end{aligned}$$

<5> **Example.** Let f be a twice continuously differentiable (actually, absolute continuity of f' would suffice) convex function, defined on a convex interval $J \subseteq \mathbb{R}$

that contains the origin. Suppose $f(0) = f'(0) = 0$. The representations

$$f(x) = x \int \{0 \le s \le 1\} f'(xs)\, ds$$
$$= x^2 \iint \{0 \le t \le s \le 1\} f''(xt)\, dt\, ds = x^2 \int_0^1 (1-t) f''(xt)\, dt,$$

establish the following facts.

(i) The function $f(x)/x$ is increasing.

(ii) The function $\phi(x) := 2f(x)/x^2$ is nonnegative and convex.

(iii) If f'' is increasing then so is ϕ.

Moreover, Jensen's inequality for the uniform distribution λ on the triangular region $\{0 \le t \le s \le 1\}$ implies that

$$\phi(x) = \lambda^{s,t} f''(xt) \ge f''\left(\lambda^{s,t} xt\right) = f''(x/3).$$

Two special cases of these results were needed in Chapter 10, to establish the Bennett inequality and to establish Kolmogorov's exponential lower bound. The choice $f(x) := e^x - 1 - x$, with $f''(x) = e^x$, leads to the conclusion that the function

$$\Delta(x) := \begin{cases} \dfrac{e^x - 1 - x}{x^2/2} & \text{for } x \ne 0 \\ 1 & \text{for } x = 0 \end{cases}$$

is nonnegative and increasing over the whole real line. The choice $f(x) := (1+x)\log(1+x) - x$, for $x \ge -1$, with $f'(x) = \log(1+x)$ and $f''(x) = (1+x)^{-1}$, leads to the conclusion that the function

$$\psi(x) := \begin{cases} \dfrac{(1+x)\log(1+x) - x}{x^2/2} & \text{for } x \ge -1 \text{ and } x \ne 0 \\ 1 & \text{for } x = 0. \end{cases}$$

is nonnegative, convex, and decreasing. Also $x\psi(x)$ is increasing on $\mathbb{R}^+$, and $\psi(x) \ge (1+x/3)^{-1}$. □

4. Relative interior of a convex set

Convex subsets of Euclidean spaces either have interior points, or they can be regarded as embedded in lower dimensional subspaces within which they have interior points.

<6> **Theorem.** *Let C be a convex subset of $\mathbb{R}^n$.*

(i) There exists a smallest subspace $\mathcal{V}$ for which $C \subseteq x_0 \oplus \mathcal{V} := \{x_0 + x : x \in \mathcal{V}\}$, for each $x_0 \in C$.

(ii) $\dim(\mathcal{V}) = n$ if and only if C has a nonempty interior.

(iii) If $\text{int}(C) \ne \emptyset$, there exists a convex, nonnegative function ρ defined on $\mathbb{R}^n$ for which $\text{int}(C) = \{x : \rho(x) < 1\} \subseteq C \subseteq \{x : \rho(x) \le 1\} = \overline{\text{int}(C)}$.

Proof. With no loss of generality, suppose $0 \in C$. Let $x_1, \ldots, x_k$ be a maximal set of linearly independent vectors from C, and let $\mathcal{V}$ be the subspace spanned by those vectors. Clearly $C \subseteq \mathcal{V}$. If $k < n$, there exists a unit vector w orthogonal to $\mathcal{V}$, and every point x of $\mathcal{V}$ is a limit of points $x + tw$ not in $\mathcal{V}$. Thus C has an empty interior.

If $k = n$, write $\bar{x}$ for $\sum_i x_i/n$. Each member of the usual orthonormal basis has a representation as a linear combination, $e_i = \sum_j a_{i,j}x_j$. Choose an $\epsilon > 0$ for which $2n\epsilon\left(\sum_i a_{i,j}^2\right)^{1/2} < 1$ for every j. For every $y := \sum_i y_i e_i$ in $\mathbb{R}^n$ with $|y| < \epsilon$, the coefficients $\beta_j := (2n)^{-1} + \sum_i a_{i,j}y_i$ are positive, summing to a quantity $1 - \beta_0 \le 1$, and $\bar{x}/2 + y = \beta_0 0 + \sum_i \beta_i x_i \in C$. Thus $\bar{x}/2$ is an interior point of C.

If $\text{int}(C) \neq \emptyset$, we may, with no loss of generality, suppose 0 is an interior point. Define a map $\rho : \mathbb{R}^n \to \mathbb{R}^+$ by $\rho(z) := \inf\{t > 0 : z/t \in C\}$. It is easy to see that $\rho(0) = 0$, and $\rho(\alpha y) = \alpha\rho(y)$ for $\alpha > 0$. Convexity of C implies that $\rho(z_1 + z_2) \le \rho(z_1) + \rho(z_2)$ for all z_i: if $z_i/t_i \in C$ then

$$\frac{z_1 + z_2}{t_1 + t_2} = \frac{t_1}{t_1 + t_2}\left(\frac{z_1}{t_1}\right) + \frac{t_2}{t_1 + t_2}\left(\frac{z_2}{t_2}\right) \in C.$$

In particular, ρ is a convex function. Also ρ satisfies a Lipschitz condition: if $y = \sum_i y_i e_i$ and $z = \sum_i z_i e_i$ then

$$\begin{aligned}\rho(y) - \rho(z) \le \rho(y - z) &= \rho\left(\textstyle\sum_i (y_i - z_i)e_i\right)\\ &\le \textstyle\sum_i \left((y_i - z_i)^+\rho(e_i) + (y_i - z_i)^-\rho(-e_i)\right)\\ &\le |y - z|\left(\textstyle\sum_i \rho(e_i)^2 \vee \rho(-e_i)^2\right)^{1/2}.\end{aligned}$$

Thus $\{\rho < 1\}$ is open and $\{\rho \le 1\}$ is closed.

Clearly $\rho(x) \le 1$ for every x in C; and if $\rho(x) < 1$ then $x_0 := x/t \in C$ for some $t < 1$, implying $x = (1 - t)0 + tx_0 \in C$. Thus $\{z : \rho(z) < 1\} \subseteq C \subseteq \{z : \rho(z) \le 1\}$. Every point x with $\rho(x) = 1$ lies on the boundary, being a limit of points $x(1 \pm n^{-1})$ from C and C^c. Assertion (iii) follows. □

If $C \subseteq x_0 \oplus \mathcal{V} \subseteq \mathbb{R}^n$, with $\dim(\mathcal{V}) = k < n$, we can identify $\mathcal{V}$ with $\mathbb{R}^k$ and C with a subset of $\mathbb{R}^k$. By part (ii) of the Theorem, C has a nonempty interior, as a subset of $x_0 \oplus \mathcal{V}$. That is, there exist points x of C with open neighborhoods (in $\mathbb{R}^n$) for which $\mathcal{N} \cap (x_0 \oplus \mathcal{V}) \subseteq C$. The set of all such points is called the ***relative interior*** of C, and is denoted by rel-int(C). Part (iii) of the Theorem has an immediate extension,

$$\text{rel-int}(C) \subseteq C \subseteq \overline{\text{rel-int}(C)},$$

with a corresponding representation via a convex function ρ defined only on $x_0 \oplus \mathcal{V}$.

5. Separation of convex sets by linear functionals

The theorems asserting existence on separating linear functionals depend on the following simple extension result.

<7> **Lemma.** *Let f be a real-valued convex function, defined on a vector space $\mathcal{V}$. Let T_0 be a linear functional defined on a vector subspace $\mathcal{V}_0$, on which $T_0(x) \le f(x)$ for all $x \in \mathcal{V}_0$. Let y_1 be a point of $\mathcal{V}$ not in $\mathcal{V}_0$. There exists an extension of T_0 to a linear functional T_1 on the subspace $\mathcal{V}_1$ spanned by $\mathcal{V}_0 \cup \{y_1\}$ for which $T_1(z) \le f(z)$ on $\mathcal{V}_1$.*

Proof. Each point z in $\mathcal{V}_1$ has a unique representation $z := x + ry_1$, for some $x \in \mathcal{V}_0$ and some $r \in \mathbb{R}$. We need to find a value for $T_1(y_1)$ for which $f(x + ry_1) \geq T_0(x) + rT_1(y_1)$ for all $r \in \mathbb{R}$. Equivalently we need a real number c such that

$$\inf_{x_0 \in \mathcal{V}_0, t>0} \frac{f(x_0 + ty_1) - T_0(x_0)}{t} \geq c \geq \sup_{x_1 \in \mathcal{V}_0, s>0} \frac{T_0(x_1) - f(x_1 - sy_1)}{s},$$

for then $T_1(y_1) := c$ will give the desired extension.

For given x_0, x_1 in $\mathcal{V}_0$ and $s, t > 0$, define $\alpha := s/(s+t)$ and $x_\alpha := \alpha x_0 + (1-\alpha)x_1$. Then, by convexity of f on $\mathcal{V}_1$ and linearity of T_0 on $\mathcal{V}_0$,

$$\frac{s}{s+t} f(x_0 + ty_1) + \frac{t}{s+t} f(x_1 - sy_1) \geq f(x_\alpha) \geq T_0(x_\alpha) = \frac{s}{s+t} T_0(x_0) + \frac{t}{s+t} T_0(x_1),$$

which implies

$$\infty > \frac{f(x_0 + ty_1) - T_0(x_0)}{t} \geq \frac{T_0(x_1) - f(x_1 - sy_1)}{s} > -\infty.$$

The infimum over x_0 and $t > 0$ on the left-hand side must be greater than or equal to the supremum over x_1 and $s > 0$ on the right-hand side, and both bounds must be finite. Existence of the desired real c follows. □

REMARK. The vector space $\mathcal{V}$ need not be finite dimensional. We can order extensions of T_0, bounded above by f, by defining $(T_\alpha, \mathcal{V}_\alpha) \succeq (T_\beta, \mathcal{V}_\mathcal{B})$ to mean that $\mathcal{V}_\beta$ is a subspace of $\mathcal{V}_\alpha$, and T_α is an extension of T_β. Zorn's lemma gives a maximal element of the set of extensions $(T_\gamma, \mathcal{V}_\gamma) \succeq (T_0, \mathcal{V}_0)$. Lemma <7> shows that $\mathcal{V}_\gamma$ must equal the whole of $\mathcal{V}$, otherwise there would be a further extension. That is, T_0 has an extension to a linear functional T defined on $\mathcal{V}$ with $T(x) \leq f(x)$ for every x in $\mathcal{V}$. This result is a minor variation on the ***Hahn-Banach theorem*** from functional analysis (compare with page 62 of Dunford & Schwartz 1958).

<8> **Theorem.** *Let C be a convex subset of $\mathbb{R}^n$ and y_0 be a point not in* rel-int(C).

(i) There exists a linear functional T on $\mathbb{R}^k$ for which $0 \neq T(y_0) \geq \sup_{x \in \overline{C}} T(x)$.

(ii) If $y_0 \notin \overline{C}$, then we may choose T so that $T(y_0) > \sup_{x \in \overline{C}} T(x)$.

Proof. With no loss of generality, suppose $0 \in C$. Let $\mathcal{V}$ denote the subspace spanned by C, as in Theorem <6>. If $y_0 \notin \mathcal{V}$, let ℓ be its component orthogonal to $\mathcal{V}$. Then $y_0 \cdot \ell > 0 = x \cdot \ell$ for all x in C.

If $y_0 \in \mathcal{V}$, the problem reduces to construction of a suitable linear functional T on $\mathcal{V}$: we then have only to define $T(z) := 0$ for $z \notin \mathcal{V}$ to complete the proof. Equivalently, we may suppose that $\mathcal{V} = \mathbb{R}^n$. Define T_0 on $\mathcal{V}_0 := \{rx_0 : r \in \mathbb{R}\}$ by $T(ry_0) := r\rho(y_0)$, for the ρ defined in Theorem <6>. Note that $T_0(y_0) = \rho(y_0) \geq 1$, because $y_0 \notin$ rel-int$(C) = \{\rho < 1\}$. Clearly $T_0(x) \leq \rho(x)$ for all $x \in \mathcal{V}_0$. Invoke Lemma <7> repeatedly to extend T_0 to a linear functional T on $\mathbb{R}^n$, with $T(x) \leq \rho(x)$ for all $x \in \mathbb{R}^n$. In particular,

$$T(y_0) \geq 1 \geq \rho(x) \geq T(x) \qquad \text{for all } x \in \overline{C} = \{\rho \leq 1\}.$$

For (ii), note that $T(y_0) > 1$ if $y_0 \notin \overline{C}$. □

<9> **Corollary.** *Let C_1 and C_2 be disjoint convex subsets of $\mathbb{R}^n$. Then there is a nonzero linear functional for which $\inf_{x \in \overline{C}_1} T(x) \geq \sup_{x \in \overline{C}_2} T(x)$.*

Proof. Define C as the convex set $\{x_1 - x_2 : x_i \in C_i\}$. The origin does not belong to C. Thus there is a nonzero linear functional for which $0 = T(0) \geq T(x_1 - x_2)$ for all $x_i \in C_i$. □

<10> **Corollary.** *For each closed convex subset F of $\mathbb{R}^n$ there exists a countable family of closed halfspaces $\{H_i : i \in \mathbb{N}\}$ for which $F = \cap_{i\in\mathbb{N}} H_i$.*

Proof. Let $\{x_i : i \in \mathbb{N}\}$ be a countable dense subset of F^c. Define r_i as the distance from x_i to F, which is strictly positive for every i, because F^c is open. The open ball $B(x_i, r_i)$ with radius r_i and center x_i is convex and disjoint from F. From the previous Corollary, there exists a unit vector ℓ_i and a constant k_i for which $\ell_i \cdot y \geq k_i \geq \ell_i \cdot x$ for all $y \in B(x_i, r_i)$ and all $x \in F$. Define $H_i := \{x \in \mathbb{R}^n : \ell_i \cdot x \leq k_i\}$.

Each x in F^c is the center of some open ball $B(x, 3\epsilon)$ disjoint from F. There is an x_i with $|x - x_i| < \epsilon$. We then have $r_i \geq 2\epsilon$, because $B(x, 3\epsilon) \supseteq B(x_i, 2\epsilon)$, and hence $x - \epsilon\ell_i \in B(x_i, r_i)$. The separation inequality $\ell_i \cdot (x - \epsilon\ell_i) \geq k_i$ then implies $\ell_i \cdot x > k_i$, that is $x \notin H_i$. □

<11> **Corollary.** *Let f be a convex (real-valued) function defined on a convex subset C of $\mathbb{R}^n$, such that* epi(f) *is a closed subset of $\mathbb{R}^{n+1}$. Then there exist $\{d_i : i \in \mathbb{N}\} \subseteq \mathbb{R}^n$ and $\{c_i : i \in \mathbb{N}\} \subseteq \mathbb{R}$ such that $f(x) = \sup_{i\in\mathbb{N}}(c_i + d_i \cdot x)$ for every x in C.*

Proof. From the previous Corollary, and the definition of epi(f), there exist $\ell_i \in \mathbb{R}^n$ and constants $\alpha_i, k_i \in \mathbb{R}$ such that

$$\infty > t \geq f(x) \text{ if and only if } k_i \geq \ell_i \cdot x - t\alpha_i \qquad \text{for all } i \in \mathbb{N}.$$

The ith inequality can hold for arbitrarily large t only if $\alpha_i \geq 0$. Define $\psi(x) := \sup_{\alpha_i > 0} (\ell_i \cdot x - k_i)/\alpha_i$. Clearly $f(x) \geq \psi(x)$ for $x \in C$. If $s < f(x)$ for an x in C then there must exist an i for which $\ell_i \cdot x - f(x)\alpha_i \leq k_i < \ell_i \cdot x - s\alpha_i$, thereby forcing $\alpha_i > 0$ and $s < \psi(x)$. □

6. Problems

[1] Let f be the convex function, taking values in $\mathbb{R} \cup \{\infty\}$, defined by

$$f(x, y) = \begin{cases} -y^{1/2} & \text{for } 0 \leq 1 \text{ and } x \in \mathbb{R} \\ \infty & \text{otherwise.} \end{cases}$$

Let T_0 denote the linear function defined on the x-axis by $T_0(x, 0) := 0$ for all $x \in \mathbb{R}$. Show that T_0 has no extension to a linear functional on $\mathbb{R}^2$ for which $T(x, y) \leq f(x, y)$ everywhere, even though $T_0 \leq f$ along the x-axis.

[2] Suppose X is a random variable for which the moment generating function, $M(t) := \mathbb{P}\exp(tX)$, exists (and is finite) for t in an open interval J about the origin of the real line. Write $\mathbb{P}_t$ for the probability measure with density $e^{tX}/M(t)$ with respect to $\mathbb{P}$, for $t \in J$, with corresponding variance $\text{var}_t(\cdot)$. Define $\Lambda(t) := \log M(t)$.

(i) Use Dominated Convergence to justify the operations needed to show that

$$\Lambda'(t) = M'(t)/M(t) = \mathbb{P}\big(Xe^{tX}/M(t)\big) = \mathbb{P}_t X,$$
$$\Lambda''(t) = \big(M(t)M''(t) - M'(t)^2\big)/M(t)^2 = \text{var}_t(X).$$

(ii) Deduce that Λ is a convex function on J.

(iii) Show that Λ achieves its minimum at $t = 0$ if $\mathbb{P}X = 0$.

[3] Let Q be a probability measure defined on a finite interval $[a, b]$. Write σ_Q^2 for its variance.

(i) Show that $\sigma_Q^2 \leq (b-a)^2/4$. Hint: Reduce to the case $b = -a$, noting that $\sigma_Q^2 \leq Q^x(x^2)$.

(ii) Suppose also that $Q^x(x) = 0$. Define $\Lambda(t) := \log\left(Q^x e^{xt}\right)$, for $t \in \mathbb{R}$. Show that $\Lambda''(t) \leq (b-a)^2/4$, and hence $\Lambda(t) \leq t^2(b-a)^2/8$ for all $t \in \mathbb{R}$.

(iii) (Hoeffding 1963) Let $X_1, \ldots, X_n$ be independent random, variables with zero expected values, and with X_i taking values only in a finite interval $[a_i, b_i]$. For $\epsilon > 0$, show that

$$\mathbb{P}\{X_1 + \ldots + X_n \geq \epsilon\} \leq \inf_{t>0} e^{-\epsilon t} \prod_i \mathbb{P}e^{tX_i} \leq \exp\left(-2\epsilon^2 / \sum_i (b_i - a_i)^2\right).$$

[4] Let P be a probability measure on $\mathbb{R}^k$. Define $M(t); = P^x\left(e^{x \cdot t}\right)$ for $t \in \mathbb{R}^k$.

(i) Show that the set $C := \{t \in \mathbb{R}^k : M(t) < \infty\}$ is convex.

(ii) Show that $\log M(t)$ is convex on rel-int(C).

[5] Let f be a convex increasing function on $\mathbb{R}^+$. Show that there exists an increasing sequence of convex, increasing functions f_n, with each f_n'' bounded and continuous, such that $0 \leq f_n(x) \leq f_{n+1}(x) \uparrow f(x)$ for each x. Hint: Approximate the right-hand derivative of f from below by smooth, increasing functions.

7. Notes

Most of the material described in this Appendix can be found, often in much greater generality, in the very thorough monograph by Rockafellar (1970).

References

Dunford, N. & Schwartz, J. T. (1958), *Linear Operators, Part I: General Theory*, Wiley.

Hoeffding, W. (1963), 'Probability inequalities for sums of bounded random variables', *Journal of the American Statistical Association* **58**, 13–30.

Rockafellar, R. T. (1970), *Convex Analysis*, Princeton Univ. Press, Princeton, New Jersey.

Appendix D

Binomial and normal distributions

SECTION 1 establishes some useful bounds for the tails of the normal distribution, then uses them to derive the perturbation inequalities needed for the proof of the main result, in Section 2

SECTION 2 describes a very precise approximation to symmetric Binomial tail probabilities via the tails of the standard normal distribution. The approximation implies existence of a very tight coupling between the Binomial and its approximating normal—the key to the KMT coupling (Chapter 10) between the empirical process and a Brownian Bridge.

SECTION 3 proves the results described in Section 2.

1. Tails of the normal distributions

The $N(0,1)$ distribution on the real line has density function $\phi(x) := \exp(-x^2/2)/\sqrt{2\pi}$ with respect to Lebesgue measure. For many limit theorems and inequalities it is only the rate of decrease of the tail probability $\bar{\Phi}(x) := \mathbb{P}\{N(0,1) > x\}$ that matters. The simplest approximation,

<1> $$\left(\frac{1}{x} - \frac{1}{x^3}\right)\phi(x) < \bar{\Phi}(x) < \frac{1}{x}\phi(x) \qquad \text{for } x > 0,$$

follows (compare with Feller 1968, Section VII.1 and Problem 7.1) by integrating from x to ∞ across the trivial inequalities

$$\left(1 - \frac{3}{t^4}\right)\phi(t) < \phi(t) < \left(1 + \frac{1}{t^2}\right)\phi(t) \qquad \text{for } t > 0.$$

Less precisely,

$$\bar{\Phi}(x) = \frac{\phi(x)}{x}\left(1 - O(x^{-2})\right) \qquad \text{as } x \to \infty.$$

When x is close to zero there are a better bounds, such as

<2> $$\bar{\Phi}(x) \le \frac{1}{2}\exp\left(-x^2/2\right) \qquad \text{for } x \ge 0,$$

an inequality that will follow from properties of the function $\rho(x) := \phi(x)/\bar{\Phi}(x)$, defined for all real x.

First note that inequality <1> provides upper and lower bounds for $1/\rho(x)$ on the positive part of the real line, which translate into

<3>
$$x < \rho(x) < x + \frac{x}{x^2 - 1} \qquad \text{for } x > 1.$$

The lower bound is also valid for $0 < x \le 1$.

To a first approximation, $\rho(x)$ increases like x. The difference $r(x) := \rho(x) - x$ is nonnegative, and it converges to zero as x tend to infinity. As the plot suggests, the function $r(\cdot)$ is actually decreasing, a property that will have pleasant consequences.

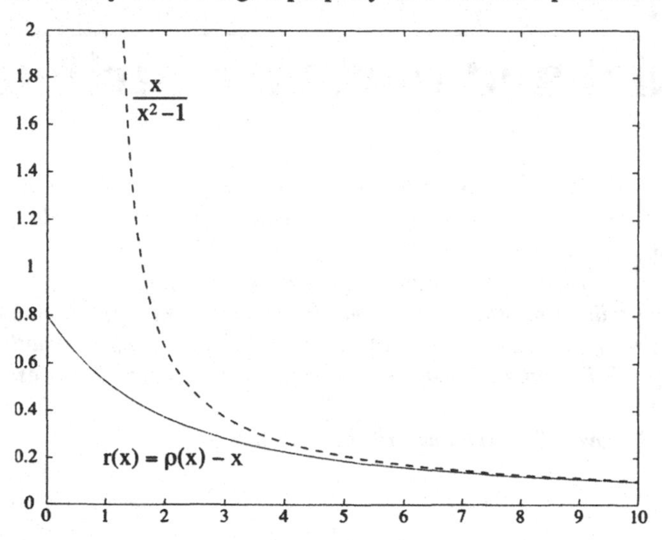

<4> **Theorem.** *The function* $\rho(x) := \phi(x)/\bar{\Phi}(x)$ *is increasing, with* $\rho(-\infty) = 0$ *and* $\rho(0) = 2/\sqrt{2\pi} \approx .7979$. *The function* $r(x) := \rho(x) - x$ *decreases to zero as* x *tends to infinity. The function* $\log \rho(x)$ *is concave, and* $\log \rho(x + \delta) < \log \rho(x) + r(x)\delta$ *for* $x \in \mathbb{R}$ *and* $\delta > 0$.

Proof. Temporarily write $G(x)$ for $1/\rho(x)$, which equals the Laplace transform of the measure μ on $\mathbb{R}^+$ with density $\exp(-z^2/2)\{z > 0\}$ with respect to Lebesgue measure:

$$G(x) = \sqrt{2\pi} \exp\left(\frac{x^2}{2}\right) \int_0^\infty \phi(z + x)\, dz = \int_0^\infty \exp\left(-xz - \frac{z^2}{2}\right) dz = \mu^z e^{-zx}.$$

For each x, define P_x to be the probability measure with density $e^{-zx}/G(x)$ with respect to μ. Dominated Convergence then lets us differentiate under the integral sign to obtain

$$G'(x)/G(x) = \mu^z \left(-ze^{-zx}\right)/G(x) = -P_x z < 0,$$
$$G''(x)/G(x) = \mu^z \left(z^2 e^{-zx}\right)/G(x) = P_x z^2.$$

From the first inequality it follows that G is decreasing, and ρ is increasing.

The derivatives of the function

$$\psi(x) := -\log \rho(x) = \log G(x) = \frac{x^2}{2} + \log \bar{\Phi}(x) + \log\left(\sqrt{2\pi}\right)$$

satisfy the equalities

$$\psi'(x) = \frac{G'(x)}{G(x)} = x - \rho(x) = -r(x),$$

$$\psi''(x) = \frac{G''(x)}{G(x)} - \left(\frac{G'(x)}{G(x)}\right)^2 = P_x z^2 - (P_x z)^2 .$$

The last expression equals the variance of the nondegenerate distribution P_x, which is strictly positive. Thus $\psi(x)$ is convex, and $r(x) = -\psi'(x)$ is decreasing. The mean-value theorem then gives

$$\log \frac{\rho(x+\delta)}{\rho(x)} = -\psi(x+\delta) + \psi(x) = \delta r(x^*) < \delta r(x),$$

where x^* lies in the interval $(x, x+\delta)$. □

REMARK. Assertion <2> is equivalent to the inequality $\rho(x) \geq \rho(0) = 2/\sqrt{2\pi}$ for $x \geq 0$, a consequence of the fact that ρ is an increasing function.

Suppose Z is $N(0,1)$ distributed. The approximation $\bar{\Phi}(x) \approx \phi(x)/x$ suggests a simple form for the conditional probability

$$\mathbb{P}\{Z > x+\delta \mid Z > x\} = \frac{\bar{\Phi}(x+\delta)}{\bar{\Phi}(x)},$$

namely, something close to $\exp(-(x+\delta)^2/2)/\exp(-x^2/2) = \exp(-x\delta - \delta^2/2)$, at least for large x and small δ. The ρ function from Theorem <4> lets us sharpen the approximation to a pair of most useful inequalities.

<5> **Corollary.** *For $x \in \mathbb{R}$ and $\delta > 0$,*

$$\exp\left(-x\delta - \frac{\delta^2}{2}\right) > \frac{\bar{\Phi}(x+\delta)}{\bar{\Phi}(x)} > \exp\left(-\rho(x)\delta - \frac{\delta^2}{2}\right).$$

Proof. For the left-hand inequality, replace $\exp(-t\delta)$ by its upper bound $\exp(-x\delta)$ in the equality

$$\bar{\Phi}(x+\delta) = \int_x^\infty \phi(t+\delta)\,dt = \int_x^\infty \phi(t)\exp(-t\delta - \delta^2/2)\,dt.$$

For the right-hand inequality write the ratio as

$$\frac{\phi(x+\delta)\rho(x)}{\phi(x)\rho(x+\delta)} = \exp\left(-x\delta - \frac{\delta^2}{2} - \log\frac{\rho(x+\delta)}{\rho(x)}\right)$$

then invoke the last assertion of Theorem <4> to bound the log term from below by $-r(x)\delta$. Remember that $\rho(x) = x + r(x)$. □

REMARK. Notice that the ratio of the upper and lower bounds equals $\exp(-r(x)\delta)$, which lies close to one if x is large and δ/x is small.

By solving for δ as a function of the ratio of tail probabilities, we obtain the perturbation inequalities needed for the proof of the main result.

<6> **Lemma.** *Let x, y, and z be related by the equality $\bar{\Phi}(z) = e^{-y}\bar{\Phi}(x)$, with $x \geq 0$.*

(i) There exists a positive constant C_1 such that both $|z - x| \leq C_1|y|$ and $|z - x - y/\rho(x)| \leq C_1|y|^2/\rho(x)$, provided $y \geq -(1+x^2)/2$.

(ii) If $x \geq 2$ and $y \geq 0$ then $0 \leq \sqrt{x^2+2y} - z \leq 2y/x^3$.

Proof. The three quantities satisfy the equation

$$g(z) = g(x) + y \qquad \text{where } g(t) := -\log \bar{\Phi}(t).$$

The function g is increasing, nonnegative, and convex, because $g'(t) = \rho(t)$, an increasing function. Inequality <2> implies that $g(t) \geq \log 2 + t^2/2$ when $t \geq 0$. In particular, under the condition on y from (i),

$$g(z) \geq \log 2 + \tfrac{1}{2}x^2 - \tfrac{1}{2}(1 + x^2) \geq \log 2 - \tfrac{1}{2} > 0,$$

implying that $z \geq -\kappa := g^{-1}\left(\log 2 - \tfrac{1}{2}\right) > -\infty$.

Write w for $z - x$. For some x_1 and x_2 between x and z,

$$y = g(z) - g(x) = w\rho(x_1) = w\rho(x) + \tfrac{1}{2}w^2\rho'(x_2).$$

From the first equality we get $|w| \leq |y|/\rho(-\kappa)$, and from the second, using the fact that $0 \leq \rho' \leq 1$, we get $|w - y/\rho(x)| \leq |y|^2/\left(2\rho(-\kappa)^2\rho(x)\right)$. Assertion (i) is proved.

For assertion (ii), note that $w \geq 0$. From Corollary <5>,

$$\rho(x)w + \tfrac{1}{2}w^2 \geq g(z) - g(x) \geq xw + \tfrac{1}{2}w^2.$$

Thus w must lie between w_1 and w_2, the positive solutions of $xw_1 + \frac{1}{2}w_1^2 = y = \rho(x)w_2 + \frac{1}{2}w_2^2$. That is,

$$w_1 := h(x) \geq w \geq h\left(x + r(x)\right) := w_2 \qquad \text{where } h(t) := -t + \sqrt{t^2 + 2y}.$$

The function $-h$ has an decreasing derivative, with

$$0 < -h'(t) = 1 - \frac{t}{\sqrt{t^2 + 2y}} = \frac{2y}{\left(t + \sqrt{t^2 + 2y}\right)\sqrt{t^2 + 2y}} \leq y/t^2,$$

which implies

$$0 \leq (x + w_1) - z \leq w_1 - w_2 \leq -r(x)h'(x) \leq \frac{x}{x^2 - 1}\frac{y}{x^2} \leq \frac{2y}{x^3} \qquad \text{if } x \geq \sqrt{2}.$$

as asserted. □

2. Quantile coupling of Binomial with normal

The quantile transformation defines an increasing function Ψ_n from $\mathbb{R}$ onto $\{0, 1, \ldots, n\}$ for which the random variable $X := \Psi_n(Y)$ has exactly a Bin$(n, 1/2)$ distribution when Y has a $N(n/2, n/4)$ distribution. More precisely, Ψ_n should take the constant value k on an interval $(\beta_{k,n}, \beta_{k+1,n}]$, for $k = 0, 1, \ldots, n$, where the cutpoints $-\infty = \beta_{0,n} < \beta_{1,n} < \ldots < \beta_{n,n} < \beta_{n+1,n} = \infty$ are determined by the requirement that $\mathbb{P}\{X \geq k\} = \mathbb{P}\{Y > \beta_{k,n}\}$ for each k. The challenge lies in locating the cutpoints.

The usual normal approximation suggests that $\mathbb{P}\{X \geq k\} \approx \mathbb{P}\{Y \geq k - 1/2\}$, that is, $\beta_{k,n} \approx k - 1/2$. Numerical calculation provides further evidence that $k - 1/2$ is indeed very close to a lower bound, at least for small values of n.

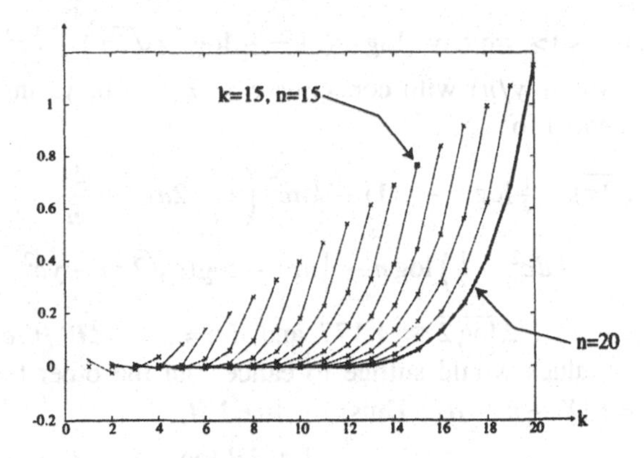

Plot of $\beta_{k,n} - (k - 1/2)$ as a function of k, for $n/2 \le k \le n$ and values of n ranging from 1 to 20. Each small cross corresponds to a different k and n. Crosses that share the same n are joined by line segments.

Notice that the plot covers only values of k greater than $n/2$, a restriction justified by the following symmetry argument. The fact that $n - X$ also has Bin(n, 1/2) distribution implies

$$\mathbb{P}\{Y > \beta_{k,n}\} = \mathbb{P}\{n - X \ge k\} = 1 - \mathbb{P}\{X \ge n - k + 1\} = 1 - \mathbb{P}\{Y > \beta_{n-k+1,n}\},$$

from which it follows that $\beta_{k,n} - n/2 = n/2 - \beta_{n-k+1,n}$, that is, $\beta_{k,n} + \beta_{n-k+1,n} = n$. Put another way, the intervals $(\beta_{1,n}, \beta_{n,n})$, $(\beta_{2,n}, \beta_{n-1,n})$, and so on, are all symmetric about $n/2$. When n is even, say $n = 2m$, the interval $(\beta_{m,n}, \beta_{m+1,n})$ is symmetric about $n/2$; so we have only to consider $k \ge m + 1 = (n+2)/2$. When n is odd, say $n = 2m + 1$, the interval $(\beta_{m,n}, \beta_{m+2,n})$ is symmetric about $n/2 = \beta_{m+1,n}$; so we have only to consider $k \ge m + 2 = (n+3)/2$.

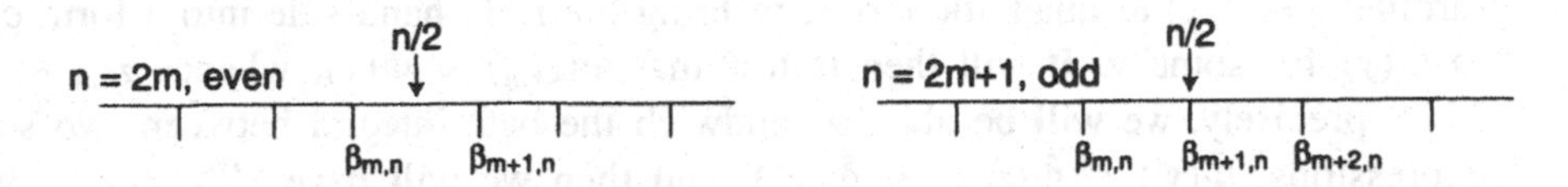

The plot also suggests that $\beta_{k,n}$ grows faster than $k - 1/2$ as k moves towards n, a suggestion supported by explicit calculations when k is close to n, for n large enough to justify simplifying approximations. It will be slightly more convenient to work with the standardized cutpoints $z_{k,n} := 2(\beta_{k,n} - n/2)/\sqrt{n}$ and the tails of the standard normal distribution. The symmetry properties and defining equalities then become $z_{n-k+1,n} = -z_{k,n}$ and

$$\mathbb{P}\{\text{Bin}(n, 1/2) \ge k\} = \bar{\Phi}(z_{k,n}) \qquad \text{for each } k. \tag{7}$$

When $z_{k,n}$ is large, it can be well approximated via inequality <1>.

<8> **Example.** For $k = n - B$, with a $B = 0, 1, 2, \dots$ fixed as n increases,

$$\mathbb{P}\{X \ge n - B\} = \left(\binom{n}{0} + \dots \binom{n}{B}\right) 2^{-n} = \exp\left(B \log n - n \log 2 + O(1)\right).$$

From inequality <1> we have $\log\bar{\Phi}(z) = -\log\left(z\sqrt{2\pi}\right) - \frac{1}{2}z^2 - O(z^{-2})$. For values $z := \sqrt{n}c\,(1 - d\ell_n + y/n)$ with constants $c > 0$, $d > 0$, y, and $\ell_n := (\log n)/n$, we have $\log\bar{\Phi}(z)$ equal to

$$-\log(c\sqrt{2\pi}) - \tfrac{1}{2}\log n - o(1) - \tfrac{1}{2}nc^2\left(1 - 2d\ell_n + \frac{2y}{n} + o(n^{-1})\right) - O(n^{-1})$$
$$= \left(dc^2 - \tfrac{1}{2}\right)\log n - \tfrac{1}{2}nc^2 - \log(c\sqrt{2\pi}) - yc^2 + o(1).$$

If we choose $c := \sqrt{2\log 2} \approx 1.177$ and $d := (1 + 2B)/(2c^2)$ then a bounded sequence of y values would suffice to cancel out the other terms, leaving a z for which $\bar{\Phi}(z) = \mathbb{P}\{X \geq n - B\}$. Thus, for fixed B,

$$z_{n-B,n} = c\sqrt{n} - \frac{1 + 2B}{2c}\frac{\log n}{\sqrt{n}} + O\left(\frac{1}{\sqrt{n}}\right)$$

and

$$\beta_{n-B,n} = \frac{n}{2} + \tfrac{1}{2}\sqrt{n}z_{n-B,n} = \frac{1+c}{2}n - \frac{1+2B}{4c}\log n + O(1).$$

□ Notice that $\beta_{n-B,n}$ exceeds $n - B$ by about $0.088n$ plus smaller order terms.

The method of proof for the main approximation result uses an elementary relationship between Binomial tails and beta integrals. If n points are placed independently according to the uniform distribution on $(0, 1)$, then $X :=$ #points in $[0, 1/2]$ has a Bin$(n, 1/2)$ distribution, the kth order statistic, T_k, has a beta distribution, and

<9> $$\mathbb{P}\{X \geq k\} = \mathbb{P}\{T_k \leq 1/2\} = \frac{n!}{(k-1)!(n-k)!}\int_0^{1/2} t^{k-1}(1-t)^{n-k}\,dt.$$

Essentially we have only to simplify the ratio of factorials by an appeal to Stirling's formula, then approximate the logarithm of the integrand by a Taylor expansion (around $t = 1/2$) to quadratic terms, to bring the right-hand side into a form close to $\bar{\Phi}(y)$ for some y. It will then follow that $\bar{\Phi}(z_{k,n}) \approx \bar{\Phi}(y)$, whence $z_{k,n} \approx y$. More precisely, we will be able to sandwich the beta integral between two such expressions, $\bar{\Phi}(y') \leq \bar{\Phi}(z_{k,n}) \leq \bar{\Phi}(y'')$, and then we will have $y' \geq z_{k,n} \geq y''$ because Ψ_n is a monotone decreasing function.

The detailed approximation looks slightly intimidating, because it needs to cover several ranges for k where different terms become important to the bounds. For k near $n/2$ it should provide a $\beta_{k,n}$ near $k - \frac{1}{2}$; for k near n it should reproduce the behavior from Example <8>; and it should make a smooth transition between the two types of behavior as k increases from $n/2$ to n.

<10> **Theorem.** *There is an increasing function $\gamma(\cdot)$ with $\gamma(0) = 1/12$ and $1 + 2\gamma(1) = c^2 := 2\log 2$ for which the standardized cutpoints*

$$z_{k,n} := 2(\beta_{k,n} - n/2)/\sqrt{n} \qquad \text{and} \qquad u_{k,n} := 2\left(k - 1/2 - n/2\right)/\sqrt{N}$$

where $N := n - 1$, are related by the approximation

$$z_{k,n} = u_{k,n}S\left(u_{k,n}/\sqrt{N}\right) + \frac{\log\left(1 - u_{k,n}^2/N\right)}{2c|u_{k,n}|} + R_{k,n} \qquad \text{where } S(\epsilon) := \sqrt{1 + 2\epsilon^2\gamma(\epsilon)},$$

and

$$-O\left(|u_{k,n}|+1\right) \le nR_{k,n} \le O\left(|u_{k,n}|+\log n\right) \qquad \text{uniformly in } \frac{n+1}{2} \le k < n.$$

For $0 < k \le (n+1)/2$, the same approximation holds but with the upper and lower bounds for $R_{k,n}$ interchanged.

The inequality $k \ge (n+1)/2$ is equivalent to $u_{k,n} \ge 0$. Equality is achieved by an integer k only when n is odd, that is, $n = 2m+1$ and $k = m+1$. By symmetry, in that case $\mathbb{P}\{X \ge m+1\} = \mathbb{P}\{X \le m\} = 1/2$, and hence $z_{m+1,n} = 0 = u_{m+1,n}$, implying $R_{m+1,n} = 0$. Thus we have only to consider the situation where $k > (n+1)/2$ for the proof.

The peculiar $\sqrt{n-1}$ standardization for $u_{k,n}$, and the exclusion of the values $k = 0$ and $k = n$ from the range, are both related to the behavior described by Example <8>. The increasing function $\gamma(\cdot)$ is defined for $0 < \epsilon < 1$ as

$$\gamma(\epsilon) = \frac{(1+\epsilon)\log(1+\epsilon) + (1-\epsilon)\log(1-\epsilon) - \epsilon^2}{2\epsilon^4} = \sum_{r=0}^{\infty} \frac{\epsilon^{2r}}{(2r+3)(2r+4)}. \tag{<11>}$$

It has an infinite derivative from the left at $\epsilon = 1$, which accounts for the behavior of the function $S(\epsilon)$ as ϵ increases to 1. More explicitly, for small positive δ,

$$S(1-\delta) = c - \frac{\delta \log(1/\delta)}{2c} + O(\delta) \qquad \text{as } \delta \to 0. \tag{<12>}$$

If $k := n - B$, for a fixed $B \ge 1$, then $u_{k,n}/\sqrt{N} = 1 - (2B)/N$ and $R_{k,n}$ is of order $O\left(n^{-1/2}\right)$. With $\delta := 2B/N$, the Theorem approximates $z_{n-B,n}$ by

$$\sqrt{N}(1-\delta)S(1-\delta) + \frac{\log(2\delta - \delta^2)}{2c\sqrt{N}(1-\delta)} = c\sqrt{N} - \frac{(1+2B)}{2c}\frac{\log N}{\sqrt{N}} + O\left(n^{-1/2}\right),$$

which differs from the expression in Example <8> by terms of order $O\left(n^{-1/2}\right)$. That is, with the $\sqrt{n-1}$ standardization we capture the extreme tail behavior correctly.

REMARK. A $\sqrt{n}$ standardization would put an extra $(\log N)/\sqrt{N}$ into the approximation, thereby slightly increasing the error bound by a factor of $\log n$. The extra factor would have no effect on the coupling bound we seek, so you could be forgiven for mostly ignoring the small difference. A much better argument for working with N instead of n, and with $k-1$ instead of k, will appear in the proof, where you will see that calculations come out much more cleanly with the slightly smaller values.

The effect of the function γ is small when $|u_{k,n}|/\sqrt{N}$ is close to zero, that is, when $|k - n/2| = o(n)$. In that case, the log term can be absorbed into the error bounds, and also $|u_{k,n}|$ is much smaller than the maximum of $\log n$ and $|u_{k,n}|^3$. The approximation then simplifies to

$$\begin{aligned} z_{k,n} &= u_{k,n}\left(1 + 2\frac{u_{k,n}^2}{N}\left(\frac{1}{12} + o(1)\right)\right)^{1/2} \pm O\left(\frac{|u_{k,n}| + \log n}{n}\right) \\ &= u_{k,n} + \frac{u_{k,n}^3}{12n}(1 + o(1)) \pm O\left(\frac{\log n}{n}\right). \end{aligned}$$

The corresponding approximation for the unstandardized cutpoints is

$$\beta_{k,n} = k - \tfrac{1}{2} + \frac{(k-n/2)^3}{3n^2}(1+o(1)) \pm O\left(\frac{\log n}{\sqrt{n}}\right) \qquad \text{for } |k - n/2| = o(n).$$

When $|k - n/2|$ is of order $o(n^{2/3})$ the cubic term contributes only a $o(1)$ to the approximation; and the cutpoint $\beta_{k,n}$ stays within $o(1)$ of $k - \frac{1}{2}$. For larger $|k - n/2|$ the cubic term accounts for the slow drift suggested by the plot of $\beta_{k,n} - (k - 1/2)$ versus k.

As a trivial consequence of the the preceding discussion, there exists a constant C_0 such that, $k - C_0 \leq \beta_{k,n}$ for $n/2 \leq k \leq n$. Also the inequality $\sqrt{x+y} \leq \sqrt{x} + \sqrt{y}$ for positive x and y, gives us the upper bound

$$z_{k,n} \leq u_{k,n} + O(u_{k,n}^2/\sqrt{n}) + O(n^{-1/2})$$

and hence (with Example <8> covering the case $k = n$),

$$\beta_{k,n} \leq \frac{n}{2} + \left(k - \tfrac{1}{2} - \frac{n}{2}\right) + O\left(\frac{(k-n/2)^2}{n}\right) + O(1) \qquad \text{for } n/2 \leq k \leq n.$$

These two inequalities give us an analog of Tusnády's inequality (from Section 10.5) for the coupling between the $X := \Psi_n(Y)$ distributed Bin$(n, 1/2)$ and the Y distributed $N(n/2, n/4)$. When $X = k \geq (n+1)/2$ we have $Y > \beta_{k,n} \geq k - C_0 = X - C_0$, and consequently $|X - n/2| \leq |Y - n/2| + C_0$. We also have, for some other constant C_1,

$$\begin{aligned} Y \leq \beta_{k+1,n} \leq k + C_1 + C_1\frac{(k-n/2)^2}{n} &= X + C_1 + C_1\frac{(X-n/2)^2}{n} \\ &\leq X + C_1 + C_1\frac{2(Y-n/2)^2 + 2C_0^2}{n} \end{aligned}$$

More succinctly, for some constant C,

<13> $$|X - n/2| \leq C + |Y - n/2| \qquad \text{and} \qquad |Y - X| \leq C + C\frac{(Y-n/2)^2}{n}.$$

By reasons of symmetry, the same inequalities also hold when $X \leq (n+1)/2$. Inequality <13> is better than needed in Chapter 10 to establish the KMT coupling.

3. Proof of the approximation theorem

From now on, all calculations will be for a fixed n. There is no harm in dropping the n from the subscripts, writing z_k instead of $z_{k,n}$, and so on. Also note that the implied constants in the $O(\cdot)$ error terms allow us to ignore finitely many values of n, and establish the Theorem only for n large enough.

The calculations will be neater when reexpressed in terms of the integers $K := k - 1$ and $N := n - 1$ and the fractions

$$\alpha := \frac{1+\epsilon}{2} = \frac{K}{N} \qquad \text{and} \qquad \beta := 1 - \alpha = \frac{1-\epsilon}{2} = \frac{N-K}{N}.$$

That is, $\epsilon = (2K/N) - 1$ and $u_k = \epsilon\sqrt{N}$. The constraint $n/2 < k \leq n-1$ corresponds to ϵ staying slightly away from two extreme values:

<14> $$1 - \frac{2}{N} \geq \epsilon = \frac{2K}{N} - 1 \geq \begin{cases} N^{-1} & \text{when } n \text{ is even,} \\ 2N^{-1} & \text{when } n \text{ is odd.} \end{cases}$$

The assertion of the Theorem <10> becomes

<15>
$$z_k = \epsilon\sqrt{N}S(\epsilon) + \frac{\log(1-\epsilon^2)}{2S(1)\epsilon\sqrt{N}} + R_k \qquad \text{where } S(\epsilon) := \left(1+2\epsilon^2\gamma(\epsilon)\right)^{1/2}$$

with $-O\left(1+\epsilon\sqrt{N}\right) \le NR_k \le O\left(\log N + \epsilon\sqrt{N}\right)$, uniformly in the range <14>.

Representation <9> becomes

$$\mathbb{P}\{X \ge k\} = \frac{n \times N!}{K!(N-K)!}\int_0^{1/2} t^K(1-t)^{N-K}\,dt.$$

The logarithm of the integrand equals N times the concave function $H(t) := \alpha\log t + \beta\log(1-t)$, for $0 < t < 1$, whose maximum occurs at α:

$$H(\alpha) = \frac{1+\epsilon}{2}\log\left(\frac{1+\epsilon}{2}\right) + \frac{1-\epsilon}{2}\log\left(\frac{1-\epsilon}{2}\right) = H\left(\tfrac{1}{2}\right) + \tfrac{1}{2}\epsilon^2 + \epsilon^4\gamma(\epsilon),$$

with $\gamma(\epsilon)$ as in <11>.

Stirling's formula (Feller 1968, Section II.9),

$$n! = \sqrt{2\pi}\exp\left((n+\tfrac{1}{2})\log n - n + \lambda_n\right) \qquad \text{with } \frac{1}{12n+1} \le \lambda_n \le \frac{1}{12n},$$

simplifies the ratio of factorials. Denote $\exp(\lambda_N - \lambda_K - \lambda_{N-K})$ by Λ_ϵ. Then

$$\begin{aligned}\frac{nN!}{K!(N-K)!} &= \frac{n}{\sqrt{2\pi}}\Lambda_\epsilon\frac{N^{N+1/2}}{K^{K+1/2}(N-K)^{N-K+1/2}}\\ &= \frac{n}{\sqrt{2\pi N}}\Lambda_\epsilon\alpha^{-N\alpha-1/2}\beta^{-N\beta-1/2}\\ &= \frac{n}{\sqrt{2\pi N\alpha\beta}}\Lambda_\epsilon\exp(-NH(\alpha)).\end{aligned}$$

Write $h(s)$ for the concave function $H(1/2-s) - H(\alpha)$. The representation becomes

$$\bar{\Phi}(z_k) = \frac{n}{\sqrt{2\pi N\alpha\beta}}\Lambda_\epsilon\int_0^{1/2}\exp(NH(t) - NH(\alpha))\,dt$$

<16>
$$= \frac{n}{N}\Lambda_\epsilon\sqrt{\frac{4N}{2\pi(1-\epsilon^2)}}\int_0^{1/2}\exp(Nh(s))\,ds.$$

The main contribution to the integral will come from a small neighborhood of $s = 0$. In this neighborhood, a Taylor expansion gives a good approximation to h:

$$\begin{aligned}h(s) &= h(0) - 2\epsilon s - 2s^2 + \frac{1}{6}s^3h'''(s^*) \qquad \text{with } 0 < s^* < s\\ &= -\epsilon^4\gamma(\epsilon) - 2\left(s+\frac{\epsilon}{2}\right)^2 + \text{REMAINDER},\end{aligned}$$

because

$$\begin{aligned}h(0) &= H(\tfrac{1}{2}) - H(\alpha) = -\tfrac{1}{2}\epsilon^2 - \epsilon^4\gamma(\epsilon),\\ h'(s) &= -2\alpha(1-2s)^{-1} + 2\beta(1+2s)^{-1} \qquad \text{whence } h'(0) = 2(\beta-\alpha) = -2\epsilon,\\ h''(s) &= -4\alpha(1-2s)^{-2} - 4\beta(1+2s)^{-2} \qquad \text{whence } h''(0) = -4.\end{aligned}$$

The third derivative is negative and decreasing,

$$h'''(s) = -16\alpha(1-2s)^{-3} + 16\beta(1+2s)^{-3} \qquad \text{whence } h'''(0) = 16(\beta - \alpha) < 0,$$
$$h^{(iv)}(s) = -96\alpha(1-2s)^{-4} - 96\beta(1+2s)^{-4} < 0 \qquad \text{so } h''' \text{ is decreasing.}$$

Thus $0 \geq$ REMAINDER $\geq \frac{1}{6}s^3 h'''(s)$ for $s \geq 0$.

Upper bound for the tail probability

Substitution of the upper bound for $h(s)$ in the representation <16> gives an upper bound for the tail probability,

$$\bar{\Phi}(z_k) \leq \frac{n}{N}\Lambda_\epsilon \sqrt{\frac{4N}{2\pi\,(1-\epsilon^2)}} \exp\left(-N\epsilon^4\gamma(\epsilon)\right) \int_0^{1/2} \exp\left(-2N\left(s+\frac{\epsilon}{2}\right)^2\right) ds.$$

We make the integral larger by increasing the upper terminal from $1/2$ to $+\infty$. The change of variable $t = 2\sqrt{N}(s + \epsilon/2)$ then simplifies the inequality to

<17>
$$\bar{\Phi}(z_k) \leq U_n(\epsilon) := \left(1 + \frac{1}{N}\right) \Lambda_\epsilon \exp\left(-N\epsilon^4\gamma(\epsilon) - \frac{1}{2}\log(1-\epsilon^2)\right) \bar{\Phi}\left(\epsilon\sqrt{N}\right).$$

Lower bound for the tail probability

To get an analogous lower bound, replace the upper terminal by some positive $\eta < 1/2$, and replace $h(s)$ by the lower bound

$$h(0) - 2\epsilon s - 2s^2 + \frac{1}{6}\eta s^2 h'''(\eta) \qquad \text{for } 0 \leq s \leq \eta.$$

There is a constant C for which

$$-h'''(\eta) = \frac{8(1+\epsilon)}{(1-2\eta)^3} - \frac{8(1-\epsilon)}{(1+2\eta)^3} \leq 12C(\epsilon + \eta) \qquad \text{for all } \eta < 1/4.$$

For small η we therefore have

$$h(s) \geq h(0) - 2\epsilon s - 2\kappa^2 s^2 \qquad \text{where } \kappa^2 := 1 + C\eta(\epsilon + \eta).$$

Substitution in <16> gives

<18>
$$\bar{\Phi}(z_k) \geq \left(1 + N^{-1}\right)\Lambda_\epsilon \exp\left(-N\epsilon^4\gamma(\epsilon) - \tfrac{1}{2}\log(1-\epsilon^2)\right)$$
$$\times \exp\left(\frac{N\epsilon^2}{2}\left(\kappa^{-2} - 1\right)\right) \times \sqrt{\frac{4N}{2\pi}} \int_0^\eta \exp\left(-2N\kappa^2\left(s + \frac{\epsilon}{2\kappa^2}\right)^2\right) ds.$$

The first factor on the right-hand side is the same as the factor for $U_n(\epsilon)$. The change of variable $t = 2\sqrt{N}\kappa(s + \epsilon/2\kappa^2)$ in the integral transforms the third factor to κ^{-1} times

<19>
$$\bar{\Phi}\left(\epsilon\sqrt{N}\kappa^{-1}\right) - \bar{\Phi}\left(\epsilon\sqrt{N}\kappa^{-1} + 2\eta\kappa\sqrt{N}\right).$$

We should try to choose η to make κ^{-1} times this difference as large as possible. Explicit maximization seems an impossible task, because the dependence on η is complicated. Lemma <5>, part (ii), gives a simpler lower bound, via the two inequalities

$$\bar{\Phi}\left(\epsilon\sqrt{N}\kappa^{-1} + 2\eta\kappa\sqrt{N}\right) \leq \bar{\Phi}\left(\epsilon\sqrt{N}\kappa^{-1}\right) \exp\left(-2N\epsilon\eta - 2N\eta^2\kappa^2\right)$$

and

$$\bar{\Phi}\left(\epsilon\sqrt{N}\kappa^{-1}\right) \geq \bar{\Phi}\left(\epsilon\sqrt{N}\right)\exp\left(\epsilon\sqrt{N}\kappa^{-1}\delta + \tfrac{1}{2}\delta^2\right) \qquad \text{where } \delta := \epsilon\sqrt{N}\left(1-\kappa^{-1}\right).$$

The difference in <19> is larger than

$$\bar{\Phi}\left(\epsilon\sqrt{N}\right)\exp\left(\epsilon\sqrt{N}\kappa^{-1}\delta + \tfrac{1}{2}\delta^2\right) \times \left(1 - \exp\left(-2N\epsilon\eta - 2N\eta^2\kappa^2\right)\right).$$

And now a miracle occurs. The exponential factor from <18> completely cancels the first exponential factor in the previous line,

$$\frac{N\epsilon^2}{2}\left(\kappa^{-2}-1\right) + \frac{\epsilon\sqrt{N}}{\kappa}\delta + \tfrac{1}{2}\delta^2 = \frac{N\epsilon^2}{2\kappa^2}\left(1-\kappa^2+2\kappa-2+(\kappa-1)^2\right) = 0,$$

leaving us with $U_n(\epsilon)\kappa^{-1}\left(1-\exp\left(-2N\epsilon\eta - 2N\eta^2\kappa^2\right)\right)$ as a lower bound for $\bar{\Phi}(z_k)$, where $U_n(\epsilon)$ denotes the upper bound from <17>.

The dependence on η now concentrates in two simpler factors. By trial and error I decided to choose η to make $2N\epsilon\eta + 2N\eta^2 = \log N$, that is, $2\eta = -\epsilon + \sqrt{\epsilon^2 + 2\ell_N}$, where $\ell_N := (\log N)/N$. Then $\kappa^2 = 1 + C\ell_N/2 \leq \exp(C\ell_N/2)$, and

<20>
$$\bar{\Phi}(z_k) \geq U_n(\epsilon)\left(1-N^{-1}\right)\exp(-C\ell_N).$$

Of course this argument assumes that n is large enough to make $\eta < 1/4$.

Inversion of the tail bounds

It remains only to replace the upper and lower bounds for $\bar{\Phi}(z_k)$ by expressions of the form $\bar{\Phi}(w)$.

First dispose of the easy case where $\epsilon \leq C_0/\sqrt{N}$ for some constant C_0. Assertion <15> then reduces to $O(N^{-1}) \leq z_k - \epsilon\sqrt{N} \leq O(\ell_N)$, and the inequalities <17> and <20> simplify to

$$\exp\left(O\left(N^{-1}\right)\right)\bar{\Phi}\left(\epsilon\sqrt{N}\right) \geq \bar{\Phi}(z_k) \geq \exp\left(-O(\ell_N)\right)\bar{\Phi}\left(\epsilon\sqrt{N}\right).$$

The asserted approximation to z_k follows immediately from Lemma <6> part (i), first with $y = O(N^{-1})$ then with $y = O(\ell_N)$.

For the remainder of the proof we may assume $\epsilon \geq C_0/\sqrt{N}$, for a constant C_0 of our choosing. Let x_k be defined by

$$\bar{\Phi}(x_k) := \exp\left(-N\epsilon^4\gamma(\epsilon) - \tfrac{1}{2}\log(1-\epsilon^2)\right)\bar{\Phi}\left(\epsilon\sqrt{N}\right).$$

Inequalities <17> and <20> can be then written

<21>
$$\exp\left(-y_k - w_k\right)\bar{\Phi}\left(x_k\right) \leq \bar{\Phi}(z_k) \leq \exp\left(-y_k\right)\bar{\Phi}\left(x_k\right),$$

where $w_k := -\log\left(1-N^{-1}\right) + C\ell_N$ and $y_k := -\log\left(1+N^{-1}\right) - \lambda_N + \lambda_K + \lambda_{N-K}$. Notice that the first three contributions to y_k are of order $O(N^{-1})$, as is the fourth term except when ϵ gets close to 1, in which case it is of order $O(1)$.

For $C_0/\sqrt{N} \leq \epsilon \leq 1 - 2/N$, for a large enough C_0, the exponent term $Y_N(\epsilon) := N\epsilon^4\gamma(\epsilon) + \tfrac{1}{2}\log(1-\epsilon^2)$ is nonnegative. For ϵ bounded away from 1, the log term is of order $O(\epsilon^2)$. In the worst case, when ϵ achieves its maximum value $1-2/N$, it is of order $O(\log N)$. From Lemma <6> part (ii),

$$\left|x_k - \sqrt{N\epsilon^2 + 2Y_N(\epsilon)}\right| \leq \frac{2Y_N(\epsilon)}{(\epsilon\sqrt{N})^3} = O\left(\epsilon/\sqrt{N}\right).$$

That is,

$$x_k = \epsilon\sqrt{N}S(\epsilon)\left(1+\frac{\log(1-\epsilon^2)}{N\epsilon^2 S(\epsilon)^2}\right)^{1/2} + O\left(\frac{\epsilon}{\sqrt{N}}\right)$$
$$= \epsilon\sqrt{N}S(\epsilon) + \frac{\log(1-\epsilon^2)}{2\epsilon\sqrt{N}S(1)} + O\left(\frac{\epsilon}{\sqrt{N}}\right),$$

the second approximation coming from the Taylor expansion $\sqrt{1+y} = 1+\frac{1}{2}y+O(y^2)$ followed by an appeal to <12>.

Notice that x_k contributes the main term in the desired approximation <15> for z_k. The y_k and w_k in <21> contribute further error terms. From Lemma <6>, part (i), the z'_k and z''_k for which

$$\bar{\Phi}(z''_k) := e^{-y_k-w_k}\bar{\Phi}(x_k) \le \bar{\Phi}(z_k) \le e^{-y_k}\bar{\Phi}(x_k) =: \bar{\Phi}(z'_k)$$

are given by

$$z'_k = x_k + O\left(y_k/\rho(x_k)\right) = x_k + O\left(1+\epsilon\sqrt{N}\right)/N,$$
$$z''_k = x_k + O\left((y_k+w_k)/\rho(x_k)\right) = x_k + O\left(\log N + \epsilon\sqrt{N}\right)/N.$$

Monotonicity of $\bar{\Phi}$ then implies $z''_k \ge z_k \ge z'_k$, completing the proof of the Theorem.

4. Notes

The material in this Appendix is an expanded version of the paper Carter & Pollard (2000), where further discussion of related literature may be found.

References

Carter, A. & Pollard, D. (2000), Tusnády's inequality revisited, Technical report, Yale University. http://www.stat.yale.edu/~pollard.

Feller, W. (1968), *An Introduction to Probability Theory and Its Applications*, Vol. 1, third edn, Wiley, New York.

Appendix E

Martingales in continuous time

SECTION 1 explains the importance of sample path properties in the study of martingales, and other stochastic processes, in continuous time. Versions are defined. A delicate measurability question, regarding first hitting times on general Borel sets, is discussed. The notion of a standard filtration is introduced. The Section summarizes some definitions and results from the start of Chapter 6.

SECTION 2 presents the extension of the Stopping Time Lemma to submartingales with right continuous sample paths.

SECTION 3 shows how to construct a supermartingale with cadlag sample paths, by microsurgery on paths. Under a regularity condition on the filtration, each submartingale with right continuous expected value is shown to have a cadlag version.

SECTION 4 presents a remarkable property of the Brownian filtration.

1. Filtrations, sample paths, and stopping times

A ***stochastic process*** is a collection of random variables $\{X_t : t \in T\}$ all defined on the same probability space $(\Omega, \mathcal{F}, \mathbb{P})$. The index set T is often referred to as "time." The theory of stochastic processes for "continuous time," where T is a subinterval of $\overline{\mathbb{R}}$, tends to be more complicated than for T countable (such as $T = \mathbb{N}$). The difficulties arise, in part, from problems related to management of uncountable families of negligible sets associated with uncountable collections of almost sure equality or inequality assertions. A nontrivial part of the continuous time theory deals with sample path properties—that is, with the behavior of a process $X_t(\omega)$ as a function of t for fixed ω—or with properties of X as a function of two variables, $X(t, \omega)$. Such properties are vital to many arguments based on approximation of processes through their values at a finite collections of times.

REMARK. I will treat the notations $X_t(\omega)$ and $X(t, \omega)$ as interchangeable. If ω is understood, I will also abbreviate to X_t or $X(t)$. The second form becomes more convenient when t is replaced by a more complicated expression: something like $X(t_0 + \tau'_{k-1})$ is much easier to read than $X_{t_0+\tau'_{k-1}(\omega)}(\omega)$.

Throughout this Appendix, T will usually denote $\mathbb{R}^+$ or a bounded interval, such as $[0, 1]$. The most desirable sample path properties are continuity, and a slightly weaker property that is known by the acronym for the French phrase meaning "continuous on the right with left limits."

<1> **Definition.** *Call a real-valued function on an interval $T \subseteq \overline{\mathbb{R}}$ **cadlag** if it is continuous from the right and has a finite limit from the left, at each point t where such assertions make sense. Write $\mathbb{D}(T)$ for the set of all cadlag real functions on T.*

For example, functions in $\mathbb{D}[0, \infty)$ are right continuous everywhere, with left limits on $(0, \infty)$. A left limit at 0 makes no sense. There is no assumption of a limit existing as $t \to \infty$. For $\mathbb{D}[0, \infty]$, however, the limit does exist as $t \to \infty$, but it need not equal the value of the function at ∞.

The measurabilty properties of the random variables provide another link with the interpretation of T as time. Typically X_t is required to be adapted to a filtration $\{\mathcal{F}_t : t \in T\}$, a family of sub-sigma-fields of $\mathcal{F}$ for which $\mathcal{F}_s \subseteq \mathcal{F}_t$ if $s < t$. That is, X_t is $\mathcal{F}_t$-measurable. If $\mathcal{F}_\infty$ is not otherwise defined, I will take it to be the sigma-field generated by $\cup_{t \in T} \mathcal{F}_t$.

The finite dimensional distributions (fidis) of a stochastic process do not completely determine the sample path properties. (See, for example, Section 9.2 for a Brownian motion with discontinuous sample paths.) Instead, good behavior of sample paths typically results from microsurgery that changes each X_t at a $\mathbb{P}$-negligible set of ω. The surgery results in a new ***version*** of the process.

<2> **Definition.** *Say that a process $\{\widetilde{X}_t : t \in T\}$ is a version of another process $\{X_t : t \in T\}$ if $\mathbb{P}\{\widetilde{X}_t \neq X_t\} = 0$ for each t in T.*

REMARK. The negligible set N_t where $\widetilde{X}_t \neq X_t$ can depend on t. If T is uncountable, we cannot dismiss $\cup_{t \in T} N_t$ as a negligible nuisance; the union might even cover the whole of Ω. An observer who sees the two processes at a fixed time would not notice any difference worth worrying about. An observer who was able to record the whole sample path, $\widetilde{X}(\cdot, \omega)$ or $X(\cdot, \omega)$, might be able to distinguish between the processes. For uncountable T it would be a much stronger requirement to insist that $\widetilde{X}(\cdot, \omega) = X(\cdot, \omega)$ *as functions on* T, for all except a negligble set of ω. Some authors (such as Métivier 1982, page 4) express the stronger property by saying that the processes are indistinguishable, or $\mathbb{P}$-equal, or equivalent up to evanescence.

The surgery that creates a new version of the process can have measurability side effects. For example, in Section 3, to obtain cadlag sample paths we will define $\widetilde{X}_t(\omega)$ by a limit of $X_s(\omega)$ values, with s ranging over rational values larger than t. On the negligible set where the limit does not exist, we will define $\widetilde{X}_t(\omega)$ in some other way. As a result of these modifications, $\widetilde{X}_t$ need not be $\mathcal{F}_t$-measurable. However it will be measurable with respect to a slightly larger sigma-field, $\widetilde{\mathcal{F}}_t := \cap_{s>t} \sigma(\mathcal{N} \cup \mathcal{F}_t)$, where $\mathcal{N} := \{A \subseteq \Omega : A \subseteq F$ for some $F \in \mathcal{F}$ with $\mathbb{P}F = 0\}$, the collection of all $\mathbb{P}$-negligible sets. Remember, $\mathbb{P}$ has a unique completion, that is, an extension to the sigma-field generated by $\mathcal{F} \cup \mathcal{N}$. It is no loss of generality to assume that $\mathcal{N} \subseteq \mathcal{F}$, but it would be a nontrivial requirement to insist, a priori, that $\mathcal{N} \subseteq \mathcal{F}_t$ for every t.

Clearly, each $\widetilde{\mathcal{F}}_t$ contains all $\mathbb{P}$-negligible sets and $\widetilde{\mathcal{F}}_t = \cap_{s>t} \widetilde{\mathcal{F}}_s$ (a property sometimes called ***right-continuity***). Such a filtration is said to be ***standard***: $\{\widetilde{\mathcal{F}}_t : t \in T\}$ is the standard filtration generated by $\{\mathcal{F}_t : t \in T\}$.

REMARK. If T happens to be a bounded interval, say $T := [0, 1]$, there is a small notational difficulty in the definition of $\widetilde{\mathcal{F}}_1$, because there are no s in T with $s > 1$. We could remedy the problem by defining $\mathcal{F}_s := \mathcal{F}_1$ for all $s > 1$, which would give $\widetilde{\mathcal{F}}_1 = \sigma(\mathcal{N} \cup \mathcal{F}_1)$. The value of $\widetilde{X}_1$ has no effect on whether the $\widetilde{X}$ process has cadlag sample paths, but we must take care to choose $\widetilde{X}_1$ to be $\widetilde{\mathcal{F}}_1$-measurable if the process is to be adapted.

Much of the power of martingale theory comes from the preservation of the defining equalities and inequalities when fixed times are replaced by suitable stopping times τ, that is, for random variables taking values in $T \cup \{\infty\}$ for which $\{\tau \leq t\} \in \mathcal{F}_t$ for each t in T. If τ takes values only in T, we define X_τ as the function with the value $X_t(\omega)$ when $\tau(\omega) = t$. If τ might take the value ∞, and $\infty \notin T$, then it is safer to work with $X_\tau\{\tau < \infty\}$, which takes the value zero when τ is infinite.

Associated with each stopping time τ is a pre-τ sigma-field $\mathcal{F}_\tau$, defined to consist of all F for which $F\{\tau \leq t\} \in \mathcal{F}_t$ for all $t \in T \cup \{\infty\}$. A random variable Z is $\mathcal{F}_\tau$-measurable if and only if $Z\{\tau \leq t\}$ is $\mathcal{F}_t$-measurable for all $t \in T \cup \{\infty\}$. If T is countable, it is easy to show that $X_\tau\{\tau < \infty\}$ is $\mathcal{F}_\tau$-measurable. For continuous time, the corresponding fact requires further assumptions about the behavior of $X(t, \omega)$ as a function of two arguments.

<3> **Definition.** *Say that a process $\{X_t : t \in \mathbb{R}^+\}$ is* ***progressively measurable*** *if its restriction to $[0, t] \times \Omega$ is $\mathcal{B}[0, t] \otimes \mathcal{F}_t$-measurable for each t in $\mathbb{R}^+$.*

Typically, good sample path properties plus adaptedness are used to deduce progressive measurability. For example, Problem [2] shows that an adapted process $\{X_t : t \in \mathbb{R}^+\}$ with right-continuous sample paths must be progressively measurable.

<4> **Theorem.** *For a given filtration, let $\{X_t : t \in \mathbb{R}^+\}$ be a progressively measurable process and τ be a stopping time. Then $X_\tau\{\tau < \infty\}$ is $\mathcal{F}_\tau$-measurable.*

Proof. Write Z for $X_\tau\{\tau < \infty\}$. Clearly $0 = Z\{\tau = \infty\}$ is $\mathcal{F}_\infty$-measurable. We need to show, for each $t \in \mathbb{R}^+$, that $Z\{\tau \leq t\}$ is $\mathcal{F}_t$-measurable. On the set $\{\tau \leq t\}$ we can replace τ by $\tau \wedge t$, giving

$$Z\{\tau \leq t\} = X(\tau(\omega) \wedge t, \omega)\{\tau(\omega) \leq t\}.$$

By definition of a stopping time, the indicator on the right-hand side is $\mathcal{F}_t$-measurable. The stopping time $\tau \wedge t$ is $\mathcal{F}_{\tau \wedge t}$-measurable, and $\mathcal{F}_{\tau \wedge t} \subseteq \mathcal{F}_t$. Thus the map $\omega \mapsto (\tau(\omega) \wedge t, \omega)$ is $\mathcal{F}_t\backslash(\mathcal{B}[0, t] \otimes \mathcal{F}_t)$-measurable. When composed with the restriction of X to $[0, t] \times \Omega$, which is $(\mathcal{B}[0, t] \otimes \mathcal{F}_t)\backslash\mathcal{B}(\mathbb{R})$-measurable, by definition of progresive measurability, it gives an $\mathcal{F}_t\backslash\mathcal{B}(\mathbb{R})$-measurable function. The random variable $Z\{\tau \leq t\}$ is a product of two $\mathcal{F}_t$-measurable functions. □

In discrete time, with an adapted process $\{X_n : n \in \mathbb{N}\}$, and a Borel set B, the ***first hitting time*** $\tau(\omega) := \inf\{n : X_n(\omega) \in B\}$ is a stopping time. (The infimum of the empty set is interpreted as $+\infty$. That is, $\tau(\omega) = +\infty$ if $X_n(\omega) \notin B$ for all n.) Clearly $\{\tau \leq n\} = \cup_{i \leq n}\{X_i \in B\} \in \mathcal{F}_n$ for each $n \in \mathbb{N}$. The analogous result for continuous time is more subtle.

For a Borel subset B we can still define $\tau(\omega) := \inf\{t \in \mathbb{R}^+ : X_t(\omega) \in B\}$. For each fixed t in $\mathbb{R}^+$, it need not even be true that $\{\tau \leq t\}$ is a union $\cup_{s \leq t}\{X_s \in B\}$, because the infimum need not be achieved; and even if the representation were valid, an uncountable union of measurable sets would be of no help. We would have more success with the representation for a strict inequality,

$$\begin{aligned}\{\tau < t\} &= \{\omega : X(s,\omega) \in B \text{ for some } s < t\} \\ &= \text{projection of } \{(s,\omega) : X(s,\omega) \in B,\ s < t\} \text{ onto } \Omega.\end{aligned}$$

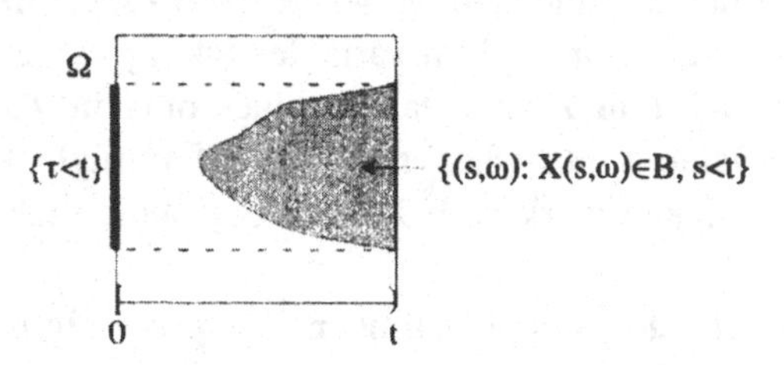

If B is an *open* set, and if the process has *right continuous* sample paths, this set takes the form of a countable union of sets $\{X_s \in B\}$, with s ranging over all rational numbers in $[0,t)$. It then follows that $\{\tau < t\} \in \mathcal{F}_t$, which is almost what we need. Indeed, we would then have

$$\{\tau \le t\} = \cap_{n\in\mathbb{N}}\{\tau < t + n^{-1}\} \in \cap_{n\in\mathbb{N}}\mathcal{F}_{t+n^{-1}} \subseteq \widetilde{\mathcal{F}}_t.$$

That is, τ is a stopping time for the standard filtration.

The argument for open B does not use the fact that $\mathcal{N} \subseteq \widetilde{\mathcal{F}}_t$. For hitting times on more general Borel sets, the negligible sets are needed. An amazing result from advanced measure theory (Dellacherie & Meyer 1978, III.1 through III.44) asserts that the projection of any $\mathcal{B}[0,t] \otimes \mathcal{F}_t$-measurable subset D of $[0,t] \times \Omega$ onto Ω is measurable with respect to the sigma field $\sigma(\mathcal{N} \cup \mathcal{F}_t)$. If X is progressively measurable, the set $D := \{(s,\omega) : X(s,\omega) \in B,\ s < t\}$ is product measurable, and hence its projection $\{\tau < t\}$ is $\sigma(\mathcal{N} \cup \mathcal{F}_t)$-measurable. It then follows that $\{\tau \le t\}$ is $\widetilde{\mathcal{F}}_t$-measurable. The complete details of the proof would take another half dozen pages of careful argument. The brave and interested reader should consult the Dellacherie-Meyer monograph. For our purposes it is enough that the general question be seen as settled, and that the role of the standard filtration be understood.

<5> **Theorem.** *If the process $\{(X_t, \mathcal{F}_t) : t \in \mathbb{R}^+\}$ is progressively measurable, and if B is a Borel subset of the real line, then $\tau := \inf\{t : X_t \in B\}$ is a stopping time for the standard filtration $\{\widetilde{\mathcal{F}}_t : t \in \mathbb{R}^+\}$ generated by $\{\mathcal{F}_t : t \in \mathbb{R}^+\}$.*

In a droll understatement of the subtlety of the underlying ideas, it is customary to refer to an assumption that the filtration be standard as "the usual conditions."

2. Preservation of martingale properties at stopping times

The basic Stopping Time Lemma from Section 6.2 asserted preservation of the submartingale property at stopping times taking values in a finite set:

Let $\{(Z_i, \mathcal{G}_i) : i = 0, 1, \ldots, N\}$ be a submartingale, and σ and τ be stopping times with $0 \le \sigma \le \tau \le N$, with N a (finite) positive integer. For each G in $\mathcal{G}_\sigma$, we have $\mathbb{P}Z_\sigma G \le \mathbb{P}Z_\tau G$. Equivalently, $Z_\sigma \le \mathbb{P}(Z_\tau \mid \mathcal{G}_\sigma)$ a.s..

The continuous time analog of the Lemma is more delicate. As with Theorem <4>, we need to make assumptions about the behavior of the martingale as a function of t. As you will see from the next Section, it is reasonable to require that the submartingale X have sample paths that are continuous from the right. Problem [2] shows that such a process is progressively measurable, so there are no measurability difficulties in working with the value of the process at stopping times.

<6> **Theorem.** *Let $\{(X_t, \mathcal{F}_t) : 0 \le t \le 1\}$ be a submartingale with right-continuous sample paths. Let $0 \le \sigma \le \tau \le 1$ be a stopping times for the filtration. Then $X_\sigma \le \mathbb{P}(X_\tau \mid \mathcal{F}_\sigma)$ almost surely. That is, $\mathbb{P}X_\sigma F \le \mathbb{P}X_\tau F$ for each F in $\mathcal{F}_\sigma$.*

Proof. We need to reduce to the case of a finite index set by using the right continuity. The method is a common one. First discretize the stopping times, defining

$$\tau_n := 0\{\tau = 0\} + \sum\nolimits_{i=1}^{2^n} \frac{i}{2^n}\left\{\frac{i-1}{2^n} < \tau \le \frac{i}{2^n}\right\} \qquad \text{for } n \in \mathbb{N}.$$

That is, τ_n is just τ rounded up to the next integer multiple of 2^{-n}. It is important that we round up, rather than down, because then $\{\tau_n \le i/2^n\} = \{\tau \le i/2^n\} \in \mathcal{F}_{i/2^n}$, which makes each τ_n a stopping time for the filtration $\{\mathcal{F}_{i/2^n} : i = 0, 1, \dots, 2^n\}$. As n increases, the sequence $\{\tau_n(\omega)\}$ decreases to $\tau(\omega)$, for each ω. The stopping times $\{\sigma_n\}$ are defined analogously.

The event F belongs to $\mathcal{F}_\sigma$, which is a sub-sigma-field of $\mathcal{F}_{\sigma_n}$. The discrete version of the Stopping Time Lemma therefore implies that $\mathbb{P}X_{\sigma_n} F \le \mathbb{P}X_{\tau_n} F$.

As n tends to infinity, right continuity ensures that $Z_n := X_{\sigma_n} \to X_\sigma$ and $Z'_n := X_{\tau_n} \to X_\tau$ along each sample path. To complete the proof, it is enough to show that $\{Z_n\}$ and $\{Z'_n\}$ are uniformly integrable. As the method of proof is the same in both cases, let us consider only the Z_n's.

The Stopping Time Lemma also implies that X_0, Z_n, Z_m, X_1 is a submartingale for each $n \ge m$. In particular, the sequence $\mathbb{P}Z_n$ decreases, to a finite limit $\kappa \ge \mathbb{P}X_0$. Given $\epsilon > 0$, there exists an m such that $\kappa \le \mathbb{P}Z_n \le \mathbb{P}Z_m \le \kappa + \epsilon$ for all $n \ge m$. For each positive constant C,

$$\begin{aligned}
\mathbb{P}|Z_n|\{|Z_n| > C\} &= \mathbb{P}Z_n\{Z_n > C\} - \mathbb{P}Z_n\{Z_n < -C\} \\
&\le \mathbb{P}X_1\{Z_n > C\} - \mathbb{P}Z_n + \mathbb{P}Z_n\{Z_n \ge -C\} && \text{submartingale} \\
&\le \mathbb{P}X_1\{Z_n > C\} - \mathbb{P}Z_m + \epsilon + \mathbb{P}Z_m\{Z_n \ge -C\} && \text{if } n \ge m \\
&\le \mathbb{P}\left(|X_1| + |Z_m|\right)\{|Z_n| > C\} + \epsilon.
\end{aligned}$$

The random variable $|X_1| + |Z_m|$ is integrable. The event $\{|Z_n| > C\}$ has probability bounded by

$$\mathbb{P}|Z_n|/C = \left(2\mathbb{P}Z_n^+ - \mathbb{P}Z_n\right)/C \le \left(2\mathbb{P}X_1^+ - \mathbb{P}X_0\right)/C,$$

which tends to zero (uniformly in n) as C tends to infinity. Uniform integrability, and the assertion of the Theorem, follow. □

<7> **Corollary.** *If $\{(M_t, \mathcal{F}_t) : 0 \le t \le 1\}$ is a martingale with right continuous sample paths then $M_\sigma = \mathbb{P}(M_1 \mid \mathcal{F}_\sigma)$ for each stopping time with $0 \le \sigma \le 1$.*

3. Supermartingales from their rational skeletons

Write S for the set of all rational numbers in $(0, 1)$. For each t in $(0, 1]$, write $\lim_{s\uparrow\uparrow t}$ to denote a limit along $\{s \in S : s < t\}$; and for each t in $[0, 1)$, write $\lim_{s\downarrow\downarrow t}$ to denote a limit along $\{s \in S : s > t\}$.

<8> **Theorem.** *Let $\{(X_t, \mathcal{F}_t) : 0 \le t \le 1\}$ be a positive supermartingale. Then there exists a $\mathbb{P}$-negligible subset N of Ω such that, for each $\omega \in N^c$:*

(i) $\sup_{s\in S} X_s(\omega) < \infty$;

(ii) for every t in $[0, 1)$, the limit $\widetilde{X}_t(\omega) := \lim_{s\downarrow\downarrow t} X_s(\omega)$ exists and is finite;

(iii) for every t in $(0, 1]$, the limit $\lim_{s\uparrow\uparrow t} X_s(\omega)$ exists and is finite.

Define $\widetilde{X}_1(\omega) := X_1(\omega)$ and $\widetilde{X}_t(\omega) := 0$ if $\omega \in N$ and $t < 1$. Then:

(iv) $\{(\widetilde{X}_t, \widetilde{\mathcal{F}}_t) : 0 \le t \le 1\}$ is a supermartingale with cadlag sample paths;

(v) $\mathbb{P}X_tF \ge \mathbb{P}\widetilde{X}_tF \ge \mathbb{P}X_{t'}F$ for all $0 \le t < t' \le 1$ and all F in $\mathcal{F}_t$;

(vi) if the map $t \mapsto \mathbb{P}X_t$ is right continuous at t_0 then $X_{t_0} = \mathbb{P}(\widetilde{X}_{t_0} \mid \mathcal{F}_{t_0})$ a.s..

Proof. Let $\{S_n\}$ be an increasing sequence of finite sets with union S. For each n, and each $x > 0$, the random variable $\tau_n := 1 \wedge \min\{s \in S_n : X_s > x\}$ is a stopping time. By the Stopping Time Lemma from Section 6.2,

$$\mathbb{P}\{\max_{s\in S_n} X_s > x\} = \mathbb{P}\{X_{\tau_n} > x\} \le \mathbb{P}X_{\tau_n}/x \le \mathbb{P}X_0/x.$$

Let n tend to infinity, then x tend to infinity to see that $\sup_{s\in S} X_s(\omega) < \infty$ except for ω in some $\mathbb{P}$-negligible set N_0.

For each pair of rational numbers $0 \le \alpha < \beta$ and $k \in \mathbb{N}$, define $A_n(k, \alpha, \beta)$ to consist of all ω for which there exist points $u_1 < v_1 < u_2 < \ldots < v_{k-1} < u_k < v_k$ in S_n for which $X(u_i, \omega) \le \alpha$ and $X(v_i, \omega) \ge \beta$ for each i. By Dubins's inequality from Section 6.3, $\mathbb{P}A_n(k, \alpha, \beta) \le (\alpha/\beta)^k$ for each n, which implies that the event $N(\alpha, \beta) := \cap_{k\in\mathbb{N}} \cup_{n\in\mathbb{N}} A_n(k, \alpha, \beta)$ has zero probability.

Define N as the union of N_0 with all $N(\alpha, \beta)$, for rational $\alpha < \beta$. For $\omega \notin N$, and each t in $[0, 1)$, we must have

$$\liminf_{s\downarrow\downarrow t} X_s(\omega) = \limsup_{s\downarrow\downarrow t} X_s(\omega) < \infty,$$

for otherwise there would be a pair of rational numbers for which $\omega \in N(\alpha, \beta) \subseteq N$. Assertion (ii) follows. The proof for assertion (iii) is similar.

Temporarily write $L_t(\omega)$ for the limit from the left. For each t in $(0, 1)$ and each $\omega \in N^c$, to each $\epsilon > 0$ there exists a $\delta > 0$ such that

$$|X_s(\omega) - \widetilde{X}_t(\omega)| \le \epsilon \qquad \text{for } s \text{ in } S \text{ with } t < s < t + \delta,$$
$$|X_s(\omega) - L_t(\omega)| \le \epsilon \qquad \text{for } s \text{ in } S \text{ with } t - \delta < s < t.$$

We must therefore have

$$|\widetilde{X}_{t'}(\omega) - \widetilde{X}_t(\omega)| \le \epsilon \qquad \text{for } t < t' < t + \delta,$$
$$|\widetilde{X}_{t'}(\omega) - L_t(\omega)| \le \epsilon \qquad \text{for } t - \delta < t' < t.$$

It follows that $\widetilde{X}(\cdot, \omega)$ is right continuous with left limit $L_t(\omega)$ at t. Similar reasoning applies at $t = 0$ and $t = 1$. Thus $\widetilde{X}(\cdot, \omega)$ is a cadlag function for every ω in N^c.

Clearly $\limsup_{s\downarrow\downarrow t} X_s$ is $\mathcal{F}_s$-measurable for each $s > t$, and hence $\widetilde{\mathcal{F}}_t$-measurable. The limit $\widetilde{X}_t$ differs from the limsup at only a negligible set of ω. The process $\{\widetilde{X}_t : 0 \le t \le 1\}$ is adapted to the standard filtration. To complete the proof of (iv), we need to show that $\mathbb{P}\widetilde{X}_t F \ge \mathbb{P}\widetilde{X}_{t'} F$ for each F in $\widetilde{\mathcal{F}}_t$, for $t < t'$. We may assume $t' < 1$. (The argument for $t' = 1$ is even simpler.) Choose sequences $\{s_n\}$ and $\{s'_n\}$ with $s'_n \downarrow\downarrow t' > s_n \downarrow\downarrow t$. By definition, $X_{s_n} \to \widetilde{X}_t$ and $X_{s'_n} \to \widetilde{X}_{t'}$, for each sample path. As in the last part of the proof of Theorem <6>, the sequences $\{X_{s_n}\}$ and $\{X_{s'_n}\}$ are uniformly integrable. (For example, put Z_n equal to $-X_{s_n}$, then argue exactly as before.) Also, because F differs only negligibly from sets in each $\mathcal{F}_{s_n}$, we have $\mathbb{P}X_{s_n} F \ge \mathbb{P}X_{s'_n} F$. In the limit we get the supermartingale inequality, $\mathbb{P}\widetilde{X}_t F \ge \mathbb{P}\widetilde{X}_{t'} F$.

For each positive constant C, the process $X_t \wedge C$ is a positive supermartingale: $\mathbb{P}(X_t \wedge C)F \ge \mathbb{P}(X_s \wedge C)F \ge \mathbb{P}(X_{t'} \wedge C)F$ if $t < s < t'$ and $F \in \mathcal{F}_t$. Invoke Dominated Convergence as s decrease to t through rational values to conclude that $\mathbb{P}(X_t \wedge C)F \ge \mathbb{P}(\widetilde{X}_t \wedge C)F \ge \mathbb{P}(X_{t'} \wedge C)F$. Let C increase to infinity to deduce (v).

The equality in (vi) is equivalent to the assertion that $\mathbb{P}X_{t_0} F = \mathbb{P}\widetilde{X}_{t_0} F$ for all F in $\mathcal{F}_{t_0}$. If for some such F and some $\epsilon > 0$ we had $\mathbb{P}X_{t_0} F \ge \epsilon + \mathbb{P}\widetilde{X}_{t_0} F$, then, via the inequality $\mathbb{P}X_{t_0} F^c \ge \mathbb{P}\widetilde{X}_{t_0} F^c$, we would get $\mathbb{P}X_{t_0} \ge \epsilon + \mathbb{P}\widetilde{X}_{t_0} \ge \epsilon + \mathbb{P}X_{t'}$, for all $t' > t_0$, which would prevent right continuity of the map $t \mapsto \mathbb{P}X_t$ at t_0. □

<9> **Corollary.** *For each integrable random variable X, there exists a cadlag version of the martingale $\mathbb{P}(X \mid \widetilde{\mathcal{F}}_t)$.*

Proof. Start from any choice of the conditional expectations $X_t := \mathbb{P}(X \mid \widetilde{\mathcal{F}}_t)$. Note that $\mathbb{P}X_t$ is constant, and hence (trivially) continuous from the right. By property (vi) and the $\widetilde{\mathcal{F}}_t$-measurability of $\widetilde{X}_t$ we have $X_t = \mathbb{P}(\widetilde{X}_t \mid \widetilde{\mathcal{F}}_t) = \widetilde{X}_t$ almost surely. □

<10> **Corollary.** *Each submartingale $\{(S_t, \widetilde{\mathcal{F}}_t)) : t \in \mathbb{R}^+\}$ with $t \mapsto \mathbb{P}S_t$ right continuous, has a cadlag version.*

Proof. For each positive integer n, invoke the Theorem for the positive martingale $M_t^{(n)} := \mathbb{P}(S_n^+ \mid \widetilde{\mathcal{F}}_t)$ and the positive supermartingale $X_t^{(n)} := M_t^{(n)} - S_t$, to get cadlag versions $\widetilde{S}_t = \widetilde{M}_t^{(n)} - \widetilde{X}_t^{(n)}$ for $n - 1 \le t \le n$. □

REMARK. The proof of the Krickeberg decomposition from Section 6.5 would carry over with only minor notational changes to continuous time. That is, each submartingale $\{(S_t, \mathcal{F}_t)) : t \in \mathbb{R}^+\}$ with $\sup_t \mathbb{P}X_t^+ < \infty$ can be expressed as the difference $M_t - X_t$ of a positive martingale M_t and a positive supermartingale X_t.

Notice that the second Corollary refers to submartingales with respect to the standard filtration. The result is not necessarily true for an arbitrary filtration. For example, if we take $\mathcal{F}_t := \sigma(\mathcal{N})$ for $t \le 1/2$ and $\mathcal{F}_t := \sigma(\mathcal{N} \cup \mathcal{F})$ for $t > 1/2$, and if A is an event with probability α not equal to zero or one, then the process $X_t(\omega) := \alpha\{t \le \frac{1}{2}\} + \{\omega \in A, t > \frac{1}{2}\}$ is a martingale for the filtration $\{\mathcal{F}_t : 0 \le t \le 1\}$. If $\{M(t)\}$ is another version of the process, then would have $M(1/2) = \alpha$ almost surely and $M(1/2+n^{-1}) \in \{0, 1\}$ almost surely, which rules out right continuity at 1/2 for almost all paths. We have $\widetilde{\mathcal{F}}_t = \mathcal{F}_t$ except at $t = 1/2$, where $\widetilde{\mathcal{F}}_{1/2} = \sigma(\mathcal{N} \cup \mathcal{F})$. The cadlag process $\widetilde{X}$ from the Lemma agrees with X except at 1/2, where $\widetilde{X}_{1/2} = A \ne X_{1/2}$ almost surely. That is, there is a single time point at which $\mathbb{P}\{\widetilde{X}_t \ne X_t\}$ is nonzero.

The general situation is similar (Doob 1953, pages 356–358): there are at most countably many t at which $\mathbb{P}\{\widetilde{X}_t \neq X_t\} > 0$. If we work only with standard filtrations, the difficulty disappears.

4. The Brownian filtration

Let $\{B_t : t \in \mathbb{R}^+\}$ be a Brownian motion with continuous sample paths. Write $\mathcal{F}_t^B$ for $\sigma\{B_s : 0 \le s \le t\}$, for $t \in \mathbb{R}^+$, the natural (or Brownian) filtration on Ω, and let $\{\widetilde{\mathcal{F}}_t^B : t \in \mathbb{R}^+\}$ be the corresponding standard filtration. The Markov property of Brownian motion implies that $\widetilde{\mathcal{F}}_t^B$ differs only negligibly from $\mathcal{F}_t^B$, and also that each martingale with respect to the standard filtration has a version with continuous sample paths. Before turning to the proof of these assertions, let me first dispose of a small detail that might cause some notational difficulties if left unsettled.

<11> **Lemma.** *The process $\{(B_t, \widetilde{\mathcal{F}}_t^B) : t \in \mathbb{R}^+\}$ is also a Brownian motion.*

Proof. We need the increment $B_t - B_s$ is independent of each F in $\widetilde{\mathcal{F}}_s^B$, when $s < t$. For each u with $s < u < t$, the event F differs negligibly from a member of $\mathcal{F}_u^B$. Thus $\mathbb{P}\exp(i\theta(B_t - B_u))\,F = \exp\left(-\theta^2(t-u)/2\right)\mathbb{P}F$, for each real θ. Invoke Dominated Convergence as u decreases to s to deduce that $\mathbb{P}\exp(i\theta(B_t - B_s))\,F$ factorizes into $\exp\left(-\theta^2(t-s)/2\right)\mathbb{P}F$, as required for a Brownian motion. □

Now recall from Chapter 9 that we may treat B as an $\mathcal{F}\backslash\mathcal{C}$-measurable map from Ω into $\mathbb{C}[0,\infty)$, the space of continuous real functions on $\mathbb{R}^+$, equipped with its cylinder sigma-field $\mathcal{C}$. The distribution of B is Wiener measure $\mathbb{W}$, a probability measure on $\mathcal{C}$. The collection of all sets $\{B \in C\}$, with $C \in \mathcal{C}$ defines a sub-sigma-field $\mathcal{F}^B$ of $\mathcal{F}$, the sigma-field generated by the Brownian motion. A random variable is $\mathcal{F}^B$-measurable if and only if it can be written in the form $f(B)$, where f is a $\mathcal{C}$-measurable map from $\mathbb{C}[0,\infty)$ into the real line. That is, $f(B)$ is a functional of the whole Brownian motion sample path. The sigma-field $\mathcal{F}_t^B$ is also generated by the map $\omega \mapsto K^t B$, where K^t is the killing functional, $(K^t x)(s) := x(t \wedge s)$, which maps $\mathbb{C}[0,\infty)$ back into itself. Thus every $\mathcal{F}_t^B$-measurable random variable is representable as $f(K^t B)$, for some $\mathcal{C}$-measurable functional f on $\mathbb{C}[0,\infty)$.

The Markov property gives an explicit representation for the conditional expectation of every $\mathbb{W}$-integrable function f of the whole Brownian motion sample path,

<12> $$\mathbb{P}(f(B) \mid \mathcal{F}_t) = \mathbb{W}^x f(K^t B + S^t x) \qquad \text{almost surely,}$$

where $(S^t x)(s)$ equals $x(t-s)$ if $t \ge s$, and is zero otherwise, as in Section 9.5.

The metric d for uniform convergence on compacta (denoted by the symbol $\xrightarrow{ucc}$) is defined on the space $\mathcal{C}[0,\infty)$ by $d(x,y) := \sum_{n\in\mathbb{N}} 2^{-n}\left(1 \wedge \sup_{0\le t\le n}|x(t) - y(t)|\right)$. The Borel sigma-field for this metric coincides with the cylinder sigma-field $\mathcal{C}$. The sigma-field $\mathcal{C}$ is also generated by the space $\mathbb{C}_{bdd}$ of all bounded, d-continuous functions on $\mathbb{C}[0,\infty)$. (Quite confusing: continuous functions on a space of continuous functions. Maybe the clumsier $\mathbb{C}_{bdd}(\mathbb{C}[0,\infty))$ would be a better symbol to save us from confusion.) The space $\mathbb{C}_{bdd}$ is dense in $\mathcal{L}^1(\mathbb{W})$.

Notice that for f in $\mathbb{C}_{bdd}$, the right-hand side of <12> is a continuous function of t, by Dominated Convergence, because $K^u B(\cdot, \omega) + S^u x \xrightarrow{ucc} K^t B(\cdot, \omega) + S^t x$ as $u \to t$, for each sample path $B(\cdot, \omega)$ and each x in $\mathbb{C}[0, \infty)$ with $x(0) = 0$. That is, *for each f in $\mathbb{C}_{bdd}$ there is a version of the martingale $\mathbb{P}(f(B) \mid \mathcal{F}_t)$ with continuous sample paths.* This fact underlies the remarkable properties of the Brownian filtration.

<13> **Theorem.** *For each t in $\mathbb{R}^+$, the standard sigma-field $\widetilde{\mathcal{F}}_t^B$ equals $\sigma(\mathcal{N} \cup \mathcal{F}_t^B)$. Each martingale for the filtration $\{\widetilde{\mathcal{F}}_t^B\}$ has a version with continuous sample paths.*

Proof. Representation <12> holds for every filtration $\{\mathcal{F}_t\}$ for which B is a Brownian motion. In particular, it holds if we choose $\mathcal{F}_t = \widetilde{\mathcal{F}}_t^B$. Remember that $\widetilde{\mathcal{F}}_t^B \subseteq \sigma(\mathcal{N} \cup \mathcal{F}_s^B) \subseteq \sigma(\mathcal{N} \cup \mathcal{F}^B)$ for every $s > t$. If X is an integrable, $\mathcal{F}_t$-measurable random variable, it differs only negligibly from an integrable random variable of the form $f(B)$, with f a $\mathcal{C}$-measurable, $\mathbb{W}$-integrable functional. In consequence,

$$X = \mathbb{P}(X \mid \mathcal{F}_t) = \mathbb{P}(f(B) \mid \mathcal{F}_t) = \mathbb{W}^x f(K^t B + S^t x) \qquad \text{almost surely.}$$

That is, X differs only negligibly from an $\mathcal{F}_t^B$-measurable random variable. The first assertion about the standard filtration follows.

Now suppose $(M_t, \mathcal{F}_t) : t \in \mathbb{R}^+\}$ is a martingale, where $\mathcal{F}_t = \widetilde{\mathcal{F}}_t^B$, as before. By Corollary <10> we may assume that it has cadlag sample paths. It is enough to prove that almost all its sample paths are continuous on each bounded subinterval of $\mathbb{R}^+$. Consider the subinterval $[0, 1]$ as a typical case.

Write M_1 as $f(B)$, with f a $\mathcal{C}$-measurable, $\mathbb{W}$-integrable functional. Approximate f in $L^1(\mathbb{W})$ norm by a sequence of functions f_n from $\mathbb{C}_{bdd}$: for each $n \in \mathbb{N}$ choose f_n so that $\mathbb{W}|f - f_n| \le 4^{-n}$. Define $M_n(t) := \mathbb{W}^x f_n(K^t B + S^t x)$, a version of the martingale $\mathbb{P}(f_n(B) \mid \mathcal{F}_t)$ with continuous sample paths. Notice that

$$\mathbb{P}|M_n(1) - M(1)| = \mathbb{P}|\mathbb{P}(f_n(B) - f(B) \mid \mathcal{F}_1)| \le \mathbb{P}|f_n(B) - f(B)| = \mathbb{W}|f_n - f|.$$

Fix a finite subset U of $[0, 1]$, define $\tau := 1 \wedge \inf\{t \in U : |M_n(t) - M(t)| > 2^{-n}\}$, a stopping time for the filtration $\{\mathcal{F}_t : t \in U\}$. Assume $1 \in U$. Then, via the Stopping Time Lemma applied to the submartingale $\{|M_n(t) - M(t)| : t \in U\}$, deduce that

$$\begin{aligned}
\mathbb{P}\{\max_{t \in U} |M_n(t) - M(t)| > 2^{-n}\} &= \mathbb{P}\{|M_n(\tau) - M(\tau)| > 2^{-n}\} \\
&\le 2^n \mathbb{P}|M_n(\tau) - M(\tau)| \\
&\le 2^n \mathbb{P}|M_n(1) - M(1)| \le 2^n \mathbb{W}|f_n - f|.
\end{aligned}$$

The last bound does not depend on U. Let U expand up to a countable dense subset of $[0, 1]$. Invoke right continuity of $M_n - M$ sample paths to deduce that

$$\mathbb{P}\{\sup_{0 \le t \le 1} |M_n(t) - M(t)| > 2^{-n}\} \le 2^{-n}.$$

By Borel-Cantelli, $\sup_{0 \le t \le 1} |M_n(t) - M(t)| \to 0$ almost surely. That is, for almost all ω, the sample path $M(\cdot, \omega)$ is a uniform limit on $[0, 1]$ of the continuous functions $M_n(\cdot, \omega)$, and hence is also continuous. □

5. Problems

[1] Let $\{N_t : 0 \le t \le 1\}$ be a family of negligible sets, for Lebesgue measure on $\mathcal{B}[0,1]$, such that $\cup_t N_t = [0,1]$. For $\omega \in \Omega := [0,1]$, define $X_t(\omega) := \{\omega \in N_t\}$. Show that $\{X_t : 0 \le t \le 1\}$ is martingale with respect to the filtration $\{\mathcal{B}[0,t] : 0 \le t \le 1\}$. Need it have cadlag sample paths?

[2] Show that an adapted process $\{X_t : t \in \mathbb{R}^+\}$ with right-continuous sample paths must be progressively measurable. Hint: For fixed t, consider approximations of the form

$$X_k(s,\omega) := X(0,\omega)\{s=0\} + \sum_{i=1}^{2^k} X(ti/2^k, \omega)\left\{t(i-1)/2^k < s \le ti/2^k\right\}.$$

Why is the X_k process $\mathcal{B}[0,t]\otimes\mathcal{F}_t$-measurable? Is it adapted? Why does it converge pointwise to X on $[0,t]\otimes\Omega$?

[3] Let $\{X_t : t \in \mathbb{R}^+\}$ be a process with continuous sample paths, adapted to a filtration $\{\mathcal{F}_t : t \in \mathbb{R}^+\}$. Define $\tau(\omega) := \inf\{t \in \mathbb{R}^+ : X_t(\omega) \in F\}$, for some closed set F. Show that $\{\tau \le t\} = \{\omega : \inf_{s\in Q_t} d(X_s(\omega), F) = 0\}$, where Q_t denotes the set of rational numbers in $[0,t]$. Deduce that τ is a stopping time for the filtration.

6. Notes

My exposition borrows ideas from Doob (1953), Métivier (1982), Breiman (1968, Chapter 14), Dellacherie & Meyer (1978, Chapter IV), and Dellacherie & Meyer (1982, Chapters V and VI).

References

Breiman, L. (1968), *Probability*, first edn, Addison-Wesley, Reading, Massachusets.

Dellacherie, C. & Meyer, P. A. (1978), *Probabilities and Potential*, North-Holland, Amsterdam.

Dellacherie, C. & Meyer, P. A. (1982), *Probabilities and Potential B: Theory of Martingales*, North-Holland, Amsterdam.

Doob, J. L. (1953), *Stochastic Processes*, Wiley, New York.

Métivier, M. (1982), *Semimartingales: A Course on Stochastic Processes*, De Gruyter, Berlin.

Appendix F

Disintegration of measures

SECTION 1 decomposes a measure on a product space into a product of a marginal measure with a kernel.

SECTION 2 specializes the decomposition to the case of a measure concentrated on the graph of a function, establishing existence of a disintegration in the sense of Chapter 5.

1. Representation of measures on product spaces

Recall from Chapter 4 how we built a measure $\mu \otimes \Lambda$, out of a sigma-finite measure μ on $(\mathcal{X}, \mathcal{A})$ and a sigma-finite kernel $\Lambda := \{\lambda_t : t \in \mathcal{T}\}$, from $(\mathcal{T}, \mathcal{B})$ to $(\mathcal{X}, \mathcal{A})$, via an iterated integral,

$$(\mu \otimes \Lambda) f := \mu^t \lambda_t^x f(x, t) \qquad \text{for } f \text{ in } \mathcal{M}^+(\mathcal{X} \times \mathcal{T}, \mathcal{A} \otimes \mathcal{B}).$$

This Section treats the inverse problem: Given a measure μ on $\mathcal{B}$ and a measure Γ on $\mathcal{A} \otimes \mathcal{B}$, when does there exist a kernel Λ for which $\Gamma = \mu \otimes \Lambda$? Such representations are closely related to the problem of constructing conditional distributions, as you saw in Chapter 5.

<1> **Theorem.** *Let Γ be a sigma-finite measure on the product sigma-field $\mathcal{A} \otimes \mathcal{B}$ of a product space $\mathcal{X} \times \mathcal{T}$, and μ be a sigma-finite measure on $\mathcal{B}$. Suppose:*

(i) $\mathcal{X}$ is a metric space and $\mathcal{A}$ is its Borel sigma-field;

(ii) the $\mathcal{T}$-marginal of Γ is absolutely continuous with respect to μ;

(iii) $\Gamma = \sum_{i \in \mathbb{N}} \Gamma_i$, where each Γ_i is a finite measure concentrating on a set $\mathcal{X}_i \times \mathcal{T}$ with $\mathcal{X}_i$ compact.

Then there exists a kernel Λ from $(\mathcal{T}, \mathcal{B})$ to $(\mathcal{X}, \mathcal{A})$ for which $\Gamma = \mu \otimes \Lambda$. The kernel is unique up to a μ-equivalence.

REMARK. The uniqueness assertion means that, if $\widetilde{\Lambda} := \{\widetilde{\lambda}_t : t \in \mathcal{T}\}$ is another kernel for which $\Gamma = \mu \otimes \widetilde{\Lambda}$, then $\lambda_t = \widetilde{\lambda}_t$, as measures on $\mathcal{A}$, for μ almost all t.

Heuristics

Suppose for the moment that Γ has a representation as $\mu \otimes \Lambda$, for some kernel Λ. If we could characterize the kernel Λ in terms of Γ and μ alone, then we could try to construct Λ for a general Γ by reinterpreting the characterization as a definition.

First note that the $\mathcal{T}$-marginal of Γ (that is, the image of Γ under the map π that projects $\mathcal{X} \times \mathcal{T}$ onto $\mathcal{T}$) must be absolutely continuous with respect to μ, because

$$(\pi\Gamma)^t g(t) = \Gamma^{x,t} g(t) = \mu^t \lambda_t^x g(t) = \mu^t \left(g(t)\lambda_t \mathcal{X}\right) \qquad \text{for } g \in \mathcal{M}^+(\mathcal{T}, \mathcal{B}).$$

If $g = 0$ a.e. $[\mu]$ then the last integral must be zero, thereby implying $(\pi\Gamma)g = 0$. Then note that, for each fixed f in $\mathcal{M}^+(\mathcal{X}, \mathcal{A})$, the iterated integral representation for Γ identifies $t \mapsto \lambda_t^x f(x)$ as a density with respect to μ of the measure Γ_f on $\mathcal{B}$ defined via an increasing linear functional,

<2> $$\Gamma_f g := \Gamma^{x,t}\left(f(x)g(t)\right) = \mu^t\left(g(t)\lambda_t^x f(x)\right) \qquad \text{for } g \in \mathcal{M}^+(\mathcal{T}, \mathcal{B}).$$

Construction of the kernel, almost

Now consider how we might reverse the argument, starting from a measure Γ on $\mathcal{A} \otimes \mathcal{B}$ for which $\pi(\Gamma)$ is absolutely continuous with respect to a sigma-finite measure μ on $\mathcal{B}$. Without loss of generality we may suppose Γ is a finite measure. The result for Γ as in the Theorem will follow by pasting together the results for each of the Γ_i.

For each f in $\mathcal{M}^+(\mathcal{T}, \mathcal{B})$, we can define (via the Radon-Nikodym Theorem from Section 3.1, even though Γ_f need not be sigma-finite) a function $\lambda(t, f)$ as the density of Γ_f with respect to μ. Here I write $\lambda(t, f)$ instead of $\lambda_t f$ to avoid an inadvertent presumption that $f \mapsto \lambda(t, f)$ is a measure. Indeed, at the moment $\lambda(t, f)$ is defined only up to a μ-equivalence; it is not even a well defined functional for each fixed t. Nevertheless, it has almost sure analogs of the properties, as a function of f, that we need. Problem [2] outlines the steps needed to prove the following result.

<3> **Lemma.** *Let Γ be a finite measure, whose $\mathcal{T}$-marginal is absolutely continuous with respect to the sigma-finite measure μ. For each fixed f in $\mathcal{M}^+(\mathcal{X}, \mathcal{A})$, the density $\lambda(t, f) := d\Gamma_f/d\mu$, with Γ_f as in <2>, is well defined and determined uniquely up to a μ-equivalence. The map $t \mapsto \lambda(t, f)$ has the following measure-like properties for all α_1, α_2 in $\mathcal{M}^+(\mathcal{T}, \mathcal{B})$ and all $f, f_1, f_2, \ldots$ in $\mathcal{M}^+(\mathcal{X}, \mathcal{A})$.*

(i) $\lambda(t, \alpha_1 f_1 + \alpha_2 f_2) = \alpha_1(t)\lambda(t, f_1) + \alpha_2(t)\lambda(t, f_2)$ a.e. $[\mu]$.

(ii) *If* $f_1 \le f_2$ *then* $\lambda(t, f_1) \le \lambda(t, f_t)$ a.e. $[\mu]$.

(iii) *If* $0 \le f_1 \le f_2 \le \ldots \uparrow f$ *then* $\lambda(t, f_n) \uparrow \lambda(t, f)$ a.e. $[\mu]$.

If it were possible to combine the negligible sets corresponding to the a.e. $[\mu]$ qualifiers on each of the assertions (i), (ii), and (iii) of the Lemma into a single μ-negligible set $\mathcal{N}$, the family of functionals $\{\lambda(t, \cdot) : t \notin \mathcal{N}\}$ would correspond to a family of measures on $\mathcal{A}$. Unfortunately, we accumulate uncountably many negligible sets as we cycle (i)—(iii) through all the required combinations of functions in $\mathcal{M}^+(\mathcal{X}, \mathcal{A})$ and positive constants.

There are two strategies in the literature for dealing with the multitude of negligible sets. The more elegant approach brings to bear the heavy measure theoretic machinery of the ***lifting theorem***. (Roughly speaking, a lifting is a map that selects a representative from each μ-equivalence classes of bounded functions in a way that eliminates the almost sure qualifiers from properties (i) and (ii) of

Lemma <3>.) The alternative is to use a separability assumption to reduce (i)—(iii) to a countable collection of requirements, each involving the exclusion of a single μ-negligible set.

REMARK. Separability methods have the longer history, but lifting is clearly the technical tool of choice when we seek more general forms for theorems about conditional probabilities, constructions of point processes, and other results involving management of uncountable families of negligible sets.

Construction of measures on compact metric spaces

I will follow the separability strategy for dealing with the host of negligible sets in Lemma <3>. The following result from Section A.4 will help us to build a measure from a set function defined initially on a countable collection of sets.

<4> **Lemma.** *Let $\mathcal{K}_0$ be a $(\emptyset, \cup f, \cap f)$ paving of subsets of $\mathcal{X}$, and let $\mathcal{K}$ denote its $\cap c$-closure. Let $\nu : \mathcal{K}_0 \to \mathbb{R}^+$ be a $\mathcal{K}_0$-tight set function that is sigma-smooth at $\emptyset$. Then ν has a unique extension to a complete, $\mathcal{K}$-regular, countably additive measure on a sigma-field $\mathcal{S} \supset \mathcal{K}_0$, defined by*

$$\nu K := \inf\{\nu K_0 : K \subseteq K_0 \in \mathcal{K}_0\} \qquad \text{for } K \in \mathcal{K},$$

$$\nu S := \sup\{\nu K : S \supseteq K \in \mathcal{K}\} \qquad \text{for } S \in \mathcal{S}.$$

The requirements of the Lemma simplify greatly when $\nu = \lambda(t, \cdot)$, with $\mathcal{K}$ as the paving of all compact subsets of a compact metric space $\mathcal{X}$, if we choose an appropriate $\mathcal{K}_0$. Problem [9] shows that there exists a countable subclass $\mathcal{K}_0$ of $\mathcal{K}$ with the following properties.

(a) The class $\mathcal{K}_0$ is a $(\emptyset, \cup f, \cap f)$ paving on $\mathcal{X}$, meaning that $\mathcal{K}_0$ contains the empty set and that it is stable under the formation of finite unions and finite intersections.

(b) For each K in $\mathcal{K}$, there exists a decreasing sequence of sets $\{K_n\}$ from $\mathcal{K}_0$ for which $K = \cap_n K_n$.

(c) For each pair of sets K_1, K_2 in $\mathcal{K}_0$ for which $K_1 \supseteq K_2$ there exists an increasing sequence of sets $\{H_n\}$ from $\mathcal{K}_0$ for which $\cup_n H_n = K_1 \backslash K_2$.

Property (b) ensures that $\mathcal{K}$ is the $\cap c$-closure of $\mathcal{K}_0$. Compactness makes the sigma-smoothness for ν automatic. The $\mathcal{K}_0$-tightness property,

$$\nu K_1 = \nu K_2 + \sup\{\nu H : K_1 \backslash K_2 \supseteq H \in \mathcal{K}_0\} \qquad \text{for } K_1 \supseteq K_2 \text{ in } \mathcal{K}_0,$$

reduces to

(A) $\nu(K \cup H) = \nu(K) + \nu(H)$ for all disjoint pairs K and H in $\mathcal{K}_0$,

(B) $\nu(K_2) + \sup_n \nu(H_n) = \nu(K_1)$, with K_1, K_2, and $\{H_n\}$ as in (b).

We now have the tools to turn Lemma <3> into a proof of Theorem <1>, first for the case of compact $\mathcal{X}$, then for the general case.

Theorem for finite Γ and compact $\mathcal{X}$

Write $\nu_t(K)$ for an arbitrarily chosen version of $\lambda(t, K)$, for each K in $\mathcal{K}_0$. There are only countably many ways to choose a pair of sets from $\mathcal{K}_0$. Taking the union of countably many μ-negligible sets, we are left with a single set N with $\mu N = 0$, for which the following properties hold when $t \notin N$:

(A)′ $\nu_t(K \cup H) = \nu_t(K) + \nu_t(H)$ for all disjoint pairs K and H in $\mathcal{K}_0$,

(B)′ $\nu_t(K_2) + \sup_n \nu_t(H_n) = \nu_t(K_1)$, with K_1, K_2, and $\{H_n\}$ as in (b).

For each such t, Lemma <4> lets us extend ν_t to a countably additive, finite measure λ_t on the Borel sigma-field of $\mathcal{X}$, for which $t \mapsto \lambda_t K$ is $\mathcal{B}$-measurable and

$$\Gamma^{x,t}\left(g(t)\{x \in K\}\right) = \mu^t\left(g(t)\lambda_t K\right) \qquad \text{for all } g \in \mathcal{M}^+(\mathcal{T}, \mathcal{B})) \text{ and } K \in \mathcal{K}_0.$$

Define λ_t as the zero measure when $t \in N$. A simple λ-cone generating class argument then shows that $t \mapsto \lambda_t A$ is $\mathcal{B}$-measurable for each Borel set A, so that $\Lambda := \{\lambda_t : t \in \mathcal{T}\}$ is a genuine kernel. A similar λ-cone argument extends the representation for Γ to all functions in $\mathcal{M}^+(\mathcal{X} \times \mathcal{T}, \mathcal{A} \otimes \mathcal{B})$, thereby confirming that $\Gamma = \mu \otimes \Lambda$.

For each K in $\mathcal{K}_0$, the function $t \mapsto \lambda_t K$ is determined up to an almost sure equivalence as a density. For any two representing kernels we would have $\lambda_t K = \widetilde{\lambda}_t K$ for $t \notin N_K$, with $\mu N_K = 0$. For t outside the μ-negligble set $\cup_{K \in \mathcal{K}_0} N_K$, a λ-class argument establishes equality of the finite measures λ_t and $\widetilde{\lambda}_t$.

General case

We may assume the compact sets $\mathcal{X}_i$ are disjoint. Invoke the special case to find kernels $\Lambda_i := \{\lambda_{t,i} : t \in \mathcal{T}\}$ for which $\Gamma_i = \mu \otimes \Lambda_i$ and $\lambda_{t,i}\mathcal{X}_i^c = 0$ for each i. Define Λ by $\lambda_t := \sum_{i \in \mathbb{N}} \lambda_{t,i}$. The uniqueness assertion reduces to the uniqueness for each $\mathcal{X}_i$, together with the fact that Γ puts zero mass outside $(\cup_i \mathcal{X}_i) \times \mathcal{T}$.

2. Disintegrations with respect to a measurable map

In Chapter 5, conditional distributions were identified as special cases of disintegrations, in the following wide sense.

<5> **Definition.** *Let $(\mathcal{X}, \mathcal{A}, \lambda)$ and $(\mathcal{T}, \mathcal{B}, \mu)$ be measure spaces, and let T be an $\mathcal{A}\backslash\mathcal{B}$-measurable map from $\mathcal{X}$ into $\mathcal{T}$. Call a kernel $\Lambda := \{\lambda_t : t \in \mathcal{T}\}$ from $(\mathcal{T}, \mathcal{B})$ to $(\mathcal{X}, \mathcal{A})$ a (T, μ)-**disintegration** of a sigma-finite measure λ if*

(i) $\lambda_t\{T \neq t\} = 0$ for μ almost all t,

(ii) $\lambda^x f(x) = \mu^t \lambda_t^x f(x)$, for each f in $\mathcal{M}^+(\mathcal{X}, \mathcal{A})$.

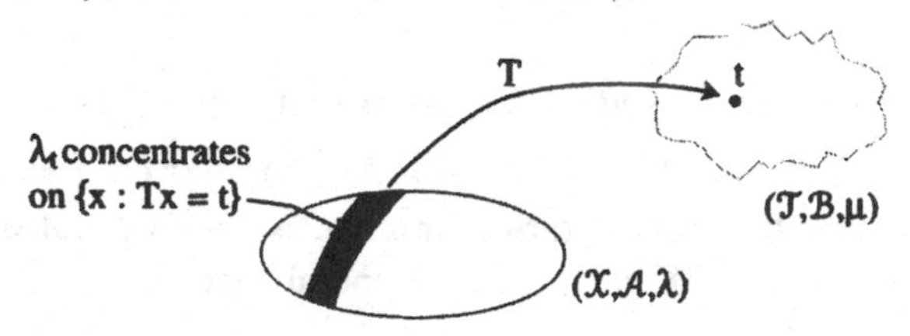

To fit disintegration into the framework of Theorem <1>, it suffices that λ be a ***Radon*** measure: a measure on the Borel sigma-field $\mathcal{B}(\mathcal{X})$ for which

$$\lambda A = \sup\{\lambda K : A \supset K, \text{with } K \text{ compact}\} \qquad \text{for } A \in \mathcal{B}(\mathcal{X}),$$

with $\lambda K < \infty$ for each compact K. A sigma-finite Radon measure must concentrate all its mass on a disjoint union of countably many compact Borel sets (Problem [3]). Every sigma-finite measure on the Borel sigma-field of a complete, separable, metric space is a Radon measure (Problem [4]), provided it gives finite measure to compact sets.

<6> **Theorem.** *Let T be an $\mathcal{A}\backslash\mathcal{B}$-measurable map from $\mathcal{X}$ into $\mathcal{T}$. Let λ be a sigma-finite Radon measure on the Borel sigma-field $\mathcal{A}$ of a metric space $\mathcal{X}$, whose image measure $T\lambda$ is absolutely continuous with respect to a sigma-finite measure μ on $\mathcal{B}$. If the set* graph$(T) := \{(x, t) \in \mathcal{X} \times \mathcal{T} : Tx = t\}$ *is $\mathcal{A} \otimes \mathcal{B}$-measurable then λ has a (T, μ)-disintegration, unique up to a μ-equivalence.*

Proof. The image measure Γ of λ under the map $x \mapsto (x, Tx)$ concentrates on graph(T), and it satisfies the conditions of Theorem <1>. The Theorem gives a kernel Λ for which $\Gamma = \mu \otimes \Lambda$, that is,

$$\lambda^x g(x, Tx) = \Gamma^{x,t} g(x, t) = \mu^t \lambda_t^x g(x, t) \qquad \text{for all } g \in \mathcal{M}^+(\mathcal{X} \otimes \mathcal{T}, \mathcal{A} \otimes \mathcal{B}).$$

Specialize to $g(x, t) \equiv f(x)$ to get the disintegration property (ii). Then take g as the indicator function of graph(T) to get $0 = \Gamma\{(x, t) : t \neq Tx\} = \mu^t \lambda_t\{T \neq t\}$, which implies the disintegration property (i). The almost uniqueness of the kernel follows from the corresponding assertion in Theorem <1>. □

3. Problems

[1] (repeated from Chapter 3) Let μ be a sigma-finite measure on $(\mathcal{T}, \mathcal{B})$. Suppose h_1 and h_2 are functions in $\mathcal{M}^+(\mathcal{T}, \mathcal{B})$ for which $\mu\left(g(t)h_1(t)\right) \leq \mu\left(g(t)h_2(t)\right)$ for all g in $\mathcal{M}^+(\mathcal{T}, \mathcal{B})$. Show that $h_1 \leq h_2$ a.e. $[\mu]$. Hint: Write $\mathcal{T} = \cup_{i \in \mathbb{N}} \mathcal{T}_i$, with $\mu \mathcal{T}_i < \infty$ for each i. Consider $g := \{t \in \mathcal{T}_i : h_2(t) < r < h_1(t)\}$ for rational r.

[2] Establish Lemma <3> by the following steps. Write γ for the $\mathcal{X}$-marginal of Γ.

(i) For fixed f in $\mathcal{M}^+(\mathcal{X}, \mathcal{A})$ and each n in $\mathbb{N}$, write $\lambda_n(t, f)$ for any choice of the Radon-Nikodym derivative $d\Gamma_{n \wedge f}/d\mu$. That is, $\lambda_n(t, f)$ is a $\mathcal{B}$-measurable function for which $\Gamma^{x,t} g(t)\left(n \wedge f(x)\right) = \mu^t g(t) \lambda_n(t, f)$ for all g in $\mathcal{M}^+(\mathcal{T}, \mathcal{B})$. Use Problem [1] to show that $0 \leq \lambda_n(t, f) \uparrow$ a.e. $[\mu]$.

(ii) Show that $\lambda(t, f) := \limsup \lambda_n(t, f)$ is a density for Γ_f with respect to μ. Use Problem [1] to show that the density is unique up to a μ-equivalence.

(iii) For fixed α_1, α_2, g in $\mathcal{M}^+(\mathcal{T}, \mathcal{B})$ and f, f_1, f_2 in $\mathcal{M}^+(\mathcal{X}, \mathcal{A})$, show that

$$\mu^t\left(g(t)\alpha_1(t)\lambda(t, f_1)\right) + \mu^t\left(g(t)\alpha_2(t)\lambda(t, f_2)\right) = \Gamma g(t)\left(\alpha_1(t) f_1(x) + \alpha_2(t) f_2(x)\right).$$

Deduce via (ii) that $\lambda(t, \alpha_1 f_1 + \alpha_2 f_2) = \alpha_1(t)\lambda(t, f_1) + \alpha_2(t)\lambda(t, f_2)$ a.e. $[\mu]$.

(iv) If $f_1, f_2 \in \mathcal{M}^+(\mathcal{X}, \mathcal{A})$ and $f_1(x) \leq f_2(x)$ a.e. $[\gamma]$, show that $\mu^t\left(g(t)\lambda(t, f_1)\right) \leq \mu^t\left(g(t)\lambda(t, f_1)\right)$ for all g in $\mathcal{M}^+(\mathcal{T}, \mathcal{B})$. Deduce that $\lambda(t, f_1) \leq \lambda(t, f_2)$ a.e. $[\mu]$.

(v) If $f_1, f_2, \ldots \in \mathcal{M}^+(\mathcal{X}, \mathcal{A})$ and $f_n(x) \uparrow f(x)$ a.e. $[\gamma]$, show that

$$\mu^t\left(g(t)\lambda(t, f_n)\right) = \Gamma\left(g(t) f_n(x)\right) \uparrow \Gamma\left(g(t) f(x)\right) = \mu^t\left(g(t)\lambda(t, f)\right)$$

for all g in $\mathcal{M}^+(\mathcal{T}, \mathcal{B})$. Deduce that $\lambda(t, f_n) \uparrow \lambda(t, f)$ a.e. $[\mu]$.

[3] Let λ be a sigma-finite Radon measure on $(\mathcal{X}, \mathcal{A})$, with $\mathcal{A}$ the Borel sigma-field. Show that there exists a sequence of disjoint compact sets $\{K_i : i \in \mathbb{N}\}$ for which $\lambda (\cup_i K_i)^c = 0$. Hint: Partition $\mathcal{X}$ into sets $\{\mathcal{X}_i : i \in \mathbb{N}\}$, each with finite measure. For each i find disjoint compact subsets $K_{i,n}$ of $\mathcal{X}_i$ for which $\lambda \left(\mathcal{X}_i \setminus \cup_{i \le n} K_{i,n}\right) < 2^{-n}$.

[4] Let λ be a finite measure on the Borel sigma-field of a complete, separable metric space $\mathcal{X}$. Write $\mathcal{K}$ for the class of all compact subsets of $\mathcal{X}$.

(i) Show that λ is tight: for each $\epsilon > 0$ there exists a K_ϵ in $\mathcal{K}$ such that $\lambda K_\epsilon^c < \epsilon$. Hint: For each positive integer n, show that the space $\mathcal{X}$ is a countable union of closed balls with radius $1/n$. Find a finite family of such balls whose union B_n has λ measure greater than $\lambda\mathcal{X} - \epsilon/2^n$. Show that $\cap_n B_n$ is compact, using the total-boundedness characterization of compact subsets of complete metric spaces (Simmons 1963, Section 25).

(ii) Deduce that $\lambda B = \sup\{\lambda K : B \supseteq K \in \mathcal{K}\}$ for each Borel set B. Hint: Compare with the Problems to Chapter 2.

(iii) Extend the result to sigma-finite measures λ on a complete, separable metric space for which $\lambda K < \infty$ for each compact K.

[5] Let T be an $\mathcal{A}\backslash\mathcal{B}$-measurable map from $(\mathcal{X}, \mathcal{A})$ into $(\mathcal{Y}, \mathcal{B})$. Suppose $\mathcal{B}$ contains all singleton sets $\{y\}$, for $y \in \mathcal{Y}$, and is countably generated: $\mathcal{B} := \sigma(\mathcal{B}_0)$ for some countable subclass $\mathcal{B}_0$. With no loss of generality you may assume that $\mathcal{B}_0$ is stable under complements. Show that graph$(T) \in \mathcal{A} \otimes \mathcal{B}$, by following these steps.

(i) Say that a set B separates points y_1 and y_2 if either $y_1 \in B$ and $y_2 \in B^c$ or $y_1 \in B^c$ and $y_2 \in B$. For fixed y_1 and y_2, show that

$$\mathcal{B}_1 := \{B \in \mathcal{B} : B \text{ does not separate } y_1 \text{ and } y_2\}$$

is a sigma-field. Deduce that $\mathcal{B}_1$ cannot contain every $\mathcal{B}_0$ set.

(ii) If $y \neq Tx$, for some $x \in \mathcal{X}$ and $y \in \mathcal{Y}$, show that $(x, y) \in (T^{-1}B^c) \otimes B$ for some $B \in \mathcal{B}_0$.

(iii) Deduce that graph$(T)^c$ is a countable union of measurable rectangles.

[6] Suppose $H : \mathcal{Y} \to \mathbb{R}$ is one-to-one. Show that $\sigma(H)$ is countably generated and contains all the singleton sets.

[7] Suppose $\mathcal{C}$ is a countably generated sigma-field on a set $\mathcal{Y}$, that contains all the singleton subsets. That is, $\mathcal{C} := \sigma(\mathcal{C}_0)$, for some countable $\mathcal{C}_0 := \{C_1, C_2, \ldots\}$. Define a real-valued, $\mathcal{C}$-measurable function $H(y) := \sum_i 3^{-i}\{y \in C_i\}$ on $\mathcal{Y}$.

(i) Show that H is one-to-one. Hint: If $H(y_1) = H(y_2)$, with $y_1 \neq y_2$, show that $\mathcal{C}$ cannot separate the points y_1, y_2, which implies that $\{y_1\} \notin \mathcal{C}$.

(ii) Equip the range $\mathcal{R} := H\mathcal{Y}$ (which might not be a Borel set) with the trace sigma-field $\mathcal{D} := \{B\mathcal{R} : B \in \mathcal{B}(\mathbb{R})\}$. Show that the map $G = H^{-1}$ from $\mathcal{R}$ onto $\mathcal{Y}$ is $\mathcal{D}\backslash\mathcal{C}$-measurable. Hint: Show that $G^{-1}C_i$ is the set of all points in $\mathcal{R}$ whose base-3 expansions have a 1 in the ith place.

(iii) Show that $\mathcal{C} = \sigma(H)$.

[8] Suppose T is a function from a set $\mathcal{X}$ into a set $\mathcal{T}$ equipped with a sigma-field $\mathcal{B}$. Suppose $S : \mathcal{X} \to \mathcal{Y}$ is $\sigma(T)\backslash\mathcal{C}$-measurable, where $\mathcal{C}$ is a countably generated sigma-field on $\mathcal{Y}$ containing all the singleton sets. Show that there exists a $\mathcal{B}\backslash\mathcal{C}$-measurable map $\gamma : \mathcal{T} \to \mathcal{Y}$ such that $S = \gamma \circ T$. Hint: Represent $H \circ S$, where H is defined in Problem [7], as $g \circ T$. Show that $\gamma = H^{-1} \circ g$ has the desired property.

[9] Let $\mathcal{X}_0$ be a countable dense subset of a compact metric space $\mathcal{X}$. Write $\overline{B}(x,r)$ for the closed ball with radius r and center x. Define $\mathcal{K}_1$ as the collection of all finite unions of balls $\overline{B}(x,r)$ with r rational and $x \in \mathcal{X}_0$. Write $\mathcal{K}_0$ for the collection of all finite intersections of sets from $\mathcal{K}_1$.

(i) Show that $\mathcal{K}_0$ is a countable $(\emptyset, \cup f, \cap f)$-paving of compact subsets of $\mathcal{X}$.

(ii) For each closed subset F of $\mathcal{X}$ show that there exists a decreasing sequence of sets $\{K_n\}$ from $\mathcal{K}_0$ for which $F = \cap_n K_n$. Hint: The union of the open balls $B(x, 1/n)$ with $x \in \mathcal{X}_0 \cap F$ covers the compact F. Find a finite subcover. Consider the union of the corresponding closed balls.

(iii) For each pair of closed subsets K_1, K_2 in $\mathcal{K}_0$ for which $K_1 \supseteq K_2$ show that there exists an increasing sequence of sets $\{H_n\}$ from $\mathcal{K}_0$ for which $\cup_n H_n = K_1\backslash K_2$. Hint: Cover the compact set $\{x \in K_1 : d(x, K_2) \geq 2/n\}$ by finitely many closed balls with radius $1/n$, then intersect their union with K_1.

4. Notes

The key idea underlying all proofs of existence of disintegrations and regular conditional distributions is that of compact approximation—existence of a class of approximating sets with properties analogous to the class of compact sets in a metric space—as a means for deducing countably additivity from finite additivity. Marczewski (1953) isolated the concept of a compact system. See Pfanzagl & Pierlo (1969) for an exposition. Bourbaki (1969, Note Historique) also gave credit to Ryll-Nardzewski for disintegration (no citation), perhaps in some point-process context. In point process theory disintegrations appear as Palm distributions—conditional distributions given a point of the process at a particular position (Kallenberg 1969).

I learned about disintegration from Dellacherie & Meyer (1978, Section III.8). They proved a result close to Theorem <6> using results about analytic sets (essentially to deduce product-measurability of graph(T)). Parthasarathy (1967, Sections V.7 and V.8) cited notes of Varadarajan for his existence proof for a disintegration. The definitive disintegration theorem is due to Pachl (1978), who established that approximation by a compact system of sets is essential for a general disintegration theorem. See Hoffmann-Jørgensen (1994, Sections 10.26–10.30) for an exposition of Pachl's method. Hoffmann-Jørgensen (1971), following Ionescu Tulcea & Ionescu Tulcea (1969), used a lifting argument to eliminate separability assumptions for existence of conditional distributions.

See the Notes to Chapter 5 for further comments about conditional distributions and disintegrations.

References

Bourbaki, N. (1969), *Intégration sur les espaces topologiques séparés*, Éléments de mathématique, Hermann, Paris. Fascicule XXXV, Livre VI, Chapitre IX.

Dellacherie, C. & Meyer, P. A. (1978), *Probabilities and Potential*, North-Holland, Amsterdam.

Hoffmann-Jørgensen, J. (1971), 'Existence of conditional probabilities', *Mathematica Scandinavica* **28**, 257–264.

Hoffmann-Jørgensen, J. (1994), *Probability with a View toward Statistics*, Vol. 2, Chapman and Hall, New York.

Ionescu Tulcea, A. I. & Ionescu Tulcea, C. T. (1969), *Topics in the Theory of Lifting*, Springer-Verlag.

Kallenberg, O. (1969), *Random Measures*, Akademie-Verlag, Berlin. US publisher: Academic Press.

Marczewski, E. (1953), 'On compact measures', *Fundamenta Mathematicae* pp. 113–124.

Pachl, J. (1978), 'Disintegration and compact measures', *Mathematica Scandinavica* **43**, 157–168.

Parthasarathy, K. R. (1967), *Probability Measures on Metric Spaces*, Academic, New York.

Pfanzagl, J. & Pierlo, N. (1969), *Compact systems of sets*, Vol. 16 of *Springer Lecture Notes in Mathematics*, Springer-Verlag, New York.

Simmons, G. F. (1963), *Introduction to Topology and Modern Analysis*, McGraw-Hill.

Index